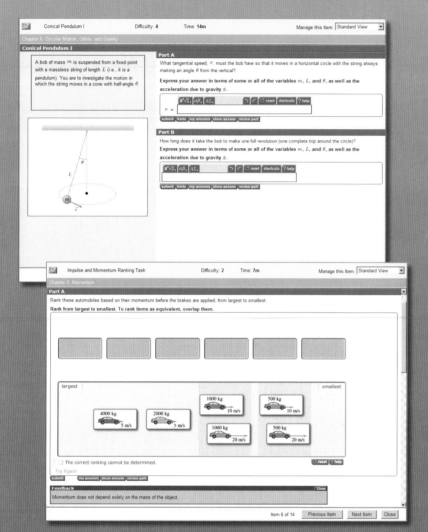

Contents

Problem-Solving Strategies and Tactics

Volume 1 (pp. 1–328) contains Chapters 1–14
Volume 2 (pp. 329–634) contains Chapters 15–26

Essential COLLEGE PHYSICS

FIRST EDITION

Volume 1
Chapters 1–14

Andrew F. Rex
University of Puget Sound

Richard Wolfson
Middlebury College

Addison-Wesley

Boston   Columbus   Indianapolis   New York   San Francisco   Upper Saddle River
Amsterdam   Cape Town   Dubai   London   Madrid   Milan   Munich   Paris   Montréal   Toronto
Delhi   Mexico City   São Paulo   Sydney   Hong Kong   Seoul   Singapore   Taipei   Tokyo

Publisher: Jim Smith
Executive Editor: Nancy Whilton
Development Director: Michael Gillespie
Development Editor: Gabriele Rennie
Senior Development Editor: Margot Otway
Editorial Manager: Laura Kenney
Senior Project Editor: Martha Steele
Editorial Assistant: Dyan Menezes
Editorial Assistant: Claudia Trotch
Media Producer: David Huth
Director of Marketing: Christy Lawrence
Executive Marketing Manager:
  Scott Dustan
*Executive Market Development
  Manager:* Scott Frost
Market Development Coordinator:
  Jessica Lyons

Managing Editor: Corinne Benson
Senior Production Supervisors: Nancy Tabor
  and Shannon Tozier
Production Service: Pre-Press PMG
Illustrations: Rolin Graphics
Text Design: Elm Street Publishing Services
Cover Design: Derek Bacchus
Manufacturing Manager: Jeff Sargent
Director, Image Resource Center: Melinda
  Patelli
Manager, Rights and Permissions: Zina Arabia
Image Permission Coordinator: Elaine Soares
Photo Research: Kristin Piljay
Text and Cover Printer and Binder: Courier,
  Kendallville
Cover Image: Mark Madeo Photography.
  "Formation", four men jumping over a wall
  at the beach in San Francisco

Credits and acknowledgments borrowed from other sources and reproduced, with permission, in this textbook appear on p. C-1.

Library of Congress Cataloging-in-Publication Data
Rex, Andrew F., 1956-
    Essential college physics / Andrew F. Rex, Richard Wolfson.—1st ed.
      p. cm.
    Includes index.
    ISBN-13: 978-0-321-61116-1 (v. 1)
    ISBN-10: 0-321-61116-0 (v. 1)
    ISBN-13: 978-0-321-61117-8 (v. 2)
    ISBN-10: 0-321-61117-9 (v. 2)
      1. Physics—Textbooks.   I. Wolfson, Richard.   II. Title.

QC23.2.R49 2009
530—dc22   2009024991

ISBN 10: 0-321-61116-0; ISBN 13: 978-0-321-61116-1 (Student edition Volume 1)

ISBN 10: 0-321-66665-8; ISBN 13: 978-0-321-66665-9 (Professional copy Volume 1)

Addison-Wesley
is an imprint of

About the Authors

Andrew F. Rex

Andrew F. Rex has been professor of physics at the University of Puget Sound since 1982. He frequently teaches the College Physics course, so he has a deep sense of student and instructor challenges. He is the author of several textbooks, including *Modern Physics for Scientists and Engineers* and *Integrated Physics and Calculus.* In addition to textbook writing, he studies foundations of the second law of thermodynamics, which has led to the publication of several papers and the widely acclaimed book *Maxwell's Demon: Entropy, Information, Computing.*

Richard Wolfson

Richard Wolfson has been professor of physics at Middlebury College for more than 25 years. In addition to his textbooks, *Essential University Physics, Physics for Scientists and Engineers,* and *Energy, Environment, and Climate,* he has written two science books for general audiences: *Nuclear Choices: A Citizen's Guide to Nuclear Technology,* and *Simply Einstein: Relativity Demystified.* His video courses for the Teaching Company include *Physics in Your Life* and *Einstein's Relativity and the Quantum Revolution: Modern Physics for Non-Scientists.*

During the three decades we have been teaching physics, algebra-based physics textbooks have grown in length, complexity, and price. We've reached the point where textbooks can be overwhelming to students, many of whom are taking physics as a requirement for another major or profession and will never take another physics class. And yet, we've also seen many students in the algebra-based course who are eager to learn how physics explains what they see in their everyday lives, how it connects to other disciplines, and how exciting new ideas in physics can be.

A Concise and Focused Book

The first thing you'll notice about this book is that it's more concise than most algebra-based textbooks. We believe it is possible to provide a shorter, more focused text that better addresses the learning needs of today's students while more effectively guiding them through the mastery of physics. The language is concise and engaging without sacrificing depth. Brevity needn't come at the expense of student learning! We've designed our text from the ground up to be concise and focused, rather than cutting down a longer book. Students will find the resulting book less intimidating and easier to use, with well coordinated narrative, instructional art program, and worked examples.

A Connected Approach

In addition to making the volume of the book less overwhelming, we've stressed connections, to reinforce students' understanding and to combat the preconception that physics is just a long list of facts and formulas.

Connecting ideas The organization of topics and the narrative itself stress the connections between ideas. Whenever possible, the narrative points directly to a worked example or to the next section. A worked example can serve as a bridge, not only to the preceding material it is being used to illustrate, but also forward by introducing a new idea that is then explicated in the following section. These bridges work both ways; the text is always looking forward and back to exploit the rich trail of connections that exist throughout physics.

Connecting physics with the real world Instead of simply stating the facts of physics and backing them up with examples, the book develops some key concepts from observations of real-world phenomena. This approach helps students to understand what physics is and how it relates to their lives. In addition, numerous examples and applications help students explore the ideas of physics as they relate to the real world. Connections are made to phenomena that will engage the students—applications from everyday life (heating a home, the physics of flight, DVDs, hybrid vehicles, and many more), from biomedicine (pacemakers, blood flow, cell membranes, medical imaging), and from cutting-edge research in science and technology (superconductivity, nanotechnology, ultracapacitors). These applications can be used to motivate interest in particular topics in physics, or they might emerge from learning a new physics topic. One thing leads to another. What results is a continuous story of physics, seen as a seamless whole rather than an encyclopedia of facts to be memorized.

Connecting words and math In the same way, we stress the connections between the ideas of physics and their mathematical expression. Equations are statements about physics—sentences, really—not magical formulae. In algebra-based physics, it's important to stress the basics but not the myriad details that cloud the issues for those new to the subject. We've reduced the number of enumerated equations, to make the essentials clearer.

- Complete edition Volumes 1–2 (shrinkwrapped) (ISBN 978-0-321-59854-7): Chapters 1–26
- Volume 1 (ISBN 978-0-321-61116-1): Chapters 1–14
- Volume 2 (ISBN 978-0-321-61117-8): Chapters 15–26
- Complete edition Volumes 1–2 (shrinkwrapped) with MasteringPhysics™ (ISBN 978-0-321-59856-1): Chapters 1–26
- Volume 1 with MasteringPhysics™ (ISBN 978-0-321-61118-5): Chapters 1–14
- Volume 2 with MasteringPhysics™ (ISBN 978-0-321-61119-2): Chapters 15–26

Connecting with how students learn   Conceptual worked examples and end-of-chapter problems are designed to help students explore and master the qualitative ideas developed in the text. Some conceptual examples are linked with numerical examples that precede or follow them, linking qualitative and quantitative reasoning skills. Follow-up exercises to worked examples ("Making the Connection") prompt students to explore further, while "Got It?" questions (short concept-check questions found at the end of text sections) help ensure a key idea is grasped before the student moves on.

Students benefit from a structured learning path—clear goals set out at the start, reinforcement of new ideas throughout, and a strategic summary to wrap up. With these aids in place, students build a solid foundation of understanding. We therefore carefully structure the chapters with learning goals, "Reviewing new concepts" reminders, and visual chapter summaries.

Connecting with how students use their textbook   Many students find using a textbook to be a chore, either because English is not their first language or because their reading skills are weak or their time limited. Even students who read with ease prefer their explanations lucid and brief, and they expect key information to be easy to find. Our goal, therefore, is a text that is clear, concise, and focused, with easy-to-find reference material, tips, and examples. The manageable size of the book makes it less intimidating to open and easier to take to class.

To complement verbal explanations in the text, the art program puts considerable information directly on the art in the form of explanatory labels and "author's voice" commentary. Thus, students can use the text and art as parallel, complementary ways to understand the material. The text tells them more, but often the illustrations will prove more memorable and will serve as keys for recalling information. In addition, a student who has difficulty with the text can turn to the art for help.

Connecting the chapters with homework   After reading a chapter, students need to be able to reason their way through homework problems with some confidence that they will succeed. A textbook can help by consistently demonstrating and modeling how an expert goes about solving a problem, by giving clear tips and tactics, and by providing opportunities for practice. Given how important it is for students to become proficient at solving problems, a detailed explanation of how our textbook will help them is provided below.

Problem-Solving Strategies

Worked examples are presented consistently in a three-step approach that provides a model for students:

Organize and Plan   The first step is to gain a clear picture of what the problem is asking. Then students gather information they need to address the problem, based on information presented in the text and considering similarities with earlier problems, both conceptual and numerical. If a student sketch is needed to help understand the physical situation, this is the place for it. Any known quantities that will be needed to calculate the answer or answers are gathered at the end of this step.

Solve   The plan is put into action, and the required steps carried out to reach a final answer. Computations are presented in enough detail for the student to see a clear path from start to finish.

Reflect   There are many things that a student might consider here. Most important is whether the answer is reasonable, in the context of either the problem or a similar known situation. This is the place to see whether units are correct or to check that symbolic answers reduce to sensible results in obvious special cases. The student may reflect on connections to other solved problems or real-life situations. Sometimes solving a problem raises a new question, which can lead naturally to another example, the next section, or the next chapter.

Conceptual examples follow a simpler two-step approach: Solve and Reflect. As with the worked examples, the Reflect step is often used to point out important connections.

Worked examples are followed by "Making the Connection," a new problem related to the one just solved, which serves as a further bridge to earlier material or the next section of text. Answers to Making the Connection are provided immediately, and thus they also serve as good practice problems—getting a second example for the price of one.

Strategy boxes follow the three-step approach that parallels the approach in worked examples. These give students additional hints about what to do in each of the three steps. "Tactic" boxes give additional problem-solving tools, outside the three-step system.

End-of-Chapter Problems

There are three types of problems:

1. *Conceptual questions,* like the conceptual worked examples, ask the students to think about the physics and reason without using numbers.

2. *Multiple-choice problems* serve three functions. First, they prepare students for their exams, in cases where instructors use that format. Second, those students who take this course in preparation for the MCAT exam or other standardized exam will get some needed practice. Third, they offer more problem-solving practice for all students.

3. *Problems* include a diversity of problem types, as well as a range of difficulty, with difficulty levels marked by one, two, or three "boxes." Problems are numerous enough to span an appropriate range of difficulty, from "confidence builders" to challenge problems. Most problems are listed under a particular section number in the chapter. General problems at the end are not tied to any section. These problem sets include multi-concept problems that require using concepts and techniques from more than one section or from an earlier chapter.

Organization of Topics

The organization of topics should be familiar to anyone who has taught College Physics. The combined Volumes 1 and 2 cover a full-year course in algebra-based physics, divided into either two semesters or three quarters.

Volume 1: Following the introductory Chapter 1, the remainder of Volume 1 is devoted to mechanics of particles and systems, including one chapter each on gravitation, fluids, and waves (including sound). Volume 1 concludes with a three-chapter sequence on thermodynamics.

Volume 2: Volume 2 begins with six chapters on electricity and magnetism, culminating and concluding with a chapter on electromagnetic waves and relativity. Following this are two chapters on optics—one on geometrical optics and one on wave optics. The final four chapters cover modern physics, including quanta, atoms, nuclei, and elementary particles.

Instructor Supplements

NOTE: For convenience, all of the following instructor supplements can also be downloaded from the "Instructor Area," accessed via the left-hand navigation bar of Mastering-Physics™ (www.masteringphysics.com).

The **Instructor Solutions Manual**, written by Brett Kraabel, Freddy Hansen, Michael Schirber, Larry Stookey, Dirk Stueber, and Robert White, provides *complete* solutions to all the end-of-chapter questions and problems. All solutions follow the Organize and Plan/Solve/Reflect problem-solving strategy used in the textbook for quantitative problems and the Solve/Reflect strategy for qualitative ones. The solutions are available by chapter in Word and PDF format and can be downloaded from the *Instructor Resource Center* (www.pearsonhighered.com/educator).

The cross-platform **Instructor Resource DVD** (ISBN 978-0-321-61126-0) provides invaluable and easy-to-use resources for your class. The contents include a comprehensive library of more than 220 applets from **ActivPhysics OnLine™**, as well as all figures, photos, tables, and summaries from the textbook in JPEG format. In addition, all the Problem-Solving Strategies, Tactics Boxes, and Key Equations are provided in editable Word as well as JPEG format. PowerPoint slides containing all the figures from the text are also included, as well as Classroom Response "Clicker" questions.

MasteringPhysics™ (www.masteringphysics.com) is a homework, tutorial, and assessment system designed to assign, assess, and track each student's progress. In addition to the textbook's end-of-chapter problems, MasteringPhysics for *Essential College Physics* also includes prebuilt assignments and tutorials.

MasteringPhysics provides instructors with a fast and effective way to assign uncompromising, wide-ranging online homework assignments of just the right difficulty and duration. The tutorials coach 90% of students to the correct answer with specific wrong-answer feedback. The powerful post-assignment diagnostics allow instructors to assess the progress of their class as a whole or to quickly identify individual students' areas of difficulty.

ActivPhysics OnLine™ (accessed through the Self Study area within www.masteringphysics.com) provides a comprehensive library of more than 420 tried and tested *ActivPhysics* applets updated for web delivery using the latest online technologies. In addition, it provides a suite of highly regarded applet-based tutorials developed by education pioneers Professors Alan Van Heuvelen and Paul D'Alessandris. The *ActivPhysics* margin icon directs students to specific exercises that complement the textbook discussion.

The online exercises are designed to encourage students to confront misconceptions, reason qualitatively about physical processes, experiment quantitatively, and learn to think critically. They cover all topics from mechanics to electricity and magnetism and from optics to modern physics. The highly acclaimed *ActivPhysics OnLine* companion workbooks help students work through complex concepts and understand them more clearly. More than 220 applets from the *ActivPhysics OnLine* library are also available on the *Instructor Resource DVD*.

The **Test Bank** contains more than 2000 high-quality problems, with a range of multiple-choice, true/false, short-answer, and regular homework-type questions. Test files are provided in both TestGen® (an easy-to-use, fully networkable program for creating and editing quizzes and exams) and Word format, and can be downloaded from www.pearson-highered.com/educator.

Student Supplements

The **Student Solutions Manuals Volume 1 (Chapters 1–14)** (ISBN 978-0-321-61120-8) and **Volume 2 (Chapters 15–26)** (ISBN 978-0-321-61128-4), written by Brett Kraabel, Freddy Hansen, Michael Schirber, Larry Stookey, Dirk Stueber, and Robert White, provide *detailed* solutions to half of the odd-numbered end-of-chapter problems. Following the problem-solving strategy presented in the text, thorough solutions are provided to carefully illustrate both the qualitative (Solve/Reflect) and quantitative (Organize and Plan/Solve/Reflect) steps in the problem-solving process.

MasteringPhysics™ (www.masteringphysics.com) is a homework, tutorial, and assessment system based on years of research into how students work physics problems and precisely where they need help. Studies show that students who use MasteringPhysics significantly increase their final scores compared to those using handwritten homework. MasteringPhysics achieves this improvement by providing students with instantaneous feedback specific to their wrong answers, simpler sub-problems upon request when they get stuck, and partial credit for their method(s) used. This individualized, 24/7 Socratic tutoring is recommended by nine out of ten students to their peers as the most effective and time-efficient way to study.

Pearson eText is available through MasteringPhysics, either automatically when MasteringPhysics is packaged with new books or as a purchased upgrade online. Allowing students access to the text wherever they have access to the Internet, Pearson eText comprises the full text, including figures that can be enlarged for better viewing. Within Pearson eText, students are also able to pop up definitions and terms to help with vocabulary and the reading of the material. Students can also take notes in Pearson eText, using the annotation feature at the top of each page.

Pearson Tutor Services (www.pearsontutorservices.com) Each student's subscription to MasteringPhysics also contains complimentary access to Pearson Tutor Services, powered by Smarthinking, Inc. By logging in with their MasteringPhysics ID and password, students will be connected to highly qualified e-instructors™ who provide additional, interactive online tutoring on the major concepts of physics. Some restrictions apply; offer subject to change.

ActivPhysics OnLine™ (accessed via www.masteringphysics.com) provides students with a suite of highly regarded applet-based tutorials (see above). The following workbooks help students work though complex concepts and understand them more clearly. The *ActivPhysics* margin icons throughout the book direct students to specific exercises that complement the textbook discussion.

ActivPhysics OnLine Workbook Volume 1: Mechanics • Thermal Physics • Oscillations & Waves (ISBN 978-0-805-39060-5)

ActivPhysics OnLine Workbook Volume 2: Electricity & Magnetism • Optics • Modern Physics (ISBN 978-0-805-39061-2)

Acknowledgments

A new full-year textbook in introductory physics doesn't just happen overnight or by accident. We begin by thanking the entire editorial and production staff at Pearson Education. The idea for this textbook grew out of discussions with Pearson editors, particularly Adam Black, whose initial encouragement and vision helped launch the project; and Nancy Whilton, who helped hone and guide this text to its current essentials state. Other Pearson staff who have rendered invaluable service to the project include Ben Roberts, Michael Gillespie, Development Manager; Margot Otway, Senior Development Editor; Gabriele Rennie, Development Editor; Mary Catherine Hagar, Development Editor; Martha Steele; Senior Project Editor; and Claudia Trotch, Editorial Assistant. In the project's early days, we were bolstered by many stimulating discussions with Jon Ogborn, whose introductory textbooks have helped improve physics education in Great Britain. In addition to the reviewers mentioned below, we are grateful to Charlie Hibbard, accuracy checker, for his close scrutiny of every word, symbol, number, and figure; to Sen-Ben Liao for meticulously solving every question and problem and providing the answer list; and to Brett Kraabel, Freddy Hansen, Michael Schirber, Larry Stookey, Dirk Stueber, and Robert White for the difficult task of writing the *Instructor Solutions Manual*. We also want to thank production supervisors Nancy Tabor and Shannon Tozier for their enthusiasm and hard work on the project; Jared Sterzer and his colleagues at Pre-Press PMG for handling the composition of the text; and Kristin Piljay, photo researcher.

Andrew Rex: I wish to thank my colleagues at the University of Puget Sound, whose support and stimulating collegiality I have enjoyed for almost 30 years. The university's staff, in particular Neva Topolski, has provided many hours of technical support throughout this textbook's development. Thanks also to student staff member Dana Maijala for her technical assistance. I acknowledge all the students I have taught over the years, especially those in College Physics classes. Seeing how they learn has helped me generate much of what you see in this book. And last but foremost, I thank my wife Sharon for her continued support, encouragement, and amazing patience throughout the length of this project.

Richard Wolfson: First among those to be acknowledged for their contributions to this project are the thousands of students in my introductory physics courses over three decades at Middlebury College. You've taught me how to convey physics ideas in many different ways appropriate to your diverse learning styles, and your enthusiasm has convinced me that physics really can appeal to a wide range of students for whom it's not their primary interest. Thanks also to my Middlebury faculty colleagues and to instructors around the world who have made suggestions that I've incorporated into my textbooks and my classrooms. It has been a pleasure to work with Andy Rex in merging our ideas and styles into a coherent final product that builds on the best of what we've both learned in our years of teaching physics. Finally, I thank my family, colleagues, and students for their patience during the intensive period when I was working on this project.

Reviewers

Chris Berven, *University of Idaho*
Benjamin C. Bromley, *University of Utah*
Michelle Chabot, *University of South Florida–Tampa*
Orion Ciftja, *Prairie View A & M University*
Joseph Dodoo, *University of Maryland–Eastern Shore*
Florence Egbe Etop, *Virginia State University*
Davene Eyres, *North Seattle Community College*
Delena Bell Gatch, *Georgia Southern University*
Barry Gilbert, *Rhode Island College*
Idan Ginsburg, *Harvard University*
Timothy T. Grove, *Indiana University–Purdue University, Fort Wayne*
Mark Hollabaugh, *Normandale Community College*
Kevin Hope, *University of Montevallo*
Joey Houston, *Michigan State University*
David Iadevaia, *Pima County Community College*
Ramanathan Jambunathan, *University of Wisconsin–Oshkosh*
Monty Mola, *Humboldt State University*
Gregor Novak, *United States Air Force Academy*
Stephen Robinson, *Belmont University*
Michael Rulison, *Ogelthorpe University*
Douglas Sherman, *San Jose State University*
James Stephens, *University of Southern Mississippi*
Rajive Tiwari, *Belmont Abbey College*
Lisa Will, *San Diego City College*
Chadwick Young, *Nicholls State University*
Sharon T. Zane, *University of Miami*
Fredy Zypman, *Yeshiva University*

Preface to the Student

Welcome to physics! Whether you're taking this course as a requirement for a pre-professional program, as a cognate for your college major, or just because you're curious, we want you to enjoy your physics experience and we hope you'll find that it's enriching and stimulating, and that it connects you with both nature and technology.

Physics is fundamental. To understand physics is to understand how the world works, both in everyday life and on scales of time and space unimaginably large and small. For that reason we hope you'll find physics fascinating. But you'll also find it challenging. Physics demands precision in thought and language, subtle interpretation of universal laws, and the skillful application of mathematics. Yet physics is also simple, because there are really only a very few basic principles to learn. Once you know those principles, you can apply them in a vast range of natural and technological applications.

We've written this book to make it engaging and readable. So read it! And read it thoroughly—*before* you begin your homework assignments. The book isn't a reference work, to be consulted only when you need to solve a particular problem or answer a particular question. Rather, it's an unfolding story of physics, emphasizing connections among different physics principles and applications, and connections to many other fields of study—including your academic major, whatever it is.

Physics is more about big ideas than it is about the nitty-gritty of equations, algebra, and numerical answers. Those details are important, but you'll appreciate them more and approach them more successfully if you see how they flow from the relatively few big ideas of physics. So look for those big ideas, and keep them in mind even as you burrow down into details.

Even though you'll need algebra to solve your physics problems, don't confuse physics with math. Math is a tool for doing physics, and the equations of physics aren't just math but statements about how the world works. Get used to understanding and appreciating physics equations as succinct and powerful statements about physical reality—not just places to "plug in" numbers.

We've written this book to give you our help in learning physics. But you can also learn a lot from your fellow students. We urge you to work together to advance your understanding, and to practice a vigorous give-and-take that will help you sharpen your intuition about physics concepts and develop your analytical skills.

Most of all, we hope you'll enjoy physics and appreciate the vast scope of this fundamental science that underlies the physical universe that we all inhabit.

Detailed Contents

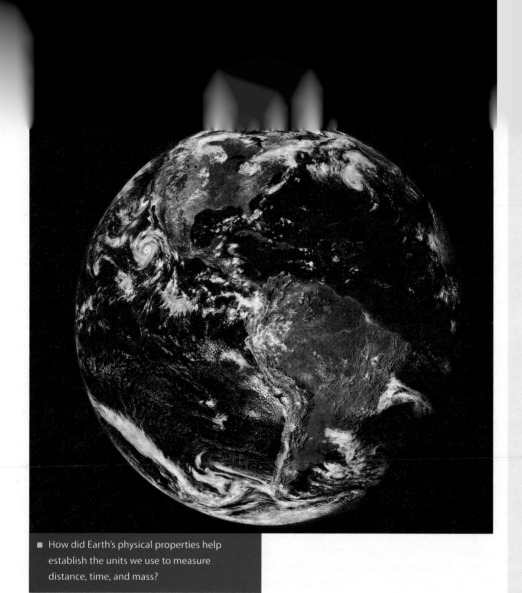

■ How did Earth's physical properties help establish the units we use to measure distance, time, and mass?

Physics provides our understanding of fundamental processes in nature. Physics is quantitative, so it's important to know *what* physical quantities are measured and *how* they're measured. Theories of physics relate different measured quantities and give us a deeper understanding of nature—the ultimate goal of physics.

This chapter introduces concepts and tools you'll need throughout your physics course. First we'll discuss distance, time, and mass and the SI unit system. We'll review scientific notation, explore SI prefixes, and explain how to convert from one unit system to another.

Then you'll see how to use dimensional analysis in physics. We'll discuss measurement, uncertainty, and the use of significant figures. Finally, we'll explain how physicists use order-of-magnitude estimates, both as a way of checking more extensive calculations and for estimating quantities that are difficult to determine exactly. With these basic concepts and tools, you'll be ready to begin the study of motion in Chapter 2.

1.1 Distance, Time, and Mass Measurements

Early in life you learned to measure **distance** and **time**. Many everyday activities require a sense of distance and time, such as meeting your friend for lunch at 12:00 noon at the restaurant a half-mile down the street.

To Learn

By the end of this chapter you should be able to

■ Recognize SI units for distance, time, and mass.

■ Use scientific notation and SI prefixes.

■ Convert among unit systems.

■ Use dimensional analysis.

■ Express results with the appropriate significant figures.

Distance and time are fundamental quantities in physics. You're familiar with your walking or driving speed being distance traveled divided by the time it takes to cover that distance. That is, for a constant speed

$$\text{speed} = \frac{\text{distance}}{\text{time}}$$

Distance and time provide the foundation for the study of motion, which will be the focus of Chapter 2 and Chapter 3 and will reappear throughout this book.

A third fundamental quantity is **mass**. You probably have some sense of mass as *how much* matter an object contains. We'll briefly touch on mass here and discuss it more in Chapter 4. A thorough understanding of mass is closely related to the study of motion, making close links between distance, time, and mass.

SI Units

Distance, time, and mass measurements go back to ancient times. People needed to know the distance from Athens to Rome, the duration of daylight hours, or how much silver was needed to trade for goods. The lack of consistent standards of measurement hindered both commerce and science in historical times.

Following the French Revolution of the late 18th century, efforts arose to develop a common system of units that was both *rational* and *natural*. It was rational in using powers of 10, rather than awkward relations such as 12 inches = 1 foot. The new system was natural in basing units on scales found in nature, which in principle anyone could measure. The foot had been based on the length of one person's foot and therefore wasn't reproducible. The new distance unit, the meter, was defined as one ten-millionth of an arc from Earth's equator to the North Pole (Figure 1.1). The gram, the unit of mass, was defined as the mass of one cubic centimeter of water. The attempt to introduce a decimal system of time—with 100 seconds per minute and so forth—proved unpopular, so we're stuck with 60-second minutes, 60-minute hours, and 24-hour days.

Those 18th century units evolved into our modern **SI system** (for Système Internationale). Definitions of the meter and gram have changed, but their values are quite close to those defined more than 200 years ago. The base units of distance, time, and mass are the **meter** (m), **second** (s), and **kilogram** (kg).

The speed of light in vacuum is a universal constant, which SI defines to be *exactly* 299,792,458 meters per second (m/s). The second is based on an atomic standard: the duration of 9,192,631,770 periods of the radiation from a particular transition in the cesium-133 atom. With units for speed (distance/time) and time defined, the meter is then defined as the distance light travels in 1/299,792,458 s.

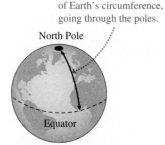

FIGURE 1.1 Arc length from the North Pole to the equator, used in the original definition of the meter.

Arc length is one-fourth of Earth's circumference, going through the poles.

North Pole

Equator

✓ **TIP**

The speed of light is defined to be an exact nine-digit number in the SI system.

Mass is still defined in terms of a prototype standard, a 1-kg lump of platinum-iridium alloy, kept at the French Bureau International des Poids et Mesures in Sèvres, France. Copies are kept throughout the world, including at the National Institute of Standards and Technology in Maryland. Using a prototype makes physicists uneasy, because its mass can change over time. Alternative and universally reproducible mass standards are under consideration.

The kilogram, meter, and second provide the basic units you'll need to study motion, force, and energy in the early chapters of this book. Later we'll introduce other SI units, such as the kelvin (K) for temperature and ampere (A) for electric current.

Reviewing New Concepts: The SI System

In the SI system,
- Distance is measured in meters (m),
- Time is measured in seconds (s), and
- Mass is measured in kilograms (kg).

EXAMPLE 1.1 **Measuring Earth**

Using Earth's mean radius R_E = 6,371,000 m, estimate the length of an arc from North Pole to equator and compare with the 18th-century definition of the meter. Approximate Earth as a perfect sphere.

ORGANIZE AND PLAN From geometry, you know that the circumference of a circle is $2\pi r$, where r is the radius. An arc from the North Pole to the equator is one-fourth of a spherical Earth's circumference (Figure 1.1).

Known: Mean radius R_E = 6,371,000 m.

SOLVE The arc distance d is one-fourth of the circumference $2\pi r$, so

$$d = \frac{2\pi R_E}{4} = \frac{\pi R_E}{2} = \frac{\pi(6,371,000 \text{ m})}{2} = 10,007,543 \text{ m}$$

This answer is remarkably close to the 10,000,000-m value in the 18th-century definition. It's higher by only

$$\frac{10,007,543 \text{ m} - 10,000,000 \text{ m}}{10,000,000 \text{ m}} \times 100\% = 0.075\%$$

REFLECT Our result is approximate, because it doesn't account for Earth's not-quite-spherical shape. In addition to irregular mountains and valleys, the planet's overall shape is slightly flattened at the poles and bulging at the equator. You might also rightly question the precision of the result, based on the number of significant figures. We'll deal with that issue in Section 1.4.

MAKING THE CONNECTION Earth's equatorial radius is 6,378,000 m. How long is an arc one-fourth of the way around the equator?

ANSWER Using the same calculation as in the example, the arc is 10,018,538 m, greater than the pole-to-equator arc. This is consistent with Earth's equatorial bulge.

Scientific Notation and SI Prefixes

Example 1.1 involved some big numbers—a common situation in physics. You'll often encounter even larger numbers, as well as some very small ones—much less than 1. Scientific notation and SI prefixes help you manage large and small numbers.

Consider Earth's mean radius R_E = 6,371,000 m. **Scientific notation** lets you present such large numbers in more compact form using powers of 10. In this case,

$$R_E = 6.371 \times 10^6 \text{ m}$$

The number multiplying the power of 10 should be at least 1 but less than 10. Thus, you express 10,000,000 m as 1×10^7 m, not 10×10^6 m. Scientific notation is also useful for very small numbers. For example, the radius of a hydrogen atom is about 0.000 000 000 053 m. In scientific notation, that's 5.3×10^{-11} m.

Scientific notation helps you appreciate the wide range of distances, times, and masses you'll encounter in physics, as shown in Tables 1.1, 1.2, and 1.3 and Figure 1.2. Note that "human-sized" scales are near the center of the range (in terms of powers of 10) on each list, at around 10^0 = 1. It's no accident that humans have defined SI units so that we can measure everyday things with numbers that aren't too large or too small. We'll return to this point later when we consider some non-SI units physicists use sometimes for extraordinarily large and small things, such as galaxies and atoms.

An alternative to scientific notation is to use SI prefixes, as shown in Table 1.4. For example, 1.25×10^5 m is more easily written as 125 km. The range of visible wavelengths of light, 4.0×10^{-7} to 7.0×10^{-7} m (see Table 1.1), is equivalently 400 nm to 700 nm. Notice that most of the standard prefixes come at intervals of 1000. Exceptions to this pattern are the prefixes c and d; for example, you're probably used to measuring short distances in cm. We'll discuss the use of centimeters, grams, and other non-SI units in Section 1.2, in the context of unit conversions.

TABLE 1.1 Selected Distances in Meters (m)

Description	Distance (m)
Distance to farthest galaxy (estimated)	1×10^{26}
Diameter of our Milky Way galaxy	9×10^{20}
Distance light travels in 1 year	9.5×10^{15}
Mean distance from Earth to Sun	1.5×10^{11}
Earth's mean radius	6.4×10^{6}
Tallest mountain on Earth	8800
Typical adult human height	1.5 to 2.0
Wavelength of visible light	4.0×10^{-7} to 7.0×10^{-7}
Diameter of hydrogen atom	1.1×10^{-10}
Size of proton (approximate)	10^{-15}

TABLE 1.2 Selected Time Intervals in Seconds (s)

Time interval	Time in seconds (s)
Age of universe	4.3×10^{17}
Age of solar system	1.6×10^{17}
One century	3.2×10^{9}
Typical college class	3200
Record time for 100-m dash	9.7
Time between sound vibrations in a "concert A" note	2.3×10^{-3}
Time between successive FM radio wave crests	9.3×10^{-9} to 1.1×10^{-8}
Time between wave crests in visible light	1.3×10^{-15} to 2.3×10^{-15}
Time for light to travel across an atom	4×10^{-19}

This galaxy is 10^{21} m across and has a mass of $\sim 10^{42}$ kg.

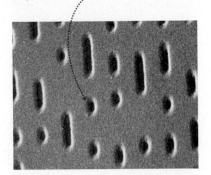

Your movie is stored on a DVD in "pits" only 4×10^{-7} m in size.

FIGURE 1.2 The study of physics ranges from the very large to the very small.

TABLE 1.3 Selected Masses in Kilograms (kg)

Typical galaxy	10^{42}
Sun	2.0×10^{30}
Earth	6.0×10^{24}
Blue whale	1.5×10^{5}
Adult human	50 to 100
Flea	10^{-5}
Dust particle	10^{-14}
Uranium atom	4.0×10^{-25}
Proton	1.7×10^{-27}
Electron	9.1×10^{-31}

TABLE 1.4 Some SI Prefixes*

Power of 10	Prefix	Abbreviation
10^{-18}	Atto	a
10^{-15}	Femto	f
10^{-12}	Pico	p
10^{-9}	Nano	n
10^{-6}	Micro	μ
10^{-3}	Milli	m
10^{-2}	Centi	c
10^{3}	Kilo	k
10^{6}	Mega	M
10^{9}	Giga	G
10^{12}	Tera	T
10^{15}	Peta	P
10^{18}	Exa	E

*For a more complete list, see Appendix B.

Scientific notation and SI prefixes are both acceptable, so you can use either. Scientific notation is handy for calculations, because you can plug the power of 10 directly into your calculator. Using SI prefixes sometimes makes comparisons more transparent. For example, if you're comparing distances of 6 mm and 30 mm, you can see that they differ by a factor of 5. In any case, it's good to know how to go between scientific notation and SI prefixes, as the following example illustrates.

✓ **TIP**

Use scientific notation or SI prefixes to express very large or very small numbers.

EXAMPLE 1.2 **An Astronomical Distance**

The mean distance from Earth to Sun is 149.6 million km. Express this distance in meters, using scientific notation.

ORGANIZE AND PLAN One million is 10^6, and the prefix k stands for 10^3, so 1 km = 10^3 m.

Known: distance d = 149.6 million km.

SOLVE Multiplying by the appropriate factors,

$$d = 149.6 \text{ million km} \times \frac{10^6}{\text{million}} \times \frac{10^3 \text{ m}}{\text{km}}$$

$$= 149.6 \times 10^9 \text{ m} = 1.496 \times 10^{11} \text{ m}$$

REFLECT One outstanding feature of SI is that only powers of 10 are involved in these kinds of conversions. In Section 1.2 we'll address the more general case when conversion factors aren't necessarily powers of 10.

MAKING THE CONNECTION Red laser light has a wavelength of 6.328×10^{-7} m. Express this value in nanometers, the unit commonly used for visible wavelengths.

ANSWER The wavelength can be written as 632.8×10^{-9} m, which is also known as 632.8 nm. Conversions between scientific notation and SI prefixes work just the same when the exponents are negative.

APPLICATION **Distance to the Moon**

Knowing the speed of light helps scientists determine the exact distance from Earth to Moon. A laser beam is directed from Earth toward a reflector that Apollo 11 astronauts placed on the Moon in 1969. Measuring the light's round-trip travel time allows the distance to be calculated. This method gives the distance (about 385,000 km on average) to within about 3 cm!

GOT IT? **Section 1.1** Rank the following masses, from greatest to least: (a) 0.30 kg; (b) 1.3 Gg; (c) 23 g; (d) 19 kg; (e) 300 g.

1.2 Converting Units

Physicists generally use SI units, and we'll follow that practice. But other units appear in everyday situations, and sometimes there are good reasons for using non-SI units in physics. For example, you know what it's like to ride in a car at 60 mi/h, but you may not have a feel for the SI equivalent of about 27 m/s. In chemistry or medicine, you'll find it convenient to measure volume in liters or cm^3 rather than m^3. Finally, physicists often find non-SI units useful when the SI value would be extremely large, such as the distance to another galaxy, or very small, such as the energy released by an atom.

Because you'll sometimes use non-SI units, it's important to be able to convert between these and their SI equivalents. You should work through the following examples and the end-of-chapter problems until you're comfortable doing conversions. It's important that you not let this mathematical detail stand in the way of your real objective, which is to learn physics!

✓ **TIP**

Learn to think in SI units. Look around at familiar distances and masses and think about how large they are in SI units.

cgs and Other Non-SI Systems

You've probably made measurements using the **cgs system**, with centimeters for distance (100 cm = 1 m), grams for mass (1000 g = 1 kg), and seconds for time. Most conversions between cgs and SI quantities only involve powers of 10. Chemists often use cgs, for example, giving 12.0 g as the mass of one mole of carbon, or 1.0 g/cm^3 as the density of water. Density (symbol ρ, the Greek letter "rho") is defined as mass/volume, or

$$\rho = \frac{m}{V}$$

Example 1.3 illustrates how to convert density from the cgs units g/cm^3 to the SI units kg/m^3.

The **English system** of units is still used in the United States outside the scientific community. Speed limits are given in miles per hour, and temperatures in degrees Fahrenheit. We'll avoid the English system, except where it might help provide you with a familiar context. Conversions between English and SI units involve conversion factors that aren't powers of 10.

Occasionally we'll meet non-SI units used in science. For example, astronomers use the light year—the distance traveled by light in 1 year—for expressing large distances, such as from the Sun to another star. Astronomers use the method of **parallax** to find the distance to nearby stars. As Earth moves around its orbit, the direction to nearby stars changes slightly, as shown in Figure 1.3. Given the angles and the known diameter of Earth's orbit, the distance to the star follows.

It's important to know which system you're using and to use it consistently. In a well-publicized incident in 1999, the Mars Climate Orbiter spacecraft was lost due to a navigation error. An investigation showed that the two scientific teams controlling the spacecraft were using two different unit systems—SI and English. Failure to convert between the two systems resulted in the spacecraft going off course and entering the Martian atmosphere with the wrong trajectory.

Tactic 1.1 shows how to convert units—you multiply the starting quantity by factors equal to 1 until the desired units replace the original ones. The following examples illustrate this strategy.

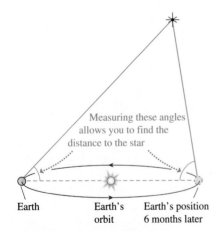

FIGURE 1.3 Measuring the distance to a star using the method of parallax.

TACTIC 1.1 **Unit Conversions**

To convert from one unit system to another, multiply the starting quantity by a fraction equal to 1, with the fraction defined by the known conversion factor. Multiplying by a factor equivalent to 1 doesn't change the quantity's physical value but trades the old unit for a new one.

Suppose you're converting 1.51 miles (mi) into meters.

- The conversion factor, found in Appendix C, is 1 mi = 1609 m. Express this as a fraction:

$$\frac{1609 \text{ m}}{\text{mi}}$$

- The fraction is equivalent to 1, because its numerator and denominator are equivalent. Multiplying the original 1.51 m by this fraction gives

$$1.51 \text{ mi} \times \frac{1609 \text{ m}}{\text{mi}} = 2430 \text{ m}$$

Notice that the mi unit canceled, leaving just meters (m).

- If necessary, repeat this process until all the old units have been traded for the desired ones.
- Once you've finished the conversion, check that the answer makes sense in terms of quantities you know or can imagine. Can you relate the final value to your experience, and if so, does the numerical value make sense?

EXAMPLE 1.3 **Density**

Gold is one of the densest pure metals, with density 19.3 g/cm^3. Convert this density to SI units.

ORGANIZE AND PLAN There are really two conversions here, for the mass unit (g to kg) and the distance unit (cm to m). Because the distance unit cm is cubed, that multiplying factor also has to be cubed.

Known: Density ρ = 19.3 g/cm^3.

SOLVE Starting with the known density and multiplying by the appropriate conversion factors,

$$\rho = 19.3 \text{ g/cm}^3 \times \frac{1 \text{ kg}}{1000 \text{ g}} \times \left(\frac{100 \text{ cm}}{1 \text{ m}}\right)^3 = 19{,}300 \text{ kg/m}^3$$

REFLECT This is a large value, but reasonable. Why? Imagine a cubical block of gold 1 m on a side. Because gold is so dense, it would be extremely massive. So 19,300 kg (about 200 times the mass of a large person) is reasonable.

MAKING THE CONNECTION Find the factor to convert density in g/cm^3 to kg/m^3. Use this factor to convert the density of water (1.0 g/cm^3) to kg/m^3.

ANSWER The results of this example show that the conversion factor is 1 g/cm^3 = 1000 kg/m^3. Therefore, the density of water is 1000 kg/m^3. Think how massive a cube of water 1 m on a side would be, and you'll see that this makes sense.

EXAMPLE 1.4 **Highway Speeds**

You're cruising down the highway at 60 mi/h. Express this speed in SI units (m/s).

ORGANIZE AND PLAN This problem requires two conversions: miles to meters and hours to seconds. Appendix C gives 1 mi = 1609 m. You can go from hours to seconds via minutes.

Known: speed = 60 mi/h; conversions: 1 mi = 1609 m; 1 h = 60 min; 1 min = 60 s.

SOLVE Multiplying 60 mi/h by the appropriate conversion factors,

$$\text{speed} = 60 \frac{\text{mi}}{\text{h}} \times \frac{1609 \text{ m}}{\text{mi}} \times \frac{1 \text{ h}}{60 \text{ min}} \times \frac{1 \text{ min}}{60 \text{ s}} = 27 \text{ m/s}$$

REFLECT Is this reasonable? The distance 27 m is about one-fourth the length of a football or soccer field, and it makes sense that a fast car could travel this distance each second. Since we'll be working in SI throughout this book, it's a good idea to get an intuitive feel for speeds in m/s. Note that we rounded the answer from about 26.82 m/s to 27 m/s. We'll discuss rounding and significant figures in Section 1.4.

MAKING THE CONNECTION Canada and many other countries express highway speeds in km/h—not an SI unit because it uses hours for time. Convert 60 mi/h to km/h.

ANSWER This conversion is simpler than that in the example, because it requires converting only mi to km and leaves the h unchanged. The answer is 97 km/h. Most cars have a km/h scale on their speedometers, along with the mi/h scale.

EXAMPLE 1.5 **Astronomical Distances—The Light Year**

A common unit in astrophysics is the light year (ly), defined as the distance traveled by light in 1 year. How many meters are in 1 ly? What is the distance in meters to the nearest star, Proxima Centauri, about 4.24 ly away?

ORGANIZE AND PLAN Speed = distance/time, so distance = speed × time. The speed of light is given in Section 1.1. We'll convert the year to seconds in steps, using days, hours, minutes, and seconds, with 365.24 days in the average year.

Known: speed of light $c = 2.998 \times 10^8$ m/s; $d = 4.24$ ly.

SOLVE For 1 light year, use time $t = 1$ y: 1 ly $= ct$. Writing this out with the appropriate conversion factors:

$$1 \text{ ly} = (2.998 \times 10^8 \text{ m/s})(1 \text{ y}) \times \frac{365.24 \text{ d}}{1 \text{ y}} \times \frac{24 \text{ h}}{1 \text{ d}} \times \frac{60 \text{ min}}{1 \text{ h}} \times \frac{60 \text{ s}}{1 \text{ min}}$$

All units except m cancel, leaving 1 ly $= 9.461 \times 10^{15}$ m. Then the distance to Proxima Centauri is

$$d = 4.24 \text{ ly} \times \frac{9.461 \times 10^{15} \text{ m}}{1 \text{ ly}} = 4.01 \times 10^{16} \text{ m}$$

REFLECT Here's a case where distances are so vast that it's not easy to see whether the result is reasonable. It's certainly large, which is good. If you had gotten an answer like 42 m, or 3×10^{-6} m, you would have known to try again!

MAKING THE CONNECTION The diameter of our Milky Way galaxy is about 9.5×10^{20} m. Express this in light years.

ANSWER Using the conversion from the example gives a diameter close to 100,000 ly, meaning it takes light 100,000 years to cross the galaxy!

GOT IT? **Section 1.2** Rank the following speeds in order from greatest to least: (a) 100 mi/h; (b) 40 m/s; (c) 135 ft/s; (d) 165 km/h.

1.3 Fundamental Constants and Dimensional Analysis

You saw in Section 1.1 that the speed of light in a vacuum is defined as $c = 299{,}792{,}458$ m/s—a quantity that, in turn, defines the meter. Note the qualifying phrase "in a vacuum"; that's because light travels more slowly in media such as air, water, or glass. The speed of light in different media is related to the refraction of light, which you'll study in Chapter 21. It makes sense that the speed of light is closely related to distance and time standards. As you'll see in Chapter 20, c is also related to fundamental constants in electricity and magnetism.

Masses of subatomic particles, such as the proton and electron in Table 1.3, are also important constants. Everything around you is made from a few basic particles. Protons and neutrons form atomic nuclei. Nuclei and electrons form atoms. Atoms join to make molecules. Molecules interact to form the solid, liquid, and gaseous substances of our world. We'll discuss the properties of liquids and solids in Chapter 10 and gases in Chapters 12–14.

We'll introduce other fundamental constants throughout the book. You can find them listed on the inside covers of this book, as well as in your scientific calculator. You should become familiar with these constants, but don't bother to memorize them.

Dimensional Analysis

Mechanics is the study of motion, and constitutes roughly the first third of this book. Distance, time, and mass are the base **dimensions** of mechanics. Other quantities in mechanics combine these three base dimensions. For example, speed is distance/time. To make dimensional comparisons easy, you can use the notation L for length, T for time, and M for mass. With this notation, the dimensions of speed (distance/time) are then written L/T.

Different **units** can describe a quantity with the same dimensions. For example, speed has dimensions L/T, but you can express speed in m/s, mi/h, fathoms per fortnight, or any other distance and time units. Units are important because they reveal the dimensions of a physical quantity. For example, a rectangle's area is the product of two lengths, so its dimensions are L^2. The corresponding SI units are m^2. If you compute an area and end up with units of m or m^3, you know you've made a mistake. After any calculation, check the units of the answer. We'll often remind you to do this in the final "Reflect" step of our problem-solving strategy.

✓**TIP**

Keep track of the units you use throughout a calculation. If your result has units inappropriate for the quantity you're trying to calculate, you've made a mistake.

You can often gain insight into a problem just by examining dimensions—a process called **dimensional analysis**. As an example, consider the *kinetic energy* of a moving body. As you'll see in Chapter 5, it has dimensions ML^2/T^2. How does kinetic energy depend on mass? Note that the dimension M appears to the first power; therefore, kinetic energy should depend linearly on mass. This leaves dimensions L^2/T^2, showing that kinetic energy depends on the *square* of the speed. Therefore, kinetic energy is proportional to mv^2, where m is mass and v is speed. We say "proportional to," because dimensional analysis can't reveal whether there are dimensionless factors involved. In this case, there's a factor of $\frac{1}{2}$, so kinetic energy is $\frac{1}{2}mv^2$.

CONCEPTUAL EXAMPLE 1.6   **Gravitational Potential Energy**

In Chapter 5, we'll introduce **potential energy**. If you hold a rock at a height h above the ground, it has potential energy, which then changes into kinetic energy (the energy of motion) as the rock falls. Potential energy depends on the height h, the rock's acceleration g (which has dimensions L/T^2), and the rock's mass m. Use dimensional analysis to find a combination of these quantities that gives the correct dimensions of potential energy, ML^2/T^2.

SOLVE First consider mass. Because the dimension M appears to the first power, the potential energy is proportional to m.

The only quantity here that includes dimensions of time is the acceleration g, with dimensions L/T^2. Because the potential energy must include time to the -2 power $(1/T^2)$, potential energy has to be proportional to g.

Where does that leave you? So far you know that potential energy is proportional to the product of m and g, which has dimensions ML/T^2. The potential energy has dimensions ML^2/T^2, so you need one more power of L. This comes from the height h, and, therefore, potential energy is proportional to the product mgh.

REFLECT Dimensional analysis reveals only how the potential energy depends on the three variables m, g, and h, and might miss numerical factors such as $\frac{1}{2}$. You'll learn in Chapter 5 that there are no missing numerical factors in this case, so potential energy $= mgh$.

1.4 Measurement, Uncertainty, and Significant Figures

In physics you often combine two or more quantities in a mathematical operation. For example, to calculate density you divide the mass by the volume. Here we'll describe how to handle the numbers in such calculations.

Measurement and Uncertainty

Measurements of physical quantities involve uncertainty. Using a millimeter ruler, you might determine the diameter of a nail to within a few tenths of a millimeter. But if you use a micrometer or calipers (Figure 1.4), you can get a result good to one hundredth of a millimeter.

Scientists distinguish between **accuracy** and **precision** of a measurement. Accuracy refers to how close a measurement is to the true or accepted value. Precision refers to the uncertainty of individual measurements, and it often follows from the "spread" of a number of repeated measurements taken using the same procedure. It's possible to be very precise yet lack accuracy. For example, if you make repeated measurements of the mass of the standard kilogram and consistently obtain values of about 1.12 kg, you have precision but not accuracy. A different instrument might be accurate but not precise, for example, if repeated measurements produce a spread of values between 0.90 kg and 1.10 kg.

FIGURE 1.4 The micrometer is used to measure the sizes of objects with great precision. An object to be measured is placed in the opening, and the spindle on the right is turned until the object is secure. Then the numerical scale registers the length of the object, typically with a precision on the order of 0.01 mm.

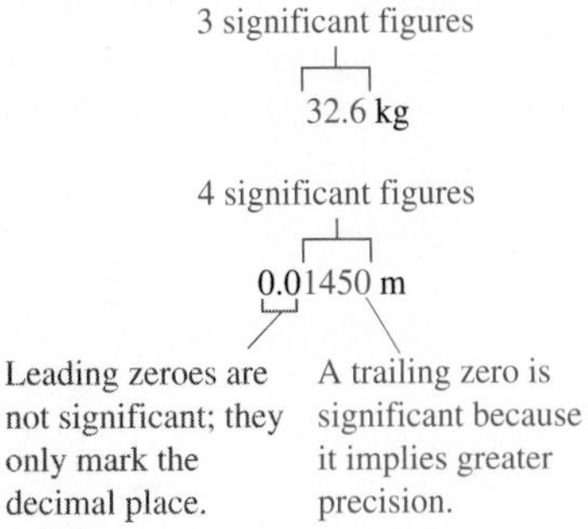

3 significant figures

32.6 kg

4 significant figures

0.01450 m

Leading zeroes are not significant; they only mark the decimal place.   A trailing zero is significant because it implies greater precision.

FIGURE 1.5 Examples of how to count significant figures.

Significant Figures

A mass measurement using the imprecise balance described above would be quoted as 1.00 kg ± 0.10 kg, meaning you claim with some confidence that the true mass is between 0.90 kg and 1.10 kg. The precision of measurement determines the number of **significant figures** in a measured quantity. For example, suppose you measure the length of a rectangular room to be 14.25 m ± 0.03 m. This length has four significant figures, because even though the last digit (the 5) is uncertain, it still conveys some information. Similarly, if you measure the width of the room to be 8.23 m ± 0.03 m, this measurement has only three significant figures, even though the uncertainty is the same.

Sometimes the number of significant figures isn't obvious, particularly when the quantity includes zeroes. Leading zeros that mark the decimal point aren't significant. Thus a 0.0015-m measurement of the thickness of a sheet of cardboard has only two significant figures. You could just as well express this as 1.5 mm or 1.5×10^{-3} m, which makes it clearer that there are two significant figures. Zeros after the decimal point, however, are significant (Figure 1.5). For example, the distance 3.600 m has four significant figures. If there were only three or two significant figures, it would be reported as 3.60 m or 3.6 m, respectively.

Suppose a car's mass is given as 1500 kg. It's not clear whether the zeros here are significant or merely mark the decimal point. Which it is depends on the precision of the scale used to weigh the car. Is it good to the nearest kilogram, or only the nearest 100 kg? In this book you can assume that all the figures shown are significant. In this case, that would mean 1500 kg has four significant figures. We would write 1.5×10^3 kg to express a measurement good to only two significant figures.

Significant Figures in Calculations

Significant figures are important in making calculations and in reporting the results. Suppose you want the area of that room described above. Based on the reported uncertainties, the area could be anywhere between 14.22 m × 8.20 m ≈ 116.6 m² and 14.28 m × 8.26 m ≈ 118.0 m². How should you report your answer?

A simplified approach, based on counting the significant digits, follows this rule:

> When multiplying or dividing two quantities, the answer should be reported with a number of significant figures equal to the smaller number of significant figures in the two factors.

Here you report three significant figures, because the width 8.23 m has only three. Thus, the answer is 14.25 m × 8.23 m = 117.2775 m² ≈ 117 m², where we've rounded to three significant figures. Note that this answer is comfortably between the extremes computed previously.

Here's another significant-figure rule:

> When adding or subtracting two quantities, the number of decimal places in the result equals the smallest number of decimal places in any of the values you started with.

Thus the sum 6.459 m + 1.15 m is rounded from 7.609 m to 7.61 m, because the term 1.15 m has only two significant figures.

Some numbers have exact values. For example, the volume of a sphere of radius r is

$$V = \frac{4}{3}\pi r^3$$

Here the numbers 4 and 3 are exact, so they don't reduce the number of significant figures in the result. The number π is also exact, even though it's irrational. You can use the value of π to as many significant figures as you like. The value of π that's built into your calculator probably carries more significant figures than you're likely to see in any measured quantity. Therefore, the number of significant figures you report for the volume of a sphere is the number of significant figures in the radius r.

✓ **TIP**

Use the value of π that's built into your calculator. If available, also use built-in values for physical constants such as the speed of light. This will give you plenty of significant figures and eliminate the chance for error in entering values manually.

EXAMPLE 1.7 **Significant Figures**

A physician uses ultrasound to measure the diameter of a fetus's head as 4.16 cm. Treating the head as a sphere, what's its volume? Discuss your use of significant figures.

ORGANIZE AND PLAN Computing the volume is straightforward. With diameter d, the radius $r = d/2$, and the volume is

$$V = \frac{4}{3}\pi r^3$$

Known: Diameter $d = 4.16$ cm.

SOLVE The radius is $r = d/2 = 2.08$ cm, so the volume is

$$V = \frac{4}{3}\pi(2.08 \text{ cm})^3 = 37.694554 \text{ cm}^3$$

where we've written all the numbers our calculator displayed. How many significant figures should be in the reported answer? The formula involves r^3, which means successive multiplication of numbers with three significant figures. By rule, the value of r^3 should be good

to three significant figures. The values of 4, 3, and π are exact, so multiplying by these factors doesn't affect the number of significant figures. Therefore, the answer should be rounded to three significant figures, or

$$V = 37.7 \text{ cm}^3$$

REFLECT It seems like you're throwing away some potentially useful information by discarding numbers in rounding. What if you need a calculated number to do another calculation? We'll address this issue in the next section.

MAKING THE CONNECTION If that fetal head has mass 37 g, what's its density?

ANSWER Density $\rho = m/V$, which gives a density of 0.98 g/cm^3. We've rounded to two significant figures, because that's all we were given for the mass. Our answer makes sense, since it's a little less than the 1 g/cm^3 density of water.

Significant Figures and Rounding

You've seen that it's often necessary to round your final answer to get the proper number of significant figures. However, every time you round, you discard potentially useful information. Therefore, you should keep more digits in intermediate calculations—as many as your calculator provides. If you need to report intermediate results, be sure to round to the appropriate number of significant figures. But keep the full number of digits as you go on to the next calculation.

✓ **TIP**

Don't round off too early when doing a calculation. Wait until the last step, when you reach the value you're reporting.

Perils of Rounding

You're finding the density of a rectangular copper block, with mass 24.75 g and sides 1.20 cm, 1.41 cm, and 1.64 cm (Figure 1.6). (a) Compute the block's volume. (b) Compute its density in two ways: first using the rounded value of the volume from part (a) and then using the non-rounded value. Compare with the known density of copper, $\rho = 8.92 \times 10^3$ kg/m^3.

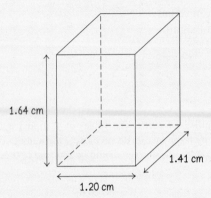

FIGURE 1.6 Our sketch for Example 1.8.

ORGANIZE AND PLAN The block is rectangular, so its volume is the product of the lengths of the three sides. Then density = mass/volume: $\rho = m/V$.

Known: sides 1.20 cm, 1.41 cm, and 1.64 cm; $m = 24.75$ g; accepted density of copper $\rho = 8.92 \times 10^3$ kg/m^3.

SOLVE (a) The block's volume is the product of its sides, so in SI:

$$V = (0.0120 \text{ m})(0.0141 \text{ m})(0.0164 \text{ m}) = 2.77488 \times 10^{-6} \text{ m}^3$$

The side lengths have three significant figures, so their product should be rounded to three figures, leaving $V = 2.77 \times 10^{-6}$ m^3.
(b) Using the rounded volume, the density becomes

$$\rho = \frac{m}{V} = \frac{0.02475 \text{ kg}}{2.77 \times 10^{-6} \text{ m}^3} = 8.935 \times 10^3 \text{ kg/m}^3$$

which rounds (again to three significant figures) to 8.94×10^3 kg/m^3. However, using the non-rounded volume from part (a),

$$\rho = \frac{m}{V} = \frac{0.02475 \text{ kg}}{2.77488 \times 10^{-6} \text{ m}^3} = 8.919 \times 10^3 \text{ kg/m}^3$$

which rounds to the accepted value 8.92×10^3 kg/m^3.

REFLECT Keeping the extra digits from the first step led to a more accurate density. Remember that the correct answer to part (a) is the rounded value, even though you use the non-rounded value subsequently in part (b).

MAKING THE CONNECTION You've found an irregularly shaped gold nugget. How can you determine its volume and density?

ANSWER Find the volume by displacement—place the sample underwater and measure the rise in water level; then weigh it and calculate density. In Chapter 10 you'll see how to measure density directly using Archimedes' principle, by weighing the sample in air and underwater.

Reviewing New Concepts: Significant Figures

- The number of significant figures in a measured quantity depends on the measurement precision.
- When multiplying or dividing quantities, the answer should be reported with a number of significant digits equal to the smaller number of significant digits in the factors.
- When adding or subtracting quantities, the number of decimal places in the result equals the smallest number of decimal places in any of the quantities.
- Keep all the digits you can in your calculator, but follow the rules for rounding when reporting an answer.

Order-of-Magnitude Estimates

Physicists often make **order-of-magnitude estimates**, giving a physical quantity to the nearest power of 10 or to within a factor of 10. Doing an order-of-magnitude estimate is useful for checking that a computation makes sense. We try to encourage this practice in the "reflect" step of our worked examples. Sometimes you don't have access to accurate values. In this case an order-of-magnitude estimate is all you can do.

For example, suppose you plan to drive across the United States with friends and want to know how much time to allow. You'll take turns driving and make brief stops for food and gas. Normal highway speed is about 100 km/h, and you guess that stops might reduce your average to 90 km/h. The distance depends on your route, but without checking maps you estimate it at 5000 km. Then with speed = distance/time, your estimated time is

$$\text{time} = \frac{5000 \text{ km}}{90 \text{ km/h}} = 56 \text{ h}$$

or about $2\frac{1}{3}$ days. There are many places where this calculation could be off. You haven't checked the exact distance, but it's surely much more than 1000 km and much less than 10,000 km. Your estimate of average speed is probably good to within about 20%. Thus, your result is likely correct to an order of magnitude. There's no way you'll complete the trip in 1 day, but it won't take 10 days, unless your car breaks down!

The next example shows that you don't have to guess at every quantity you use in an order-of-magnitude estimate. Feel free to consult books or web sources for numbers you need.

EXAMPLE 1.9 **How Many Atoms?**

Estimate the number of atoms in a human body. Assume a typical body mass of 70 kg.

ORGANIZE AND PLAN This estimate requires some knowledge of the body's composition, because atoms have vastly different masses. You've probably heard that more than half the human body consists of water. A quick check of web resources gives estimates of 60% to 70% H_2O. So a first rough guess might be that the body is two-thirds hydrogen and one-third oxygen.

It's hard to do much better than that. Except for the water, you consist largely of organic molecules comprising carbon, hydrogen, and oxygen. Another quick resource check shows that almost 99% of the body is made up of these three atoms. So of the one-third of the body that is not water, much of it is also hydrogen and oxygen. We'll estimate the number of atoms by assuming that two-thirds are hydrogen and one-third are oxygen.

Known: Body mass = 70 kg.

SOLVE The periodic table (Appendix D) gives the mass of a hydrogen atom as about 1 u = 1.66×10^{-27} kg, and oxygen as 16 u. Then the mass of a water molecule (H_2O) is 18 u, so the average mass of the

three atoms is $(18 \text{ u})/3 = 6 \text{ u}$. The number of atoms can then be found through unit conversions, starting with the 70-kg body mass:

$$70 \text{ kg} \times \frac{1 \text{ u}}{1.66 \times 10^{-27} \text{ kg}} \times \frac{1 \text{ atom}}{6 \text{ u}} = 7 \times 10^{27} \text{ atoms}$$

REFLECT This is an example of an estimate where the actual number might vary. Two 70-kg people could have different portions of bone, muscle, and fat. But the point of the order-of-magnitude estimate is to get within a factor of 10, and it's unlikely that one 70-kg person has 10 times the number of atoms of another. Given the numbers used in this calculation, the results are likely accurate to well within a factor of 10.

MAKING THE CONNECTION A typical body actually contains about 63% hydrogen, 24% oxygen, and 12% carbon. Would the use of this more accurate data change our order-of-magnitude estimate?

ANSWER Each carbon atom has a mass of 12 u. The weighted average of these three atomic masses is still about 6 u, so our order-of-magnitude estimate won't change.

GOT IT? Section 1.4 An athlete's mass is reported as 102.50 kg. How many significant figures are being reported? (a) 2; (b) 3; (c) 4; (d) 5.

Chapter 1 in Context

This chapter introduced some of the basic tools you'll need to do quantitative physics. The *SI system* is preferred for most measurements and computations, and you should begin to develop a feel for the distances and masses measured in SI units. Occasionally, some non-SI systems are used, so it's important to know how to do *unit conversions*. Other tools we've introduced include *dimensional analysis*, the proper use of *significant figures*, *rounding*, and *estimation*.

Looking Ahead With basic tools in place, you will begin your study of motion—kinematics—in Chapters 2 and 3. Chapter 2 covers motion in one dimension and introduces the concepts of displacement, velocity, and acceleration. Chapter 3 extends that study to two-dimensional motion and introduces vectors for describing motion in more than one dimension. Chapter 4 then introduces forces, which are responsible for changes in motion.

CHAPTER 1 SUMMARY

Distance, Time, and Mass Measurements

(Section 1.1) The **SI unit system** sets our standards of measure. Distance is measured in **meters (m), time** in **seconds (s)**, and **mass in kilograms (kg)**.

Large and small SI quantities are expressed using **scientific notation** or **SI prefixes**.

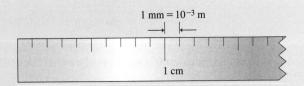

Converting Units

(Section 1.2) **Scientific notation and SI prefixes** both present very small or very large numbers in compact form:

Mean radius of Earth = 6,371,000 m = 6.371×10^6 m = 6.371 Mm

Some common non-SI systems are the **cgs system** (centimeters, grams, seconds) and the **English system**, still widely used in the United States.

Conversions between cgs and SI quantities only involve powers of 10. **Speed** and **density** are two common examples of such conversions.

cgs to SI: 100 cm/s = 1 m/s   1 g/cm³ = 1000 kg/m³

Some relationships: Speed = $\dfrac{\text{distance}}{\text{time}}$ Density $\rho = \dfrac{m}{V}$

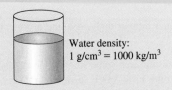

Water density:
$1 \text{ g/cm}^3 = 1000 \text{ kg/m}^3$

Fundamental Constants and Dimensional Analysis

(Section 1.3) **Fundamental constants** include the speed of light, c, and the masses of subatomic particles such as electrons and protons.

Dimension refers to a specific physical quantity and how that quantity depends on distance, time, and mass. **Units** reveal the dimensions of the physical quantity.

Dimensional analysis can be used to analyze problems without using numbers.

Some fundamental constants:

c = 299,792,458 m/s m(proton) = 1.67×10^{-27} kg
m(electron) = 9.11×10^{-31} kg

Notation for dimensional analysis:
L for length   **T** for time   **M** for mass.

Accuracy describes how close a measurement is to the true or accepted value.

Precision describes the repeated reliability of individual measurements.

Measurement, Uncertainty, and Significant Figures

(Section 1.4) The number of **significant figures** reflects the precision of measurement. **Rounding** to get the correct number of significant figures may reduce accuracy, so should be avoided until the final answer.

Rounding to the nearest 0.1m:
42.682 m → 42.7 m

Order-of-magnitude estimates give physical quantities to within a factor of 10. Estimates are useful for determining whether a computed or reported quantity makes sense.

Determining significant figures:

Distance = 0.0015 m = 1.5 mm = 1.5×10^{-3} m. This quantity has two significant figures. Zeros used solely to mark the decimal point are not significant.

NOTE: Problem difficulty is labeled as ■ straightforward to ■■■ challenging. Problems labeled BIO are of biological or medical interest.

Conceptual Questions

1. Astronomers sometimes measure distances in **astronomical units**, where 1 AU = 1.496×10^{11} m, the mean distance from Earth to Sun. Why is this useful for distances within our solar system?

2. Describe a situation in which you might find it practical to use a non-SI unit.

3. What are the disadvantages of using a prototype (a piece of metal) for the standard kilogram?

4. **Nanotechnology** (which includes building small-scale machines and the electronic circuit boards inside computers) has gained a lot of attention recently. Why do you think the word "nanotechnology" was chosen?

5. Explain the difference between dimensions and units.

6. Radio signals used to communicate with spacecraft travel at the speed of light. What problems do you think might arise if scientists want to send a series of signals to maneuver a spacecraft into orbit around Saturn?

7. Can you add or subtract quantities with different units? Can you multiply or divide quantities with different units?

8. How many significant figures are in the number $\sqrt{2}$?

Multiple-Choice Problems

9. Ten million kg can be written as (a) 10^{13} g; (b) 10^{10} g; (c) 10^{7} g; (d) 10^{6} g.

10. The SI unit of speed is (a) mi/h; (b) km/h; (c) km/s; (d) m/s.

11. The age of the universe is approximately 13.7 Gy. This is approximately (a) 4.3×10^{11} s; (b) 4.3×10^{14} s; (c) 4.3×10^{17} s; (d) 4.3×10^{20} s.

12. The density of one kind of steel is 8.25 g/cm^3. Expressed in SI units, this is (a) 0.825 kg/m^3; (b) 825 kg/m^3; (c) 8250 kg/m^3; (d) $82,500 \text{ kg/m}^3$.

13. A car is speeding down the road at 85 mi/h. In SI units, this is (a) 38 m/s; (b) 40 m/s; (c) 42 m/s; (d) 44 m/s.

14. A car completes a 500-mile race in 3 hours, 8 minutes. Its average speed is (a) 63 m/s; (b) 67 m/s; (c) 71 m/s; (d) 75 m/s.

15. One planet has four times the surface area of another. What is the ratio of their volumes? (a) 2; (b) 4; (c) 8; (d) 16.

16. How many significant figures are in the quantity 16.500 m? (a) 2; (b) 3; (c) 4; (d) 5.

17. How many significant figures are in the quantity 0.0053 kg? (a) 2; (b) 3; (c) 4; (d) 5.

18. Expressed with the correct number of significant figures, what is the volume of a rectangular room that measures 12.503 m by 10.60 m by 9.5 m? (a) 1300 m³; (b) 1260 m³; (c) 1259 m³; (d) 1259.1 m³.

Problems

Section 1.1 Distance, Time, and Mass Measurements

19. ■ Express the following in scientific notation: (a) 13,950 m; (b) 0.0000246 kg; (c) 0.000 000 0349 s; (d) 1,280,000,000 s.

20. ■ Express the quantities in the preceding problem using SI units and prefixes with no powers of 10.

21. ■■ One metric ton is defined to be 1000 kg. How many kilograms are in a megaton?

22. ■■ Express the speed of light in units of μm/fs.

23. ■■ Earth's mean radius is 6.371 Mm. (a) Assuming a uniform sphere, what's Earth's volume? (b) Using Earth's mass of 5.97×10^{24} kg, compute Earth's average density. How does your answer compare with the 1000-kg/m³ density of water?

24. ■■ The average distance to the Moon is 385,000 km. How much time does it take a laser beam, traveling at the speed of light, to go from Earth to the Moon and back?

25. ■■ The summit of Mount Everest is 8847 m above sea level. What fraction of Earth's radius is that? Express your answer as a decimal using scientific notation.

26. ■■ Use the quantities in Tables 1.1, 1.2, and 1.3 to express the following quantities with SI prefixes and no powers of 10: (a) the distance traveled by light in 1 year; (b) the time since the formation of the solar system; (c) the mass of a typical dust particle.

Section 1.2 Converting Units

27. ■ A cheetah runs at 70 mi/h. Express this in m/s.

28. ■ The density of aluminum is 2.70 g/cm³. Express this in kg/m³.

29. ■ Basketball star Yao Ming is 7 feet, 6 inches tall. What's that in meters?

30. ■■ In 2004, Lance Armstrong won the 3395-km Tour de France with a time of 83 hours, 36 minutes, and 2 seconds. What was Armstrong's average speed, in m/s?

31. ■■ One place in the Hoh Rainforest in the state of Washington receives an average annual rainfall of 200 inches. What's this in meters?

32. ■■ The winner of the Kentucky Derby runs the 1.25-mile race in 2 minutes, 2.0 s. What's the horse's average speed in m/s? Compare with the speed of a human sprinter running (for a shorter distance!) at 10.0 m/s.

33. ■■ Early astronomers often used Earth's diameter as a distance unit. How many Earth diameters make up the distance (a) from Earth to Moon; (b) from Earth to Sun?

34. ■■ One year is approximately 365.24 days. (a) How many seconds are in 1 year? (b) A reasonable approximation to your answer in part (a) is $\pi \times 10^7$. By what percentage does this value differ from your answer?

35. ■■ One mole of atoms contains Avogadro's number, 6.02×10^{23} atoms. The mass of one mole of carbon atoms is exactly 12 g. What is the mass of a carbon atom, in kg?

36. ■■ A water molecule has mass 3.0×10^{-26} kg. How many molecules are in 1 liter (1000 cm³) of water?

37. ■■■ Derive the following conversion factors: (a) mi to km; (b) kg to μg; (c) km/h to m/s; (d) ft³ to m³.

38. ■■ It's claimed that a typical college class lasts about one microcentury. Express one microcentury in minutes. Comment.

39. ■■ Earth's equatorial radius is 6378 km. A spacecraft is in circular orbit 100 km above the equator. If the spacecraft orbits every 86.5 minutes, what's its speed?

40. ■■■ You're planning to drive across the border from the United States to Canada. Suppose the currency exchange rate is $1.00 US = $1.07 CDN. Are you better off buying gasoline at $4.30 per gallon in the United States or $1.36 per liter in Canada?

41. ■■■ Astronomers use the astronomical unit (abbreviated AU), equal to 1.496×10^{11} m, which is the mean distance from Earth to the Sun. Find the distances of the following planets from the Sun

in AU: (a) Mercury, 5.76×10^{10} m; (b) Mars, 2.28×10^{11} m; (c) Jupiter, 7.78×10^{11} m; Neptune, 4.50×10^{12} m.

42. ■■■ Astronomers define a **parsec** (short for parallax second) to be the distance at which 1 astronomical unit subtends an angle of 1 second. The astronomical unit (AU) was defined in the preceding problem, and in angular measure 1 degree = 60 minutes and 1 minute = 60 seconds of arc. Find the conversion factors that relate (a) parsecs to AU; (b) parsecs to m.

Section 1.3 Fundamental Constants and Dimensional Analysis

43. ■■ Planet A has twice the radius of planet B. What is the ratio of their (a) surface areas and (b) volumes? (Assume spherical planets.)

44. ■ (a) What are the dimensions of density? (b) What are the SI units of density?

45. ■ How much time does it take for light to travel (a) from Moon to Earth; (b) from Sun to Earth; (c) from Sun to the planet Neptune?

46. ■■ Newton's second law of motion (Chapter 4) says that the acceleration of an object of mass m subject to force F depends on both m and F. The dimensions of acceleration are L/T^2, and the dimensions of force are ML/T^2. Apart from dimensionless factors, how does acceleration depend on mass and force?

47. ■■ A spring hangs vertically from the ceiling. A mass m on the end of the spring oscillates up and down with period T, measured in s. The stiffness of the spring is described by the spring constant k, with units of kg/s^2. Apart from dimensionless factors, how does the period of oscillation depend on k and m? (You'll study this system in Chapter 7.)

48. ■■ The period (time for a complete oscillation) of a simple pendulum depends on the pendulum's length L and the acceleration of gravity g. The dimensions of L are L, and the dimensions of g are L/T^2. Apart from dimensionless factors, how does the period of the pendulum depend on L and g?

49. ■■ A ball is dropped from rest from the top of a building of height h. The speed with which it hits the ground depends on h and the acceleration of gravity g. The dimensions of h are L, and the dimensions of g are L/T^2. Apart from dimensionless factors, how does the ball's speed depend on h and g?

Section 1.4 Measurement, Uncertainty, and Significant Figures

50. ■ How many significant figures are in each of the following? (a) 130.0 m; (b) 0.04569 kg; (c) 1.0 m/s; (d) 6.50×10^{-7} m.

51. ■ How many significant figures are in each of the following? (a) 0.04 kg; (b) 13.7 Gy (the age of the universe); (c) 0.000 679 mm/s; (d) 472.00 s.

52. ■ Find the area of a rectangular room measuring 9.7 m by 14.5 m. Express your answer with the correct number of significant figures.

53. ■■ Find the area of a right triangle with sides 15.0 cm, 20.0 cm, and 25.0 cm. Express your answer with the correct number of significant figures.

54. ■■ A brand of steel has density 8194 kg/m^3. (a) Find the volume of a 14.00-kg piece of this steel. (b) If this piece is spherical, what's its radius?

55. ■■ Using calipers, you find that an aluminum cylinder has length 8.625 cm and diameter 1.218 cm. An electronic pan balance shows that its mass is 27.13 g. Find the cylinder's density.

56. ■■■ (a) A diver jumps from a board 1 m above the water. Make an order-of-magnitude estimate of the diver's time of fall and speed when she hits the water. (b) Repeat your estimates for a dive from a 10-m tower.

57. ■■ Estimate the number of heartbeats in an average lifetime.

58. ■■■ Estimate (a) the number of atoms and (b) the number of protons in planet Earth.

59. ■■ Make an order-of-magnitude estimate of the thickness of one page in this book.

60. ■■■ The brilliant 20th-century physicist Enrico Fermi, who worked for a time at the University of Chicago, made a classic estimate—the number of piano tuners in Chicago. (a) Try to repeat Fermi's estimate. State carefully the assumptions and estimates you're making. (b) Try a similar estimate of the number of auto repair shops in the Los Angeles metropolitan area.

General Problems

61. ■■ Take planets Venus and Earth to be spherical. Earth is slightly larger than Venus, with mass larger by a factor of 1.23 and radius by a factor of 1.05. (a) Which of the planets has the larger average density? (b) Find the ratio of their densities.

62. BIO ■■ **Fetus growth.** A child is born after 39 weeks in its mother's womb. (a) If the child's birth mass is 3.3 kg, how much mass on average does the fetus gain each day in the womb? (b) Assuming the fetal density is 1020 kg/m^3, what is the average volume gained each day?

63. ■■ Assume Saturn to be a sphere (ignore the rings!) with mass 5.69×10^{26} kg and radius 6.03×10^7 m. (a) Find Saturn's mean density. (b) Compare Saturn's density with that of water, 1000 kg/m^3. Is the result surprising? Note that Saturn is composed mostly of gases.

64. ■■ Compute the number of minutes in one 365-day year. Note: The answer is used in the song "Seasons of Love" from the musical *Rent*.

65. BIO ■■ **Horse race.** In 1973 the horse Secretariat set a record time of 2 minutes, 24 seconds in the 1.5-mile Belmont Stakes. (a) What was Secretariat's average speed, in SI units? (b) Find the ratio of Secretariat's speed to that of a human sprinter who runs the 100-m dash in 9.8 seconds.

66. ■■■ Eratosthenes, a Greek living in Egypt in the 3rd century BCE, estimated Earth's diameter using the following method. On the first day of summer, he noted that the Sun was directly overhead at noon in Syene. At the same time in Alexandria, 5000 stades northward, the Sun was 7.2° from overhead. One stade is about 500 feet. (a) Find Earth's diameter in stades and in meters. (b) Compare with today's accepted value, 12.7 Mm.

67. ■■ Estimate how many atoms are in your 0.500-L bottle of water (density 1000 kg/m^3).

68. BIO ■■■ **Blood flow.** The flow rate of a fluid is expressed as volume flowing per time. (a) What are the dimensions of flow rate, in terms of the dimensions M, L, and T? (b) What are its SI units? (c) Suppose a typical adult human heart pumps 5.0 L of blood per minute. Express this rate in SI. (d) If the heart beats 70 times per minute, what volume of blood flows through the heart in each beat?

69. BIO ■■■ **Oxygen intake.** Air has density 1.29 kg/m^3 at sea level and comprises about 23% oxygen (O_2) by mass. Suppose an adult human breathes an average of 15 times per minute, and each breath takes in 400 mL of air. (a) What mass of oxygen is inhaled each day? (b) How many oxygen molecules is this? (Note: The mass of one oxygen molecule is 32 u.)

70. ■■ In baseball, home plate and first, second, and third bases form a square 90 feet on a side. (a) Find the distance in meters across a

diagonal, from first base to third or home plate to second. (b) The pitcher throws from a point 60.5 feet from home plate, along a line toward second base. Does the pitcher stand in front of, on, or behind a line drawn from first base to third?

Answers to Chapter Questions

Answer to Chapter-Opening Question
Originally, the meter was one ten-millionth of the distance from the North Pole to the equator, the second was 1/86,400 of 1 day (Earth's rotation period), and the kilogram was 1000 times the mass of 1 cm^3 of water (water covers most of the planet). Today the units are defined differently but are close to these original suggestions.

Answers to GOT IT? Questions
Section 1.1   (b) > (d) > (e) = (a) > (c)
Section 1.2   (d) > (a) > (c) > (b)
Section 1.4   (d) 5

The server tosses the tennis ball straight up. What's the ball's acceleration as it goes up and down in free fall?

Chapter 1 gave you basic tools for quantitative physics. In Chapters 2 and 3 you'll learn to describe the motion of objects—a branch of physics called *kinematics*. The key quantities here are position, velocity, and acceleration. Kinematics only *describes* motion, without reference to causes. In Chapter 4 we'll visit the "cause" question, exploring the relation between forces and *changes* in motion. This branch of physics is *dynamics*, and it's governed by Newton's laws of motion.

Chapter 2 considers only motion in one dimension. There are many examples of motion that's one dimensional or nearly so. Studying one-dimensional motion will familiarize you with important concepts of kinematics, particularly velocity and acceleration. In Chapter 3 we'll expand these concepts for motion in more than one dimension.

2.1 Position and Displacement

Frames of Reference

When your friend asks for directions to your house, you might say something like: "Starting from your house, go one block east on Graham Street; turn right on McLean and go three blocks south; then look for the big white house on the northeast corner of Chestnut and McLean." Such directions assume a common *frame of reference*—in this case, a starting point, a unit of distance (city blocks), and knowledge of north, south, east, and west.

To Learn

By the end of this chapter you should be able to

- Distinguish displacement and distance.
- Distinguish speed and velocity.
- Understand average velocity and instantaneous velocity.
- Understand acceleration and its relation to velocity and position in one-dimensional motion.
- Solve problems involving constant acceleration in one dimension (including free fall).

An agreed-on reference frame is essential in describing motion. Physicists normally use Cartesian coordinates with SI units. The two-dimensional Cartesian system in Figure 2.1a should be familiar from math classes. That system could be used to describe two-dimensional motion, such as that trip between houses or the flight of a baseball, Motion in three dimensions—like an airplane's flight—requires a third axis, as shown in Figure 2.1b.

Coordinate systems are only artifacts we use to describe the physical world. Therefore, you're free to choose a coordinate system that suits your situation. That means choosing the **origin**—the point where the axes meet, and the zero of each coordinate—and the orientations of the coordinate axes. If you're doing an experiment with colliding pucks on an air table, you might use a two-dimensional coordinate system with a typical placement of the x- and y-coordinate axes (Figure 2.2a). To describe the flight of a soccer ball—also two-dimensional—a convenient choice has the x-axis horizontal and the y-axis vertical (Figure 2.2b). You could put the origin on the ground or at the height where the ball is kicked. But what about a skier going down a smooth slope? You might be tempted to make the x-axis horizontal and the y-axis vertical again. Although that choice isn't "wrong," a better choice might be to put the x-axis along the slope (Figure 2.2c). That choice makes the skier's motion entirely along the x-axis, so it's one dimensional.

The remainder of this chapter considers only one-dimensional motion. This allows us to introduce concepts of kinematics without worrying about a second or third dimension. There are many "real-world" situations in which motion is confined to one dimension, at least to an excellent approximation. Drop a rock, for example, and it falls straight down. You'll learn about freely falling objects in Section 2.5. Because it's simpler and yet covers real-world situations, one-dimensional motion is a good place to start your study of kinematics.

FIGURE 2.1 Two examples of coordinate systems.

(a) Two-dimensional coordinate system—could be used to represent motion in two dimensions.

(b) Three-dimensional coordinate system—could be used for motion in three dimensions.

Objects and Point Particles

Real objects—such as cars, stars, people, and baseballs—take up space and occupy more than one point. When we locate an object at a point on our coordinate system, we're treating the object as a **point particle**, with all its important properties (mass, for example, or electric charge) concentrated in a single point.

The fact that real objects aren't point particles isn't as great an issue as you might think. Later we'll show how to describe an object's motion in terms of a special point that represents a kind of average location. For now, though, you can consider any fixed point of reference you want—for example, the tip of a person's nose or the front of a car's hood (Figure 2.3)—as the point used to fix position.

TIP

You can use a point particle to locate a solid object at a single position.

Displacement and Distance

Walking down your street, you can describe your position using a single coordinate axis. You're free to put the axis wherever it's convenient, and to call it what you want. Here we've called it the x-axis, and made the sensible choice to put the origin ($x = 0$) at the start (in this case, literally the origin of your journey!), with the $+x$-direction being the direction you're walking (Figure 2.4). None of these choices affects physical reality, but they may make the mathematical description easier. For example, our choice of the $+x$-direction avoids negative positions.

Motion involves change in position. Physicists call the change in position **displacement**, and in one dimension it's represented by the symbol Δx. (The Greek uppercase delta, Δ, generally means "the change in . . .") Moving from some initial position x_0 to a final position x results in a displacement:

$$\Delta x = x - x_0 \quad \text{(Displacement in one dimension; SI unit: m)} \quad (2.1)$$

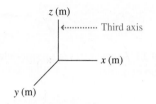
(a) Coordinate system for pucks colliding on air table

(b) Coordinate system for kicked ball

(c) Coordinate system for skier on slope

FIGURE 2.2 The coordinate axes we chose for the three situations.

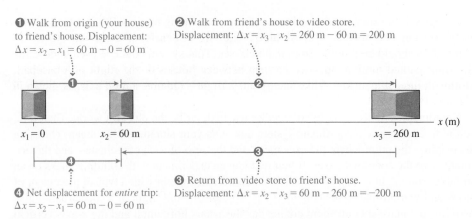

❶ Walk from origin (your house) to friend's house. Displacement: $\Delta x = x_2 - x_1 = 60 \text{ m} - 0 = 60 \text{ m}$

❷ Walk from friend's house to video store. Displacement: $\Delta x = x_3 - x_2 = 260 \text{ m} - 60 \text{ m} = 200 \text{ m}$

$x_1 = 0$ $x_2 = 60 \text{ m}$ $x_3 = 260 \text{ m}$ x (m)

❸ Return from video store to friend's house. Displacement: $\Delta x = x_2 - x_3 = 60 \text{ m} - 260 \text{ m} = -200 \text{ m}$

❹ Net displacement for *entire* trip: $\Delta x = x_2 - x_1 = 60 \text{ m} - 0 = 60 \text{ m}$

FIGURE 2.4 A trip illustrating displacement in one dimension.

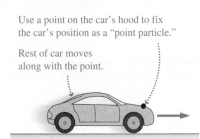

Use a point on the car's hood to fix the car's position as a "point particle."

Rest of car moves along with the point.

FIGURE 2.3 A real object (a car) modeled as a point particle.

For the walk of Figure 2.4, it's simple to compute displacements for each of the three segments shown. The figure also shows that positive displacements correspond to motion in the $+x$-direction, and negative displacements to motion in the $-x$-direction.

Note that the net **distance** traveled—here, the total number of meters you walked—is always positive and is not necessarily the same as displacement. Walking to your friend's house, then to the video store, then back to your friend's house gives a displacement $(x_2 - x_1)$ of only 60 m, but the total distance you walked is 260 m + 200 m = 460 m. This distinction between distance and displacement will be important when we define average velocity and average speed in Section 2.2.

✓ **TIP**

Displacement (change in position) is not necessarily the same as distance traveled.

CONCEPTUAL EXAMPLE 2.1   **Displacement and Distance**

Grand Island, Nebraska, is 160 km west of Lincoln. The road between these cities is essentially straight. For a round trip between Grand Island and Lincoln, find the displacement and the total distance traveled.

SOLVE There's no coordinate system specified, so your first task is to choose one. With the road running west-east, a good choice is to put $x = 0$ at Grand Island, with the $+x$-axis pointing east (Figure 2.5).

For a round trip, the final position and initial position are the same. Therefore, from the definition of displacement, $\Delta x = x - x_0 = 0$. The displacement is zero. On the other hand, the distance traveled is the sum of the two 160-km segments, out and back, or 320 km.

REFLECT How can displacement be zero when you've taken a trip? It's because displacement means the *net* change in position. If you return to your starting point, there's no net change in position—no matter what distance you've covered.

Grand Island   Lincoln

$x = 0$ $x = 160 \text{ km}$ x

FIGURE 2.5 Our sketch for Conceptual Example 2.1.

GOT IT? Section 2.1 Which one or more of the following are true about displacement and distance in one-dimensional motion? (a) Distance traveled can never be negative. (b) Distance traveled is always equal to displacement. (c) Distance traveled is always less than displacement. (d) Distance traveled is always more than displacement. (e) Distance traveled is greater than or equal to displacement.

2.2 Velocity and Speed

Describing Motion—Average Velocity and Average Speed

You developed an intuitive feel for motion long before taking physics. You sense the difference between driving at 30 miles per hour and going 60. Those are measures of **speed**, an important quantity in describing motion. You've probably also used the term **velocity**, which is related to speed but not exactly equivalent. Here we'll define these terms carefully and show how they're used in one-dimensional motion.

Consider a 100-m footrace. A video shows one runner's position at 1.0-s intervals, recorded in Figure 2.6a. Figure 2.6b shows a graph of this position-versus-time data. The data and graph show that this world-class runner finished the race in 10.0 s. But the graph also contains a wealth of information showing just how the runner got from start to finish.

Using the data (Figure 2.6a) and graph (Figure 2.6b), you can analyze the runner's motion during different parts of the race. Figure 2.6c is a **motion diagram**, showing a moving object's position at equal time intervals. It shows that the runner travels much farther during the second time interval than the first. A measure of the runner's progress during each time interval is **average velocity**, defined as the object's displacement Δx divided by the time interval Δt during which the displacement takes place. Symbolically,

$$\bar{v}_x = \frac{\Delta x}{\Delta t} \qquad \text{(Average velocity for motion in one dimension; SI unit: m/s)} \qquad (2.2)$$

where the bar over the v signifies an average value. The physical dimensions of velocity are distance divided by time, or m/s in SI. Because displacement can be positive, negative, or zero, so can average velocity. Positive velocity corresponds to displacement in the $+x$-direction, negative to displacement in the $-x$-direction.

Time, t (s)	Position, x (m)
0.0	0.0
1.0	4.7
2.0	13.6
3.0	23.4
4.0	34.0
5.0	45.0
6.0	56.0
7.0	67.0
8.0	78.0
9.0	89.0
10.0	100.0

(a) The data

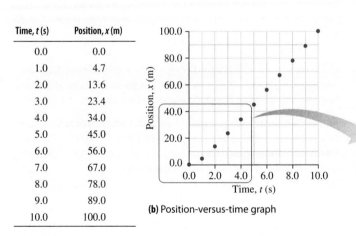

(b) Position-versus-time graph

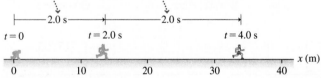

(c) Motion diagram for the first 4 seconds

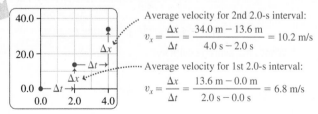

Average velocity for 2nd 2.0-s interval:
$$v_x = \frac{\Delta x}{\Delta t} = \frac{34.0\ \text{m} - 13.6\ \text{m}}{4.0\ \text{s} - 2.0\ \text{s}} = 10.2\ \text{m/s}$$

Average velocity for 1st 2.0-s interval:
$$v_x = \frac{\Delta x}{\Delta t} = \frac{13.6\ \text{m} - 0.0\ \text{m}}{2.0\ \text{s} - 0.0\ \text{s}} = 6.8\ \text{m/s}$$

(d) How to compute average velocity

FIGURE 2.6 Analyzing the runner's motion in a sprint race.

As an example, consider again the runner's motion in Figure 2.6. Using Equation 2.2, you can show that the runner's average velocity is much higher for the second 2-s interval of the race than it is for the first 2 s. As shown in Figure 2.6d, the average velocity is 6.8 m/s for the first 2 s and 10.2 m/s for the next 2 s.

The **average speed** for one-dimensional motion is defined as

$$\bar{v} = \frac{\text{distance traveled}}{\Delta t} \tag{2.3}$$

Recall from Section 2.1 that distance is always positive. Therefore, average speed is always positive, and it's not necessarily the same as average velocity.

EXAMPLE 2.2 **To the Store**

Consider your trip to the video store in Figure 2.4. It has three parts: (1) You walk from your house to the store in 3 minutes, 20 seconds; (2) You spend 5 minutes in the store; (3) You walk back to your friend's house in 2 minutes, 5 seconds. (a) Find your average velocity and average speed for each part of the trip. (b) Find your average velocity and average speed for the entire trip.

ORGANIZE AND PLAN Our sketch is shown in Figure 2.7. Average velocity is $\bar{v}_x = \Delta x/\Delta t$ (Equation 2.2), and average speed is $\bar{v}_x = $ distance traveled/Δt (Equation 2.3). To get the speed and velocity in SI units, we'll convert the times into seconds.

SOLVE For part (1), the time is 3 minutes, 20 seconds = 180 s + 20 s = 200 s. Then the average velocity is

$$\bar{v}_x = \frac{\Delta x}{\Delta t} = \frac{260 \text{ m}}{200 \text{ s}} = 1.3 \text{ m/s}$$

The average speed for part (1) is

$$\bar{v} = \frac{\text{distance traveled}}{\Delta t} = \frac{260 \text{ m}}{200 \text{ s}} = 1.3 \text{ m/s}$$

For part (2), $\Delta t = 5$ min = 300 s. However, displacement and distance are zero, so both the average speed and average velocity are zero.

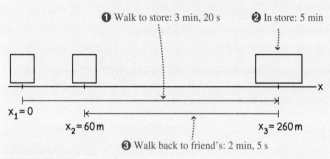

❶ Walk to store: 3 min, 20 s ❷ In store: 5 min

$x_1 = 0$
$x_2 = 60$ m
$x_3 = 260$ m
❸ Walk back to friend's: 2 min, 5 s

FIGURE 2.7 What's the average velocity?

For part (3), $\Delta t = 120$ s + 5 s = 125 s. Traveling in the $-x$-direction, the displacement is -200 m, and the distance is 200 m. Therefore, the average velocity for part (3) is

$$\bar{v}_x = \frac{\Delta x}{\Delta t} = \frac{-200 \text{ m}}{125 \text{ s}} = -1.6 \text{ m/s}$$

and the average speed is

$$\bar{v} = \frac{\text{distance traveled}}{\Delta t} = \frac{200 \text{ m}}{125 \text{ s}} = 1.6 \text{ m/s}$$

For the entire trip, the time is $\Delta t = 200$ s + 300 s + 125 s = 625 s. Earlier we found that the displacement is $\Delta x = 60$ m, while the total distance is 460 m. So the average velocity is

$$\bar{v}_x = \frac{\Delta x}{\Delta t} = \frac{60 \text{ m}}{625 \text{ s}} = 0.096 \text{ m/s}$$

and the average speed for the trip is

$$\bar{v} = \frac{\text{distance traveled}}{\Delta t} = \frac{460 \text{ m}}{625 \text{ s}} = 0.74 \text{ m/s}$$

REFLECT Note that average velocity and average speed are the same for some intervals but not others. Also, the average velocity and average speed for the entire trip are quite different, due to the reversal of direction at the video store.

MAKING THE CONNECTION What information does your average velocity of 1.3 m/s for part (1) of the trip convey about how fast you were walking at each moment of that part?

ANSWER From the average velocity alone, it's impossible to know how fast you were walking at each moment. You may have walked at a steady pace, or stopped intermittently for traffic. "At each moment" implies looking at a much shorter time interval.

Instantaneous Velocity

Look again at the runner's data showing position at 1-s intervals (Figure 2.6a). Using that data, you can compute the average velocity for any of those intervals. But what happens throughout each interval? Is it possible, for example, to say anything about the runner's velocity at $t = 1.6$ s? That would require more data—namely, positions at smaller time intervals.

Suppose you time the runner with a video-capture system capable of measuring positions to within 1 mm (0.001 m) and time intervals of 0.01 s. Imagine starting at some fixed time, say 1.00 s, and looking at the average velocity over shorter and shorter time intervals, all starting at $t = 1.00$ s. Table 2.1 shows some representative data in the interval between 1.00 s and 2.00 s that let you consider intervals ranging from one full second down to 0.01 s. For example, in the interval from 1.00 s to 2.00 s, the average velocity is

$$\bar{v}_x = \frac{\Delta x}{\Delta t} = \frac{13.629 \text{ m} - 4.711 \text{ m}}{2.00 \text{ s} - 1.00 \text{ s}} = 8.92 \text{ m/s}$$

as shown in the table.

Note that as the time interval shrinks, the average velocity seems to approach a specific value—in this case, about 6.8 m/s. If you had data for even smaller intervals and more precise positions, you would see that the average velocity does approach one value, in the limit as the time interval Δt approaches zero. That value is the **instantaneous velocity**. For one-dimensional motion, instantaneous velocity v_x is defined mathematically as

$$v_x = \lim_{\Delta t \to 0} \frac{\Delta x}{\Delta t} \quad \text{(Instantaneous velocity for motion in one dimension; SI unit: m/s)} \quad (2.4)$$

If you've taken calculus, you'll recognize Equation 2.4 as the definition of the **derivative**, with the shorthand notation $v_x = dx/dt$. We won't use calculus in this book, so you won't be asked to compute exact values of instantaneous velocity. It's important to remember, though, that a moving object has an instantaneous velocity at each moment in time.

TABLE 2.1 Data Table as Described, Including Average Velocity

Time t (s)	Position x (m)	Average velocity (m/s) with $\Delta t = t - 1.00$ s
1.00	4.711	
1.01	4.779	6.80
1.02	4.848	6.85
1.05	5.056	6.90
1.10	5.411	7.00
1.20	6.151	7.20
1.50	8.611	7.80
2.00	13.629	8.92

✓**TIP**

In lab you might do the experiment of Example 2.3 using video capture or a motion detector.

EXAMPLE 2.3   **Falling!**

In lab, a spark timer marks the position of a falling steel ball at 0.01-s intervals (Figure 2.8a). The ball starts from rest at time $t = 0$ at a position on the tape designated $x = 0$. Some data (position and time measurements) from dots farther down the tape are given in Figure 2.8b. Use the data to estimate the falling ball's instantaneous velocity at time $t = 0.60$ s.

ORGANIZE AND PLAN Instantaneous velocity is $v_x = \lim_{\Delta t \to 0} \Delta x / \Delta t$

(Equation 2.4). With data at finite intervals, you can't compute this limit exactly, but you can estimate it from the average velocity, using small intervals starting at the desired time (0.60 s). The average velocity over a time interval Δt is $\bar{v}_x = \Delta x / \Delta t$ (Equation 2.2).

There are plenty of time intervals to use. Starting at $t = 0.60$ s, the average velocity can be computed for intervals from 0.01 to 0.05 s. The value these average velocities approach as the time interval gets smaller is the best estimate of the ball's instantaneous velocity at $t = 0.60$ s.

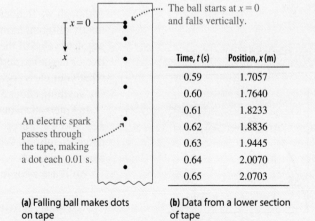

The ball starts at $x = 0$ and falls vertically.

$x = 0$

x

An electric spark passes through the tape, making a dot each 0.01 s.

Time, t (s)	Position, x (m)
0.59	1.7057
0.60	1.7640
0.61	1.8233
0.62	1.8836
0.63	1.9445
0.64	2.0070
0.65	2.0703

(a) Falling ball makes dots on tape

(b) Data from a lower section of tape

FIGURE 2.8 Analyzing the motion of a falling body.

cont'd.

SOLVE The computed average velocities are:

Initial time (s)	Final time (s)	Δx (m)	Average velocity (m/s)
0.60	0.61	0.0593	5.93
0.60	0.62	0.1196	5.98
0.60	0.63	0.1805	6.02
0.60	0.64	0.2430	6.08
0.60	0.65	0.3063	6.13

The average velocities approach roughly 5.9 m/s as the time interval approaches zero, so 5.9 m/s is our best estimate of the ball's velocity at $t = 0.60$ s.

REFLECT Look at the interval 0.59−0.60 s, for which the average velocity computes to 5.83 m/s. Having average velocities of 5.83 m/s just before 0.60 s and 5.93 m/s just after 0.60 s validates our estimate of 5.9 m/s for the instantaneous velocity at $t = 0.60$ s.

MAKING THE CONNECTION The table shows that the ball's average velocity over the interval 0.60−0.65 s is about 6.13 m/s. Is the instantaneous velocity at $t = 0.65$ s less than, equal to, or greater than 6.13 m/s?

ANSWER The instantaneous velocity at $t = 0.65$ s should be greater than 6.13 m/s, because the ball's velocity increases as it falls. If the average for the 0.60–0.65 s interval is 6.13 m/s, and the instantaneous velocity is increasing, then the instantaneous velocity must be greater than 6.13 m/s at the end. As a check, you can compute that the average velocity during the interval 0.64–0.65 s is 6.33 m/s, which is significantly greater than 6.13 m/s.

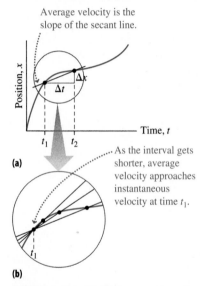

(a)

Average velocity is the slope of the secant line.

As the interval gets shorter, average velocity approaches instantaneous velocity at time t_1.

(b)

FIGURE 2.9 Graphical interpretation of average velocity and instantaneous velocity.

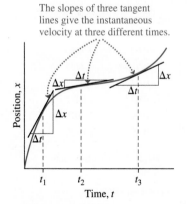

The slopes of three tangent lines give the instantaneous velocity at three different times.

FIGURE 2.10 Instantaneous velocity on a position-versus-time graph.

✓**TIP**
You can estimate instantaneous velocity numerically by shrinking the time interval.

Graphical Interpretation of Average and Instantaneous Velocity

A position-versus-time graph offers a useful view of both average and instantaneous velocity. The graph in Figure 2.9a shows a line (called a **secant line**) drawn between two points. The slope of the secant line is just the rise (Δx) divided by the run (Δt). But $\Delta x/\Delta t$ is also the average velocity for the time interval Δt. Therefore, *on a position-versus-time graph, the slope of the secant line gives the average velocity for a time interval.*

Instantaneous velocity is the average velocity in the limit as the time interval shrinks toward zero. Moving the endpoints of the secant line closer shrinks the interval, until the points coincide and the secant becomes the tangent (Figure 2.9b). Figure 2.10 shows the result: *on a position-versus-time graph, the slope of the tangent line gives the instantaneous velocity at a given time.*

Instantaneous Speed

Although we frequently interchange "speed" and "velocity" in everyday usage, they're distinctly different terms in physics. For one-dimensional motion, **instantaneous speed** is the absolute value of the instantaneous velocity. Although instantaneous velocity may be positive or negative, depending on direction, instantaneous speed is always positive. Speed is what your car's speedometer measures—it tells you how fast you're going, but it doesn't say anything about direction.

In kinematics, instantaneous velocity and speed appear more frequently than their averages. So we'll drop "instantaneous" and just use "velocity" and "speed" for the instantaneous quantities. Whenever we want the average, we'll explicitly say "average" and put a bar over the v.

We'll use v for instantaneous speed. Symbolically,

$$v = |v_x| \quad \text{(Instantaneous speed for motion in one dimension; SI unit: m/s)} \quad (2.5)$$

Be careful and consistent with notation, and you won't confuse velocity and speed. The symbol v without a subscript always means speed, but the subscripted v_x means velocity.

Reviewing New Concepts

- Average velocity over an interval is displacement divided by the time interval.
- Instantaneous velocity is the average velocity in the limit as the time interval approaches zero.
- Average speed over an interval is distance traveled divided by the time interval.
- Instantaneous speed is the absolute value of instantaneous velocity.

CONCEPTUAL EXAMPLE 2.4   Velocity: Positive, Negative, or Zero?

Figure 2.11 graphs the position of a car going forward and backward along a straight line, which we take to be the x-axis. Identify the time(s) when the car's (instantaneous) velocity is positive, negative, and/or zero.

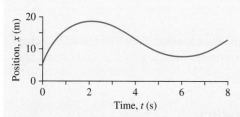

FIGURE 2.11 Position-versus-time graph for the car.

SOLVE The instantaneous velocity at any point is the slope of the tangent line. We've identified the answers on the graph (Figure 2.12).

REFLECT Positive velocity corresponds to increasing values of position x, and negative velocity corresponds to decreasing x.

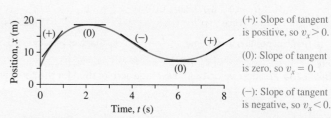

(+): Slope of tangent is positive, so $v_x > 0$.

(0): Slope of tangent is zero, so $v_x = 0$.

(−): Slope of tangent is negative, so $v_x < 0$.

FIGURE 2.12 The car's velocity related to slope of the tangent line.

MAKING THE CONNECTION   When does the car have its maximum velocity in the $+x$-direction?

ANSWER Answering this question means measuring the slope of the steepest rising tangent line. It looks like the maximum slope occurs at the beginning of the graph (near $t = 0$).

GOT IT? Section 2.2   Which velocity-versus-time graph goes with the position-versus-time graph shown at right?

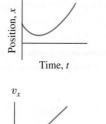

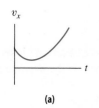

(a)

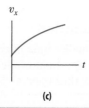

(b)

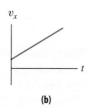

(c)

(d)

2.3 Acceleration

Changing Velocity

Look again at the footrace shown in Figure 2.6. The runner's velocity changes throughout the race, and you could find its value by drawing tangent lines to the graph at as many points as you wish. Because the velocity is changing, the runner is **accelerating**.

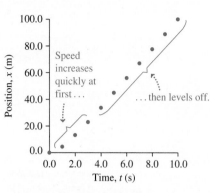

(a) Graph of position versus time

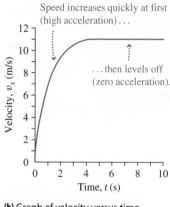

(b) Graph of velocity versus time

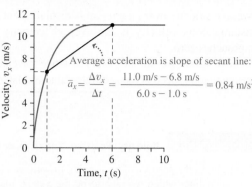

(c) How to find average acceleration on a graph of velocity versus time

FIGURE 2.13 Position, velocity, and average acceleration for the sprinter's run.

Any change in velocity (*not* just an increase) involves **acceleration**. The rest of this chapter shows how acceleration, velocity, and position are used together to understand one-dimensional motion.

Average and Instantaneous Acceleration

Acceleration is the rate of change of velocity, just as velocity is the rate of change of position. The **average acceleration** over a time interval Δt is defined for one-dimensional motion as the change in velocity divided by the time interval:

$$\bar{a}_x = \frac{\Delta v_x}{\Delta t} \quad \begin{array}{l}\text{(Average acceleration for motion} \\ \text{in one dimension; SI unit: m/s}^2\text{)}\end{array} \quad (2.6)$$

A good way to visualize acceleration is to graph velocity versus time. The result for our runner is shown in Figure 2.13b. The average acceleration is the slope of a secant line between two points on the graph (Figure 2.13c). Its units can be expressed as (meters per second) per second, showing explicitly that acceleration is a rate of change of velocity (m/s) with respect to time (s). However, it's customary to combine the two occurrences of seconds to form the more compact unit m/s^2. Although we'll always write m/s^2 (read "meters per second squared"), you may sometimes find it useful to think (m/s)/s.

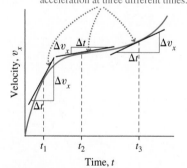

The slopes of three tangent lines give the instantaneous acceleration at three different times.

FIGURE 2.14 Instantaneous acceleration on a velocity-versus-time graph.

Instantaneous Acceleration

As with velocity, taking smaller time intervals brings the average acceleration closer to the instantaneous acceleration. **Instantaneous acceleration** is defined as the average acceleration in the limit as the time interval Δt approaches zero:

$$a_x = \lim_{\Delta t \to 0} \frac{\Delta v_x}{\Delta t} \quad \begin{array}{l}\text{(Instantaneous acceleration for} \\ \text{motion in one dimension; SI unit: m/s}^2\text{)}\end{array} \quad (2.7)$$

In analogy with instantaneous velocity (Section 2.2), **the slope of the tangent line on a velocity-versus-time graph gives the instantaneous acceleration at that time** (Figure 2.14). Once again, we'll omit "instantaneous" and use just "acceleration" for the instantaneous value.

Acceleration can be positive, negative, or zero. Suppose you're driving along a city street (the *x*-axis, in the +*x*-direction), with velocity versus time given in Figure 2.15a. The graph shows that positive acceleration corresponds to increasing velocity, zero acceleration to constant velocity, and negative acceleration to decreasing velocity. Careful! That's *velocity*, not *speed*. Reverse the trip of Figure 2.15a and head in the −*x*-direction; now your displacements and velocities are negative, as shown in Figure 2.15b. What about your acceleration? At the start of this trip, your speed—the magnitude of your velocity—is *increasing*, but the velocity is getting *more negative*, and thus the acceleration is negative, too.

Sometimes people use "decelerate" to mean a decrease in speed, but as Figure 2.15b shows, that can be confusing. Therefore, we'll avoid the term "decelerate" and will instead describe precisely any changes in velocity and in speed, which, as you've seen, are not necessarily the same.

Reviewing New Concepts

To summarize, here are some important ideas about acceleration:

- Acceleration is zero whenever velocity is constant. It doesn't matter whether that velocity is positive, negative, or zero.
- When an object is traveling in the +*x*-direction, positive acceleration results in higher speeds, and negative acceleration in lower speeds.
- When an object is traveling in the −*x*-direction, positive acceleration results in lower speeds, and negative acceleration in higher speeds.

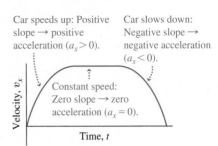

(a) Car traveling in +*x*-direction

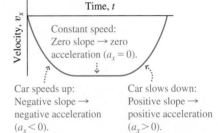

(b) Car traveling in −*x*-direction

FIGURE 2.15 Acceleration related to the velocity-versus-time graph.

EXAMPLE 2.5 **Drag Race**

Velocity data for a drag race car are given below. The velocity is measured at five points along the track—20 m, 100 m, 200 m, 300 m, and 402 m, and the time at each point is the total time since the race started. (The race covers 0.25 miles, about 402 m.) Find the average acceleration for each of the five time intervals and for the entire race.

Time (label)	Time (s)	Position (m)	Velocity v_x (m/s)
t_0	0	0	0
t_1	0.843	20	45.3
t_2	2.147	100	91.9
t_3	3.069	200	122.1
t_4	3.852	300	135.2
t_5	4.539	402	143.8

ORGANIZE AND PLAN Average acceleration is change in velocity divided by change in time. The positions given play no role in computing acceleration. Average acceleration is $\bar{a}_x = \Delta v_x / \Delta t$. For reference, we've labeled the times t_0, t_1, etc. Thus, for the interval t_0 to t_1, the average acceleration is

$$\bar{a}_x = \frac{\Delta v_x}{\Delta t} = \frac{v_x(t_1) - v_x(t_0)}{t_1 - t_0}$$

and so on for the other time intervals.

SOLVE Using the method developed above (under "Organize and Plan"), the results are as follows.

Time interval	Average acceleration $\bar{a}_x$ (m/s²)	Time interval	Average acceleration $\bar{a}_x$ (m/s²)
t_0 to t_1	53.7	t_3 to t_4	16.7
t_1 to t_2	35.7	t_4 to t_5	12.5
t_2 to t_3	32.8		

For the entire race, the average acceleration is

$$\bar{a}_x = \frac{\Delta v_x}{\Delta t} = \frac{143.8 \text{ m/s} - 0.0 \text{ m/s}}{4.539 \text{ s} - 0.0 \text{ s}} = 31.7 \text{ m/s}^2$$

REFLECT The data show that the average acceleration is greatest at the beginning of the race. As the race car accelerates, it becomes more difficult to increase the velocity, mainly due to aerodynamic drag (discussed in Chapter 4). The average acceleration for the entire race is somewhere between the highest and lowest interval values, as should be expected.

MAKING THE CONNECTION How does the race car's acceleration compare to that of your car?

ANSWER Car manufacturers cite the time for acceleration from zero to 60 mi/h (26.8 m/s); a typical time for a mid-sized car is about 7 s, giving an average acceleration of 3.8 m/s²—nearly a factor of 10 lower than the race car!

GOT IT? Section 2.3   Tell whether the acceleration implied by the graph is positive, negative, or zero in each of the time intervals: (a) A to B; (b) B to C; (c) C to D.

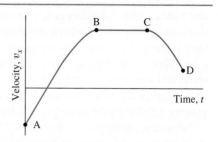

2.4 One-Dimensional Motion with Constant Acceleration

Now we consider the special case of constant acceleration in one dimension. Examples of constant or nearly constant acceleration are common and include objects in free fall near Earth's surface—a situation important enough that we'll devote Section 2.5 to it. You'll encounter other examples of constant acceleration throughout the book, including charged particles moving in uniform electric fields (Chapter 15).

Keep in mind that constant acceleration is a special case, and even then is normally an approximation. For example, a falling object is subject to aerodynamic drag, which gradually changes its acceleration. But describing constant acceleration is mathematically straightforward and offers insights into motion, so we'll spend some time on this special case.

The Kinematic Equations—Predicting the Future

We'll now derive three relationships that not only describe motion under constant acceleration, but also let you predict an object's future position and velocity as long as its acceleration remains constant. These relationships are the **kinematic equations** for constant acceleration.

A constant acceleration a_x causes an object's velocity to change from v_{x0} to v_x during the time interval 0 to t (Figure 2.16). With constant acceleration, there's no difference between instantaneous acceleration a_x and average acceleration $\bar{a}_x$. Therefore, by definition of average acceleration,

$$\bar{a}_x = a_x = \frac{v_x - v_{x0}}{t - 0}$$

Rearranging to solve for v_x gives the first kinematic equation:

$$v_x = v_{x0} + a_x t \tag{2.8}$$

Equation 2.8 lets you predict future velocities. Suppose you're driving at 60 mi/h (26.8 m/s) when you let up on the gas and accelerate at a constant -1.8 m/s^2 for 4.5 s. At the end of that time, your velocity is

$$v_x = v_{x0} + a_x t = 26.8 \text{ m/s} + (-1.8 \text{ m/s}^2)(4.5 \text{ s}) = 18.7 \text{ m/s}$$

In addition to predicting future velocities, Equation 2.8 provides a general relationship among velocity, initial velocity, acceleration, and time. If you know any three, you can find the fourth. The next example illustrates this point.

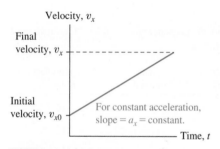

FIGURE 2.16 Velocity versus time for constant acceleration.

EXAMPLE 2.6   **Stopping Time**

In the situation just described, how much time does it take to come to a stop (starting from 26.8 m/s)? Assume that the acceleration continues for as long as is needed to stop.

ORGANIZE AND PLAN Figure 2.17 shows the motion diagram for the slowing car. Because it's slowing, successive frames in the diagram show that the car travels shorter distances in later time intervals.

Equation 2.8 relates the initial and final velocities, the constant acceleration, and time for one-dimensional motion. With the other three quantities given, you can find the time. Time appears in the equation $v_x = v_{x0} + a_x t$ (Equation 2.8).

Known: $v_{x0} = 26.8$ m/s; $v_x = 0$ m/s (because you want to *stop*); $a_x = -1.8$ m/s^2.

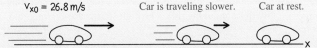

$v_{x0} = 26.8$ m/s   Car is traveling slower.   Car at rest.

FIGURE 2.17 Motion diagram for a car gradually slowing to rest.

SOLVE Solving for t and inserting numerical values gives

$$t = \frac{v_x - v_{x0}}{a_x} = \frac{0 \text{ m/s} - 26.8 \text{ m/s}}{-1.8 \text{ m/s}^2} = 15 \text{ s}$$

REFLECT Fifteen seconds seems like a reasonable stopping time from highway speed. Note the cancellation of units to yield seconds, and the cancellation of the minus signs, giving a positive time. This shows why it's important to have the correct signs on velocity and acceleration.

MAKING THE CONNECTION   Could you have stopped sooner, say, in 4 s?

ANSWER You can solve Equation 2.8 for the acceleration required to go from 26.8 m/s to 0 m/s in any desired time. For 4.0 s, the acceleration is -6.7 m/s^2. A typical car has maximum braking acceleration between -7 m/s^2 and -9 m/s^2. So 4 s is reasonable, but you couldn't do much better.

Another kinematic equation follows from average velocity. Refer again to Figure 2.16. For any such straight-line graph, the average over an interval is just the average of the initial and final values, which in this case means

$$\bar{v}_x = \tfrac{1}{2}(v_x + v_{x0})$$

If you start at $t = 0$, $\Delta t = t$, and Equation 2.2 gives

$$\bar{v}_x = \frac{\Delta x}{t}$$

Equating these two expressions for $\bar{v}$ gives

$$\frac{\Delta x}{t} = \tfrac{1}{2}(v_x + v_{x0})$$

Substituting v_x from Equation 2.8 gives

$$\frac{\Delta x}{t} = \tfrac{1}{2}(v_{x0} + a_x t + v_{x0}) = v_{x0} + \tfrac{1}{2}a_x t$$

Finally, multiply by t to get the displacement Δx:

$$\Delta x = v_{x0}t + \tfrac{1}{2}a_x t^2$$

or

$$x = x_0 + v_{x0}t + \tfrac{1}{2}a_x t^2 \qquad (2.9)$$

This is the second kinematic equation, giving displacement as a function of acceleration, initial velocity, and time. Equation 2.9 is another "predict the future" result—if you know an object's initial velocity and constant acceleration, you can predict its position at a later time.

EXAMPLE 2.7 **Roller Coaster!**

The "Xcelerator" roller coaster at Knott's Berry Farm uses a hydraulic drive to accelerate its trains at ground level from rest to 82.0 mi/h in 2.30 s. Assuming constant acceleration, find (a) the acceleration and (b) the roller coaster's displacement during the acceleration period.

ORGANIZE AND PLAN Again, drawing a motion diagram (Figure 2.18) helps you visualize the situation. Because the train's speed is increasing, the distance traveled increases in successive time intervals. In Equation 2.8, $v_x = v_{x0} + a_x t$, all the quantities except the acceleration a_x are known. You can solve for a_x, then use Equation 2.9 for displacement: $\Delta x = x - x_0 = v_{x0}t + \frac{1}{2}a_x t^2$.

Known: $v_{x0} = 0$; $v_x = 82.0$ mi/h; $t = 2.30$ s.

We'll need to convert $v_x = 82.0$ mi/h to m/s so we can use SI throughout.

SOLVE (a) First, the conversion:

$$82.0 \text{ mi/h} \times \frac{1609 \text{ m}}{1 \text{ mi}} \times \frac{1 \text{ h}}{3600 \text{ s}} = 36.6 \text{ m/s}$$

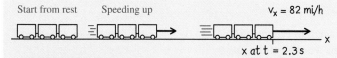

Start from rest   Speeding up $v_x = 82$ mi/h

x at t = 2.3 s

FIGURE 2.18 Motion diagram for accelerating roller coaster.

Solving Equation 2.8 for acceleration and inserting numerical values gives

$$a_x = \frac{v_x - v_{x0}}{t} = \frac{36.6 \text{ m/s} - 0 \text{ m/s}}{2.3 \text{ s}} = 15.9 \text{ m/s}^2$$

(b) Equation 2.9 then gives the displacement

$$\Delta x = v_{x0}t + \frac{1}{2}a_x t^2 = (0 \text{ m/s})(2.30 \text{ s}) + \frac{1}{2}(15.9 \text{ m/s}^2)(2.30 \text{ s})^2 = 42.1 \text{ m}$$

REFLECT For comparison, the acceleration of a commercial jet on take-off is about 3 m/s². This roller coaster's acceleration is some five times larger, which should definitely get the rider's attention! But that's just the start of things. The thrilling experience you get on the roller coaster depends in large part on the fact that the acceleration changes rapidly throughout the ride. The high speeds are exciting, too.

MAKING THE CONNECTION What are the velocity and displacement of the roller coaster at the halfway time, $t = 1.15$ s?

ANSWER Using Equations 2.8 and 2.9 with $a_x = 15.9$ m/s² gives $v_x = 18.3$ m/s and $\Delta x = 10.5$ m after 1.15 s. Note that the velocity is half the final velocity, because under constant acceleration velocity changes linearly with time. However, the displacement is only one-fourth of the final displacement, because displacement increases as t^2. This is clear from the motion diagram in Figure 2.18.

APPLICATION

Air Bags

Unrestrained occupants of a car undergoing a frontal collision suffer extreme negative accelerations when they hit the dashboard or steering wheel, dropping from driving speed to rest in a very short time. An air bag, deploying as shown, increases the time for the driver's body to come to rest, reducing the magnitude of the acceleration and thus minimizing injury. The air bag itself is triggered by an acceleration sensor that detects the abnormally large acceleration of the collision.

In principle, kinematic Equations 2.8 and 2.9 provide all the information needed to describe one-dimensional motion under constant acceleration, because they give both velocity and position as functions of time. But sometimes you don't know the time, so it's helpful to have a third equation that doesn't include time. We can get that equation by solving Equation 2.8 for time t,

$$t = \frac{v_x - v_{x0}}{a_x}$$

Substitute into Equation 2.9:

$$\Delta x = v_{x0}\left(\frac{v_x - v_{x0}}{a_x}\right) + \frac{1}{2}a_x\left(\frac{v_x - v_{x0}}{a_x}\right)^2$$

and rearrange to get

$$v_x^2 = v_{x0}^2 + 2a_x\Delta x \tag{2.10}$$

This is our third kinematic equation. It's useful because it relates an object's initial and final velocities to its acceleration and displacement, without any reference to time. To summarize, the three kinematic equations are:

Kinematic equations for constant acceleration:		
$v_x = v_{x0} + a_x t$	(Predicts velocity; SI unit: m/s)	(2.8)
$x = x_0 + v_{x0}t + \frac{1}{2}a_x t^2$	(Predicts position; SI unit: m)	(2.9)
$v_x^2 = v_{x0}^2 + 2a_x\Delta x$	(Relates final and initial velocities, acceleration, and displacement)	(2.10)

The next few examples illustrate uses of these equations, but they're by no means exhaustive. The problems at the end of the chapter offer further opportunities to explore kinematics.

✓ TIP

When acceleration is constant, use the three kinematic equations (2.8, 2.9, and 2.10) to relate position, velocity, acceleration, and time.

EXAMPLE 2.8 Long-Distance Runner

A long-distance runner goes at a steady 4.9 m/s for most of a race. Near the end she accelerates at a constant 0.30 m/s^2 for 5.0 s. (a) What's her velocity at the end of the acceleration period? (b) How far did she run during this period?

ORGANIZE AND PLAN Both parts of this problem require kinematic equations. For part (a), initial velocity, acceleration, and time are used to get final velocity. In part (b), the same information is used to find displacement.

Equation 2.8 relates velocity to initial velocity, acceleration, and time: $v_x = v_{x0} + a_x t$. Equation 2.9 gives displacement from the same quantities: $\Delta x = v_{x0} t + \frac{1}{2} a_x t^2$.

Known: $a_x = 0.30 \text{ m/s}^2$; $v_{x0} = 4.9 \text{ m/s}$; $t = 5.0 \text{ s}$.

SOLVE (a) Equation 2.8 gives the velocity v_x at the end of the 5.0-s acceleration period:

$$v_x = v_{x0} + a_x t = 4.9 \text{ m/s} + (0.30 \text{ m/s}^2)(5.0 \text{ s}) = 6.4 \text{ m/s}$$

(b) Using the given quantities in Equation 2.9 yields the displacement:

$$\Delta x = v_{x0} t + \tfrac{1}{2} a_x t^2 = (4.9 \text{ m/s})(5.0 \text{ s})$$
$$+ \tfrac{1}{2}(0.30 \text{ m/s}^2)(5.0 \text{ s})^2 = 28 \text{ m}$$

So the runner went 28 m while she accelerated to 6.4 m/s.

REFLECT In both cases, the units combine to give the correct units for the answers: m/s in part (a) and m in part (b). Given the relatively small acceleration, it seems reasonable that the runner covers 28 m while speeding up. Given the answer to part (a), you could just as well have solved part (b) using Equation 2.10, which doesn't involve time. In kinematics, there are often multiple ways to solve problems.

MAKING THE CONNECTION How would the answers change if the runner had *slowed* at the same rate for 5.0 s?

ANSWER All that changes is the sign of the acceleration. Thus $a_x = -0.30 \text{ m/s}^2$, which, following the same method, yields a final velocity $v_x = 3.4 \text{ m/s}$ and displacement $\Delta x = 21 \text{ m}$.

EXAMPLE 2.9 How Long a Runway?

A Boeing 777 aircraft takes off at 295 km/h after accelerating from rest at 2.80 m/s^2. What's the minimum runway length required?

ORGANIZE AND PLAN We're not given the acceleration time, so Equation 2.10, which doesn't involve time, is the appropriate kinematic equation. We can solve Equation 2.10, $v_x^2 = v_{x0}^2 + 2a_x \Delta x$, for displacement Δx in terms of the given quantities.

Known: $a_x = 2.80 \text{ m/s}^2$; $v_{x0} = 0 \text{ m/s}$; $v_x = 295 \text{ km/h} = 81.9 \text{ m/s}$, where we converted 295 km/h to m/s for consistency with the other quantities.

SOLVE Solve Equation 2.10 for displacement and insert numerical values:

$$\Delta x = \frac{v_x^2 - v_{x0}^2}{2a_x} = \frac{(81.9 \text{ m/s})^2 - (0.0 \text{ m/s})^2}{2(2.80 \text{ m/s}^2)} = 1.20 \text{ km}$$

REFLECT The units came out correctly for displacement (meters, which we converted to kilometers to express in three significant figures), and the distance—a little under a mile—seems reasonable. Of course, a runway needs to be substantially longer to give a margin of safety. Most commercial airport runways are between 2 km and 4 km long.

MAKING THE CONNECTION At landing, the 777's velocity is about the same as at takeoff, and after touchdown its acceleration is normally between -2.0 m/s^2 and -2.5 m/s^2. How does this affect consideration of a safe runway length?

ANSWER The magnitude of the landing acceleration (2.0 m/s^2 to 2.5 m/s^2) is slightly smaller than the takeoff acceleration. Since acceleration appears in the denominator in our solution for this example, that means a slightly greater displacement. This suggests a longer runway is necessary to ensure safety on both takeoff and landing.

EXAMPLE 2.10 **Landing on Mars**

It's 2051, and you're taking your grandchildren to Mars. At 47.8 km above the surface, your spacecraft is dropping vertically at 325 m/s. (a) What constant acceleration is needed for a "soft landing"—that is, one with zero velocity? (b) With that acceleration, how much time will it take to reach the surface from 47.8 km?

ORGANIZE AND PLAN Our sketch is shown in Figure 2.19, with the motion directed along the *x*-axis and the spacecraft starting at $x = 0$. You aren't given time in part (a), so again Equation 2.10 is the appropriate kinematic equation for this part. Once you have the acceleration, either of the other kinematic equations can be used to find the landing time. We'll solve Equation 2.10, $v_x^2 = v_{x0}^2 + 2a_x\Delta x$, for the acceleration a_x in terms of the other quantities given in the problem.

Known: $v_{x0} = 325$ m/s; $v_x = 0.0$ m/s; $\Delta x = 47.8$ km.

SOLVE (a) Solve Equation 2.10 for a_x and insert numerical values:

$$a_x = \frac{v_x^2 - v_{x0}^2}{2\Delta x} = \frac{(0.0 \text{ m/s})^2 - (325 \text{ m/s})^2}{2(4.78 \times 10^4 \text{ m})} = -1.10 \text{ m/s}^2$$

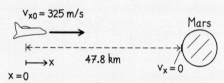

FIGURE 2.19 Spacecraft approaching Mars.

(b) You could use either of the other kinematic equations to find the landing time. However, Equation 2.8, $v_x = v_{x0} + a_xt$, is simpler to use, because it's linear in *t* whereas Equation 2.9 has t^2 and so is quadratic. Solving Equation 2.8 for *t* gives

$$t = \frac{v_x - v_{x0}}{a_x} = \frac{0.0 \text{ m/s} - 325 \text{ m/s}}{-1.10 \text{ m/s}^2} = 295 \text{ s}$$

In other words, you'll take just under 5 (exciting!) minutes to reach the surface of Mars.

REFLECT Once again, note the proper combination of units throughout. The acceleration should be manageable; it's a lot less than we've seen for jet aircraft and braking cars, although it lasts longer.

MAKING THE CONNECTION How much time would it take to go from Earth to Mars if you traveled the entire distance at 325 m/s (the starting velocity for the landing approach in this problem)?

ANSWER The answer depends on the actual distance of the trip, which varies considerably depending on the relative positions of Earth and Mars in their orbits. The shortest possible distance is the difference in the radii of the two planets' orbits. Using the data in Appendix E, that difference is about 7.8×10^{10} m. Thus the time would be $(7.8 \times 10^{10} \text{ m})/(325 \text{ m/s}) = 2.4 \times 10^8$ s, or almost 8 years. Actually, spacecraft take 5–10 months to reach Mars.

PROBLEM-SOLVING STRATEGY 2.1 **Solving Kinematics Problems with Constant Acceleration**

ORGANIZE AND PLAN
- Visualize the situation. Make a schematic diagram with coordinate system as needed.
- Be sure that acceleration is constant.
- Determine what you know, including numerical values. Make sure you've assigned correct signs to all quantities—acceleration, velocity, and position.
- Identify what you're trying to find.
- Plan how to use the given information to solve for unknown(s). Consider: Is time involved, as a given quantity or an unknown to be found? If so, use the first two kinematic equations (2.8 and 2.9). If time is neither given nor sought, use the third kinematic equation (2.10).

SOLVE
- Gather the information you're given.
- Combine and solve equations for the unknown quantity.
- Insert numerical values and solve, making sure to use appropriate units.

REFLECT
- Check dimensions and units of answer. Are they reasonable?
- If problem relates to something familiar, think about whether answer makes sense.

2.5 Free Fall

Gravity—A Case of Constant Acceleration

Gravity is one of the most obvious forces shaping our everyday experience. What isn't as obvious is that objects in **free fall**—moving under the influence of gravity alone—experience the same constant acceleration, regardless of size or mass. The reason that's not obvious is air resistance, which influences falling objects. Drop a coin and sheet of paper simultaneously, and the coin accelerates much faster. That's because air affects the paper significantly, but not the coin. Wad the paper into a tight ball to minimize air resistance, and you'll see it fall much more like the coin.

Air resistance had ancient people convinced that heavier objects fall faster. It was Galileo Galilei (1564–1642) who proved otherwise. Galileo did careful experiments with balls rolling down inclined planes. He found that different balls had the same constant acceleration for a given angle of inclination. Extrapolating to the case of vertical drops, Galileo argued that the acceleration should have the same constant value for all objects. Legend has it that Galileo proved his result by dropping two different cannon balls (one much heavier than the other) from the Tower of Pisa. Although there's doubt about whether Galileo actually did this, he wrote with uncanny accuracy about what *would* have happened. Galileo wrote that if the effects of air resistance could be eliminated, a feather and penny dropped simultaneously would fall together. Figure 2.20 shows a modern experiment confirming Galileo's hypothesis.

Near Earth's surface, the gravitational acceleration of falling bodies is about $9.8 \, \text{m/s}^2$. The actual value varies with location and elevation, ranging from about $9.78 \, \text{m/s}^2$ to $9.83 \, \text{m/s}^2$. In this book, we'll consistently use $9.80 \, \text{m/s}^2$ as a convenient value. The acceleration due to gravity is important enough to deserve its own symbol, g, to distinguish it from other kinds of acceleration. When you insert numbers in a calculation involving free fall, use $g = 9.80 \, \text{m/s}^2$.

This chapter is limited to one-dimensional motion. But in anticipation of Chapter 3's expansion to two dimensions, we'll now adopt a coordinate system with the x-axis horizontal and the y-axis upward. Again, this is just a convention; the laws of physics, after all, don't depend on your choice of coordinate axes! In our new coordinates, one-dimensional motion under the influence of gravity is motion in the y-direction.

The Kinematic Equations Revisited

Since gravity produces constant acceleration, our kinematic equations still hold. We just have to change the label x to y to reflect our new coordinate system. And we know the acceleration a_y in this case: it's $-g$ (that is, $-9.80 \, \text{m/s}^2$). The minus sign is because we've chosen the $+y$-axis to point upward, so acceleration is in the $-y$-direction. Making these changes in our kinematic equations yields the equations for free fall in one dimension:

Kinematic equations for free fall:

$v_y = v_{y0} - gt$	(Predicts velocity; SI unit: m/s)	(2.11)
$y = y_0 + v_{y0}t - \frac{1}{2}gt^2$	(Predicts position; SI unit: m)	(2.12)
$v_y^2 = v_{y0}^2 - 2g\Delta y$	(Relates final and initial velocity, acceleration, and displacement; SI unit: $(\text{m/s})^2$ or m^2/s^2)	(2.13)

Remember when doing calculations that g is a *positive number* ($9.80 \, \text{m/s}^2$). We've already taken the downward direction into account by using $a_y = -g$. Equations 2.11, 2.12, and 2.13 let you solve free-fall problems with the same general approaches you used for other one-dimensional, constant acceleration problems in Section 2.4.

✓**TIP**

The kinematic equations apply to free fall with constant downward acceleration of gravity g.

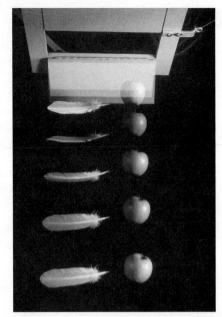

FIGURE 2.20 In a laboratory experiment, a feather and apple are dropped simultaneously in a vacuum chamber. The time lapse photo shows that they accelerate at the same rate.

Falling Faster than g

The Tower of Terror ride at Disney's California Adventure puts passengers on a mock elevator that's pulled down with acceleration about $1.3g$ (about $13 \, \text{m/s}^2$). The resulting sensation is unusual: Because you're fastened to your seat, and your body would naturally accelerate downward at g, it's as if you're being pulled *upward* relative to the falling car. Part of the excitement comes from this unusual feeling, as well as not knowing just what the acceleration will be from one moment to the next!

EXAMPLE 2.11 **Tower of Pisa Revisited**

The top floor of the Tower of Pisa is 58.4 m above the ground. You duplicate Galileo's purported experiment by dropping two balls from the Tower. (a) What's their velocity when they strike the ground? (b) How much time does it take them to fall?

ORGANIZE AND PLAN Figure 2.21 shows a motion diagram for one ball as it accelerates downward. The first question doesn't involve time, which suggests using the third kinematic equation (2.13): $v_y^2 = v_{y0}^2 - 2g\Delta y$. Once you know the final velocity, you can use either of the others to find the time. As you saw earlier, that will be easier with the equation that's linear in time; here that's Equation 2.11: $v_y = v_{y0} - gt$.

Known: $g = 9.80 \text{ m/s}^2$; $v_{y0} = 0 \text{ m/s}$ (dropped from rest); $\Delta y = -58.4 \text{ m}$ (note the negative, a *drop* of 58.4 m, from higher to lower y).

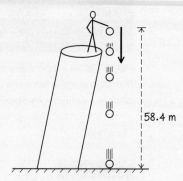

FIGURE 2.21 Motion diagram for Galileo's experiment.

SOLVE (a) Solving Equation 2.13 for v_y means taking a square root—a step that always yields both positive and negative values:

$$v_y = \pm\sqrt{v_{y0}^2 - 2g\Delta y}$$
$$= \pm\sqrt{(0 \text{ m/s})^2 - 2(9.80 \text{ m/s}^2)(-58.4 \text{ m})}$$
$$= \pm 33.8 \text{ m/s}$$

Of the two possible answers ($+33.8 \text{ m/s}$ and -33.8 m/s, the negative one makes sense here. That's because the $+y$-axis is vertically upward, so the balls' velocity is in the $-y$-direction. So the answer is $v_y = -33.8 \text{ m/s}$. The *speed*—the absolute value of velocity—is $+33.8 \text{ m/s}$.

(b) Solving Equation 2.11 for time gives

$$t = \frac{v_{y0} - v_y}{g} = \frac{0 \text{ m/s} - (-33.8 \text{ m/s})}{9.80 \text{ m/s}^2} = 3.45 \text{ s}$$

REFLECT The time seems reasonable for a drop from a high tower—the equivalent of a 15- to 20-story building. Note that signs are important! You really needed that minus sign on v_y to get a positive answer for the time.

MAKING THE CONNECTION What happens (qualitatively) to the answers if air resistance affects the dropped balls?

ANSWER Air resistance reduces the balls' speed, so they'll hit the ground with a speed less than 33.8 m/s. The time, correspondingly, will be greater than 3.45 s.

EXAMPLE 2.12 **Stomp Rocket**

The "stomp rocket" is a toy consisting of a spring-loaded mechanism that shoots a toy rocket straight up. If a rocket is launched from the ground at 12.6 m/s, (a) what maximum height does it reach? (b) What's the rocket's velocity when it's at half the height found in part (a)?

ORGANIZE AND PLAN The initial velocity is given, and the final velocity is implied: At the top of any free-fall flight, the velocity is zero. The final height is the displacement Δy for the flight. For part (b), half the maximum height gives a new displacement, which can be used to find the unknown velocity.

The displacement Δy appears with the initial and final velocities in Equation 2.13: $v_y^2 = v_{y0}^2 - 2g\Delta y$. Because you're never asked about time, use this same equation in both parts.

Known: $g = 9.80 \text{ m/s}^2$; $v_{y0} = 12.6 \text{ m/s}$.

SOLVE (a) At the top of its flight, the rocket is momentarily at rest ($v_y = 0 \text{ m/s}$). Solving Equation 2.13 for displacement Δy then gives the rocket's maximum height:

$$\Delta y = \frac{v_{y0}^2 - v_y^2}{2g} = \frac{(12.6 \text{ m/s})^2 - (0 \text{ m/s})^2}{2(9.80 \text{ m/s}^2)} = 8.10 \text{ m}$$

(b) Half the maximum height is 4.05 m, which becomes the new displacement Δy. With the same initial velocity, the final velocity is

$$v_y = \pm\sqrt{v_{y0}^2 - 2g\Delta y}$$
$$= \pm\sqrt{(12.6 \text{ m/s})^2 - 2(9.80 \text{ m/s}^2)(4.05 \text{ m})}$$
$$= \pm 8.91 \text{ m/s}$$

cont'd.

As in the preceding example, the square root has two possible values, in this case +8.91 m/s and −8.91 m/s. This time, they're both correct! The problem asked for the rocket's velocity at 4.05 m, but didn't specify whether the rocket was going up or down. On the way up, it passes 4.05 m with speed 8.91 m/s. On the way down, it passes 4.05 m with the same speed. In free-fall situations, the speed at a given height is always the same going up as going down, as long as we can neglect air resistance.

REFLECT The displacement in part (a) sounds reasonable; it's about rooftop level for a two-story house. Notice that the speed in part (b) is more than half the initial speed, even though the height is half the maximum. That's because the rocket travels faster during the first part of its flight, when the gravitational acceleration hasn't had much time to slow it.

MAKING THE CONNECTION Find the time for the rocket to reach its maximum height and the times at which it passes through the half-maximum height of 4.05 m.

ANSWER In each case, the equation $v_y = v_{y0} - gt$ gives the answer. At maximum height, $v_y = 0$, so $t = 1.29$ s. When the rocket passes the halfway mark going up, $v_y = +8.91$ m/s, which gives $t = 0.38$ s; when the rocket is going back down, $v_y = -8.91$ m/s, which gives $t = 2.19$ s. Note that the two halfway points are both about 0.8 s from the time the rocket reaches its peak. Motion in free fall is symmetric: It takes the same amount of time to go up and down through the same vertical distance. You could also get *both* answers by solving Equation 2.12 as a quadratic equation in t, with $y = 4.05$ m.

GOT IT? Section 2.5   Which diagram represents velocity (v_y) versus time for a rock thrown straight upward from the ground?

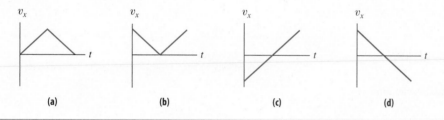

(a)   (b)   (c)   (d)

Chapter 2 in Context

This chapter provided an introduction to motion (*kinematics*), based on the concepts *position*, *velocity*, and *acceleration*. You've seen how these are related, with velocity and acceleration defined as the rates of change of position and velocity, respectively. You've explored the special case of *constant acceleration*, including *free fall* near Earth, and you know how to use *kinematic equations* to solve problems involving constant acceleration.

Looking Ahead: More kinematics, then dynamics. Chapter 2 was restricted to one-dimensional motion, to help you concentrate on the key concepts and relationships. In Chapter 3, we'll extend the concepts of position, velocity, and acceleration to motion in two dimensions. Then in Chapter 4 we'll discuss forces, and show how they're responsible for acceleration. The study of forces and motion—called dynamics—will bring you a deeper understanding of physics.

CHAPTER 2 SUMMARY

Position and Displacement

(Section 2.1) **Displacement** is the net change in an object's position. Total **distance traveled** is the sum of individual distances without regard to direction.

Displacement of an object from an initial position x_0 to a final position x:

$$\Delta x = x - x_0$$

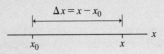

Velocity and Speed

(Section 2.2) **Average velocity** is the change in position of an object divided by the corresponding time interval.

Instantaneous velocity is the limit of average velocity as the time interval Δt approaches zero.

Average velocity: $\bar{v}_x = \dfrac{\Delta x}{\Delta t}$

Instantaneous velocity: $v_x = \lim\limits_{\Delta t \to 0} \dfrac{\Delta x}{\Delta t}$

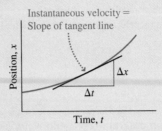

Acceleration

(Section 2.3) **Average acceleration** is the change in velocity of an object divided by the time interval.

Instantaneous acceleration is the limit of average acceleration as the time interval Δt approaches zero.

Average acceleration: $\bar{a}_x = \dfrac{\Delta v_x}{\Delta t}$

Instantaneous acceleration: $a_x = \lim\limits_{\Delta t \to 0} \dfrac{\Delta v_x}{\Delta t}$

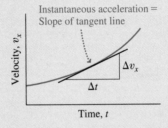

One-Dimensional Motion with Constant Acceleration and Free Fall

(Sections 2.4 and 2.5) **Kinematic equations** relate position, velocity, acceleration, and time for the case of constant acceleration.

Objects in **free fall**—under the influence of gravity alone—experience constant acceleration, regardless of size or mass, because they are under the influence of gravity and no other forces.

Kinematic equations for constant acceleration:

$$v_x = v_{x0} + a_x t \quad x = x_0 + v_{x0}t + \tfrac{1}{2}a_x t^2 \quad v_x^2 = v_{x0}^2 + 2a_x \Delta x$$

Kinematic equations for free fall:

$$v_y = v_{y0} - gt \quad y = y_0 + v_{y0}t - \tfrac{1}{2}gt^2$$

NOTE: Problem difficulty is labeled as ■ straightforward to ■■■ challenging. Problems labeled BIO are of biological or medical interest.

Conceptual Questions

1. In one-dimensional motion, when are displacement and distance traveled the same? When are they different?
2. In one-dimensional motion, when are average speed and average velocity the same? When are they different?
3. If the acceleration of an object is zero, can its velocity be negative?
4. If the acceleration of an object is negative, can its velocity be zero? Can its velocity be positive? Explain.
5. Can an object have zero velocity yet nonzero acceleration? Give an example.
6. Galileo studied uniform acceleration by rolling balls down inclined ramps. He observed that a ball starting from rest would travel through 1, 3, 5, 7, 9, etc. units of distance during successive, equal time intervals. Explain why this observation is consistent with constant acceleration.
7. Given these graphs of velocity versus time for one-dimensional motion, construct graphs of position versus time and acceleration versus time.

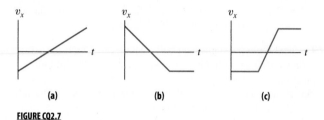

(a)   (b)   (c)

FIGURE CQ2.7

8. Your car's maximum acceleration is about 3.0 m/s². For about how long can it maintain that acceleration?
9. A car going 25 m/s passes another going 20 m/s in the same direction. What can you say about the accelerations of the two cars?
10. A ball is launched straight up from the ground and falls straight back down. At any given position along the path, which of these quantities are the same and which are different for the upward and downward motions: velocity, speed, and acceleration?
11. What can you say about an object's displacement during some time interval if its average velocity is zero during that interval? What can you say about the displacement if the average acceleration is zero?
12. *Zeno's paradox* says that when going from a starting point to your destination, you first cover half the distance, then half the remaining distance (one-fourth the original distance), then half the remaining distance (one-eighth the original distance), and so on. Since each step takes you only halfway to your goal, you should never reach any destination. How would you refute Zeno's paradox?

Multiple-Choice Problems

13. If a trip from Earth to Moon (about 385,000 km) takes 2.5 days, the average speed for the trip is (a) 1.8 m/s; (b) 29.7 m/s; (c) 1800 m/s; (d) 27,000 m/s.
14. BIO A cheetah can maintain its top speed of 32 m/s for about 35 s. During that time it travels about (a) 750 m; (b) 850 m; (c) 1000 m; (d) 1100 m.
15. For the first 1200 m of a 1500-m race, a runner's average speed is 6.14 m/s. To finish in under 4 minutes, the runner's average speed for the remainder of the race must be at least (a) 6.73 m/s; (b) 7.14 m/s; (c) 8.05 m/s; (d) 8.29 m/s.
16. A runner sprints in the $+x$-direction at 9.2 m/s for 100 m, stops, reverses direction, and jogs backward at 3.6 m/s for 50 m. The average velocity for the entire run is (a) 2.0 m/s; (b) 4.0 m/s; (c) 6.4 m/s; (d) 10.9 m/s.
17. In one-dimensional motion, displacement (a) can never be negative; (b) can be positive, negative, or zero; (c) is the same as distance traveled; (d) can be greater than distance traveled.
18. What's your average speed when you run at 4.0 m/s for 60 m, then at 6.0 m/s for another 60 m? (a) 4.8 m/s; (b) 5.0 m/s; (c) 5.2 m/s; (d) 5.4 m/s.
19. The average acceleration to boost a spacecraft from 1250 m/s to 1870 m/s in 35 s is (a) 53.4 m/s²; (b) 35.7 m/s²; (c) 17.7 m/s²; (d) 9.80 m/s².
20. Starting from rest, a go-cart undergoes constant acceleration −1.4 m/s². When its velocity reaches −10 m/s, its displacement from its starting point is (a) −7.14 m; (b) −14.3 m; (c) −35.7 m; (d) −100 m.
21. An arrow with speed 21.4 m/s embeds itself 3.75 cm into a target before stopping. While in the target, its constant acceleration was (a) −570 m/s²; (b) −1140 m/s²; (c) −6100 m/s²; (d) −12,200 m/s².
22. An object is dropped from rest off the 442-m Sears Tower in Chicago. Its fall time is (a) 45.1 s; (b) 19.1 s; (c) 9.5 s; (d) 4.7 s.
23. A ball dropped from rest from height h reaches the ground with a speed v. If the drop height is changed to $2h$, its speed at the ground is (a) $4v$; (b) $2v$; (c) $\sqrt{2}v$; (d) v.
24. You're driving at 12.8 m/s, and are 16.0 m from an intersection when you see a stoplight turn yellow. What acceleration do you need in order to just stop before the intersection? (a) −0.8 m/s²; (b) −5.1 m/s²; (c) −7.4 m/s²; (d) −10.2 m/s².
25. For an object traveling in a straight line with constant acceleration, the graph of its velocity versus time is (a) a horizontal line; (b) a diagonal line; (c) a parabola.
26. A ball is thrown straight up from the ground at 10 m/s, and simultaneously a ball is thrown straight down from a 10-m-high ledge at 10 m/s. At what height do the two balls pass? (a) 2.5 m; (b) 3.2 m; (c) 3.8 m; (d) 5.0 m.

Problems

Section 2.1 Position and Displacement
27. ■ In the example described by Figure 2.4, what are the displacement and total distance traveled for a round trip from your friend's house to the video store?
28. ■ Using the data in Conceptual Example 2.1, find the displacement for a trip from Lincoln to Grand Island. (Choose a coordinate system with the $+x$-axis pointing east.)
29. ■■ Using the data in Conceptual Example 2.1, find the displacement and distance traveled for (a) 3, (b) $3\frac{1}{2}$, and (c) $3\frac{3}{4}$ round trips from Grand Island to Lincoln.

Section 2.2 Velocity and Speed
30. ■ Find the average speed (in m/s) of a runner who completes each of the following races in the time given: (a) a marathon (41 km) in 2 hours, 25 minutes; (b) 1500 m in 3 minutes, 50 seconds; (c) 100 m in 10.4 s.

31. ■ How much time does light from the Sun take to reach Earth? Use data from Appendix E.

32. ■ The best major league fastball goes about 44 m/s. (a) How much time does this pitch take to travel 18.4 m to home plate? (b) Make the same calculation for a "change-up" thrown at a speed of 32 m/s.

33. ■■ If you run at 4.0 m/s for 100 m, then at 5.0 m/s for another 100 m, what's your average speed?

34. ■■ (a) Find your average speed if you go 10 m/s for 100 s and then 20 m/s for 100 s. (b) Find your average speed if you go 10 m/s for 1000 m and then 20 m/s for 1000 m. (c) Why are your answers different?

35. ■■ You are flying from Seattle to Anaheim with a connection in Oakland. The distance from Seattle to Oakland is 1100 km, and Oakland to Anaheim is 550 km. If both airplanes average 800 km/h and the layover in Oakland is 80 min, find (a) the total time for the trip and (b) your average speed.

36. ■■ In a 2-km crew race, boat 1 starts at 4.0 m/s for the first 1500 m, but slows to 3.1 m/s for the rest of the race. Boat 2 goes a steady 3.6 m/s for the first 1200 m and then 3.9 m/s for the remainder. Who wins?

37. ■ A plane flies east at 210 km/h for 3.0 h, then turns around and flies west at 170 km/h for 2.0 h. Taking the +x-axis to point east, find the plane's average velocity and average speed for the trip.

38. ■■ A runner is planning for a 10-km race. She can maintain a steady speed of 4.10 m/s for as much time as needed before ending the race with a 7.80-m/s sprint. If she wants to finish in 40 min or less, how far from the finish should she begin to sprint?

39. ■■ A dogsled goes straight at 9.5 m/s for 10 h. Then the dogs rest for the remainder of the day. What's the average velocity for the entire 24-h day?

40. ■■ A car travels a straight road at 100 km/h for 30 min then at 60 km/h for 10 min. It then reverses and goes at 80 km/h for 20 min. Find the average velocity and average speed for the entire trip.

41. ■■■ You're checking your car's speedometer. With cruise control engaged, the speedometer reads a constant 60 mi/h. (a) Using highway mileposts, you clock 4 min, 45 s to go 5 m. If the mileposts are accurate, what's the error in your speedometer reading? (b) How much time should you take to cover each mile if your true speed is 65 mi/h?

42. ■■■ On a windless day, a bird can fly at a constant 10 m/s. (a) It flies 10 km east and then home again. How much time does the round trip take? (b) A 5.0-m/s wind from the west gives the bird a groundspeed of 15 m/s flying east and 5 m/s flying west. Find the time for the round-trip flight under these conditions. (c) Compare your answers to parts (a) and (b). Why aren't they the same?

43. BIO ■■ **Chasing a zebra.** A cheetah running at 30 m/s is pursuing a zebra going in a straight line at 14 m/s. If the zebra has a 35-m head start, how much time does it take for the cheetah to catch up?

44. ■■ A parachutist free falls 440 m in 10.0 s. She then opens her chute and drops the remaining 1350 m. If her average velocity for the entire trip is 3.45 m/s, what's her average velocity while the chute is open?

45. ■■ In 1675, the Danish astronomer Olaf Römer used observations of the eclipses of Jupiter's moons to estimate that it took the light about 22 min to cross the 299-million-km diameter of Earth's orbit. Use Römer's data to compute the speed of light, and compare that with today's value of 3.00×10^8 m/s.

Use the graph in Figure P2.46 to complete the next two problems.

46. ■■ Find the average velocity over each 2.0-s time interval, e.g., 0.0–2.0 s, 2.0–4.0 s.

47. ■■■ Construct a graph of velocity versus time for the entire time interval.

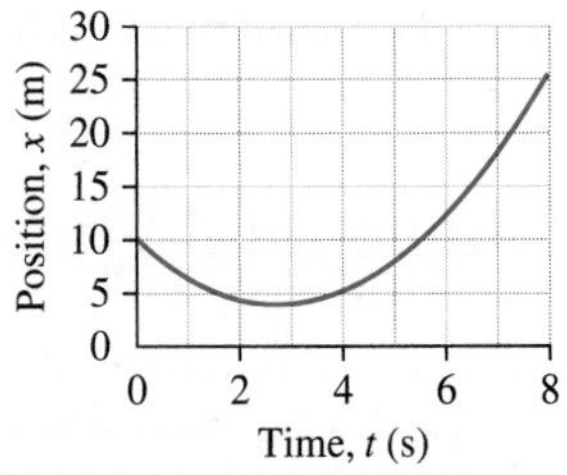

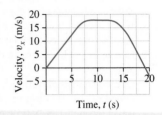

FIGURE P2.46

Section 2.3 Acceleration

For the next four problems, refer to Figure P2.48, a graph of velocity versus time for a car starting from rest on a straight road.

48. ■■ Find the average acceleration for each of the four intervals (0–5 s, 5–10 s, etc.).

49. ■■ Construct a graph of instantaneous acceleration from $t = 0$ to $t = 20$ s.

FIGURE P2.48

50. ■■ Draw a motion diagram for this trip.

51. ■■ Where is the acceleration (a) greatest, (b) least, and (c) zero? (d) Compute the greatest and least acceleration.

52. ■ For the sprinter described in Section 2.2, what's the average acceleration for the first half of the race?

The next three problems deal with a stock car, which starts from rest at time $t = 0$ with velocity (m/s) increasing for 4.0 s, according to the function $v_x = 1.4t^2 + 1.1t$.

53. ■ (a) Find the car's velocity at the end of the 4.0-s interval. (b) Find the average acceleration for this interval.

54. ■■ Graph the velocity as a function of time. When is the acceleration greatest and when is it least?

55. ■■ Estimate the instantaneous acceleration at $t = 2.0$ s.

56. ■■ Draw a motion diagram for the car trip graphed in Figure 2.15a.

57. ■■ Draw a motion diagram for the car trip graphed in Figure 2.15b.

58. BIO ■■ **Acceleration in blood flow.** Figure P2.58 shows a pattern of coronary artery blood flow rates (cm/s) in a patient successfully treated for a myocardial infarction (heart attack). The upper and lower peaks represent the diastolic and systolic phases of the heartbeat. Estimate the average acceleration of the blood between the peaks of these phases.

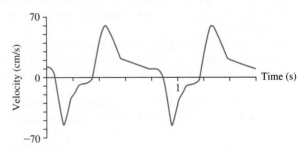

FIGURE P2.58

Section 2.4 One-Dimensional Motion with Constant Acceleration

59. ■ You're driving at 50 km/h, when the traffic light 40 m away turns yellow. Find (a) the constant acceleration required to stop at the light and (b) the stopping time. Is the acceleration reasonable?

60. BIO ■ **Animal and human acceleration.** (a) What constant acceleration is required for a cheetah to go from rest to its top speed of 90 km/h in 3.0 s? (b) Repeat the calculation for a human who takes 2.0 s to reach a top speed of 10 m/s.

61. ■ ■ A golfer putts her golf ball straight toward the hole. The ball's initial velocity is 2.52 m/s, and it accelerates at a rate of -0.65 m/s^2. (a) Will the ball make it to the hole, 4.80 m away? (b) If your answer is yes, what's the ball's velocity when it reaches the hole? If your answer is no, how close does it get before stopping?

62. ■ A rocket sled accelerates at 21.5 m/s^2 for 8.75 s. (a) What's its velocity at the end of that time? (b) How far has it traveled?

63. ■ ■ A car going initially with a velocity 13.5 m/s accelerates at a rate of 1.9 m/s^2 for 6.2 s. It then accelerates at a rate of -1.2 m/s^2 until it stops. (a) Find the car's maximum speed. (b) Find the total time from the start of the first acceleration until the car is stopped. (c) What's the total distance the car travels?

64. ■ ■ ■ One car is going 50 km/h, another 100 km/h. Both have brakes that provide -3.50 m/s^2 accelerations. (a) Find the stopping time for each car. (b) Find the stopping distance for each car. (c) Use your answer to part (a) to find the ratio of the stopping times, and use your answer to part (b) to find the ratio of the stopping distances.

65. ■ ■ A bullet going 310 m/s strikes a 5.0-cm-thick target. (a) What constant acceleration is required if the bullet is to stop within the target? (b) What's its acceleration if the bullet emerges from the target at 50 m/s?

66. ■ ■ A fully loaded 737 aircraft takes off at 250 km/h. If its acceleration is a steady 3.0 m/s^2, how long a runway is required? How much time does it take the plane to reach takeoff?

67. ■ ■ ■ A car is speeding at 75 mi/h (33.4 m/s). A police cruiser starts in pursuit from rest when the car is 100 m past the cruiser. At what rate must the cruiser accelerate to catch the speeder before the state line, 1.2 km away from the speeding car?

68. ■ ■ An x-ray tube accelerates electrons from rest at 5×10^{14} m/s^2 through a distance of 15 cm. Find (a) the electrons' velocity after this acceleration and (b) the acceleration time. (Such high accelerations are possible because electrons are extremely light.)

69. ■ A jet touches down at 310 km/h (86.1 m/s). Find the (constant) acceleration required to stop the aircraft 1000 m down the runway.

70. ■ ■ ■ You're approaching an intersection at 50 km/h (13.9 m/s). You see the light turn yellow when you're 35 m from the intersection. Assume a reaction time of 0.6 s before braking begins and a braking acceleration of -3.0 m/s^2. (a) Will you be able to stop before the intersection? (b) The yellow light stays on for 3.4 s before turning red. If you continue at 50 km/h without braking, will you make it through the 9.5-m-wide intersection before the light turns red?

71. ■ ■ Disney's "Rockin' Roller Coaster" accelerates in a straight line from rest to 60 mi/h in 2.8 s. (a) What is its (constant) acceleration? (b) How far does it travel during the first 2.8 s?

72. BIO ■ ■ **Brain injuries in auto accidents.** Brain injuries generally occur any time the brain's acceleration reaches 100g for even a short time. Consider a car running into a solid barrier. With an airbag, the driver's head moves through a distance of 20 cm while the airbag stops it. Without an airbag, the head continues forward until the seatbelt stops the torso, causing the head to stop in a distance of only 5.0 cm. For each case, find the maximum speed with which the car can strike the barrier without causing brain injury.

Section 2.5 Free Fall

73. ■ Jerry knocks a flowerpot off its third-story ledge, 9.5 m above the ground. If it falls freely, how fast is the flowerpot moving when it crashes to the sidewalk?

74. ■ ■ After solving a difficult physics problem, an excited student throws his book straight up. It leaves his hand at 3.9 m/s from 1.5 m above the ground. (a) How much time does it take until the book hits the floor? (b) What's its velocity then?

75. ■ The Jurassic Park ride at Universal Studios theme park drops 25.6 m straight down essentially from rest. Find the time for the drop and the velocity at the bottom.

76. ■ A batter pops the baseball straight up at 19.5 m/s. The catcher loses sight of the ball, so the first baseman has to rush in to catch it. How much time does he have?

77. ■ ■ A rock is launched straight up from the ground at 16.5 m/s. Graph the rock's velocity and position versus time from launch until it reaches the ground.

78. ■ ■ The gravitational acceleration of bodies dropped near the Moon's surface is about 1.6 m/s^2. Find the times for an object dropped from rest to fall 1.0 m on the Moon and on Earth.

79. ■ ■ The first astronaut to reach Mars decides to measure the gravitational acceleration by dropping a rock from a 45.2-m-high cliff. If the rock falls for 5.01 s, what's g_{Mars}?

80. ■ ■ A tennis ball gun launches tennis balls at 18.5 m/s. It's pointed straight up and launches one ball; 2.0 s later, it launches a second ball. (a) At what time (after the first launch) are the two balls at the same height? (b) What is that height? (c) What are both balls' velocities at that point?

81. ■ ■ ■ A world-class volleyball player can jump vertically 1.1 m from a standing start. (a) How long is the player in the air? (b) Graph the athlete's position versus time. (c) Use your graph to explain why an athlete might appear to "hang" in the air near the top of the jump.

82. ■ ■ To escape a fire, you jump from a 2.5-m-high window ledge. To cushion your landing, you begin with legs straight and end with your legs bent into a crouch 55 cm below your normal height. Find your (constant) acceleration upon striking the ground.

83. ■ ■ ■ A rocket accelerates straight up from the ground at 12.6 m/s^2 for 11.0 s. Then the engine cuts off and the rocket enters free fall. (a) Find its velocity at the end of its upward acceleration. (b) What maximum height does it reach? (c) With what velocity does it crash to Earth? (d) What's the total time from launch to crash?

84. ■ ■ ■ A ball thrown straight up from the ground passes a window 5.6 m up. An observer looking out the window sees the ball pass the window again, going down, 3.2 s later. Find (a) the velocity with which the ball was initially thrown and (b) the total time for the round trip, from the time the ball was thrown until it reaches the ground.

85. ■ ■ ■ In lab, a student measures an acceleration of 3.50 m/s^2 for a ball rolling down a 30-degree incline. Then the ball rolls up a 45-degree incline, reaching the same height from which it was released (Figure P2.85). Find the ball's acceleration along the second ramp.

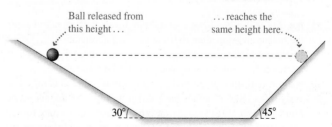

FIGURE P2.85

86. ■ ■ ■ A helicopter rises vertically with a constant upward acceleration of 0.40 m/s^2. As it passes an altitude of 20 m, a wrench slips out the door. (a) How soon and (b) at what speed does the wrench hit the ground?

87. ■■■ For the situation in the preceding problem, let $t = 0$ be the moment when the wrench slips from the helicopter. Draw graphs of position and velocity versus time for the wrench, from until it hits the ground.

88. **BIO** ■■ **Jumping flea.** For its size, the flea can jump to amazing heights—as high as 30 cm straight up, about 100 times the flea's length. (a) For such a jump, what takeoff speed is required? (b) How much time does it take the flea to reach maximum height? (c) The flea accomplishes this leap using its extremely elastic legs. Suppose its upward acceleration is constant while it thrusts through a distance of 0.90 mm. What's the magnitude of that acceleration? Compare with g.

89. **BIO** ■■ **Falling cat.** Young cats develop a "righting reflex" that enables them to land on their feet after a fall. Upon landing they absorb the impact by extending their feet and then crouching after their feet touch the ground. (a) Find the speed with which a cat reaches the ground after a fall from a 6.4-m-high window. (b) After this cat touches the ground, it comes to rest with a constant acceleration as it crouches through a distance of 14 cm. Find the acceleration during the crouching maneuver.

General Problems

90. ■■■ Starting from rest on your bicycle, you go in a straight line with acceleration $2.0\ \text{m/s}^2$ for 5.0 s. Then you pedal with a constant velocity for another 5.0 s. (a) What's your final velocity? (b) What is the total distance cycled? (c) Draw graphs of position and velocity versus time for the entire trip.

91. ■■ You're driving at a legal $13.4\ \text{m/s}$, and you're 15.0 m from an intersection when you see a stoplight turn yellow. (a) What acceleration do you need to stop at the intersection? (b) What's the corresponding stopping time? (c) Repeat part (a), but now assume a reaction time of 0.60 s before you brake.

92. ■■■ The graph in Figure GP2.92 shows velocity versus time for a ball projected upward along an incline. Take the $+x$-axis directed upward along the incline. (a) Describe what you would see if you were watching this ball. (b) How far up the ramp does the ball get from its initial position? (c) Graph the acceleration and position versus time for the rolling ball.

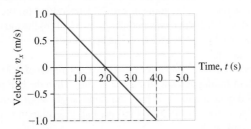

FIGURE GP2.92

93. ■■ You're driving down a straight highway at $25\ \text{m/s}$. You apply the brakes and stop after 10.0 s of constant acceleration. (a) Graph (a) your velocity and (b) your position, both versus time. (c) Draw a motion diagram, showing your car at 2.0-s intervals.

94. ■■■ A ball rolls up an incline with initial speed $2.40\ \text{m/s}$. Exactly 6.0 s later, it passes back down through the launch point. Draw a graph of (a) velocity and (b) position versus time for the entire trip, with $x = 0$ being the launch point and the positive x-axis pointing up the ramp. (c) Find the ball's average velocity and average speed for the entire trip.

95. ■■ You throw a baseball straight up at $12.1\ \text{m/s}$. (a) Find the time(s) when the ball is 5.20 m above its launch point. (b) Find the velocity at each time you found in part (a).

96. ■■■ A student throws a baseball straight up from 1.50 m above the ground with a speed of $11.0\ \text{m/s}$. Simultaneously, another student on top of the 12.6-m-high physics building throws a baseball straight down at $11.0\ \text{m/s}$. When and where do the two balls meet?

97. ■■■ A "moving sidewalk" in an airport runs at a constant $1.0\ \text{m/s}$. From opposite ends of the 50-m-long sidewalk, two old friends begin running toward one another, each with a speed of $4.0\ \text{m/s}$ relative to the moving sidewalk. Where do they meet, relative to the fixed ends of the sidewalk?

98. **BIO** ■■■ **Cheetah chase.** A cheetah can accelerate from rest to 60 mph in 3.0 s. (a) Find the cheetah's (assumed constant) acceleration in SI units. (b) Although they're fast, cheetahs tire quickly. A gazelle running at a constant $20\ \text{m/s}$ has a 25-m head start on a resting cheetah. The cheetah runs toward the gazelle, accelerating from rest to 60 mph in 3.0 s and then maintaining that speed for 10 s before tiring. Does the cheetah catch the gazelle?

99. ■■ Runner A leads runner B by 85.0 m in a distance race, and both are running at $4.45\ \text{m/s}$. Runner B accelerates at $0.10\ \text{m/s}^2$ for the next 10 s and then runs with constant velocity. How much total time elapses before B passes A?

100. ■■ At the edge of a 12-m-tall building, two children throw rocks at $10\ \text{m/s}$, one upward and one downward. (a) Find the time until each rock hits the ground. (b) Find the speed of each when it hits.

101. ■■■ A train passes through a station at a constant $11\ \text{m/s}$. On a parallel track sits another train at rest. At the moment the first train passes, the second begins to accelerate at $1.5\ \text{m/s}^2$. When and where do the trains meet again?

102. ■■■ Two 110-m-long trains are traveling at $22.5\ \text{m/s}$, going in opposite directions on parallel tracks. (a) How much time elapses from the moment the front ends of the trains pass to when the rear ends pass? (b) Repeat part (a), but this time suppose that when the front ends pass, one train begins to accelerate at $1.0\ \text{m/s}^2$.

103. ■■ Attempting to waste the last 4.8 s of a basketball game, a player on the team that's ahead throws the ball straight up. He releases it 1.6 m above the ground, and an opposing player catches it at the same height on the way down. The ball isn't permitted to touch the roof, 17.2 m above the ground. Did the winning team run out the clock, or will the opponent have time to make a shot?

Answers to Chapter Questions

Answer to Chapter-Opening Question
The ball's acceleration is constant throughout its flight: $9.80\ \text{m/s}^2$, directed straight down.

Answers to GOT IT? Questions
Section 2.1   (a) Distance traveled can never be negative. (e) Distance traveled is greater than or equal to displacement.

Section 2.2   (d)

Section 2.3   (a) positive (b) zero (c) negative

Section 2.5   (d)

■ A well-struck golf ball's flight is two-dimensional. How can you predict the ball's time in flight and distance traveled?

Chapter 3 extends our study of kinematics to motion in a plane. We'll begin by reviewing trigonometry, and then introduce vectors to describe motion in two dimensions. You'll see how position, displacement, velocity, and acceleration are vector quantities. Then we'll consider two special cases: the motion of projectiles near Earth and uniform circular motion.

3.1 Trigonometry Review

Trigonometry is used throughout physics, so here's a quick review. If you're familiar with sine, cosine, and tangent functions and right triangles, you can skim or skip this section.

Trigonometry is especially important in two-dimensional motion. We'll describe such motion using Cartesian coordinate systems, with x and y axes perpendicular. This 90° angle means you'll see many right triangles, to which trigonometry applies.

Defining the Trig Functions and Inverse Trig Functions

Figure 3.1 shows a right triangle with sides a, b, and c. The sine, cosine, and tangent of the angle θ are defined as:

$$\sin\theta = \frac{\text{opposite}}{\text{hypotenuse}} = \frac{a}{c} \quad \cos\theta = \frac{\text{adjacent}}{\text{hypotenuse}} = \frac{b}{c} \quad \tan\theta = \frac{\text{opposite}}{\text{adjacent}} = \frac{a}{b} \quad (3.1)$$

To Learn

By the end of this chapter you should be able to

- Distinguish scalars and vectors.
- Understand vectors in component form and as a magnitude and direction.
- Add and subtract vectors, both analytically and graphically.
- Describe position, velocity, and acceleration vectors.
- Understand and analyze projectile motion.
- Understand and analyze uniform circular motion.

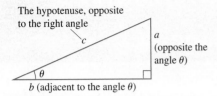

The hypotenuse, opposite to the right angle

c

a (opposite the angle θ)

θ

b (adjacent to the angle θ)

FIGURE 3.1 Right triangle used to define the trigonometric functions.

Each of these relationships involves three quantities: two sides and a trig function of the angle θ. If you know any two, you can find the third. For example, if $c = 12$ cm and $\theta = 30°$, then $a = c \sin(30°) = (12 \text{ cm})(0.50) = 6.0$ cm. Your calculator computes trig functions, and it accepts angle measurements in degrees, radians, and sometimes other units. In Chapter 7 you'll see why radians are useful for circular motion. Until then we'll use degree measurements, so for now be sure your calculator is in the "degree" mode.

If you know any two sides of a right triangle, you can use **inverse trigonometric functions** to find the angle θ. We'll write $\sin^{-1}\theta$, $\cos^{-1}\theta$, and $\tan^{-1}\theta$ for the inverse functions; others may use $\arcsin\theta$, $\arccos\theta$, and $\arctan\theta$. For the triangle in Figure 3.1, the inverse functions are

$$\theta = \sin^{-1}\left(\frac{a}{c}\right) \quad \theta = \cos^{-1}\left(\frac{b}{c}\right) \quad \theta = \tan^{-1}\left(\frac{a}{b}\right) \tag{3.2}$$

For example, if $a = 5.2$ cm and $c = 9.5$ cm, then

$$\theta = \sin^{-1}\left(\frac{a}{c}\right) = \sin^{-1}\left(\frac{5.2 \text{ cm}}{9.5 \text{ cm}}\right) = \sin^{-1}(0.547) = 33°$$

The Pythagorean theorem relates the three sides of a right triangle. Again in reference to Figure 3.1, it states

$$a^2 + b^2 = c^2 \tag{3.3}$$

If you know any two sides, the Pythagorean theorem lets you find the third.

EXAMPLE 3.1 **Sun Angle**

At noon on the day of the spring equinox, a 1.85-m-tall man standing on level ground casts a 1.98-m-long shadow. What's the Sun's elevation angle above the horizon?

ORGANIZE AND PLAN As usual, it helps to draw a diagram (Figure 3.2); it shows a right triangle, the two known sides, and the unknown angle θ. With the opposite and adjacent sides known, the inverse tangent gives the angle:

$$\theta = \tan^{-1}(a/b)$$

Known: opposite side $a = 1.85$ m; adjacent side $b = 1.98$ m.

SOLVE Inserting the values gives

$$\theta = \tan^{-1}\left(\frac{a}{b}\right) = \tan^{-1}\left(\frac{1.85 \text{ cm}}{1.98 \text{ cm}}\right) = \tan^{-1}(0.934) = 43.0°$$

REFLECT This answer seems about right. It's a good idea to become familiar with trig functions of some benchmark angles (see

Table 3.1). One of these is $\tan(45°) = 1$. In this problem, $\tan\theta = 0.934$—a little less than 1, so the angle is a little under 45°. If the man's height and shadow length were equal, the angle would be exactly 45°.

TABLE 3.1 Trig Functions of 0°, 30°, 45°, 60°, and 90°*

Angle	sin	cos	tan
0°	0	1	0
30°	1/2	$\sqrt{3}/2 \approx 0.866$	$1/\sqrt{3} \approx 0.577$
45°	$1/\sqrt{2} \approx 0.707$	$1/\sqrt{2} \approx 0.707$	1
60°	$\sqrt{3}/2 \approx 0.866$	1/2	$\sqrt{3} \approx 1.73$
90°	1	0	Undefined

*Note that for trig functions given by irrational numbers, the decimal equivalents are given to three significant figures.

MAKING THE CONNECTION Given that the Sun is directly above the equator at noon on the equinox, find the (northern) latitude of the observer in this example.

ANSWER For an observer on the equator, a Sun-Earth line would be vertical. For every degree northward, the Sun moves one degree from that line. In this example, the Sun makes a 43° angle with the horizon, so it's 47° from vertical. Thus the latitude is 47°.

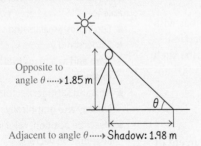

Opposite to angle θ ·····▸ 1.85 m

θ

Adjacent to angle θ ·····▸ Shadow: 1.98 m

FIGURE 3.2 Finding the Sun's elevation.

GOT IT? **Section 3.1** Rank in order the values of the following trig functions, from least to greatest: (a) $\tan 60°$; (b) $\cos 90°$; (c) $\sin 0°$; (d) $\sin 90°$; (e) $\cos 180°$; (f) $\tan 120°$.

3.2 Scalars and Vectors

Scalars are physical quantities specified by a single number (with appropriate units). For example, the volume of water in a swimming pool (192.4 m³), your body temperature (37.0°C), and the ratio of shadow length to the man's height in Example 3.1 (1.07) are all scalars. The first two also require physical units, while the third is dimensionless.

A **vector** is a physical quantity that must be specified by two or more numbers. Vectors occur throughout physics. Position in a plane is one example and is at the heart of kinematics. In two-dimensional motion, position, displacement, velocity, and acceleration are all vector quantities.

Position Vectors

Figure 3.3a shows a map of a town. You describe your position by naming the appropriate streets, saying "I'm at the corner of 2nd Avenue and C Street." Or you could use Cartesian coordinates (x, y) as shown. The pair of numbers describing your coordinates, taken together as a single physical quantity, constitutes your **position vector**. It's the need for *two* numbers (or three in three dimensions) that makes position a vector.

A pair of Cartesian coordinates is one way to represent the position vector. An equivalent representation is graphical, as shown in Figure 3.3b. Graphically, the vector is an arrow from the origin to the position you're describing.

 TIP

A position vector can be represented two ways: as a pair of Cartesian coordinates and graphically, as an arrow from the origin to the object being located.

Vector Notation

We'll indicate vector quantities with an arrow over a symbol; for example, we'll call the position vector $\vec{r}$. The arrow shows this is a vector, not a scalar. **Whenever you're writing a vector quantity, you must use that arrow!** Some books use boldface letters (e.g., **r**) to denote vectors. But the arrow is easier to distinguish when you're writing on paper or blackboard, so it's preferable.

 TIP

Because scalars and vectors are different kinds of quantities, **never** write an equation that has a vector on one side and a scalar on the other!

In the example above, the quantities $x = 200$ m and $y = 100$ m are the two **components** of the position vector. There are two common notations for keeping track of a vector's components. Math books often separate components with a comma and surround them with brackets:

$$\vec{r} = \langle 200 \text{ m}, 100 \text{ m} \rangle$$

In this book we'll use notation common in physics, based on **unit vectors**. In unit vector notation, we write

$$\vec{r} = 200 \text{ m}\,\hat{\imath} + 100 \text{ m}\,\hat{\jmath}$$

The unit vectors $\hat{\imath}$ and $\hat{\jmath}$ ("eye hat," "jay hat") designate the x- and y-directions, respectively. They're vectors each having a magnitude of one dimensionless unit along the x- and y-coordinate axes. So the unit vectors carry information about direction but not about the size or dimensions of a physical vector quantity. The notation $\vec{r} = 200 \text{ m}\,\hat{\imath} + 100 \text{ m}\,\hat{\jmath}$ shows that you reach position $\vec{r}$ by starting at the origin, then going 200 m in the x-direction ($\hat{\imath}$) and 100 m in the y-direction ($\hat{\jmath}$).

Thus, a general position vector in the plane is

$$\vec{r} = x\hat{\imath} + y\hat{\jmath} \tag{3.4}$$

where x and y are Cartesian coordinates. Note that x and y don't have arrows over them. **The individual components of a vector are single numbers and are therefore scalars.**

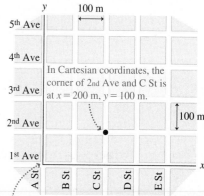

In Cartesian coordinates, the corner of 2nd Ave and C St is at $x = 200$ m, $y = 100$ m.

From the origin (0 m, 0 m), the x-axis points east and the y-axis north.

(a)

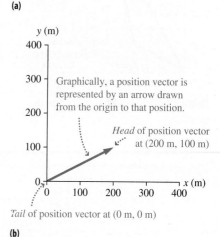

Graphically, a position vector is represented by an arrow drawn from the origin to that position.

Head of position vector at (200 m, 100 m)

Tail of position vector at (0 m, 0 m)

(b)

FIGURE 3.3 Two ways to represent a position relative to a set of coordinate axes.

Vector components can be positive, negative, or zero. On our map, the position six blocks due south of the origin is $\vec{r} = 0$ m $\hat{\imath} - 600$ m $\hat{\jmath}$ (meaning $x = 0$ m, $y = -600$ m). Three blocks west and five blocks south of the origin, the position is $\vec{r} = -300$ m $\hat{\imath} - 500$ m $\hat{\jmath}$ (meaning $x = -300$ m, $y = -500$ m).

✓ TIP

Each vector component is a scalar.

Reviewing New Concepts

To summarize, here are important ideas about position vectors in two dimensions:

- A position vector has two components (x, y), corresponding to the two Cartesian coordinates.
- Graphically, a position vector is an arrow from the origin to that position.
- Symbolically, you can use unit vectors $\hat{\imath}$ and $\hat{\jmath}$ to represent a vector, with the x-component attached to $\hat{\imath}$ and the y-component attached to $\hat{\jmath}$.

Displacement Vectors

The definition of displacement is the same in two dimensions as in one (see Chapter 2): **displacement is change in position**. In two dimensions, however, displacement, like position, is a vector (see Figure 3.4 and Equation 3.5).

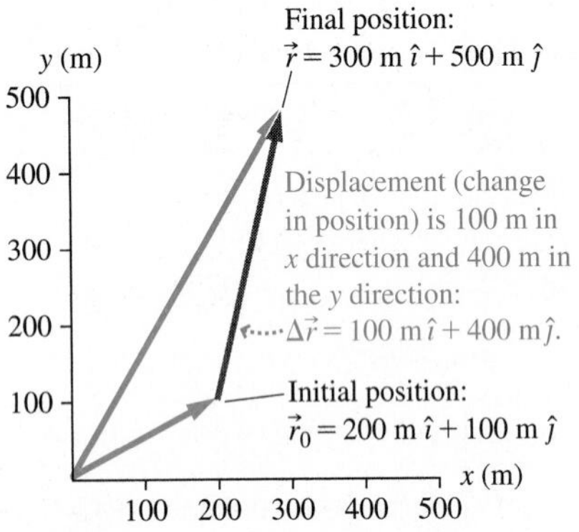

$$\Delta \vec{r} = \vec{r} - \vec{r}_0 \quad \text{(Displacement; SI unit: m)} \quad (3.5)$$

✓ TIP

It's conventional to put the vector symbol over the r but not the Δ to represent the change in a vector. As before, we'll use the subscript 0 to designate "initial."

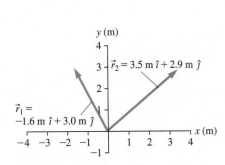

FIGURE 3.4 Displacement is change in position.

Equation 3.5 requires that you subtract two vectors to find displacement. Before proceeding, we'll need to discuss vector addition and subtraction.

Adding and Subtracting Vectors

Adding and subtracting vectors is straightforward:

The sum of two vectors is a vector whose components are the sums of the corresponding components. Symbolically, the sum of vectors $\vec{r}_1 = x_1\hat{\imath} + y_1\hat{\jmath}$ and $\vec{r}_2 = x_2\hat{\imath} + y_2\hat{\jmath}$ is

$$\vec{r}_1 + \vec{r}_2 = (x_1 + x_2)\hat{\imath} + (y_1 + y_2)\hat{\jmath} \quad \text{(Vector sum)} \quad (3.6)$$

The difference of two vectors is analogous:

$$\vec{r}_2 - \vec{r}_1 = (x_2 - x_1)\hat{\imath} + (y_2 - y_1)\hat{\jmath} \quad \text{(Vector difference)} \quad (3.7)$$

Consider, for example, position vectors $\vec{r}_1 = -1.6$ m $\hat{\imath} + 3.0$ m $\hat{\jmath}$ and $\vec{r}_2 = 3.5$ m $\hat{\imath} + 2.9$ m $\hat{\jmath}$ shown in Figure 3.5. Their sum follows from Equation 3.6:

$$\vec{r}_1 + \vec{r}_2 = (-1.6 \text{ m} + 3.5 \text{ m})\hat{\imath} + (3.0 \text{ m} + 2.9 \text{ m})\hat{\jmath} = 1.9 \text{ m } \hat{\imath} + 5.9 \text{ m } \hat{\jmath}$$

and their difference from Equation 3.7:

$$\vec{r}_2 - \vec{r}_1 = (3.5 \text{ m} - (-1.6 \text{ m}))\hat{\imath} + (2.9 \text{ m} - 3.0 \text{ m})\hat{\jmath} = 5.1 \text{ m } \hat{\imath} - 0.1 \text{ m } \hat{\jmath}$$

FIGURE 3.5 Vectors $\vec{r}_1$ and $\vec{r}_2$.

✓ TIP

Add vectors by adding the components. Subtract vectors by subtracting the components. Don't forget units!

Displacement is change in position—the difference between two position vectors, as given by Equation 3.5: $\Delta\vec{r} = \vec{r} - \vec{r}_0$. For the example of Figure 3.4, the displacement is

$$\Delta\vec{r} = \vec{r} - \vec{r}_0 = 300 \text{ m }\hat{\imath} + 500 \text{ m }\hat{\jmath} - (200 \text{ m }\hat{\imath} + 100 \text{ m }\hat{\jmath})$$
$$= (300 \text{ m} - 200 \text{ m})\hat{\imath} + (500 \text{ m} - 100 \text{ m})\hat{\jmath} = 100 \text{ m }\hat{\imath} + 400 \text{ m }\hat{\jmath}$$

This is just the result shown in Figure 3.4.

Vector addition shows that displacement $\Delta\vec{r}$ is the vector you need to add to an initial position to get to a final position, as shown in Figure 3.4. Verbally, "initial plus change equals final." In the example we've been considering,

$$\vec{r} = \vec{r}_0 + \Delta\vec{r} = 200 \text{ m }\hat{\imath} + 100 \text{ m }\hat{\jmath} + (100 \text{ m }\hat{\imath} + 400 \text{ m }\hat{\jmath})$$
$$= (200 \text{ m} + 100 \text{ m})\hat{\imath} + (100 \text{ m} + 400 \text{ m})\hat{\jmath} = 300 \text{ m }\hat{\imath} + 500 \text{ m }\hat{\jmath}$$

Vector Magnitude and Direction

Physically, vector quantities have both direction and **magnitude**—the size of the quantity. We use the symbol r for the magnitude of a vector $\vec{r}$—that is, the same variable without the arrow. We designate the **direction angle** by the angle θ counterclockwise from the x-axis.

Figure 3.6 shows that the vector $\vec{r} = x\hat{\imath} + y\hat{\jmath}$ has a magnitude r and direction given by:

$$r = \sqrt{x^2 + y^2} \qquad \text{(Vector magnitude)} \tag{3.8a}$$

$$\theta = \tan^{-1}\left(\frac{y}{x}\right) \qquad \text{(Vector direction)} \tag{3.8b}$$

✓ TIP

The magnitude of the vector $\vec{r}$ is represented by the symbol r, with no arrow.

Solving for θ requires care, because tangent is a multi-valued function. When you calculate the inverse tangent function, your calculator outputs the "principal value," which lies between $-90°$ and $+90°$. If the vector you're considering points in the second or third quadrant, add 180° to the calculator's output. Figure 3.7 illustrates this for the vector $\vec{r} = -1.23 \text{ m }\hat{\imath} - 2.35 \text{ m }\hat{\jmath}$. If you use the components $x = -1.23 \text{ m}$, $y = -2.35 \text{ m}$ in your calculator, you get

$$\theta = \tan^{-1}\left(\frac{y}{x}\right) = \tan^{-1}\left(\frac{-2.35 \text{ m}}{-1.23 \text{ m}}\right) = 62.4°$$

Because this vector points in the third quadrant, add 180° to get $\theta = 62.4° + 180° = 242.4°$.

You've seen how to use right-triangle geometry to find a vector's direction and magnitude. You can also go the other way, getting components from direction and magnitude. In Figure 3.6, the components x and y are right-triangle sides adjacent and opposite the angle θ, respectively. Equation 3.1 shows that $\cos\theta = x/r$ and $\sin\theta = y/r$. Therefore, the vector components are

$$x = r\cos\theta \text{ and } y = r\sin\theta \qquad \text{(x- and y-components of a position vector $\vec{r}$)} \tag{3.9}$$

Components and magnitude/direction are equivalent ways to express a vector. Either pair fully describes a vector in two dimensions; Equations 3.8 and 3.9 let you go back and forth between the two descriptions.

Graphical Interpretation of Vector Arithmetic

The representation of vectors as arrows leads to a graphical view of vector addition and subtraction (Figure 3.8). Adding two vectors, in this case $\vec{r}_1$ and $\vec{r}_2$, is equivalent to placing

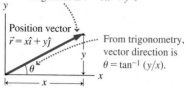

From Pythagorean theorem, vector magnitude is $r = \sqrt{x^2 + y^2}$.

From trigonometry, vector direction is $\theta = \tan^{-1}(y/x)$.

FIGURE 3.6 Magnitude and direction of a position vector.

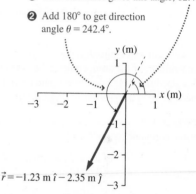

❶ Your calculator gives this angle, 62.4°.

❷ Add 180° to get direction angle $\theta = 242.4°$.

$\vec{r} = -1.23 \text{ m }\hat{\imath} - 2.35 \text{ m }\hat{\jmath}$

FIGURE 3.7 Adjusting your calculator's output for second and third quadrant direction angles.

Two vectors

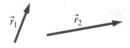

Adding the vectors

❶ Place $\vec{r}_1$ and $\vec{r}_2$ head to tail (in either order).

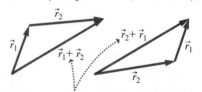

❷ The vector sum points from the *tail* of one vector to the *head* of the other. Note that $\vec{r}_1 + \vec{r}_2 = \vec{r}_2 + \vec{r}_1$.

Subtracting $\vec{r}_1$ from $\vec{r}_2$

❶ Place the vectors tail to tail.

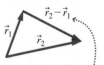

❷ $\vec{r}_2 - \vec{r}_1$ points from the *head* of $\vec{r}_1$ to the *head* of $\vec{r}_2$.

FIGURE 3.8 How to add and subtract vectors graphically.

the tail of $\vec{r}_2$ at the head of $\vec{r}_1$. **You get the vector sum $\vec{r}_1 + \vec{r}_2$ by drawing an arrow from the tail of $\vec{r}_1$ to the head of $\vec{r}_2$.** The difference $\Delta\vec{r} = \vec{r}_2 - \vec{r}_1$ is the vector you add to $\vec{r}_1$ to get $\vec{r}_2$. Graphically, that means **you draw $\Delta\vec{r}$ from the head of $\vec{r}_1$ to the head of $\vec{r}_2$.**

The graphical representation vector arithmetic has the advantage that you can visualize the situation. Its disadvantage is that adding and subtracting graphically isn't as accurate as using components. Knowing both methods will contribute to your understanding of the many vector situations that arise throughout physics.

CONCEPTUAL EXAMPLE 3.2 **Adding and Subtracting Vectors Graphically**

Consider again the two vectors discussed earlier:

$$\vec{r}_1 = -1.6 \text{ m } \hat{\imath} + 3.0 \text{ m } \hat{\jmath} \text{ and } \vec{r}_2 = 3.5 \text{ m } \hat{\imath} + 2.9 \text{ m } \hat{\jmath}$$

Use graphical methods to find the sum $\vec{r}_1 + \vec{r}_2$ and the difference $\vec{r}_2 - \vec{r}_1$.

SOLVE The drawings show the graphical solutions, constructed using our rules: For the sum, connect the vectors head to tail (Figure 3.9a). For the difference, join the tails and draw the difference vector from the head of $\vec{r}_1$ to the head of $\vec{r}_2$ (Figure 3.9b).

REFLECT Let's compare the graphical and component methods. Earlier, using components, we found

$$\vec{r}_1 + \vec{r}_2 = 1.9 \text{ m } \hat{\imath} + 5.9 \text{ m } \hat{\jmath}$$
$$\vec{r}_2 - \vec{r}_1 = 5.1 \text{ m } \hat{\imath} - 0.1 \text{ m } \hat{\jmath}$$

Do the graphical solutions agree? Look at the vector $\vec{r}_1 + \vec{r}_2$ obtained graphically; its x-component is about 2 m and its y-component about 6 m. Similarly, $\vec{r}_2 - \vec{r}_1$ has an x-component of about 5 m and y-component just under zero. The graphical sum and difference match the numerical answers, within the accuracy of our drawings.

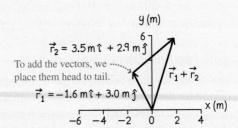

(a)

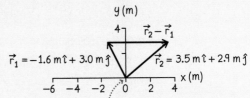

(b)

FIGURE 3.9 (a) Adding $\vec{r}_1$ and $\vec{r}_2$ graphically; (b) Finding $\vec{r}_2 - \vec{r}_1$ graphically.

TACTIC 3.1 **Vector Addition and Subtraction**

VECTOR ADDITION
- Vectors add by component. To find the sum of two vectors, add their x-components and y-components separately.
- Graphically, add vectors $\vec{r}_1$ and $\vec{r}_2$, by placing the tail of $\vec{r}_2$ on the head of $\vec{r}_1$. Then the sum $\vec{r}_1 + \vec{r}_2$ is the vector from the tail of $\vec{r}_1$ to the head of $\vec{r}_2$.

VECTOR SUBTRACTION
- To find the difference of two vectors using components, subtract their x-components and y-components separately.
- Graphically, place the vectors tail to tail. Draw the difference $\vec{r}_2 - \vec{r}_1$ from the head of $\vec{r}_1$ to the head of $\vec{r}_2$.

Multiplying Vectors by Scalars

We frequently multiply vectors by scalars. For example, the vector 1.9 m $\hat{\imath}$ is the product of the vector $\hat{\imath}$ and the scalar 1.9 m. In general, the distributive law applies, so:

$$3(2 \text{ m } \hat{\imath} + 5 \text{ m } \hat{\jmath}) = 6 \text{ m } \hat{\imath} + 15 \text{ m } \hat{\jmath}$$

More generally, with $\vec{r} = x\hat{\imath} + y\hat{\jmath}$,

$$a\vec{r} = ax\,\hat{\imath} + ay\,\hat{\jmath} \qquad \text{(Multiplication of a vector by a scalar)} \qquad (3.10)$$

Graphically, multiplying a vector by a positive scalar a changes the vector's magnitude by a factor of a but doesn't change its direction (see Figure 3.10). For example, if $\Delta\vec{r}$ represents your displacement for a trip, then $2.5\Delta\vec{r}$ represents a trip in the same direction, but 2.5 times farther. Multiplying by a negative scalar a reverses the vector's direction and also changes its magnitude by a factor of the absolute value of a. Multiplying a vector by -1 reverses the direction while keeping the magnitude unchanged. Finally, if the scalar has dimensions, they multiply any existing vector dimensions. You'll see examples in the next section.

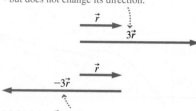

Multiplying $\vec{r}$ by 3
• increases its magnitude by a factor of 3
• but does not change its direction.

Multiplying $\vec{r}$ by -3
• increases its magnitude by a factor of 3
• and *reverses* its direction.

FIGURE 3.10 Multiplying a vector by a positive scalar and by a negative scalar.

Reviewing New Concepts

- To add vectors in two dimensions, add the components.
- To subtract vectors in two dimensions, subtract the components.
- Graphically form the vector sum $\vec{r}_1 + \vec{r}_2$ by placing the tail of $\vec{r}_2$ at the head of $\vec{r}_1$. Then the sum is the vector drawn from the tail of $\vec{r}_1$ to the head of $\vec{r}_2$.
- Graphically form the difference by placing the tails of $\vec{r}_1$ and $\vec{r}_2$ together. Then the difference $\Delta\vec{r} = \vec{r}_2 - \vec{r}_1$ is the vector drawn from the head of $\vec{r}_1$ to the head of $\vec{r}_2$.
- To multiply a vector by a scalar, multiply each component of the vector by that scalar.

GOT IT? Section 3.2   What's the sum of $\vec{r}_1$ and $\vec{r}_2$?

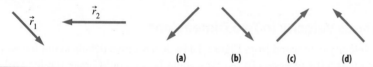

(a)   (b)   (c)   (d)

3.3 Velocity and Acceleration in Two Dimensions

In Chapter 2, you saw how in one dimension an object's change in position—its displacement—over time defines its velocity, and then in turn how its change in velocity over time defines its acceleration. These definitions are the same for two-dimensional motion. The only difference is that displacement in two dimensions is a vector, so velocity and acceleration are also vectors.

Average Velocity in Two Dimensions

Figure 3.11 shows the motion of a leopard stalking its prey. Here we've chosen coordinates with the $+x$-axis eastward and the $+y$-axis northward. In Chapter 2, we defined average velocity as displacement Δx divided by the corresponding time interval Δt. Expanding that definition to two-dimensional motion with displacement $\Delta\vec{r}$ gives

$$\overline{\vec{v}} = \frac{\text{displacement}}{\text{time}} = \frac{\Delta\vec{r}}{\Delta t} \qquad \begin{array}{l}\text{(Average velocity in} \\ \text{two-dimensional motion; SI unit: m/s)}\end{array} \qquad (3.11)$$

Note that the vector $\Delta\vec{r}$ is divided by the scalar Δt; equivalently, it's multiplied by $1/\Delta t$—the process of scalar multiplication discussed in Section 3.2. Because Δt is always positive, the resulting average velocity $\overline{\vec{v}}$ is a vector in the same direction as $\Delta\vec{r}$. Its units are

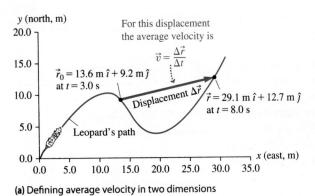

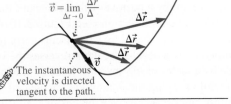

(a) Defining average velocity in two dimensions

(b) Approaching the instantaneous velocity by shortening the time interval to zero

FIGURE 3.11 Average velocity and instantaneous velocity for two-dimensional motion.

those of displacement (m) divided by those of time (s)—or m/s, as we expect for velocity. For the positions and times marked in Figure 3.11a, the leopard's average velocity is

$$\vec{v} = \frac{\Delta \vec{r}}{\Delta t} = \frac{(29.1 \text{ m } \hat{\imath} + 12.7 \text{ m } \hat{\jmath}) - (13.6 \text{ m } \hat{\imath} + 9.2 \text{ m } \hat{\jmath})}{8.0 \text{ s} - 3.0 \text{ s}}$$

$$= \frac{(29.1 \text{ m} - 13.6 \text{ m})\hat{\imath} + (12.7 \text{ m} - 9.2 \text{ m})\hat{\jmath}}{5.0 \text{ s}}$$

$$= \frac{15.5 \text{ m } \hat{\imath} + 3.5 \text{ m } \hat{\jmath}}{5.0 \text{ s}} = 3.1 \text{ m/s } \hat{\imath} + 0.7 \text{ m/s } \hat{\jmath}$$

Instantaneous Velocity in Two Dimensions

The average velocity calculated from Figure 3.11a doesn't give details about the leopard's motion during that 5.0-s interval. To see those details, we could break the leopard's trek into shorter time intervals, as in Figure 3.11b. As for one-dimensional motion, the limit of arbitrary small intervals gives the instantaneous velocity $\vec{v}$:

$$\vec{v} = \lim_{\Delta t \to 0} \frac{\Delta \vec{r}}{\Delta t} \quad \text{(Instantaneous velocity in two dimensions; SI unit: m/s)} \quad (3.12)$$

The approach to this limit in Figure 3.11b shows that the leopard's instantaneous velocity vector is tangent to the trajectory. Instantaneous velocity tells you how an object is moving at a particular moment in time, so it's more informative than the average velocity over an interval. Just as in one-dimensional motion, we'll drop "instantaneous" and use simply "velocity" to mean instantaneous velocity.

Determining Velocity, Speed, and Direction in Two Dimensions

As with any two-dimensional vector, you can express the velocity vector $\vec{v}$ either in components or with magnitude and direction. Calling the velocity components v_x and v_y, we can write:

$$\vec{v} = v_x \hat{\imath} + v_y \hat{\jmath}$$

You can see, for example, that the leopard's velocity in Figure 3.11 has both x- and y-components. As with any vector, the individual components v_x and v_y are scalars; they combine with the unit vectors to make a two-dimensional vector (Figure 3.12).

As with any vector, the Pythagorean theorem gives the magnitude (v) of the velocity vector $\vec{v}$:

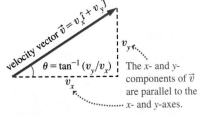

FIGURE 3.12 Velocity components v_x and v_y.

$$v = \sqrt{v_x^2 + v_y^2} \quad \text{(Speed; SI unit: m/s)} \quad (3.13)$$

The magnitude v has a special name: **speed**. Although you may interchange "speed" and "velocity" in everyday usage, in physics they're distinct. Velocity $\vec{v}$ is a vector, with speed v its magnitude—a scalar.

✓ **TIP**

Velocity and **speed** are distinct but closely related. Velocity is a vector, with magnitude and direction. Speed, a scalar, is the magnitude of velocity.

The direction of velocity follows from trigonometry, as it did for position. Figure 3.12 shows that

$$\theta = \tan^{-1}\left(\frac{v_y}{v_x}\right) \quad \text{(Direction in two-dimensional motion)} \tag{3.14}$$

Components and magnitude/direction are equivalent ways of expressing the same velocity, just as they are with any vectors. In two dimensions, each approach requires two numbers—either the two components or the magnitude and direction angle. Equations 3.13 and 3.14 take you from components to magnitude/direction. The triangle in Figure 3.12 shows that you can reverse the process, getting components from magnitude/direction:

$$v_x = v \cos\theta \quad v_y = v \sin\theta \quad \begin{array}{l}\text{(Velocity components in}\\ \text{two dimensions; SI unit: m/s)}\end{array} \tag{3.15}$$

✓ **TIP**

If you know a vector's components, Equations 3.13 and 3.14 give you the vector's magnitude and direction. If you know the magnitude and direction, Equation 3.15 gives you the components.

EXAMPLE 3.3   **A Speedy Leopard**

To track our leopard's motion, scientists lay out a grid with the $+x$-axis east and $+y$-axis north. At one instant they clock the leopard at 17.6 m/s in a direction 60.0° from the $+x$-axis. (a) Find the components of the leopard's velocity. (b) The cat changes velocity, keeping the same x-component v_x but changing its y-component to 10.6 m/s. Find its new speed and direction.

ORGANIZE AND PLAN This problem involves going back and forth between the magnitude/direction and component representations of vectors. It's helpful to visualize each case by sketching the vector triangle.

For part (a), you're given speed and direction, so we'll use Equation 3.15 to get the components (Figure 3.13a):

$$v_x = v \cos\theta, \ v_y = v \sin\theta$$

For part (b), you'll need the inverse operation, provided by Equations 3.13 and 3.14 and Figure 3.13b:

$$v = \sqrt{v_x{}^2 + v_y{}^2}, \ \theta = \tan^{-1}\left(\frac{v_y}{v_x}\right)$$

Known: $v = 17.6$ m/s and $\theta = 60.0°$.

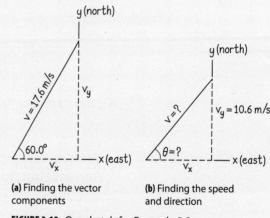

(a) Finding the vector components

(b) Finding the speed and direction

FIGURE 3.13 Our sketch for Example 3.3.

SOLVE (a) Inserting the numerical values,

$$v_x = v \cos\theta = (17.6 \text{ m/s}) \cos(60.0°) = 8.80 \text{ m/s}$$
$$v_x = v \sin\theta = (17.6 \text{ m/s}) \sin(60.0°) = 15.2 \text{ m/s}$$

cont'd.

(b) The x-component of velocity remains $v_x = 8.80$ m/s, but now the y-component is $v_y = 10.6$ m/s. The new speed and direction are

$$v = \sqrt{v_x^2 + v_y^2} = \sqrt{(8.80 \text{ m/s})^2 + (10.6 \text{ m/s})^2}$$
$$= 13.8 \text{ m/s}$$

$$\theta = \tan^{-1}\left(\frac{v_y}{v_x}\right) = \tan^{-1}\left(\frac{10.6 \text{ m/s}}{8.80 \text{ m/s}}\right) = 50.3°$$

Now the leopard is running at 13.8 m/s, at 50.3° from the x-axis.

REFLECT Reducing v_y from 15.2 m/s to 10.6 m/s reduced the speed by only 1.4 m/s (from 15.2 m/s to 13.8 m/s). It also reduced the

angle by about 10°. In compass terms, the leopard has turned a bit eastward.

MAKING THE CONNECTION Again the leopard starts with velocity $v_x = 8.80$ m/s, $v_y = 15.2$ m/s, but then changes to $v_x = -8.80$ m/s, $v_y = -15.2$ m/s. How do speed and direction change?

ANSWER The speed doesn't change. Mathematically, the velocity vector got multiplied by -1, changing the direction by 180° (to $60° + 180° = 240°$), but leaving the magnitude unchanged.

CONCEPTUAL EXAMPLE 3.4 **Velocity and Speed**

Can an object's speed remain constant while its velocity changes? If so, give an example. If not, explain why not.

SOLVE Yes. Constant speed with changing velocity is possible if components v_x and v_y vary such that $v = \sqrt{v_x^2 + v_y^2}$ remains constant.

A good example is a car rounding a curve at constant speed (Figure 3.14). The speedometer remains steady but the velocity isn't constant, because the *direction* of motion is changing. We'll discuss circular motion in Section 3.5.

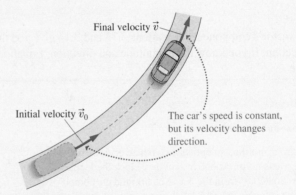

Final velocity $\vec{v}$

Initial velocity $\vec{v}_0$

The car's speed is constant, but its velocity changes direction.

FIGURE 3.14 Constant speed with changing velocity.

REFLECT The converse of this situation is not possible. If an object's speed is changing, then its velocity is changing, too.

Acceleration in Two Dimensions

Acceleration occurs when velocity changes. Once again, we'll define average and instantaneous acceleration as we did for one-dimensional motion in Chapter 2.

Average acceleration $\vec{a}$ during a time interval Δt is the change in velocity divided by the time interval. Symbolically,

$$\vec{a} = \frac{\Delta \vec{v}}{\Delta t} \quad \text{(Average acceleration in two-dimensional motion; SI unit: m/s}^2\text{)} \tag{3.16}$$

For example, in Example 3.3, the leopard's velocity changed from $\vec{v}_0 = 8.80$ m/s $\hat{\imath}$ + 15.2 m/s $\hat{\jmath}$ to $\vec{v} = 8.80$ m/s $\hat{\imath}$ + 10.6 m/s $\hat{\jmath}$. If that change took 2.0 s, the average acceleration was

$$\overline{\vec{a}} = \frac{\Delta\vec{v}}{\Delta t} = \frac{(8.80\text{ m/s }\hat{\imath} + 10.6\text{ m/s }\hat{\jmath}) - (8.80\text{ m/s }\hat{\imath} + 15.2\text{ m/s }\hat{\jmath})}{2.0\text{ s}}$$

$$= \frac{(8.80\text{ m/s} - 8.80\text{ m/s})\hat{\imath} + (10.6\text{ m/s} - 15.2\text{ m/s})\hat{\jmath}}{2.0\text{ s}}$$

$$= \frac{(0\text{ m/s})\hat{\imath} + (-4.6\text{ m/s})\hat{\jmath}}{2.0\text{ s}} = -2.3\text{ m/s}^2\,\hat{\jmath}$$

In this case, the average acceleration is entirely in the $-y$-direction. This makes sense, given that the y-component of velocity decreased while the x-component remained unchanged.

Instantaneous acceleration $\vec{a}$ is the average acceleration in the limit as the time interval approaches zero. Thus

$$\vec{a} = \lim_{\Delta t \to 0} \frac{\Delta\vec{v}}{\Delta t} \quad \text{(Instantaneous acceleration in two-dimensional motion; SI unit: m/s}^2) \quad (3.17)$$

Again, we'll drop "instantaneous" and simply use "acceleration" for instantaneous acceleration.

3.4 Projectile Motion

A **projectile** is an object launched with some initial velocity that then flies through the air under the influence of gravity. Our ancestors hurled sticks and stones at their prey; those were projectiles. The invention of cannons made projectile motion an important military application. College students are more likely to see projectiles on their athletic fields. Think of all the ball games that require skillful launching of projectiles!

The Projectile's Constant Acceleration

Picking a coordinate system is the first step in analyzing two-dimensional motion. The logical choice for projectile motion is a horizontal x-axis and a vertical y-axis. A good choice for the origin is the projectile's launch point, regardless of whether that's ground level or not. Figure 3.15 shows the coordinate system and the projectile's curved trajectory.

Our modern understanding of projectile motion began with Galileo, who described the horizontal and vertical components of a projectile's motion separately. We'll follow this idea by considering vector components of position (x and y), velocity (v_x and v_y), and acceleration (a_x and a_y). By tracking the individual components as they change with time, you can understand projectile motion.

 TIP

Think of the x- and y-components of a projectile's flight separately.

Figure 3.16 provides insight into Galileo's reasoning. The photo shows two balls launched simultaneously, one dropped from rest and the other projected horizontally. Note that the vertical motions of the two balls are identical, as each falls under the influence of gravity. Thus, the vertical motion of any projectile is the free-fall motion you studied in Chapter 2: a constant vertical acceleration $a_y = -g = -9.80$ m/s^2. There's no horizontal acceleration ($a_x = 0$); the horizontal velocity component remains constant.

Figure 3.17 shows graphically how a projectile's constant downward acceleration results in its curved path. The downward acceleration affects the vertical velocity component but not the horizontal component. The vertical component shrinks gradually to zero, then

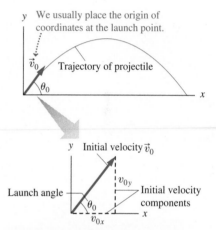

FIGURE 3.15 The path (trajectory) of a projectile.

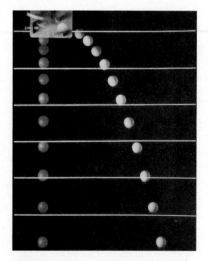

FIGURE 3.16 Strobe photograph of two balls released at the same time, one dropped from rest and the other projected horizontally. The horizontal lines show that the balls accelerate vertically downward at the same rate.

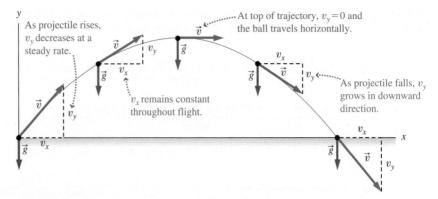

FIGURE 3.17 Motion diagram for a projectile, showing constant velocity in the horizontal direction and free fall in the vertical direction, with constant downward acceleration $\vec{g}$.

reverses and grows toward more negative values. The trajectory is, in fact, a parabola, although we won't prove this.

Here we're ignoring the effect of air resistance on projectiles. Just as with free fall (Section 2.5), that's a good approximation for some objects, but not others. It works well for Galileo's cannon balls, but not ping-pong balls. We'll discuss air resistance in Chapter 4.

Kinematic Equations for Projectiles

With both components of acceleration constant ($a_x = 0$ and $a_y = -g$), we can apply the one-dimensional constant acceleration equations separately for the horizontal and vertical motions. Table 3.2 shows the results.

TABLE 3.2 Kinematic Equations in Two Dimensions

Kinematic equations for x		For projectile, with $a_x = 0$	
$x = v_{0x}t + \frac{1}{2}a_x t^2$	(2.9)	$x = v_{0x}t$	(3.18a)
$v_x = v_{0x} + a_x t$	(2.8)	$v_x = v_{0x}$	(3.18b)
$v_x^2 = v_{0x}^2 + 2a_x \Delta x$	(2.10)	$v_x = v_{0x}$	
Kinematic equations for y		**For projectile, with $a_y = -g$**	
$y = v_{0y}t + \frac{1}{2}a_y t^2$		$y = v_{0y}t - \frac{1}{2}gt^2$	(3.19a)
$v_y = v_{0y} + a_y t$		$v_y = v_{0y} - gt$	(3.19b)
$v_y^2 = v_{0y}^2 + 2a_y \Delta y$		$v_y^2 = v_{0y}^2 - 2g\Delta y$	(3.19c)

✓ **TIP**

Table 3.2 gives the kinematic equations for projectile motion. These results assume launch from the origin, so $x_0 = y_0 = 0$.

The equations in Table 3.2 are powerful tools that let you predict a projectile's future position (x and y) and velocity (v_x and v_y) given its initial velocity. The following examples show some ways you can use these equations.

EXAMPLE 3.5   Fore!

A golfer swings her 6-iron, sending the ball off the ground at 30° with speed 39.0 m/s. (a) Find the x- and y-components of the initial velocity. (b) Find both velocity components 1.00 s after the ball was hit.

ORGANIZE AND PLAN Figure 3.18 is our sketch of the initial velocity vector $\vec{v}$ and its components v_{0x} and v_{0y}. You can see from the diagram, or you can apply Equation 3.15, to get

$$v_{0x} = v_0 \cos(30°)$$
$$v_{0y} = v_0 \sin(30°)$$

With the initial velocity components known, the velocity components at any later time are $v_x = v_{0x}$ (Equation 3.18b), because the horizontal component doesn't change, and $v_y = v_{0y} - gt$ (Equation 3.19b), accounting for gravity.

Known: $v_0 = 39.0$ m/s and $\theta_0 = 30°$.

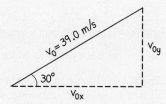

FIGURE 3.18 Our sketch showing the launch angle and velocity components.

SOLVE (a) Using numerical values of initial speed and angle gives

$$v_{0x} = v_0 \cos(30°) = (39.0 \text{ m/s}) \cos(30°) = 33.8 \text{ m/s}$$
$$v_{0y} = v_0 \sin(30°) = (39.0 \text{ m/s}) \sin(30°) = 19.5 \text{ m/s}$$

(b) The horizontal component doesn't change, so after 1.00 s, $v_x = v_{0x} = 33.8$ m/s. The vertical component becomes $v_y = v_{0xy} - gt = 19.5$ m/s $- (9.80 \text{ m/s}^2)(1.00 \text{ s}) = 9.7$ m/s.

REFLECT After 1.00 s the ball is still going up, because $v_y > 0$. But the ball's downward acceleration will continue to decrease v_y, with $v_y = 0$ at the peak of the trajectory. After that the ball will fall as it continues to accelerate downward.

MAKING THE CONNECTION At the peak of its trajectory, the vertical component of the ball's velocity is zero. What's its acceleration at this point?

ANSWER The acceleration *always* has magnitude g and points downward. The peak is no different. The fact that the ball instantaneously has no vertical motion is irrelevant. Just before the peak, the ball was moving upward. Just after, it's moving downward. So its vertical velocity component is changing continually, even at the instant when the vertical component is zero.

CONCEPTUAL EXAMPLE 3.6   Velocity Components

Suppose you know a projectile's initial velocity components v_{0x} and v_{0y}. Assuming a launch from the ground, with flight over level ground, what are v_x and v_y when the projectile lands?

SOLVE The x-component never changes, so $v_x = v_{0x}$ (Figure 3.19). The vertical motion is the same as for one-dimensional free fall. In Section 2.5 you saw that the speed in free fall at a given height is the same going up and down. Thus the vertical velocity component coming down is just the negative of the component going up. So when the projectile lands, $v_y = -v_{0y}$.

REFLECT The flight of a projectile is symmetric about its midpoint (the highest point), as the next example illustrates. That means a projectile takes the same time to rise to its peak as to fall back to its initial height.

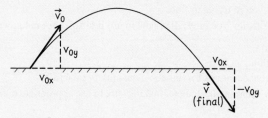

FIGURE 3.19 The projectile's velocity components at the beginning and end of its flight.

EXAMPLE 3.7 **How High, How Far, and How Much Time?**

A batter hits a "pop up," with the ball leaving the bat at 23.8 m/s and 60° above the horizontal. (a) How much time does the fielder have to reach the ball? (b) How high does the ball go? (c) How far does it go horizontally? (This is the **range** of the projectile.) Assume that the ball is caught at the same height at which it left the bat.

ORGANIZE AND PLAN Knowing the initial speed and angle lets you find the initial velocity components (Figure 3.20 or Equation 3.15). The other information follows from that initial velocity and the kinematic equations (Table 3.2).

As in Example 3.5, the initial velocity components are $v_{0x} = v_0 \cos(60°)$, $v_{0y} = v_0 \sin(60°)$. Conceptual Example 3.6 showed that when the ball returns, $v_y = -v_{0y}$. Let T be the total time of flight. You can find T from Equation 3.19b, using $v_y = -v_{0y}$: $-v_{0y} = v_{0y} - gT$. Conceptual Example 3.6 also showed that due to the symmetry of the motion, the ball reaches its maximum height at time $t = T/2$. Then, with $t = T/2$, the maximum height follows from Equation 3.19a: $y = v_{0y}t - \frac{1}{2}gt^2$ (with $t = T/2$). Equation 3.18a will give the range: $x = v_{0x}T$.

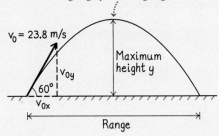

The projectile's motion is symmetric, so the ball spends half its flight time going up and half going down.

$v_0 = 23.8$ m/s

Maximum height y

v_{0y}

60°

v_{0x}

Range

FIGURE 3.20 The launch and maximum height of the baseball.

Known: $v_0 = 23.8$ m/s; launch angle $\theta = 60°$.

SOLVE (a) The initial velocity components are

$$v_{0x} = v_0 \cos(60°) = (23.8 \text{ m/s}) \cos(60°) = 11.9 \text{ m/s}$$

$$v_{0y} = v_0 \sin(60°) = (23.8 \text{ m/s}) \sin(60°) = 20.6 \text{ m/s}$$

Solving our equation $-v_{0y} = v_{0y} - gT$ for T,

$$T = \frac{v_{0y} - (-v_{0y})}{g} = \frac{20.6 \text{ m/s} - (-20.6 \text{ m/s})}{9.80 \text{ m/s}^2} = 4.20 \text{ s}$$

(b) The maximum height is reached at half this time (see Figure 3.20), or $t = 2.10$ s. At that time, the height is

$$y = v_{0y}t - \frac{1}{2}gt^2 = (20.6 \text{ m/s})(2.10 \text{ s}) - \frac{1}{2}(9.80 \text{ m/s}^2)(2.10 \text{ s})^2$$

$$= 21.7 \text{ m}$$

(c) Using the flight time from (a), the ball's range is

$$x = v_{0x}t = (11.9 \text{ m/s})(4.20 \text{ s}) = 50.0 \text{ m}$$

REFLECT The time, height, and distance seem about right for baseball. The distance isn't far past second base, so a fielder should be able to reach the ball in 4.2 s!

***

MAKING THE CONNECTION Our solution neglected the ball's initial height. If that height was 0.5 m, and given that the fielder has until the ball reaches the ground to make the catch, how does the answer to part (a) change?

ANSWER The 4.20-s answer is the time to return to the ball's launch height. To drop another 0.5 m (with $v_y \approx -v_{0y} = -20.6$ m/s) takes about 0.024 s—so the extra time is insignificant.

***

PROBLEM-SOLVING STRATEGY 3.1 **Solving Projectile Motion Problems**

ORGANIZE AND PLAN
- Draw a diagram and identify any points of interest in the projectile's flight.
- Identify what you already know, including numerical values.
- Determine what you're trying to find.
- Review the information you have; plan how to use it to solve for unknown(s).

SOLVE
- Gather information you have.
- Combine and solve equations for unknown quantity or quantities.
- Insert numerical values and calculate, making sure to use appropriate units.

REFLECT
- Does the answer(s) have the correct units?
- If the problem relates to projectile flight on a familiar scale, do the results make sense?

CONCEPTUAL EXAMPLE 3.8 **Range and Slope**

A golfer always hits the ball at the same launch speed and angle above the horizontal. Discuss the distance traveled when the terrain slopes uphill, relative to the ball's range on level ground. Repeat for terrain sloping downward.

SOLVE Think about the ball's parabolic trajectory: While in flight it's the same regardless of the ground's slope. Figure 3.21a shows that the trajectory is interrupted on the uphill slope, so the ball doesn't travel as far. The opposite is true for the downhill slope, so the ball goes a bit farther (Figure 3.21b).

REFLECT Good golfers take slope into account, knowing that they get less distance on uphill shots than on downhill shots. They also need to consider the effects of wind speed and direction.

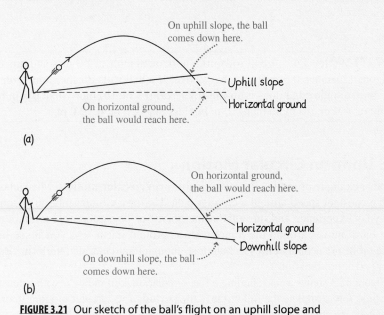

(a)

(b)

FIGURE 3.21 Our sketch of the ball's flight on an uphill slope and on a downhill slope.

EXAMPLE 3.9 **A Hilly Golf Course**

Consider again the golf shot in Example 3.5, with $v_0 = 39.0$ m/s and $\theta = 30°$. Now, however, the golfer tees off from a 2.90-m-high hill surrounded by level ground. How far does the ball travel horizontally before striking the ground?

ORGANIZE AND PLAN As usual, we'll consider x- and y-components of motion separately. The quantity you want to find—horizontal distance—is given by Equation 3.18a: $x = v_{0x}t$. We already found v_{0x} and v_{0y} in Example 3.5. To find the ball's position (x) when it reaches the ground, we need its flight time. That follows from the kinematic equation for the y-component of the ball's position:

$$y = v_{0y}t - \frac{1}{2}gt^2$$

Everything here is known except t, so we can solve for t when the ball hits the ground. Since our equations assume the

launch point is the origin (Figure 3.22), and the ground is 2.90 m below that point, we'll take $y = -2.90$ m.

Known: $v_{0x} = 33.8$ m/s, $v_{0y} = 19.5$ m/s (Example 3.5), $y = -2.90$ m (when the ball lands).

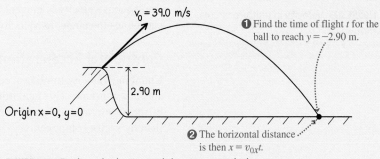

FIGURE 3.22 Finding the horizontal distance traveled.

cont'd.

SOLVE Inserting known values into the equation for y:

$$y = v_{0y}t - \frac{1}{2}gt^2$$

$$-2.90 \text{ m} = (19.5 \text{ m/s})t - \frac{1}{2}(9.80 \text{ m/s}^2)t^2$$

This is a quadratic equation for t. The quadratic formula gives solutions $t = 4.12$ s and $t = -0.144$ s. Only the positive time is meaningful in this example, so the flight time is $t = 4.12$ s. Then the horizontal distance is

$$x = v_{0x}t = (33.8 \text{ m/s})(4.12 \text{ s}) = 139 \text{ m}$$

REFLECT Time and distance both seem reasonable for a golf ball. Comparing the results for the same ball hit on level ground (see Making the Connection below) shows that the elevated launch increases both time and distance.

MAKING THE CONNECTION What are the time and distance for a ball launched with the same initial velocity on level ground?

ANSWER Working through the problem the same way (but with final height $y = 0$), you'll find $t = 3.98$ s and $x = 135$ m. These answers make sense (see Conceptual Example 3.8). Here the ball spends 140 ms less time in the air and falls 4 m short of what it did in Example 3.9.

GOT IT? Section 3.4 A ball rolls off a horizontal table at 1.0 m/s and hits the ground a horizontal distance 0.5 m from table's edge. If a ball rolls off the same table at 2.0 m/s, how far from the edge does it hit the floor? (a) 0.5 m; (b) more than 0.5 m but less than 1.0 m; (c) 1.0 m; (d) more than 1.0 m but less than 2.0 m; (e) 2.0 m.

3.5 Uniform Circular Motion

Another example of motion in a plane is **uniform circular motion**, when an object moves with a constant speed around a circular path. Uniform circular motion is common in our everyday experience and throughout the universe. To a good approximation, this is the way Earth moves around the Sun, or Moon around Earth. (As you'll see in Chapter 9, these orbits are actually ellipses, but close enough to circular that uniform circular motion is a good approximation.)

Figure 3.23 shows a car rounding a flat, circular track of radius R with a constant speed v. The time T to complete the full circle is the **period** of its circular motion. In one period the car travels the circle's circumference, $2\pi R$, so its speed is

$$v = \frac{\text{distance}}{\text{time}} = \frac{2\pi R}{T}$$

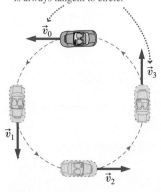

Speed is constant but direction changes. Direction of velocity $\vec{v}$ is always tangent to circle.

FIGURE 3.23 Velocity in uniform circular motion.

✓ **TIP**

Use the uppercase T for period, to distinguish period from a variable time t.

In uniform circular motion, *speed* is constant, but *velocity* is continuously changing because, as Figure 3.23 shows, the *direction* of the velocity changes. Remember that the instantaneous velocity is tangent to the trajectory—in this case, tangent to the circle. As the car rounds the circle, its velocity is different at every point because its direction changes.

Centripetal Acceleration

Changing velocity implies acceleration. How do you find acceleration in this case, when speed is constant? This is a little harder than with projectile motion, because the direction of the acceleration is itself changing. We'll approach the problem geometrically rather than by vector components.

Our approach is based on the definition of acceleration (Equation 3.17):

$$\vec{a} = \lim_{\Delta t \to 0} \frac{\Delta \vec{v}}{\Delta t}$$

Again imagine that car, or any object traveling in a circle of radius R with constant speed v. We'll look at small changes in velocity $\Delta \vec{v}$ over correspondingly small time intervals Δt, then consider the limit as Δt approaches zero. Figure 3.24 shows the vector subtraction to find $\Delta \vec{v} = \vec{v} - \vec{v}_0$, the velocity change as the object moves a small distance around the circle.

Figure 3.25 shows that our object travels an arc of length $v\Delta t$ moving at constant speed v during the interval Δt. Here's where the approximation and limiting process come in. For short time intervals, the circular arc can be approximated as a straight line, as shown in the bottom portion of Figure 3.25. That line and the two radii form an isosceles triangle that is similar to the velocity triangle in Figure 3.24, because the angles $\Delta \theta$ are the same. Recall from geometry that the sides of similar triangles have equal ratios. In this case, those ratios are (from Figures 3.24 and 3.25):

$$\frac{\Delta v}{v} = \frac{v\Delta t}{R}$$

This approximation becomes exact as Δt approaches zero. Therefore, in that limit, the magnitude of the acceleration is

$$a = \lim_{\Delta t \to 0} \frac{\Delta v}{\Delta t} = \frac{v^2}{R}$$

Acceleration is a vector, with magnitude and direction. What's the direction of the acceleration for uniform circular motion? Look again at Figure 3.24. In the limit as Δt approaches zero, the angle $\Delta \theta$ also approaches zero, and $\Delta \vec{v}$ becomes perpendicular to both $\vec{v}$ and $\vec{v}_0$. With the velocity tangent to the circle, this means that $\Delta \vec{v}$ points inward along the radius. Because $\vec{a} = \lim_{\Delta t \to 0} \Delta \vec{v}/\Delta t$, the acceleration $\vec{a}$ also points radially inward.

Physicists call the acceleration in uniform circular motion **centripetal acceleration** (meaning "center seeking") and use the symbol $\vec{a}_r$. The subscript r stands for *radial*, because the centripetal acceleration is radially inward. To summarize, centripetal acceleration points toward the center of the circle and has a magnitude

$$a_r = \frac{v^2}{R} \quad \text{(Magnitude of centripetal acceleration; SI unit: m/s}^2) \quad (3.20)$$

Although we've discussed uniform circular motion in the context of full circles, Equation 3.20 applies to *any* motion that involves circular arcs traversed at constant speed. It doesn't matter, for example, whether a race car travels around a test track or your own car rounds a circular turn at constant speed.

Reviewing New Concepts: Centripetal Acceleration

An object in uniform circular motion (constant speed v around a circle of radius R) has a centripetal acceleration, which

- Is directed toward the center of the circle, and

- Has a magnitude $a_r = \dfrac{v^2}{R}$.

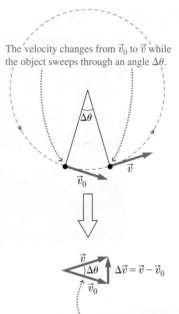

The velocity changes from $\vec{v}_0$ to $\vec{v}$ while the object sweeps through an angle $\Delta \theta$.

$\Delta \vec{v} = \vec{v} - \vec{v}_0$

Because the velocity vector is tangent to the circle, it changes direction by an angle $\Delta \theta$ while the object sweeps through the same angle $\Delta \theta$ around the circle.

FIGURE 3.24 Changing velocity for an object in uniform circular motion.

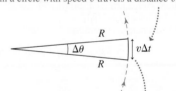
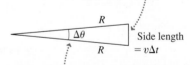

Over a time interval Δt, an object moving in a circle with speed v travels a distance $v\Delta t$.

As $\Delta t \to 0$, the arc approaches a straight line . . .

Side length $= v\Delta t$

. . . forming an isosceles triangle whose short side approximates $v\Delta t$.

FIGURE 3.25 Angle and arc length for an object in uniform circular motion, in the limit as $\Delta t \to 0$.

Space Station Gravity: Saving Muscle and Bone

Astronauts in orbiting spacecraft (Chapter 9) experience apparent weightlessness, which can result in deterioration of bone and muscle on long missions. Spinning a cylindrical spacecraft creates "artificial gravity," which can prevent physiological problems. Astronauts on the inside walls are accelerated toward the center with acceleration v^2/R, where v is the speed at that point and R is the spacecraft's radius. Spin the spacecraft at the right rate, and you can match Earth's g!

EXAMPLE 3.10 **Test Track**

Engineers determine that a race car's maximum centripetal acceleration on a 1.2-km-radius circular track is $0.45g$. How fast can the car safely go?

ORGANIZE AND PLAN Given acceleration and radius, speed is the only unknown in Equation 3.20: $a_r = v^2/R$. You've been given the *maximum* centripetal acceleration, which corresponds to the *maximum* safe speed v. Solving for speed,

$$v = \sqrt{a_r R}$$

Speed is positive, so you want the positive root.

Known: Acceleration $a_r = 0.45g = 0.45(9.80 \text{ m/s}^2) = 4.4 \text{ m/s}^2$; radius $R = 1.2$ km $= 1200$ m.

SOLVE Inserting the numerical values,

$$v = \sqrt{a_r R} = \sqrt{(4.4 \text{ m/s}^2)(1200 \text{ m})} = 73 \text{ m/s}$$

REFLECT Notice that with SI units as the units for a_r and R, the speed comes out in m/s. The speed 73 m/s is just over 160 miles per hour, fast but reasonable for a race car.

MAKING THE CONNECTION What limits the car's centripetal acceleration?

ANSWER As you'll learn in Chapter 4, friction between tires and road keeps the car from skidding off its circular path. Water or oil reduces friction and limits the safe speed. Banking the track, on the other hand, reduces the necessary frictional force and makes for safer cornering.

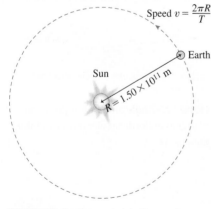

FIGURE 3.26 Earth moving in its nearly circular orbit.

Any object in uniform circular motion has centripetal acceleration, which you can find knowing the object's speed and the radius of the circle. As we noted, Earth's orbit is nearly circular. With radius 1.50×10^{11} m (Figure 3.26) and period 1 year ($= 3.15 \times 10^7$ s), Earth's orbital speed is

$$v = \frac{2\pi R}{T} = \frac{2\pi(1.50 \times 10^{11}\text{m})}{3.15 \times 10^7\text{s}} = 2.99 \times 10^4 \text{ m/s}$$

That's about 30,000 m/s, or 30 km/s. Then Earth's centripetal acceleration is

$$a_r = \frac{v^2}{R} = \frac{(2.99 \times 10^4 \text{ m/s})^2}{1.50 \times 10^{11} \text{ m}} = 5.96 \times 10^{-3} \text{ m/s}^2$$

Despite Earth's high orbital speed, our centripetal acceleration is low. That's because Earth moves only about 1 degree in angle each day around its orbit, so its direction doesn't change rapidly.

GOT IT? Section 3.5   A car rounds a circular track at a constant speed. Which of the following changes will double its centripetal acceleration? (a) Doubling the car's speed; (b) cutting its speed in half; (c) doubling the track's radius; (d) cutting the radius in half.

Chapter 3 in Context

Chapter 2 introduced the concepts of position, velocity, and acceleration in one dimension. Chapter 3 used *vectors* to extend these concepts into two dimensions. It takes two numbers to specify vector quantities in two dimensions, while a single number suffices for scalars. You'll see many examples of scalars and vectors throughout this book.

We then considered two special cases of motion in a plane: *projectile motion* and *uniform circular motion*. Projectile motion is best understood by considering separately the horizontal and vertical components of motion. In uniform circular motion, acceleration has a constant magnitude but a direction that changes continuously so it's always toward the center of the circle—*centripetal acceleration*.

Looking Ahead: Kinematics, dynamics, and circular motion Up to this point we've considered kinematics, which only *describes* motion. In Chapter 4 we'll begin the study of dynamics, showing how acceleration is related to forces. You'll learn about everyday forces, which can be thought of as pushes and pulls, and gravity, which affects objects near Earth and throughout the universe. Our brief introduction to circular motion will be expanded when you study rotations (Chapter 8) and the orbits of planets and satellites (Chapter 9).

CHAPTER 3 SUMMARY

Scalars and Vectors

(Section 3.2) **Scalars** are quantities described by one number (and appropriate units).

Vectors are physical quantities described by more than one number, or by magnitude and direction.
Vectors can be added either by component or graphically.

Vector components: $x = r\cos\theta$ and $y = r\sin\theta$

Adding vectors by components:

$$\vec{r}_1 + \vec{r}_2 = (x_1 + x_2)\hat{\imath} + (y_1 + y_2)\hat{\jmath}$$

Subtracting vectors by components:

$$\vec{r}_2 - \vec{r}_1 = (x_2 - x_1)\hat{\imath} + (y_2 - y_1)\hat{\jmath}$$

Velocity and Acceleration in Two Dimensions

(Section 3.3) In two dimensions, position, velocity, and acceleration are vectors.

Average velocity is the displacement, a vector $\Delta\vec{r}$, divided by the time, the scalar Δt.

Average acceleration is the change in velocity divided by the change in time.

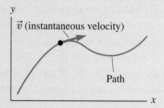

Instantaneous velocity in two dimensions: $\vec{v} = \lim\limits_{\Delta t \to 0} \dfrac{\Delta\vec{r}}{\Delta t}$

Instantaneous acceleration in two dimensions: $\vec{a} = \lim\limits_{\Delta t \to 0} \dfrac{\Delta\vec{v}}{\Delta t}$

Projectile Motion

(Section 3.4) A projectile's acceleration is constant, vertically downward with magnitude g.

The **kinematic equations for projectiles** relate position, velocity, and time.

Kinematic equations for projectiles:

Horizontal: $x = v_{0x}t \quad v_x = v_{0x}$

Vertical: $y = v_{0y}t - \frac{1}{2}gt^2 \quad v_y = v_{0y} - gt \quad v_y^2 = v_{0y}^2 - 2g\Delta y$

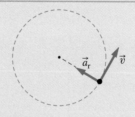

Uniform Circular Motion

(Section 3.5) In **uniform circular motion**, **speed** is constant and **velocity** is tangent to the circle.

The **centripetal acceleration** is directed toward the center of the circle.

The relationship between the **speed** v and **period** T: $v = \dfrac{2\pi R}{T}$

Centripetal acceleration: $a_r = \dfrac{v^2}{R}$

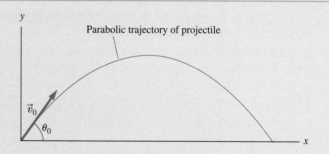

NOTE: Problem difficulty is labeled as ■ straightforward to ■ ■ ■ challenging. Problems labeled BIO are of biological or medical interest.

Conceptual Questions

1. Identify each of the following as a vector or scalar: (a) The surface area of a floor; (b) the position of a point on Earth's surface; (c) centripetal acceleration; (d) the number of pages in this book.
2. A complete topological/weather map of a state contains the following quantities for various locations. Identify each as vector or scalar: (a) position (latitude and longitude); (b) altitude; (c) wind speed; (d) wind velocity; (e) temperature; (f) barometric pressure.
3. Is the magnitude of a vector a scalar or a vector? Is the direction of a vector (in two dimensions) a vector or a scalar?
4. In two-dimensional motion, are average speed and average velocity ever the same? Explain.
5. If both components of a vector are doubled, does the vector's magnitude double? Explain.
6. Suppose you multiply a vector by -1. How does this affect its (a) components; (b) magnitude; (c) direction?
7. Can you add a position vector to a velocity vector? Why or why not?
8. Can an object have constant speed yet changing velocity? Can it have changing speed yet constant velocity?
9. Does an object's average velocity during a time interval have to be equal to its instantaneous velocity at some time during the interval? Explain why or give a counter example.
10. You want to throw a ball over a wall using the smallest possible launch speed. Should the launch angle be less than, equal to, or greater than 45°?
11. You have a rubber-band slingshot that you want to fire from the top of a building and reach the greatest possible horizontal distance. Should the launch angle be less than, equal to, or greater than 45°?
12. A car rounds a circular track at constant speed. While continuing on the same track, its speed then begins to increase. What can you say about the magnitude and direction of the car's acceleration, compared with when it had a constant speed?
13. A propeller blade turns at a constant rate. Compare the centripetal acceleration at the blade tip with a point halfway between the rotation axis and the tip.

Multiple-Choice Problems

14. A 1.60-m-tall person is standing on level ground, with the Sun 30° above the horizon. The length of the person's shadow is (a) 0.92 m; (b) 1.85 m; (c) 2.77 m; (d) 3.20 m.
15. If you drive 30 km east, then 25 km north, how far are you from your start? (a) 34 km; (b) 39 km; (c) 46 km; (d) 55 km.
16. The magnitude of the position vector $\vec{r} = 1.7\,\text{m}\,\hat{\imath} - 4.9\,\text{m}\,\hat{\jmath}$ is (a) -3.2 m; (b) 6.6 m; (c) 26.9 m; (d) 5.2 m.
17. The position vector $\vec{r} = -2.3\,\text{m}\,\hat{\imath} - 4.0\,\text{m}\,\hat{\jmath}$ makes a direction angle of (a) 60°; (b) 210°; (c) 240°; (d) 330°.
18. If you bicycle 240 m in the $+x$-direction, then 360 m in the $-y$-direction in a total time of 1 min, your average velocity is (a) 10 m/s; (b) 7.2 m/s; (c) 4.0 m/s $\hat{\imath} - 6.0$ m/s $\hat{\jmath}$; (d) 8.0 m/s $\hat{\imath} + 12.0$ m/s $\hat{\jmath}$.
19. Your car's instantaneous velocity is $11.9\,\text{m/s}\,\hat{\imath} + 19.5\,\text{m/s}\,\hat{\jmath}$. Your speed is (a) 22.8 m/s; (b) 25.6 m/s; (c) 31.4 m/s; (d) 33.6 m/s.

20. A running dog's velocity changes from 2.13 m/s $\hat{\imath} - 1.91$ m/s $\hat{\jmath}$ to 1.25 m/s $\hat{\imath} + 2.03$ m/s $\hat{\jmath}$ in a time of 5.0 s. The dog's average acceleration during that time interval is (a) -0.88 m/s² $\hat{\imath} + 3.94$ m/s² $\hat{\jmath}$; (b) -0.18 m/s² $\hat{\imath} + 0.79$ m/s² $\hat{\jmath}$; (c) -0.18 m/s² $\hat{\imath} + 0.02$ m/s² $\hat{\jmath}$; (d) 1.13 m/s² $\hat{\imath} + 0.79$ m/s² $\hat{\jmath}$.
21. The maximum range (on level ground) of a projectile launched at 25.5 m/s is (a) 2.8 m; (b) 47.0 m; (c) 66.4 m; (d) 250 m.
22. A projectile launched at 10.8 m/s at 50° above the horizontal reaches a maximum height of (a) 6.0 m; (b) 4.4 m; (c) 3.5 m; (d) 2.9 m.
23. A projectile is launched horizontally from a 12.4-m-tall building at 14.0 m/s. How far (horizontally) from the base of the building does it strike the ground? (a) 22.3 m; (b) 17.7 m; (c) 12.4 m; (d) 10.9 m.
24. A golfer hits the ball from a deep sand bunker onto a nearby green at a higher elevation. How does the ball's speed (v) when it hits the green compare to its speed (v_0) when it left the bunker? (a) $v > v_0$; (b) $v = v_0$; (c) $v < v_0$; (d) cannot be determined from the information given.
25. A car rounds a circular track of radius 875 m. What's its maximum speed if its centripetal acceleration must not exceed 3.50 m/s²? (a) 250 m/s; (b) 55.3 m/s; (c) 47.5 m/s; (d) 27.6 m/s.
26. The Moon orbits Earth in a nearly circular orbit of radius 3.84×10^8 m and period 27.3 days. What is the Moon's centripetal acceleration? (a) 2.72×10^{-3} m/s²; (b) 1.49×10^{-2} m/s²; (c) 0.108 m/s²; (d) 9.80 m/s².

Problems

Section 3.1 Trigonometry Review
27. ■ What are all the angles in a right triangle with sides 3 m, 4 m, and 5 m?
28. ■ The hypotenuse of a right triangle measures 12.5 cm. If the three angles are 30°, 60°, and 90°, what are the lengths of the other two sides?
29. ■ ■ Standing on level ground, a person casts a shadow 1.12 m long when the Sun is 55° above the horizon. How tall is the person?
30. ■ ■ On level ground, you stand 25.0 m from the base of a tree and determine that the treetop is at 42° above the horizontal, as measured from ground level. How tall is the tree?
31. ■ A right triangle has hypotenuse 25.0 cm and one side 19.8 cm. (a) Find the third side. (b) Find the three angles of the triangle.
32. ■ ■ When driving west toward the Rocky Mountains, you see a peak 3.1° above the horizon. Your GPS gives your current elevation as 1580 m, and according to the map, you're 25.0 km (horizontally) from the peak. Find the peak's elevation.
33. ■ ■ A sign on a mountain road reads "Caution: 6 degree downgrade for next 5.5 km." By how much will your elevation have changed after you complete that stretch of road?
34. ■ ■ It's 35 km by road from town to a ski resort. The resort's elevation is 1150 m above town. What's the road's average inclination angle?
35. ■ ■ On the map, let the $+x$-axis point east and the $+y$-axis north, with direction angles measured counterclockwise from the $+x$-axis. What direction angle should you head if your destination is (a) 4.5 km north and 2.3 km west; (b) 9.9 km west and 3.4 km south; (c) 1.2 km east and 4.0 km south?

Section 3.2 Scalars and Vectors

36. ■■ You walk 1250 m east and then 900 m south in a total time of 20 min. Compute your (a) displacement and (b) average velocity in m/s.

Problems 37–42 concern the position vectors $\vec{r}_1 = 2.39\text{ m }\hat{\imath} - 5.07\text{ m }\hat{\jmath}$ and $\vec{r}_2 = -3.56\text{ m }\hat{\imath} + 0.98\text{ m }\hat{\jmath}$.

37. ■ Find the magnitude and direction of both vectors.

38. ■ Find $\vec{r}_1 + \vec{r}_2$.

39. ■ Find $\vec{r}_1 - \vec{r}_2$.

40. ■■ Repeat the preceding two problems using graphical methods and compare with the results you computed exactly.

41. ■■ Find the angle between the two vectors. *Hint:* Find the angle each makes with the +x-axis.

42. ■■ (a) Compute the vector $-\vec{r}_2$. (b) Use components to show that $\vec{r}_1 - \vec{r}_2$ is equivalent to $\vec{r}_1 + (-\vec{r}_2)$. (c) Show graphically that $\vec{r}_1 - \vec{r}_2$ is equivalent to $\vec{r}_1 + (-\vec{r}_2)$.

43. ■ Find the x- and y-components of these vectors in the x-y plane: (a) $\vec{A}$ has magnitude 6.4 m, direction angle 80°; (b) $\vec{B}$ has magnitude 13 m, direction 30°; (c) $\vec{C}$ has magnitude 10 m and points in the −y-direction.

44. ■■ Position vector $\vec{r}$ has magnitude of 13.0 m and direction angle 210°. (a) Find its components. (b) Find the components, magnitude, and direction of the vector $-2\vec{r}$.

45. ■■ Add the following vectors, first graphically, then using components: $\vec{R}$ has magnitude 6.0 m and points in the +x-direction, and $\vec{S}$ has a magnitude 9.0 m and direction angle 60°.

46. ■■ You drive due east at 85 km/h for 50 min. Then you follow a road going 60° south of east, going 90 km/h for another 30 min. Find your total displacement.

47. ■■■ A roller coaster goes 26 m along a section of track sloping 10° below the horizontal, followed by a 15-m section sloping upward at 6°, followed by an 18-m section of level track. Find the net displacement.

Section 3.3 Velocity and Acceleration in Two Dimensions

48. ■■ A circular racetrack with 250 m radius lies in the x-y plane and is centered at the origin. A car rounds the track counterclockwise starting at the point (250 m, 0). Find the total distance traveled and the displacement after (a) one-quarter lap; (b) one-half lap; (c) one complete lap.

49. ■■ The car in the preceding problem travels at constant speed and completes one lap in 55 s. Assume the same starting point (250 m, 0). Find (a) the car's speed, (b) its average velocity for one-half lap, and (c) its instantaneous velocity at the end of one lap.

50. ■■ A runner completes an 800-m race with two laps of the 400-m track in 1 min, 54.3 s. Find the runner's average velocity and average speed for the race.

51. ■■ A snail crawling on a sheet of graph paper goes from the origin to the point $x = 5.6$ cm, $y = 4.3$ cm in 1 min. Find its average velocity and average speed.

52. ■ On the map, let the +x-axis point east and the +y-axis north. (a) An airplane flies at 810 km/h northwestward direction (i.e., midway between north and west). Find the components of its velocity. (b) Repeat for the case when the plane flies due south at the same speed.

53. ■■ If the plane in the preceding problem took 45 s to execute the turn from a northwest heading to a south heading, what was its average acceleration during the turn?

54. ■■ You drive 100 km/h east for 30 min, then 80 km/h north for 40 min. Find your average speed and average velocity for the entire trip.

55. ■■ In this problem let the +x-axis be horizontal and the +y-axis straight up. A rocket is launched from rest on the ground. After 55 s, its speed is 950 m/s in a direction 75° above the horizontal. Find (a) the components of the rocket's velocity vector at that time and (b) its average acceleration for the first 55 s of flight.

56. ■■ A dog running in the +y-direction at 6.7 m/s hears its owner call from behind, and 2.5 s later it's running at the same speed in the opposite direction. Compute (a) the dog's change in velocity and (b) its average acceleration.

57. ■■ A baseball traveling horizontally at 32 m/s is struck by the bat, giving it a speed of 36 m/s in the opposite direction. (a) Find the change in the ball's velocity. (b) If the ball was in contact with the bat for 0.75 ms, what was its average acceleration? (Give both the magnitude and direction.)

58. ■■ A car rounds a 1.25-km-radius circular track at 90 km/h. Find the magnitude of the car's average acceleration after it has completed one-fourth of the circle.

59. ■■ Take the +x-axis horizontal and the +y-axis upward. A skier glides down a 7.5° slope with acceleration 1.15 m/s² along the slope. (a) Find the components of the acceleration. (b) If the skier started from rest, find her velocity and speed after 10.0 s.

60. ■■ You drive west at 75 km/h for 20 min, then follow a road that goes 40° south of west, driving at 90 km/h for another 40 min. Find (a) your total displacement and (b) your average speed and average velocity for the trip.

61. ■■■ A billiard ball approaches a side cushion at a 45° angle with speed 1.80 m/s. It rebounds at a 45° angle, as shown in Figure P3.61, with the same speed. (a) Find the change in the ball's velocity. (b) More realistically, the ball might lose a little speed, rebounding at 1.60 m/s. Now what's its velocity change?

FIGURE P3.61

62. ■■ Meteorologists track a storm using radar. The radar shows a storm centered 35 km west of town. Ninety minutes later, it's 25 km north of town. Assuming the storm moved with constant velocity, find that velocity.

Section 3.4 Projectile Motion

63. ■■ A baseball hit just above the ground leaves the bat at 27 m/s at 45° above the horizontal. (a) How far away does the ball strike the ground? (b) How much time is the ball in the air? (c) What is its maximum height?

64. ■ Make a scale drawing of the trajectory of the baseball in the preceding problem. Let the motion be in the x-y plane, with x horizontal and y vertical.

65. ■■ A ball rolls off a horizontal table at 0.30 m/s and strikes the ground 0.15 m horizontally from the base of the table. (a) How high is the table? (b) If another ball rolls off the same table at 0.60 m/s, how far from the base of the table does it hit?

66. ■■ You hit a tennis ball from just above the ground. It leaves the racket at 20 m/s going 15° upward. (a) Show that the ball lands before reaching the baseline at the other end of the court, 25 m away. (b) If you struck the ball harder but at the same angle, what's the maximum initial speed it could have and still land in the court?

67. ■■■ A projectile is fired horizontally at 13.4 m/s from the edge of a 9.50-m-high cliff and strikes the ground. Find (a) the horizontal distance it traveled; (b) the elapsed time; and (c) its final velocity.

68. ■■ In a cathode-ray tube, electrons are projected horizontally at 1.2×10^6 m/s and travel a horizontal distance 8.5 cm across the tube. (a) How long are the electrons in flight? (b) Find the vertical distance the electrons fall (under the influence of gravity) during their flight.

69. ■■ A firefighter aims a hose upward at a 75° angle above the horizontal. If the water emerges from the hose 1.5 m above the ground at a speed of 22 m/s, what height does it reach? *Hint:* Think of the water stream as composed of individual droplets subject to gravity in flight.

70. ■■ A long jumper maintains a constant horizontal velocity component of 7.50 m/s while jumping upward with vertical component 3.85 m/s. Determine her horizontal range.

71. ■■ Guns on World War II battleships could shell targets 15 km away. Find (a) the minimum launch speed needed to achieve that range and (b) the shell's flight time under these conditions.

72. ■■ The fastest major league pitcher throws a ball at 45.0 m/s. If he throws the ball horizontally, how far does it drop vertically on the 18.4-m trip to home plate?

73. ■■ In 1971 astronaut Alan Shepard took a golf club to the Moon, where the acceleration due to gravity is about $g/6$. If he struck a ball with a speed and launch angle that would result in a 120-m horizontal range on Earth, how far would it go?

74. ■■■ A rifle has muzzle velocity 425 m/s. If you're 50.0 m from a target the same height as the rifle, at what angle above the horizontal should you aim in order to hit the target?

75. BIO ■■ **Leaping gazelle.** A gazelle attempts to leap a 2.1-m fence. Assuming a 45° takeoff angle, what's the minimum speed for the jump?

76. ■■■ A football kicker is trying to make a field goal from 45.0 m. If he kicks at 40° above the horizontal, what minimum speed is needed to clear the 3.05-m-high crossbar?

77. ■■■ A soccer player takes a penalty kick 11.0 m from the goal. To score, she has to shoot the ball under the 2.44-m-high crossbar. If she kicks at 19.8 m/s, what range of launch angles will result in a goal? *Note:* The ball is permitted to strike the ground before reaching the goal.

78. ■■■ A soccer player 20.0 m from the goal stands ready to score. In the way stands a goalkeeper, 1.70 m tall and 5.00 m out from the goal, whose crossbar is at 2.44 m high. The striker kicks the ball toward the goal at 18 m/s. Determine whether the ball makes it over the goalkeeper and/or over the goal for each of the following launch angles (above the horizontal): (a) 20°; (b) 25°; (c) 30°.

79. ■■■ In the preceding problem, for what range of angles does the player score a goal—meaning the ball passes above the goalkeeper but below the crossbar?

Section 3.5 Uniform Circular Motion

80. ■ Compute the centripetal acceleration of a point on Earth's equator, given Earth's 24-h rotation period. (See Appendix E for needed data.) Compare your answer with $g = 9.80$ m/s^2.

81. ■■ Compute the centripetal acceleration of a point on Earth's surface at 38° north latitude. Compare with the answer to the preceding problem.

82. ■ You're designing a highway so that the maximum centripetal acceleration on a curve is no more than 1.0 m/s^2. What is the minimum curvature radius to accommodate a 100-km/h maximum speed?

83. ■■ In a loop-the-loop roller coaster, the minimum centripetal acceleration at the top of the loop is 9.8 m/s^2 if the car is not to fall off the track. Why? For a 7.3-m-radius loop, what minimum speed must the car have at the top?

84. ■■ In 1974, physicist Gerard K. O'Neill proposed a cylindrical space station that would rotate about its central axis to simulate gravity along the outer surfaces. For a 1.1-km-radius cylinder, find the time for one revolution if "gravity" at the surface is to be 9.8 m/s^2.

85. ■ Charged particles round a 1.2-km-radius circular particle accelerator at nearly the speed of light. Find (a) the period and (b) the centripetal acceleration of the charged particles.

86. ■■ Using astronomical data in Appendix E, find the centripetal acceleration of (a) Venus and (b) Mars in their orbits (assumed circular) around the Sun. Compare your results with Earth's acceleration found in Section 3.5.

87. ■ A washing machine drum has a diameter of 46 cm and spins at 500 revolutions per minute. Find the centripetal acceleration at the drum's surface.

88. BIO ■■ **Extreme accelerations.** To simulate the extreme accelerations during launch, astronauts train in a large centrifuge. If the centrifuge diameter is 10.5 m, what should be its rotation period to produce a centripetal acceleration of (a) $4g$; (b) $6g$?

89. ■ A neutron star has 12-km radius and rotation period 1.0 s. What's the centripetal acceleration on its surface at the equator?

90. ■■ A satellite is in a 91.5-min-period circular orbit 350 km above Earth's surface. Find (a) the satellite's speed; (b) its centripetal acceleration.

91. ■■ A cyclist goes counterclockwise around a circular track with a constant speed. Which of the diagrams in Figure P3.91 correctly shows the cyclist's velocity and acceleration vectors when the bicycle is on the right edge of the track?

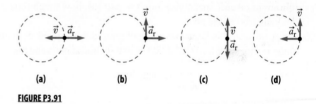

(a)   (b)   (c)   (d)

FIGURE P3.91

General Problems

92. ■■ You're sitting in your yard, 10.5 m from your 7.20-m-high house. The Sun is on the opposite side of the house. What angle must it be above the horizon for you to be in sunshine?

93. ■■ A running track consists of two straight 80.0-m segments and two semicircles of radius 38.2 m as shown in Figure GP3.93. (a) Find the distance a runner travels in one complete lap. (b) Find the displacement of a runner who covers half a lap, from A to B.

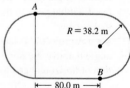

FIGURE GP3.93

94. ■■ A golf ball is hit on level ground at 25 m/s and 32° above the horizontal. What is its velocity (a) at the peak of its flight and (b) when it lands? (c) How far does it travel horizontally?

95. ■■■ Repeat the preceding problem for a ball launched from a 7.2-m-tall hill.

96. BIO ■ ■ **Motion of blood.** Figure GP3.96 outlines the human circulatory system. Given the scale shown, estimate (a) the distance traveled by and (b) the net displacement of a red blood cell as it moves from the right hand to the left foot.

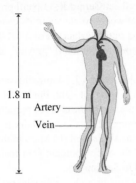

1.8 m

Artery ——

Vein ——

FIGURE GP3.96

97. ■ ■ ■ A radar signal detects a cruise boat 10 km due east of your position, traveling northward at 25 km/h. Your speedboat can go 40 km/h. (a) In what direction should you head to intercept the cruise boat? (b) How much time will it take to reach it? (c) Where will you intercept it?

98. ■ ■ ■ A golfer hits a ball from a 1.50-m-deep sand bunker, with speed 13.5 m/s at an angle of 55°. (a) How far does it travel horizontally before reaching the ground? (b) What is its speed when it strikes the ground? Compare with its initial speed. (c) What would be the horizontal range of an identically launched ball on level ground? Compare with your answer to part (a).

99. BIO ■ ■ **Extreme accelerations.** Kangaroos have made leaps with ranges of up to 12.8 m on level ground. (a) What is the minimum takeoff speed required for such a jump? (b) For the minimum takeoff speed you found in part (a), how much time is the kangaroo in the air?

100. ■ ■ A projectile is launched on level ground with speed v_0 at an angle θ above the horizontal. Show that its horizontal range is

$$R = \frac{v_0^2 \sin(2\theta)}{g}$$

Hint: You may need the trig identity $2 \sin\theta \cos\theta = \sin(2\theta)$.

101. ■ ■ Use the equation derived in the preceding problem to answer the following: (a) What launch angle gives maximum range, for a given initial speed? (b) For a given initial speed, which other launch angle gives the same range as a 22° launch angle? (c) What happens to the range for a 90° launch? Why does this make sense?

102. ■ ■ ■ A projectile is launched from the origin with speed v_0 at an angle θ above the horizontal. Show that its altitude y as a function of horizontal position x is

$$y = (\tan\theta)x - \left(\frac{g}{2v_0^2 \cos^2(\theta)}\right)x^2$$

103. ■ ■ Use the results of the preceding problem to explain why the projectile's trajectory is a parabola.

104. ■ ■ ■ A basketball player stands in the corner of the court at the three-point line, 6.33 m from the basket, with the hoop 3.05 m above the floor. (a) If the player shoots the ball from a height of 2.00 m at 30° above the horizontal, what should the launch speed be to make the basket? (b) How much would the launch speed have to increase to make the ball travel 40 cm farther and miss the hoop entirely?

105. ■ ■ Moon orbits Earth in 27.3 days, with orbital radius 384,000 km. Find the centripetal acceleration of the Moon in its orbit, and compare the result with g. (In Chapter 9, we'll show how Isaac Newton used this result to help understand how gravity works.)

106. ■ ■ ■ An archer standing a horizontal distance d from a tree fires an arrow toward an apple that hangs at height h. At the moment the arrow is fired, the apple falls. Show that as long as the arrow's initial speed is large enough to travel a horizontal distance d, it will hit the apple.

107. BIO ■ ■ **Medical centrifuge.** Medical technicians use small centrifuges to isolate blood cells. A typical unit holds six test tubes, rotates at 3380 revolutions per minute, and produces a centripetal acceleration of $1600g$. How far are the test tubes from the rotation axis?

Answers to Chapter Questions

Answer to Chapter-Opening Question

If you know the ball's initial velocity, the kinematic equations for constant acceleration allow you to find the time of flight. The horizontal velocity component of a projectile is constant, so once you know the time of flight the horizontal range is just that horizontal velocity component multiplied by the time of flight.

Answers to GOT IT? Questions

Section 3.1 $\tan 120° < \cos 180° < \sin 0° = \cos 90° < \sin 90° < \tan 60°$

Section 3.2 (a)

Section 3.4 (c) 1.0 m

Section 3.5 (d) Cutting the radius in half

■ Will the falling skydiver continue to accelerate indefinitely until she opens her chute?

Chapters 2 and 3 introduced *kinematics*—the study of motion. Here we'll turn to *dynamics*—the study of forces that cause changes in motion. The concept of force will give you a deeper understanding not only of motion but also of mass.

We'll introduce Newton's three laws of motion, which form the foundation of dynamics. Newton's second law relates the force on an object to that object's mass and acceleration, thus providing the link between dynamics and kinematics. We'll then introduce some specific forces: normal force, tension/compression forces, gravitational force, and frictional and drag forces. Finally, we'll consider forces that result in uniform circular motion.

4.1 Force and Mass

In the preceding two chapters, you learned key relationships among position, velocity, and acceleration. These kinematic relationships are fundamental to your understanding of physics, and they certainly provide insight into how things move. But they don't tell us *why* things move as they do. To answer that "why" question, we'll now turn our attention to **forces**. The study of forces is called **dynamics**, and the combination of force and motion—kinematics and dynamics—is **mechanics**.

To Learn

By the end of this chapter you should be able to
- Understand force and mass.
- Describe Newton's three laws of motion.
- Relate the net force to changes in motion.
- Describe the effects of friction and drag forces.
- Recognize the role of forces in circular motion.

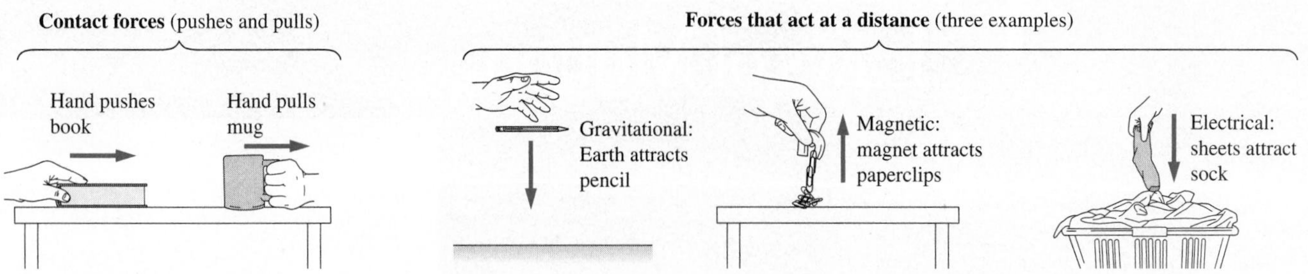

Contact forces (pushes and pulls)

Hand pushes book

Hand pulls mug

Forces that act at a distance (three examples)

Gravitational: Earth attracts pencil

Magnetic: magnet attracts paperclips

Electrical: sheets attract sock

FIGURE 4.1 Some examples of forces. Notice that a force always involves an interaction between two objects—hand and book, Earth and pencil, and so on.

What Is Force?

Your everyday experience gives you a feel for kinematic quantities including velocity and acceleration—although you've seen that physics often requires more precision, as in the distinction between velocity and speed. Experience also gives you some intuitive understanding of forces and their effects. But again, be cautious: Some of your experience with forces actually results in misconceptions that we'll overcome by focusing with the precise lens of physics.

You probably associate force with "push" or "pull." These are the most obvious forces—you push a book or pull a mug across a table (Figure 4.1). Other forces are less visible. Drop your pencil, and the force of gravity pulls it toward Earth—even though the two interacting objects, Earth and pencil, aren't in contact. Physicists use the term "action at a distance" to describe such noncontact forces. Gravity isn't the only action-at-a-distance force. Others include magnetism—the invisible pull of a magnet on a paper clip, for example—and the electric force that makes your socks attract and cling together when you pull them from the dryer and that, more profoundly, holds atoms and molecules together. You'll learn much more about electric and magnetic forces in Chapters 15–18.

As the preceding examples suggest, forces always involve *interaction* between *two objects*. A push requires something to do the pushing and something to be pushed. Gravity requires that two objects attract each other. A magnet needs another magnet or magnetic substance to act upon. The idea that a force acts mutually between a pair of objects shows up in the third of Newton's laws, which we'll introduce in the next section.

Another important fact about force is that it's a vector quantity. Every force has a strength (magnitude) and a direction. You can't describe a force completely without accounting for both—and that makes force a vector.

Reviewing New Concepts

To summarize, here are some important ideas about forces:
- Force can act by contact between two objects, as in a push or pull, or it can act over a distance, as with gravity, electricity, and magnetism.
- The application of force involves an interaction between two objects.
- Force is a vector, with magnitude and direction.

Mass—Quantity of Matter and Resistance to Force

Mass is another familiar concept, which you may think of as measuring the amount of matter in an object. A stick of butter has a mass just over 100 grams; your mass is probably between 50 and 100 kilograms. Because it's expressed as a single number, mass is a scalar.

But there's another meaning of mass: **resistance to change in motion**. This resistance to change is called **inertia**, so sometimes the words *inertia* and *mass* are used interchangeably in this context. We'll fine-tune the concept of inertia as resistance to change in motion after introducing Newton's second law in Section 4.2.

Net Force and Force Diagrams

Often multiple forces act on an object, such as the ball in Figure 4.2. The sum of these forces is the **net force**, and it's the net force that's responsible for the change in an object's motion. Symbolically, the net force is the vector sum of all forces acting on an object:

$$\vec{F}_{net} = \vec{F}_1 + \vec{F}_2 + \cdots + \vec{F}_n$$

Like any other vector, forces add by component or graphically, as you learned in Chapter 3. Figure 4.2 shows how to find the net force—in this case, the sum of the three forces acting on the ball.

Force diagrams help you visualize the forces acting on an object. Consider a book at rest on a table (Figure 4.3a). Two forces act on the book: gravity, pulling down, and the table, pushing up. Figure 4.3b shows the corresponding force diagram. The forces of gravity and the table are labeled $\vec{w}$ and $\vec{n}$, respectively. (The symbol $\vec{w}$ stands for weight and $\vec{n}$ for normal force, meaning a force perpendicular to the table. We'll explain these forces in detail later.)

The net force on the book is the vector $\vec{F}_{net} = \vec{w} + \vec{n}$. We've already noted that a net force causes a *change* in an object's motion. But here the book is sitting at rest, its motion unchanging. Therefore, the net force is zero. That's why we drew the vectors $\vec{w}$ and $\vec{n}$ with the same length in the force diagram. Their directions, however, are opposite, so their vector sum is zero.

Some textbooks call the force diagram a *free-body diagram*. We prefer "force diagram" to emphasize that the diagram should contain only force vectors and not velocity or acceleration vectors, which belong in a *motion diagram*.

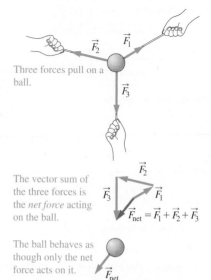

FIGURE 4.2 The net force on an object is the sum of all the individual forces.

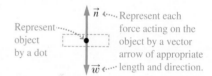

(a) Sketch of situation

(b) Force diagram for book

FIGURE 4.3 Constructing a force diagram.

TACTIC 4.1 **Drawing Force Diagrams**

- Diagram the physical situation, as in Figure 4.3a.
- Identify all forces acting on the object under study.
- On a separate force diagram (see Figure 4.3b), sketch the object and draw vectors representing forces acting on it, following the conventions below. (Complex problems might require multiple force diagrams for multiple objects.)
  - Use a dot to represent the object. Place all force vector tails on the dot. (Exception: two forces acting in the same direction may be placed tip to tail to show their sum.)
  - Make sure that vector directions and magnitudes are as accurate as the problem allows, to help visualize how the forces will add. (Sometimes a force is unknown; then you'll have to guess its length or direction.)
  - Label each vector with the proper symbol.

4.2 Newton's Laws of Motion

Newton's First Law

Consider again that book resting on the table. Give it a push and it slides across the table. Assume the push was brief, and the book is sliding on its own, as shown in Figure 4.4a. The frictional force between table and book causes the book to slow down—an example of a force causing *change* in motion. (We'll discuss friction later in this chapter.) In the corresponding force diagram (Figure 4.4a), there are now three forces acting on the book: the normal force, gravity, and friction. The sum of the forces is the net force on the book. The forces $\vec{w}$ and $\vec{n}$ are unchanged from when the book was at rest, so their sum remains zero. In this case, therefore, the frictional force $\vec{f}$ equals the net force $\vec{F}_{net}$.

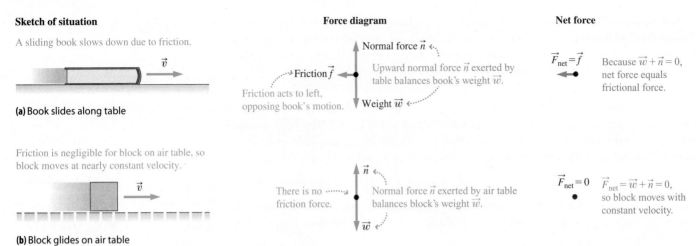

FIGURE 4.4 The book in (a) experiences a net force that changes its motion (it slows down). The block in (b) experiences zero net force, so its motion does not change (constant velocity).

One of the great misconceptions of ancient times—and one that persists among students new to physics—is the notion that a force is required to sustain motion. (This misconception is often credited to the Greek philosopher Aristotle, but he was by no means solely responsible for it.) The example of the sliding book illustrates why this misconception is common: It doesn't take long for the book to stop sliding after your push ends.

But picture a small block sliding across an air table (Figure 4.4b). There's very little friction, so to a good approximation the frictional force is zero. Then the only forces acting on the block are the normal force and gravity. The net force is $\vec{F}_{net} = \vec{w} + \vec{n} = 0$, and the block moves with constant velocity. This illustrates that a net force causes *changes* in motion, not motion itself. Force *is not required for motion, and an object subject to zero net force has constant velocity—that is, unchanging motion.* That constant velocity may be nonzero (as for the sliding block in Figure 4.4b), or it may be zero (as for the book at rest on the table in Figure 4.3).

The preceding example illustrates Newton's first law of motion, stated formally as the following:

Newton's first law of motion: If the net force on an object is zero, then the object maintains constant velocity.

A more colloquial form of Newton's first law describes explicitly the cases of zero and nonzero velocity: "An object at rest remains at rest, and an object in motion remains in motion with constant velocity, unless acted upon by a net force."

✓ **TIP**

Velocity is a vector quantity. *Constant* velocity means that both the magnitude and direction of the velocity are constant.

Historically, Galileo first provided convincing arguments for what we now call Newton's first law. Because the term "inertia" describes the tendency for objects to remain at rest or in uniform motion, Newton's first law is also known as the "law of inertia."

CONCEPTUAL EXAMPLE 4.1 **Newton's First Law in Practice**

An astronaut on the International Space Station does an experiment to test Newton's first law. She attaches a ball to a piece of string and whirls it around in a circle. Eventually the string breaks. Describe the subsequent path of the ball. *Note:* You can ignore gravity, which is not evident in the station's reference frame because it's in free fall.

SOLVE The string provides a net force on the ball that continually changes the direction of the ball's motion, keeping it in a circular path. Once the string breaks, it no longer provides that force, leaving zero net force on the ball. According to Newton's first law, it then moves with constant velocity, meaning straight-line motion at constant speed.

What is that constant velocity? During its circular motion, the ball's velocity was tangent to the circle (see Section 3.3). So the ball's straight-line path follows whatever the velocity vector was at the moment the string broke (Figure 4.5), until it runs into the wall of the station or some other force acts on it.

REFLECT You could do this experiment on Earth, although after the string breaks, the ball will fall under the influence of gravity. Observe carefully, though, and you'll see the ball start to head off in a straight line.

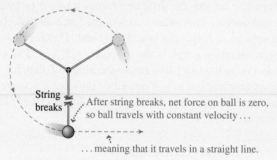

FIGURE 4.5 The ball's motion.

Newton's Second Law of Motion

A tennis ball comes quickly toward you; you pull back your racket and swing. In physics terms, you want your racket to apply a force to the ball, changing its velocity and sending it back toward your opponent.

How is velocity change related to force? An answer comes from measuring an object's acceleration (rate of velocity change) as a function of the object's mass and the net force applied. Experiments yield the following results:

- The object's acceleration is proportional to the net force acting on it;
- The object's acceleration is inversely proportional to its mass.

These results combine in a single equation:

$$\vec{a} = \frac{\vec{F}_{net}}{m}$$

This experimentally based result is Newton's second law. We'll restate it in its more familiar form, with the net force on one side of the equation.

Newton's second law of motion: An object's acceleration and the net force acting on it are directly proportional:

$$\vec{F}_{net} = m\vec{a}$$

Because mass is a positive scalar, this proportionality means that acceleration and net force vectors are always in the same direction.

Newton's second law is powerful: Given the net force on an object of known mass, you can determine its acceleration ($\vec{a} = \vec{F}_{net}/m$). Then you can use kinematics to predict the object's motion. Conversely, measuring an object's acceleration determines the net force acting on it.

Newton's second law shows that the SI units for force are $kg \cdot m/s^2$. In SI, such combinations are often redefined as a new unit. In this case, the **newton** (N) is the SI unit of force, with

$$1\,N = 1\,kg \cdot m/s^2$$

Mass, Inertia, and Newton's Law

Newton's second law lets us expand the idea of mass as inertia, or resistance to force. Imagine applying different forces, horizontally, to a puck on a frictionless air table, making it accelerate. Figure 4.6a shows how you might use a spring scale to apply the force—with the bonus that the scale also measures the magnitude of the force. You could simultaneously measure the puck's acceleration using video capture or a motion detector.

Figure 4.6b graphs the results for different forces applied to pucks with different masses. Here we're considering only magnitudes of the force and acceleration vectors, so we'll write a scalar version of Newton's law: $F_{net} = ma$. This form shows that the slope of the force-versus-acceleration graph is the mass, $m = F_{net}/a$. The graph for the puck with the larger mass ($m = 0.04\,kg$) has twice the slope of that for the smaller mass ($m = 0.02\,kg$). The experiment we've just described provides a new way to think dynamically about mass as measuring an object's response to force, rather than simply describing the quantity of matter.

Weight and Gravitational Acceleration

The force of gravity—also called **weight**—acts on all objects. Near Earth's surface, the weight of an object with mass m is

$$\vec{w} = m\vec{g} \qquad \text{(Weight of an object with mass } m\text{; SI unit: N)} \qquad (4.1)$$

Be careful to distinguish mass and weight. Mass, a scalar quantity measured in kilograms, is an intrinsic property of an object, independent of its location. Weight, a vector with units of newtons, depends on the object's mass and its location. The magnitude (g) of the acceleration vector $\vec{g}$ varies over Earth's surface and decreases with increasing altitude.

Equation 4.1 shows why (neglecting air resistance) all bodies have the same gravitational acceleration. Following Galileo, suppose you drop 5.0-kg and 1.0-kg rocks simultaneously. The magnitudes of their weights are

$$w_5 = (5.0\,kg)(9.8\,m/s^2) = 49\,N$$

and

$$w_1 = (1.0\,kg)(9.8\,m/s^2) = 9.8\,N$$

In free fall, weight is the only force on each, giving a net force $F_{net} = w$. Turning Equation 4.1 around to solve for each rock's acceleration magnitude a gives

$$a_5 = \frac{F_{net}}{m} = \frac{w}{m} = \frac{49\,N}{5.0\,kg} = 9.8\,m/s^2$$

and

$$a_1 = \frac{F_{net}}{m} = \frac{w}{m} = \frac{9.8\,N}{1.0\,kg} = 9.8\,m/s^2$$

At a given location, the acceleration is always the same (g), independent of mass.

Spring scale measures force applied to block.

(a) Measuring the applied force

For a block of a given mass, a plot of net force vs. acceleration is a straight line.

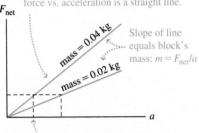

F_{net}

mass $= 0.04\,kg$

Slope of line equals block's mass: $m = F_{net}/a$

mass $= 0.02\,kg$

a

To obtain a desired acceleration, the more massive block requires a larger net force.

(b) Net force versus acceleration

FIGURE 4.6 Mass as resistance to force.

Inertial Reference Frames

A word of caution about Newton's first and second laws: They only hold in reference frames moving with constant velocity. Because Newton's first law is about inertia, reference frames with constant velocity—that is, zero acceleration—are called **inertial reference frames**. Accelerated frames are **noninertial**.

Imagine you're on an airplane, awaiting takeoff. You're challenging authority, leaving your tray table down with a glass of juice on it. With the plane at rest on the runway, the glass maintains constant velocity (in this case zero, relative to the ground), obeying Newton's first law. Then the engines roar, and you accelerate down the runway (Figure 4.7a). The glass slides toward you and spills in your lap! What happened? Newton's first law was no longer valid in the noninertial reference frame of the accelerating plane. In that frame, the glass seemed to accelerate toward you without a net force—in clear violation of Newton's first law. It's better to describe the situation in the inertial reference frame of the ground; there the plane accelerated forward, leaving the unattached glass behind. The glass's inertia—its tendency to stay put—kept it at rest with respect to the ground while airplane, tray, and passengers accelerated.

You're given another glass of juice once the plane reaches its constant cruising velocity. Now you're back in an inertial frame of reference. Here Newton's first law holds, and the glass stays put (Figure 4.7b). But watch out for turbulence, because any change in velocity eliminates the plane's inertial status. The juice's inertia—now its tendency to retain that constant cruising velocity—will make it spill again!

Newton's Third Law

You've seen that forces require pairs of interacting objects: hammer and nail, magnet and paper clip, Earth and a falling body, and so on. Newton's third law illustrates an associated pairing of forces. You may have heard the third law stated as "For every action, there is an equal and opposite reaction." This terminology is unfortunate, first because *action* is a physical quantity different from force, and second because two vectors aren't equal if they're opposite! A better statement of Newton's third law is

> **Newton's third law of motion:** When two objects (A and B) interact, the force $\vec{F}_{AB}$ that object A exerts on object B is equal in magnitude and opposite in direction to the force $\vec{F}_{BA}$ that B exerts on A. Symbolically,
>
> $$\vec{F}_{AB} = -\vec{F}_{BA}$$
>
> Note the double subscripts, which we'll consistently use when describing a force pair. The first subscript always indicates the object exerting the force, the second the object on which the force is exerted. Thus, $\vec{F}_{AB}$ means the force that A exerts on B.

TIP

Whenever you are dealing with a force pair, use a double subscript. The order of the subscripts in $\vec{F}_{AB}$ means the force A exerts on B.

Newton's third law describes how forces come in pairs. One object can't exert a force on another object without itself experiencing a force of equal magnitude in the opposite direction. Stand and push against a wall. The wall pushes back on you, with a magnitude equal to the magnitude of your push. Newton's third law is valid whether or not there's contact between the interacting bodies. An object in free fall experiences the action-at-a-distance force of Earth's gravity, directed downward. By Newton's third law, the body exerts on Earth a force of equal magnitude and opposite direction—in this case, upward. Because Earth's mass is so great ($\sim 6 \times 10^{24}$ kg), its acceleration is negligible compared with the falling object's 9.80 m/s^2.

Glass accelerates toward you when plane accelerates down runway.

(a) Noninertial reference frame

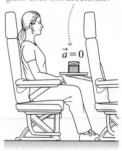

When plane's velocity is constant, glass does not accelerate.

(b) Inertial reference frame

FIGURE 4.7 Motion in noninertial and inertial reference frames. Acceleration vectors $\vec{a}$ are measured in the passenger's noninertial reference frame.

APPLICATION

Rocket!

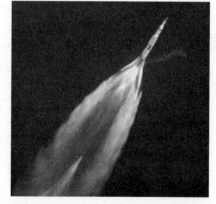

The rocket provides a good example of Newton's third law. The rocket exerts a force on the hot gases produced in its engine, ejecting them at a high speed. Those gases exert a force on the rocket, of equal magnitude and opposite direction, and that's what accelerates the rocket.

CONCEPTUAL EXAMPLE 4.2   **Newton on Ice**

Two skaters, with masses $m_A = 50$ kg and $m_B = 80$ kg, start from rest on frictionless ice and push off against each other with constant force. Describe their motion during and after the push.

SOLVE   By Newton's third law, the skaters experience forces of equal magnitude but opposite direction, so they'll accelerate in opposite directions. Because the force magnitudes are equal $(F_{AB} = F_{BA})$, but the skaters' masses aren't $(m_A \neq m_B)$, their acceleration magnitudes are also unequal (Figure 4.8a). Newton's second law gives

$$a_A = F_{BA}/m_A$$
$$a_B = F_{AB}/m_B$$

Because $m_A < m_B$, the lighter skater has greater acceleration, and so has greater speed when the skaters separate. After separation, they coast apart with these constant (and unequal) speeds (Figure 4.8b).

REFLECT   This example illustrates the notion of inertia. The more massive skater has more inertia, and so is not moving as fast at the end.

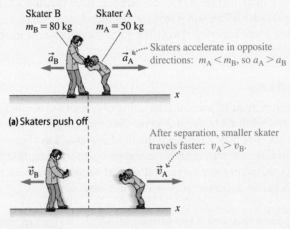

(a) Skaters push off

(b) Motion after separation

FIGURE 4.8 Motion of the skaters.

EXAMPLE 4.3   **Newton on Ice—Quantitative**

The skaters of Conceptual Example 4.2 push with a constant 200-N force. (a) Find each skater's acceleration during the push. (b) If they push for 0.40 s, what are their velocities after separation?

ORGANIZE AND PLAN   The ideas are the same as in Conceptual Example 4.2. Newton's second law gives the accelerations, and then the rules of kinematics relate acceleration to the final velocities.

We'll use coordinates in which the smaller skater (A) moves in the $+x$-direction (Figure 4.9a). As in the conceptual example, the skaters' accelerations have magnitudes

$$a_A = F_{BA}/m_A$$
$$a_B = F_{AB}/m_B$$

In our coordinates, the acceleration directions are $+x$ for skater A, $-x$ for B. With an acceleration a_x over time t, kinematics gives the final velocity $v_x = v_{0x} + a_x t$ (Figure 4.9b). Here $v_{0x} = 0$ for both. Their masses are $m_A = 50$ kg and $m_B = 80$ kg.

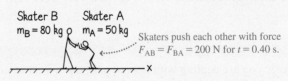

(a) Finding the accelerations

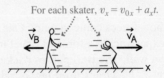

(b) Finding the velocities

FIGURE 4.9 Computing acceleration and velocity.

Known: $m_A = 50$ kg; $m_B = 80$ kg; force magnitude $= 200$ N.

cont'd.

SOLVE (a) Skater A accelerates in the $+x$-direction at

$$a_A = \frac{F_{BA}}{m_A} = \frac{200\ \text{N}}{50\ \text{kg}} = 4.0\ \text{m/s}^2$$

and B accelerates in the $-x$-direction at

$$a_B = \frac{F_{AB}}{m_B} = \frac{200\ \text{N}}{80\ \text{kg}} = 2.5\ \text{m/s}^2$$

(b) For A, $a_x = 4.0\ \text{m/s}^2$, so the kinematic equation gives

$$v_x = v_{0x} + a_x t = 0\ \text{m/s} + (4.0\ \text{m/s}^2)(0.40\ \text{s}) = 1.6\ \text{m/s}$$

For B, $a_x = -2.5\ \text{m/s}^2$ (that is, $2.5\ \text{m/s}^2$ in the $-x$-direction), so

$$v_x = v_{0x} + a_x t = 0\ \text{m/s} + (-2.5\ \text{m/s}^2)(0.40\ \text{s}) = -1.0\ \text{m/s}$$

REFLECT As predicted in Conceptual Example 4.2, the lighter skater's speed is greater. Note that final speeds are inversely proportional to mass.

MAKING THE CONNECTION Is there an equation that relates the four final quantities (two masses and two speeds) in this problem?

ANSWER Yes. The products of the masses and speeds are equal. That is, $m_A v_A = m_B v_B$. In Chapter 6, you'll see how the concept of *momentum* governs this equality.

CONCEPTUAL EXAMPLE 4.4   **Newton's Third Law on the Air Track**

You push glider A along a horizontal air track, and A in turn pushes a less massive glider B ($m_B < m_A$). The gliders remain in contact as you push. Considering only the horizontal forces, sketch the situation and draw force diagrams for your hand and for the two gliders. Identify the forces that constitute pairs according to Newton's third law.

SOLVE Figure 4.10a is the sketch. Hand H pushes glider A from the left with force $\vec{F}_{HA}$ (Figure 4.10b). By Newton's third law, the glider pushes back on the hand with force $\vec{F}_{AH}$ of equal magnitude but opposite direction. $\vec{F}_{HA}$ and $\vec{F}_{AH}$ constitute one force pair. Similarly, glider A pushes to the right on B with force $\vec{F}_{AB}$. By Newton's third law, B pushes to the left on A with force $\vec{F}_{BA}$ of equal magnitude. $\vec{F}_{AB}$ and $\vec{F}_{BA}$ are another third-law pair.

In each pair, the force magnitudes are equal, as the third law requires. However, two forces from different pairs may have different magnitudes. In this case, $F_{HA} = F_{AH} > F_{AB} = F_{BA}$.

REFLECT Note that the two forces of a third-law pair *always* act on *different* objects. Therefore, you'll never find the two forces of a pair in the same force diagram.

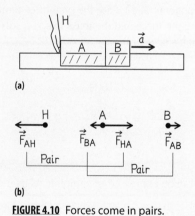

(a)

(b)

FIGURE 4.10 Forces come in pairs.

✓ **TIP**

The two forces in a third-law force pair act on different objects, so they never appear on a single object's force diagram.

GOT IT? Section 4.2   A rocket of weight $\vec{w}$ is accelerating straight up just after launch. The exhaust gases exert a force $\vec{F}_{gas}$ on the rocket. Which is the correct force diagram for the rocket?

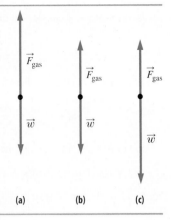

4.3 Applications of Newton's Laws

Newton's Second Law in Component Form

Solving quantitative problems in dynamics usually involves Newton's second law:

$$\vec{F}_{net} = m\vec{a}$$

If two vectors are equal, so are their respective components. For motion in the *x-y* plane, the *x*-component of $\vec{F}_{net}$ equals the *x*-component of $m\vec{a}$, and the *y*-component of $\vec{F}_{net}$ equals the *y*-component of $m\vec{a}$. Symbolically:

$$F_{net,x} = ma_x \quad \text{(x-component)} \tag{4.2}$$

$$F_{net,y} = ma_y \quad \text{(y-component)} \tag{4.3}$$

Here we've "traded" a single vector equation $\vec{F}_{net} = m\vec{a}$ for two scalar equations, which are generally easier to solve. We'll illustrate with a simple example (Figure 4.11), in which a 0.160-kg hockey puck is pushed across the ice (which we take to be the *x-y* plane) with net force

$$\vec{F}_{net} = 1.10\ \text{N}\,\hat{\imath} + 1.25\ \text{N}\,\hat{\jmath}$$

As always, the *x*- and *y*-components of $\vec{F}_{net}$ are the quantities multiplying the unit vectors $\hat{\imath}$ and $\hat{\jmath}$, respectively. Knowing the force and the mass lets you solve Equations 4.2 and 4.3 for the components of the puck's acceleration:

$$a_x = \frac{F_{net,x}}{m} = \frac{1.10\ \text{N}}{0.160\ \text{kg}} = 6.88\ \text{m/s}^2$$

$$a_y = \frac{F_{net,y}}{m} = \frac{1.25\ \text{N}}{0.160\ \text{kg}} = 7.81\ \text{m/s}^2$$

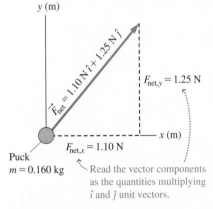

FIGURE 4.11 Force components.

✓ **TIP**

Like other vectors, force can be expressed in components or magnitude/direction.

Newton's second law works just as well with magnitude/direction instead of components. In this example, the net force has magnitude $F_{net} = \sqrt{(1.10\ \text{N})^2 + (1.25\ \text{N})^2} = 1.67\ \text{N}$, at angle $\theta = \tan^{-1}(1.25\ \text{N}/1.10\ \text{N}) = 48.7°$ above the +*x*-axis. Then by Newton's second law $F_{net} = ma$, and the acceleration's magnitude is

$$a = \frac{F_{net}}{m} = \frac{1.67\ \text{N}}{0.160\ \text{kg}} = 10.4\ \text{m/s}^2$$

in the same direction. You should convince yourself that this is equivalent to our result using vector components.

TIP

Remember from Chapter 3 the notation for vectors and magnitudes: F_{net} means the magnitude of the vector $\vec{F}_{net}$, and a means the magnitude of $\vec{a}$.

This example illustrates a typical class of dynamics problem: Given an object's mass and the force(s) acting on it, find its acceleration. Of course, there are other variations on this theme. In this section, we'll present some examples of a range of dynamics problems you might encounter. In doing so, we'll introduce some common types of force.

PROBLEM-SOLVING STRATEGY 4.1   **Solving Dynamics Problems**

ORGANIZE AND PLAN
- Draw a diagram to visualize the situation.
- Choose a suitable coordinate system.
- Identify object(s) of interest and the forces acting; draw force diagram(s).
- Determine what you know, including given numerical values.
- Determine what you're trying to find.
- Review information you have; plan how to use it to solve for unknown(s).

SOLVE
- Combine and solve equations for unknown quantity.
- Insert numerical values and calculate answers; use appropriate units.

REFLECT
- Check dimensions, units of answer. Are they reasonable?
- If problem relates to something familiar, consider whether answer makes sense.

The Normal Force

The *normal force* (symbol $\vec{n}$), a contact force, is the force that a surface exerts on an object that's on the surface. The term "normal" means that the force is always directed perpendicular to the surface, whether it's horizontal or inclined.

Figure 4.12 shows a book lying on a table. The two forces acting on the book are its weight $\vec{w} = m\vec{g}$ and the normal force $\vec{n}$. With the book at rest, Newton's first law says that the net force is zero:

$$\vec{F}_{net} = \vec{w} + \vec{n} = 0$$

In the coordinate system shown, the y-components of the normal force and the gravitational force are, respectively, $+n$ and $-mg$. Their sum is zero, so

$$n + (-mg) = 0$$

Rearranging,

$$n = mg \tag{4.4}$$

This simple example shows that the normal force points straight up with magnitude mg, just balancing the book's weight. Thus the book remains at rest, accelerating neither upward nor downward.

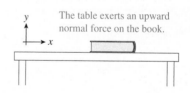

Force diagram

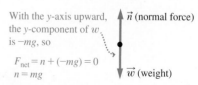

FIGURE 4.12 The normal force for an object resting on a horizontal surface.

TIP

In this example the normal force $\vec{n}$ and weight $\vec{w}$ do *not* constitute a third-law pair. The normal force involves an interaction between book and table, while the weight (gravitational force) involves book and Earth.

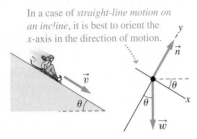

In a case of *straight-line motion on an incline*, it is best to orient the *x*-axis in the direction of motion.

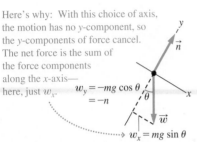

Here's why: With this choice of axis, the motion has no *y*-component, so the *y*-components of force cancel. The net force is the sum of the force components along the *x*-axis— here, just w_x.

$w_y = -mg \cos \theta$
$= -n$

$w_x = mg \sin \theta$

FIGURE 4.13 Why an inclined coordinate system is best for straight-line motion on an incline.

Motion on an Incline

Figure 4.13 shows a child sledding down an essentially frictionless slope at angle θ. (We'll consider friction for this scenario in Section 4.4.) You can find the sled's acceleration by using Newton's second law.

Until now, we've routinely adopted coordinate systems with *x*-axis horizontal and *y*-axis vertical. But that's not the best choice in this case, where the sled follows a straight line along the incline. Here, it's better to define the coordinate system as in Figure 4.13, with the $+x$-axis along the incline and the $+y$-axis perpendicular.

The advantage of this coordinate choice is that the motion you're trying to understand becomes one-dimensional (along the *x*-axis) and is therefore easier to analyze. *Must* you use this coordinate system? No. Remember that coordinate systems are just constructions we make for convenience. So you *could* use a conventional horizontal/vertical system, but that would make the math harder.

In our inclined coordinate system, the normal force points in the $+y$-direction, while the gravitational force has components in both the *x*- and *y*-directions. As always, the magnitude of the gravitational force is mg. Figure 4.13 shows that the angle between the force $\vec{w}$ and the negative *y*-axis is the same as the incline angle θ. That's because rotating the plane from horizontal through an angle θ rotates the coordinate axes through the same angle.

✓ TIP

Choose an appropriate coordinate system for motion along an incline. Usually this means taking the *x*-axis down the incline and the *y*-axis perpendicular.

Now you can find the sled's acceleration. Adding the normal and gravitational forces gives the net force, which has the following components:

x-component: $F_{net,x} = mg \sin \theta = ma_x$

y-component: $F_{net,y} = n - mg \cos \theta = ma_y$

where the components of the gravitational force follow from Figure 4.13. What you see here are the two components of Newton's second law. The mass m cancels from the *x* equation, leaving

$$a_x = g \sin \theta \quad \text{(Acceleration along incline)} \quad (4.5)$$

Thus the sled slides down the slope with acceleration $g \sin \theta$. On a 5° slope, that's $a_x = g \sin(5°) = 0.85$ m/s². As long as that acceleration doesn't last too long, the sled's speed will remain reasonably safe.

What about the *y*-components? You didn't need them to find the acceleration, but they do yield some useful information. With no acceleration in the *y*-direction, the *y*-components sum to zero: $n - mg \cos \theta = 0$, or $n = mg \cos \theta$. So for an object on a frictionless incline, subject only to gravity and the normal force, the normal force has a lower magnitude $n = mg \cos \theta$ than its value $n = mg$ on a horizontal surface. This will be important in Section 4.4, when we consider frictional forces.

✓ TIP

Don't assume that the normal force $n = mg$. The magnitude of the normal force depends on the situation. It's perpendicular to the surface and not necessarily vertical.

CONCEPTUAL EXAMPLE 4.5 **Slopes: Gradual and Steep**

You've just seen that an object slides down a frictionless slope with acceleration of magnitude $g \sin\theta$. Analyze this result dimensionally, and check whether it makes sense for the smallest and largest possible slope angles.

SOLVE Dimensionally, the equation agrees perfectly. Trig functions (here, $\sin\theta$) are dimensionless. Because g is the acceleration due to gravity, the dimensions of the acceleration a_x are m/s^2, as they should be.

The limiting cases here are $\theta = 0°$ (a flat slope; Figure 4.14a) and $\theta = 90°$ (a vertical cliff; Figure 4.14b). With $\theta = 0°$, the acceleration equation becomes $a_x = g \sin\theta = 0$, as we expect for a flat surface. With $\theta = 90°$, the acceleration becomes $a_x = g \sin 90° = g$. Right again: "sliding" down a vertical cliff is just free fall, because the surface doesn't exert any force.

REFLECT Dimensional analysis and the analysis of extreme cases are good ways to check a solution. They won't *prove* that the solution is right, but if everything checks out, there's a good chance you're right.

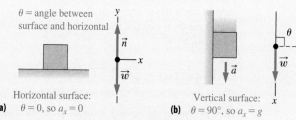

(a) Horizontal surface: $\theta = 0$, so $a_x = 0$

(b) Vertical surface: $\theta = 90°$, so $a_x = g$

FIGURE 4.14 Special cases for the sliding sled.

Tension

Another common force is *tension* (symbol $\vec{T}$), the force transmitted through elongated, stretchable structures such as strings, ropes, and, in the body, muscles and tendons. As you lift your laptop computer from the desk, for example, it's primarily the tension in your biceps that supports the computer.

Figure 4.15 shows a worker pulling a crate across the floor. The rope tension is most immediately responsible for the crate's motion; in turn, the worker's muscles exert forces on the rope. You can analyze the crate's motion if you determine the net force, then use Newton's second law to find its acceleration. We'll neglect friction in this example, which is plausible because the crate is on a dolly that's easy to roll.

The force diagram in Figure 4.15 shows three forces acting on the crate. The crate doesn't move vertically, so the vertical (y) force components sum to zero:

$$F_{\text{net},y} = n + (-mg) = ma_y = 0$$

which gives $n = mg$.

Horizontally (x-direction),

$$F_{\text{net},x} = T = ma_x$$

where as usual T means the magnitude of the vector $\vec{T}$. Therefore, the crate's acceleration is in the $+x$-direction, given by

$$a_x = \frac{T}{m}$$

If the worker can muster a pulling force of 420 N (about half his weight), and the crate's mass is 120 kg, this gives an acceleration

$$a_x = \frac{T}{m} = \frac{420\,\text{N}}{120\,\text{kg}} = 3.5\,\text{m/s}^2$$

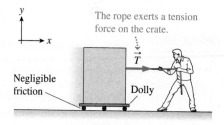

The rope exerts a tension force on the crate.

Negligible friction

Dolly

Force diagram for crate

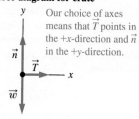

Our choice of axes means that $\vec{T}$ points in the $+x$-direction and $\vec{n}$ in the $+y$-direction.

FIGURE 4.15 Using tension to move a crate.

Crate, Revisited

In our crate example the rope was horizontal. More realistically, the worker might stand upright and pull at an upward angle, say 25.0° above horizontal. Assuming the same 120-kg crate and rope tension of 420 N, find (a) the normal force between floor and crate and (b) the crate's acceleration.

ORGANIZE AND PLAN Following Problem-Solving Strategy 4.1, begin with a diagram (Figure 4.16). Again you can identify three forces acting on the crate and draw a force diagram. Next, write the equations for the force components:

$$F_{net,x} = ma_x$$
$$F_{net,y} = ma_y = 0$$

The crate doesn't move vertically, so the y-component of the acceleration is zero.

Known: $T = 420$ N; $m = 120$ kg; $\theta = 25.0°$.

SOLVE (a) Summing forces in the y-direction,

$$F_{net,y} = T \sin \theta + n - mg = ma_y = 0$$

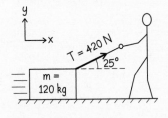
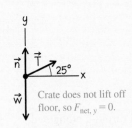

FIGURE 4.16 Forces on the crate.

Solving for n,

$$n = mg - T \sin \theta$$
$$n = (120 \text{ kg})(9.80 \text{ m/s}^2) - (420 \text{ N}) \sin (25.0°)$$
$$= 999 \text{ N}$$

(b) Summing forces in the x-direction,

$$F_{net,x} = T \cos \theta = ma_x$$

so

$$a_x = \frac{T \cos \theta}{m} = \frac{(420 \text{ N}) \cos (25.0°)}{120 \text{ kg}} = 3.17 \text{ m/s}^2$$

REFLECT Both the normal force and acceleration are different with the rope no longer horizontal. The upward component of rope tension helps support the crate, so less normal force is required. Also, changing the rope angle decreases the tension's horizontal component, reducing the acceleration.

MAKING THE CONNECTION In a more realistic situation, why is it easier to pull with the rope angled upward?

ANSWER Part of the answer involves anatomy and physiology. Standing upright makes it easier to use back and leg muscles. Another, more subtle, part of the answer involves friction. Later in this chapter, you'll see that the magnitude of the frictional force is proportional to the magnitude of the normal force. Thus, reducing the normal force helps overcome friction.

Weighing in on the Normal Force

What a scale actually measures is the normal force between the scale and whatever is on it. Suppose you (mass 65 kg) stand on a scale in an elevator. What's the scale reading when the elevator is (a) accelerating upward at 2.25 m/s², (b) moving with constant velocity, and (c) accelerating downward at 2.25 m/s²?

ORGANIZE AND PLAN There are two forces acting on you: the normal force upward, and your weight downward (Figure 4.17). The sum of these forces provides acceleration, as given by Newton's second law. With the +y-axis upward, the vertical component of Newton's second law becomes

$$F_{net,y} = n + (-mg) = ma_y$$

Solving for the magnitude of the normal force n,

$$n = ma_y + mg$$

or

$$n = m(a_y + g)$$

Known: $m = 65$ kg.

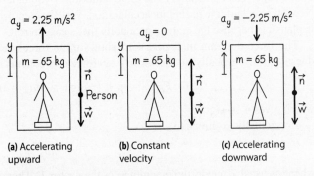

FIGURE 4.17 Moving in an elevator car.

SOLVE (a) Accelerating upward (Figure 4.17a), $a_y = 2.25$ m/s². Therefore,

$$n = m(a_y + g) = (65 \text{ kg})(2.25 \text{ m/s}^2 + 9.80 \text{ m/s}^2) = 783 \text{ N}$$

(b) With constant velocity (Figure 4.17b), $a_y = 0$, and the normal force is

$$n = m(a_y + g) = (65 \text{ kg})(9.80 \text{ m/s}^2) = 637 \text{ N}$$

cont'd.

(c) With downward acceleration (Figure 4.17c), $a_y = -2.25\ \text{m/s}^2$, and

$$n = m(a_y + g) = (65\ \text{kg})(-2.25\ \text{m/s}^2 + 9.80\ \text{m/s}^2)$$
$$= 491\ \text{N}$$

REFLECT Even without a scale, you can feel this difference as the elevator starts up, cruises at constant speed, and slows to a stop at the top of its trip. The normal force—the scale reading—is sometimes called *apparent weight*. You certainly feel heavier or lighter while the elevator is accelerating. But your true weight—the force gravity exerts on you—remains $mg = 637\ \text{N}$ regardless of the elevator's motion.

MAKING THE CONNECTION Describe the changes in your apparent weight as you ride an elevator from top to bottom of a tall building.

ANSWER The trip starts with a downward acceleration, making you feel lighter. For most of the ride, you have constant velocity and you feel your normal weight. Stopping at the bottom means accelerating upward, so you feel heavier. Note that the direction of the *velocity* doesn't matter at all. You feel heavier accelerating upward, whether it's at the start of an upward journey or the end of a downward one.

GOT IT? Section 4.3 For the sled on the incline in Figure 4.13, which force has greater magnitude: (a) The normal force; (b) the weight; (c) neither—they have the same magnitude?

4.4 Friction and Drag

So far we've neglected friction and drag forces, yet these are present in most real-world situations—sometimes to the extent that neglecting them leads to inaccurate or even absurd solutions. Now that you have seen how to solve motion problems using Newton's laws, it's time to include friction and drag.

Both friction and drag oppose an object's motion. **Frictional forces** result from an interaction between the object and a surface it contacts. Examples include a hockey puck sliding across ice, your book sliding across a table, and your shoes gripping the floor as you walk. **Drag forces** affect objects moving through fluids, such as a skydiver or a swimmer.

Causes of Friction

Friction results ultimately from electrical forces on atoms in two contacting surfaces. Although surface roughness plays a role, friction is present even when surfaces appear smooth. We'll discuss the fundamental nature of the electric force later in this chapter and in more detail in Chapter 15.

There are three important types of friction. **Kinetic friction** (or sliding friction) acts between an object and a surface it slides over. *Rolling* friction involves a round object rolling over a surface. **Static friction** acts when an object is at rest on a surface and may keep the object from sliding. We'll show how to quantify each type of friction, and then we'll analyze the effect of friction like any other force, using Newton's laws.

Kinetic Friction

Figure 4.18 shows a book sliding across a tabletop. Three forces act on the book: its weight $\vec{w}$, the normal force $\vec{n}$, and the force of kinetic friction $\vec{f}_k$. Kinetic friction acts to oppose motion, so we've drawn it opposite the book's velocity. We'll use lowercase vectors $\vec{f}$ and magnitudes f for frictional forces, to distinguish them from other forces.

Friction involves complicated interactions at the microscopic scale, but these interactions result, approximately, in a simple relationship: The magnitude of the frictional force f_k is proportional to the magnitude of the normal force n:

$$f_k = \mu_k n \quad \text{(Force of kinetic friction; SI unit: N)} \quad (4.6)$$

The reason f_k is proportional to n is illustrated in Figure 4.19. In Equation 4.6, μ_k (the Greek letter "mu") is the **coefficient of kinetic friction**. Mathematically, μ_k is a constant expressing the proportionality between the frictional force f_k and the normal force n. The coefficient μ_k is dimensionless, and its value depends on the roughness of the surfaces. Smooth, slippery surfaces have low values of μ_k (typically less than 0.2), whereas rough or sticky surfaces may have μ_k values of 1 or more. Table 4.1 lists some typical values.

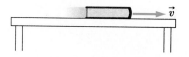

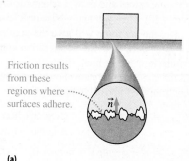

FIGURE 4.18 Kinetic friction acts on a sliding object.

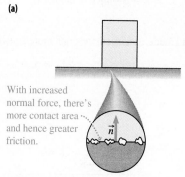

FIGURE 4.19 Illustrating the relationship between frictional force and normal force.

Synovial joints (such as the knee) have cartilage between bones to reduce friction and ease movement—see Table 4.1. Worn cartilage results in pain and loss of mobility. In serious cases physicians replace the knee joint, using low-friction plastics in place of cartilage. A typical coefficient of kinetic friction for the prosthetic knee is 0.05–0.10, which is higher than that of a healthy joint but lower than the coefficient for the diseased joint it's replacing.

TABLE 4.1 Approximate Coefficients of Friction for Selected Materials

Materials	Coefficient of kinetic friction μ_k	Coefficient of static friction μ_s
Rubber on concrete (dry)	0.80	1.0
Rubber on concrete (wet)	0.25	0.30
Wood on snow (snowboard/ski)	0.06	0.12
Steel on steel (dry)	0.60	0.80
Steel on steel (oiled)	0.05	0.10
Wood on wood	0.20	0.50
Steel on ice (speed skating)	0.006	0.012
Teflon on Teflon	0.04	0.04
Human synovial joints	0.003	0.10

EXAMPLE 4.8 **Sliding Away?**

You slide your textbook across a lab bench, starting it at 1.8 m/s in the $+x$-direction. The coefficient of kinetic friction between book and bench is 0.19. (a) What's the book's acceleration? (b) Will the book reach the edge of the bench, 1.0 m away?

ORGANIZE AND PLAN Like any dynamics problem, the strategy is to identify the forces and draw a schematic diagram and force diagram (Figure 4.20). Then you apply Newton's laws to solve the problem. In this case, the net force will give the acceleration (Newton's second law, $\vec{F}_{net} = m\vec{a}$). With the acceleration known, kinematic equations determine how far the book slides.

In the coordinate system shown, the components of Newton's law follow from the force diagram:

$$F_{net,x} = -f_k = ma_x \qquad \text{(Negative because the force is in the } -x\text{-direction)}$$

$$F_{net,y} = n + (-mg) = ma_y = 0 \qquad \text{(Zero because there's no vertical motion)}$$

With kinetic friction, there's one more equation (Equation 4.6):

$$f_k = \mu_k n$$

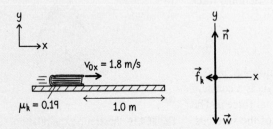

FIGURE 4.20 Our sketch for Example 4.8.

Known: $\mu_k = 0.19$, initial speed $= 1.8$ m/s.

SOLVE (a) With $a_y = 0$, the y equation gives $n = mg$. Using this result in Equation 4.6 for the frictional force, the x equation becomes

$$ma_x = -f_k = -\mu_k n = -\mu_k mg$$

The mass m cancels, leaving

$$a_x = -\mu_k g = -(0.19)(9.80 \text{ m/s}^2) = -1.9 \text{ m/s}^2$$

As expected, the acceleration due to friction is in the $-x$-direction, opposite the book's velocity.

(b) You now have acceleration and distance, so the kinematic equation $v_x^2 = v_{0x}^2 + 2a_x(x - x_0)$ is useful. You want to know how far the book goes—that's displacement $x - x_0$. Solving for $x - x_0$ gives

$$x - x_0 = \frac{v_x^2 - v_{0x}^2}{2a_x} = \frac{(0 \text{ m/s})^2 - (1.8 \text{ m/s})^2}{2(-1.9 \text{ m/s}^2)} = 0.85 \text{ m}$$

The book stops short of the edge, saving your valuable investment!

REFLECT The answers don't depend on the book's mass for much the same reason that a falling object's acceleration is independent of mass. With more mass, there's more normal force, and thus more frictional force. But it takes more force to stop a greater mass, so the acceleration remains the same.

MAKING THE CONNECTION What's being assumed here about the coefficient of friction μ_k? Is this a good assumption?

ANSWER The implicit assumption is that the μ_k is the same over the entire path. Whether that's a good assumption depends on how uniform the table surface is. If the book hits a rough or smooth patch, μ_k will change, and so will the acceleration.

EXAMPLE 4.9 **Sledding, Revisited**

Section 4.3 analyzed a sled on a frictionless 5° slope. Suppose, more realistically, that there's kinetic friction, with $\mu_k = 0.035$. Find the sled's acceleration and compare with the frictionless case.

ORGANIZE AND PLAN With motion on an incline, you've seen that it's best to take the x-axis pointing downslope (Figure 4.21). With the sled's motion in the $+x$-direction, the force of kinetic friction is then in the $-x$-direction.

The plan is to add the force components and use Newton's second law to find the acceleration. Referring to the force diagram (Figure 4.21),

$$F_{net,x} = mg \sin \theta - f_k = ma_x$$
$$F_{net,y} = n - mg \cos \theta = ma_y$$

There's also the friction relation: $f_k = \mu_k n$.

Known: $\mu_k = 0.035$; slope = 5°.

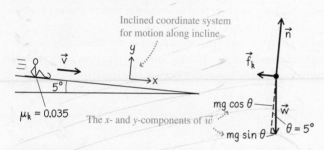

Inclined coordinate system for motion along incline

$\mu_k = 0.035$

The x- and y-components of $\vec{w}$

$mg \cos \theta$

$mg \sin \theta$ $\theta = 5°$

FIGURE 4.21 Sliding down a 5° incline.

SOLVE There's no motion in the y-direction, so $a_y = 0$, and the y-equation gives $n - mg \cos \theta = 0$, or

$$n = mg \cos \theta$$

Using this result, the x-equation becomes

$$ma_x = mg \sin \theta - f_k$$
$$ma_x = mg \sin \theta - \mu_k n$$
$$ma_x = mg \sin \theta - \mu_k mg \cos \theta$$

The mass m cancels, leaving

$$a_x = g \sin \theta - \mu_k g \cos \theta$$

That's the general result for an incline with friction. Using numbers for this example, $a_x = g \sin \theta - \mu_k g \cos \theta = (9.80 \text{ m/s}^2)(\sin 5°) - (0.035)(9.80 \text{ m/s}^2)(\cos 5°) = 0.51 \text{ m/s}^2$. That's significantly lower than the frictionless case, where $a_x = 0.85 \text{ m/s}^2$.

REFLECT Once again, the mass canceled. Although we've included friction, we've neglected the drag force of air resistance. For serious sled racers (such as Olympic athletes), mass becomes important, as drag forces have less effect on more massive riders.

MAKING THE CONNECTION Could the coefficient of friction ever be high enough to give zero acceleration?

ANSWER Sure! Setting $a_x = 0$ in this example gives $\mu_k = 0.087$—not unreasonable for steel runners on snow. With zero acceleration you could give your sled an initial push and then coast down the hill with constant velocity.

Rolling Friction

Roll a ball across a level floor and it eventually stops. The culprit is **rolling friction**, a process physically quite different from kinetic friction. One source of rolling friction is deformation, as in your car's tires (Figure 4.22). The forces of deformation work against the motion, just as with kinetic friction. Another source of rolling friction is because the point of contact on a rolling object is momentarily at rest, resulting in molecular bonds between the two surfaces. The force required to break those bonds shows up as rolling friction.

Quantitatively, rolling friction is similar to kinetic friction. The frictional force $\vec{f}_r$ is directed opposite the rolling body's velocity, with magnitude f_r proportional to the normal force:

$$f_r = \mu_r n \qquad \text{(Force of rolling friction: SI unit: N)} \qquad (4.7)$$

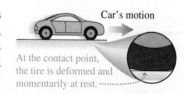

Car's motion

At the contact point, the tire is deformed and momentarily at rest.

FIGURE 4.22 Sources of rolling friction.

where μ_r is the **coefficient of rolling friction**. The big difference between kinetic and rolling friction is that for comparable surfaces, rolling friction is far smaller. That's why the wheel was such a profound invention. Consider how far your car rolls (in neutral, without braking) compared with how quickly it stops in a skid. Quantitatively, μ_k is about a factor of 40 larger than μ_r. For rubber on dry concrete, μ_r is about 0.02, but μ_k is 0.80.

Static Friction

There are two forces acting on a book at rest on your desk: gravity and the normal force (Figure 4.12). They sum to zero, so by Newton's second law the book doesn't accelerate. Now you apply a small horizontal force—yet the book still doesn't budge (Figure 4.23).

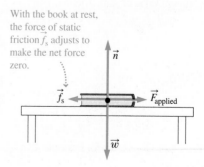

With the book at rest, the force of static friction $\vec{f}_s$ adjusts to make the net force zero.

$\vec{n}$

$\vec{f}_s$ $\vec{F}_{applied}$

$\vec{w}$

FIGURE 4.23 The force of static friction.

Antilock Brakes

Because the contact point of a rolling wheel is momentarily at rest, friction associated with changes in the wheel's speed is static friction—as long as the wheel is rolling. But slam the brakes hard, especially on an older car, and the wheel locks. It's no longer rolling, but sliding, and now kinetic friction acts. Because $\mu_k < \mu_s$, the result is a longer stopping distance. Today's antilock braking systems use computers to control brakes so the wheels just avoid locking. This maintains the larger static friction and gives a shorter stopping distance. More significantly, it keeps a locked wheel from precipitating an out-of-control skid.

The reason is **static friction**, a force that adjusts to keep the sum of all forces (including static friction) equal to zero.

Push hard enough, and the book accelerates—suggesting that there's a maximum magnitude for the force of static friction. Again, that maximum value is proportional to the normal force. These facts about static friction are incorporated in the following *inequality*:

$$f_s \leq \mu_s n \quad \text{(Force of static friction; SI unit: N)} \quad (4.8)$$

Equation 4.8 gives the *magnitude* of $\vec{f}_s$; its direction is whatever makes the net force on the object zero.

Static friction results from attractive forces between atoms in the contacting surfaces. Those forces are generally stronger than those for moving surfaces, and thus μ_s is generally larger than μ_k for the same surfaces. Table 4.1 includes both coefficients.

EXAMPLE 4.10 **Measuring Static Friction**

One way to measure the coefficient of static friction between two surfaces (say, a coin and a board) is to lay the coin on the board and slowly incline the board from the horizontal. At the moment the coin begins to slide, measure the board's inclination angle. Suppose you find that the coin slips when the angle reaches 23°. What is the coefficient of static friction between coin and board?

ORGANIZE AND PLAN First, sketch the situation and use your sketch to draw the force diagram (Figure 4.24). Notice that the force of static friction is up the incline. That's because the sum of the other forces ($\vec{w}$ and $\vec{n}$) points down the incline. The force of static friction is just enough to give zero net force on the coin. The force of static friction reaches its maximum value at the moment the coin slips, so $f_s = \mu_s n$.

We've adopted the inclined coordinate system appropriate to such problems, with the $+x$-axis parallel to the slope. With the coin at rest, each component of the net force is zero. Then components of Newton's law are $F_{net,x} = mg \sin \theta - f_s = ma_x = 0$ and $F_{net,y} = n - mg \cos \theta = ma_y = 0$. There's also the relation $f_s = \mu_s n$ for the maximum value of static friction.

Known: Slipping begins when $\theta = 23°$.

SOLVE Putting $f_s = \mu_s n$ into the x-equation,

$$mg \sin \theta - \mu_s n = 0$$

The y-equation gives $n = mg \cos \theta$; substituting into the equation above gives

$$mg \sin \theta - \mu_s mg \cos \theta = 0$$

The weight mg cancels, and solving for μ_s gives

$$\mu_s = \tan \theta$$

With $\theta = 23°$, the numerical result is

$$\mu_s = \tan(23°) = 0.42$$

REFLECT The result is dimensionally correct, because the functions and friction coefficients μ are both dimensionless. The numerical result seems reasonable—it's within the values given in Table 4.1.

MAKING THE CONNECTION Once the coin begins to slide, does it continue to accelerate?

ANSWER Yes. Because $\mu_k < \mu_s$, once kinetic friction takes over there's a nonzero net force on the coin, so it keeps accelerating.

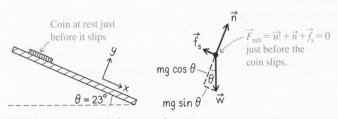

FIGURE 4.24 Our sketch for Example 4.10.

Moving with Friction

Although friction slows moving objects, it also makes common types of motion possible. Walking requires the force of static friction between your foot and the ground (Figure 4.25). You push back with your momentarily stationary foot—that's why it's *static* friction—and by Newton's third law the ground pushes you forward. For good traction $\mu_s > 0.5$ is desirable. In sports played on slippery grass, dirt, or mud, cleats or spikes improve traction by digging into the surface.

Driving is similar. The engine drives the wheel, pushing the tire's momentarily stationary contact point backward; the third-law response is a forward push from the road on the car. Braking is the opposite, with the frictional force forward and the road force backward.

FIGURE 4.25 The force of static friction is essential in walking.

Drag Forces

Drag forces are "fluid friction," retarding objects moving through fluids such as air or water. Dive into a pool, and the water's drag force keeps you from crashing into the bottom. Air drag limits your cycling speed and lowers automobile fuel mileage. In both cases aerodynamic design decreases drag.

Drag forces aren't constant, but depend on an object's speed—in some cases linearly $(F_{drag} \propto v)$ and in others quadratically $(F_{drag} \propto v^2)$. Either way, it's impossible to solve Newton's second law without calculus.

TIP

An important difference between kinetic friction and fluid drag is that the drag force strengthens as speed increases.

There's one aspect of drag forces we can describe without calculus. Consider a skydiver falling vertically (Figure 4.26). The drag force depends on speed, so it increases as the skydiver accelerates downward. At some point, the magnitude of the drag force equals that of the gravitational force, so the net force is zero. At that point there's no acceleration, and the skydiver falls with constant speed.

That constant speed is the **terminal speed** (symbol $v_{terminal}$). A skydiver's terminal speed is around 50 to 80 m/s, depending on the body's orientation. Falling "spread eagle" gives greater drag and decreases terminal speed. Opening the parachute greatly increases drag, dropping terminal speed enough for a safe landing.

Early in fall, upward drag force is less than skydiver's weight, so skydiver accelerates.

Later in fall, drag force equals weight, so skydiver's velocity is constant.

FIGURE 4.26 Drag forces on a skydiver.

GOT IT? Section 4.4 A batter hits a baseball with initial velocity directed 45° above the horizontal. When the ball is at the peak of its flight, what is the direction of the drag force: (a) upward; (b) downward; (c) horizontal?

Reviewing New Concepts

To summarize, we've identified four distinct forms of friction:

- Kinetic friction, experienced by sliding objects, with a force $f_k = \mu_k n$;
- Rolling friction, with force $f_r = \mu_r n$;
- Static friction, holding objects in place with a variable force $f_s \leq \mu_s n$;
- Drag on objects moving through fluids, with velocity-dependent force opposite the direction of motion.

4.5 Newton's Laws and Uniform Circular Motion

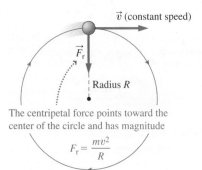

$\vec{v}$ (constant speed)

$\vec{F}_r$

Radius R

The centripetal force points toward the center of the circle and has magnitude

$$F_r = \frac{mv^2}{R}$$

FIGURE 4.27 Uniform circular motion requires centripetal force.

Recall from Section 3.5 (Equation 3.20) that an object moving at speed v in uniform circular motion with radius R has acceleration $a_r = v^2/R$ directed toward the center of the circle. According to Newton's second law, $\vec{F}_{net} = m\vec{a}$, so there must be a net force of magnitude

$$F_r = \frac{mv^2}{R} \qquad \text{(Centripetal force; SI unit: N)} \qquad (4.9)$$

directed toward the center of the circle (Figure 4.27). Because it's center-directed, the force causing uniform circular motion is called **centripetal force**.

It's important to recognize that centripetal force is *not* another category of force, like normal force, tension, or gravity. Rather, it's a name for the *net force*—the sum of all the forces acting on an object—when that net force results in circular motion.

Even though an object in uniform circular motion has constant speed, it doesn't have constant *velocity*, because its direction is changing. So it's accelerating and, by Newton's second law, there *must* be a net force acting on it—that's the centripetal force.

✓**TIP**

When drawing force diagrams for circular motion, *do not* include a separate force vector for the centripetal force. Show all the physical forces acting; their vector sum—the net force—is the centripetal force.

EXAMPLE 4.11 **A Whirling Puck**

A 0.525-m string connects a 0.325-kg puck to a peg at the center of a frictionless air table. If the string tension is 25.0 N, find the puck's centripetal acceleration and its speed.

ORGANIZE AND PLAN Our diagram (Figure 4.28) is drawn from above the table, showing the circular path and the string. The force diagram is a side view with the puck approaching, to show the three forces acting on the puck.

The normal force balances gravity, so string tension alone provides the net force. For uniform circular motion, the net force—here the tension—is the centripetal force, so

$$T = F_r = ma_r = \frac{mv^2}{R}$$

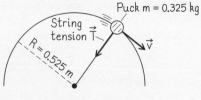

Puck $m = 0.325$ kg

String tension $\vec{T}$

$\vec{v}$

$R = 0.525$ m

Top view

Tension directed toward center of circle

$\vec{n}$

$\vec{T}$

$\vec{w}$

Force diagram
(view as puck approaches)

FIGURE 4.28 The puck's circular motion.

This expression involves both unknowns: acceleration a_r and speed v. The radius R is the length of the string.

Known: $R = 0.525$ m; $m = 0.325$ kg; $T = 25.0$ N.

SOLVE Solving for the centripetal acceleration gives

$$a_r = \frac{T}{m} = \frac{25.0\text{ N}}{0.325\text{ kg}} = 76.9\text{ m/s}^2$$

Then $a_r = v^2/R$, so

$$v = \sqrt{a_r R} = \sqrt{(76.9\text{ m/s}^2)(0.525\text{ m})} = 6.35\text{ m/s}$$

REFLECT With this speed the period is $2\pi R/v = 0.52$ s—just about two revolutions per second and entirely reasonable for this situation.

MAKING THE CONNECTION Supposing there's a *small* coefficient of friction between puck and air table, sketch the net force on the puck.

ANSWER The drawing has the tension from the string plus a force of kinetic friction opposite the puck's velocity. The net force is no longer toward the center, so this can't be uniform circular motion. That makes sense, because friction should slow the puck.

EXAMPLE 4.12 **Rounding a Curve**

The coefficient of static friction between a car's tires and a flat road is 0.84. Find the maximum speed on a turn of radius 240 m.

ORGANIZE AND PLAN Figure 4.29 is our diagram. The force diagram is a view from in front of the car, showing the weight and normal force vertical and only friction horizontal. Therefore, friction provides the entire centripetal force. Given that force and the radius, you can find the speed. Maximum speed occurs when static friction is at its maximum, $f_s = \mu_s n$.

Known: $R = 240$ m; $\mu_s = 0.84$.

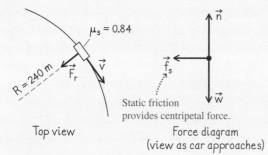

Top view

Force diagram
(view as car approaches)

FIGURE 4.29 Static friction keeps the car on the track.

SOLVE The vertical forces balance, so, as before, $n = mg$. Therefore, $f_s = \mu_s n = \mu_s mg$. Because friction provides the centripetal force mv^2/R, the horizontal component of Newton's law becomes

$$\mu_s mg = mv^2/R$$

which can be solved for the maximum speed v:

$$v = \sqrt{\mu_s Rg} = \sqrt{(0.84)(240 \text{ m})(9.80 \text{ m/s}^2)} = 44 \text{ m/s}$$

REFLECT That's a high speed—nearly 100 miles per hour. Exceed it, and you slide into a higher-radius path, going off the road or into the other lane.

MAKING THE CONNECTION Why does *static* friction govern this situation, rather than rolling or kinetic friction?

ANSWER Rolling friction and kinetic friction both act opposite the direction of motion. Static friction acts to *prevent* motion. Although the car is moving, it's not sliding or rolling in the *radial* direction. So it's *static* friction that acts to keep the tire from slipping radially.

EXAMPLE 4.13 **A Banked Curve**

Highway and racetrack curves are often banked, so the normal force provides the centripetal force and cars don't have to rely on friction. The Daytona International Speedway has one of the steepest banked curves, with maximum angle 31° on a curve of radius 320 m. What is the maximum speed for this curve, assuming no friction?

ORGANIZE AND PLAN Figure 4.30 shows the physical situation and the force diagram. Absent friction, there are only two forces acting on the car: the normal force $\vec{n}$ and the car's weight $\vec{w}$. The car moves in a horizontal circle, so the vertical component of $\vec{n}$ balances the weight, while its horizontal component provides the centripetal force. Writing the two components of Newton's law lets you relate centripetal force to speed.

Known: $R = 320$ m, $\theta = 31°$.

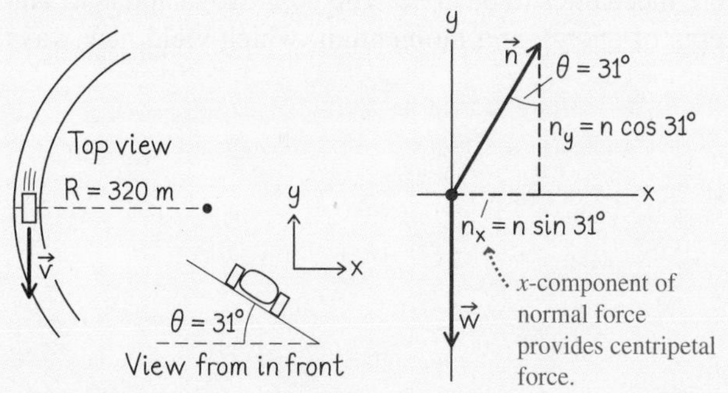

y

$\vec{n}$ $\theta = 31°$

$n_y = n \cos 31°$

x

$n_x = n \sin 31°$

... x-component of normal force provides centripetal force.

Top view
R = 320 m

$\vec{v}$

$\theta = 31°$
View from in front

$\vec{w}$

FIGURE 4.30 Analyzing the car on a banked track.

SOLVE The horizontal (x) component of Newton's law is $F_r = ma_r$, so

$$n \sin \theta = \frac{mv^2}{R}$$

while the vertical component (y) is

$$n \cos \theta + (-mg) = 0$$

Solving this second equation for n,

$$n = \frac{mg}{\cos \theta}$$

and substituting into the x-equation:

$$\frac{mv^2}{R} = n \sin \theta = \frac{mg \sin \theta}{\cos \theta}$$

or

$$\frac{mv^2}{R} = mg \tan \theta$$

The masses cancel, and the maximum speed v becomes

$$v = \sqrt{Rg \tan \theta} = \sqrt{(320 \text{ m})(9.80 \text{ m/s}^2) \tan(31°)} = 43 \text{ m/s}$$

cont'd.

Again, that's about 100 miles per hour. Race cars actually go about twice this speed, because friction between tires and road also contributes to the centripetal force. You'll have the chance to explore the combined effects of friction and a banked track in some end-of-chapter problems.

MAKING THE CONNECTION How does the maximum speed change as the banking angle increases?

ANSWER Our analysis gives $v = \sqrt{Rg\tan\theta}$. The tangent function increases with angle θ, approaching infinity as $\theta \rightarrow 90°$. The higher the speed, the higher the banking angle needed.

GOT IT? Section 4.5 A string holds an air table puck in uniform circular motion as in Example 4.11. For the cases shown, rank the string tension in order from lowest to highest.

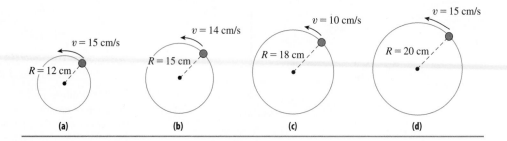

(a)   (b)   (c)   (d)

Chapter 4 in Context

This chapter introduced *Newton's three laws of motion*—the basis of dynamics—and illustrated their use in common situations. We've examined *mass as inertia* (resistance to force) in relation to Newton's first and second laws. We've explored the dynamics of circular motion. But we've barely scratched the surface. The descriptions we've given of such forces as *friction*, *normal forces*, and *tension* are useful in problem solving, but they don't probe the deeper nature of force. We'll address those issues in appropriate contexts throughout the rest of this book. Here's a brief preview.

Looking Ahead: Fundamental forces Physicists believe there are four *fundamental* forces: gravitational, electromagnetic, strong, and weak. Gravity is familiar in everyday experience, and you've already seen it in many examples. We'll do more with gravity in Chapter 9. Electricity and magnetism are both manifestations of a unified *electromagnetic* force, covered in Chapters 15–20.

The other two forces don't show up in your everyday experience, but are important at the scale of the atomic nucleus. The *strong* force binds quarks to make protons, neutrons, and other subatomic particles. A residual effect is to bind protons and neutrons into nuclei. The *weak* force governs the beta decay process in the nucleus. We'll discuss strong and weak forces in Chapter 25.

For now though, there's more mechanics to be done. The next two chapters (5 and 6) introduce the important concepts of energy and momentum, which yield new ways to analyze motion.

Force and Mass

(Section 4.1) **Force** is an interaction between two objects, such as a push or pull, or action at a distance.

Mass (inertia) is resistance to change in motion.

Net force: $\vec{F}_{net} = \vec{F}_1 + \vec{F}_2 + \cdots + \vec{F}_n$

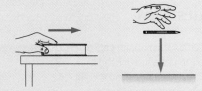

Newton's Laws of Motion

(Section 4.2) **Newton's first law:** Zero net force implies constant velocity.

Newton's second law: Net force is proportional to acceleration.

Newton's third law: Forces come in pairs, with equal magnitudes and opposite directions.

Newton's second law: $\vec{F}_{net} = m\vec{a}$

Newton's third law: $\vec{F}_{AB} = -\vec{F}_{BA}$

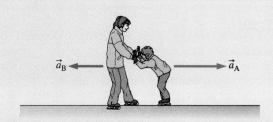

Applications of Newton's Laws

(Section 4.3) Given an object's mass and the force(s) acting on it, you can find its acceleration.

Newton's second law in components: $F_{net,x} = ma_x \quad F_{net,y} = ma_y$

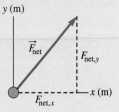

Friction and Drag

(Section 4.4) **Frictional forces** result from interactions between an object and the surface it rests on or moves across.

Drag forces retard the motion of an object moving through a fluid, such as air or water.

Force of kinetic friction: $f_k = \mu_k n$

Force of rolling friction: $f_r = \mu_r n$

Force of static friction: $f_s \leq \mu_s n$

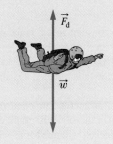

Newton's Laws and Uniform Circular Motion

(Section 4.5) **Centripetal force,** the net force on an object in **uniform circular motion**, is toward the center of the circle.

Centripetal force: $F_r = \dfrac{mv^2}{R}$

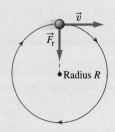

NOTE: Problem difficulty is labeled as ■straightforward to ■ ■ ■challenging. Problems labeled BIO are of biological or medical interest.

Conceptual Questions

1. *The Rules of Golf* specifies a maximum golf ball mass of 45.93 g. Why would it be unfair to use a heavier ball?
2. Explain the difference between mass and weight.
3. Describe how each of the following changes over time from the moment a skydiver opens her parachute to just before she reaches the ground: gravitational force, drag force, net force, speed, and acceleration.
4. A slab of concrete rests on an incline. Identify all the forces on the slab, and draw a force diagram.
5. Explain why it's desirable to bank racetrack curves with the steeper angle on the outer part of the track.
6. You enter an elevator car on the ground floor and stand on a scale. Describe the scale readings as the car accelerates upward, then moves upward at a constant speed, and then slows to a stop at the top floor. Repeat for the downward trip.
7. Billiard ball A strikes an identical, stationary ball B (Figure CQ4.7). After the collision, ball B's velocity is directed 30° to A's initial motion. Assuming constant forces acting during the brief collision, what was the direction of the force acting on ball A? Explain your reasoning.

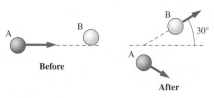

FIGURE CQ 4.7

8. You're in a car rounding a curve at a constant speed. Is the car's interior an inertial or noninertial reference frame? Explain.
9. Draw a force diagram showing gravitational and drag forces on a baseball at a point when the ball's velocity is directed 25° above horizontal.
10. Sketch the parabolic trajectory of a projectile launched from level ground without air resistance. On the same diagram, estimate the trajectory of a projectile launched with the same velocity but subject to a drag force from the air.
11. How does friction help you walk? Is it kinetic friction or static friction?
12. You're pulling a heavy box of books across the floor. If there's friction between box and floor, why is it better to pull at some angle above the horizontal, rather than horizontally? Is the same true if there's no friction?
13. A skydiver falls with constant velocity after opening her parachute. Is the force the parachute exerts on her greater than, less than, or equal in magnitude to her weight? Explain your reasoning.
14. A junior-high science student says that since an object with twice the mass has twice the weight, it should fall twice as fast, because acceleration is proportional to force. What's your response?
15. A roller coaster goes over a hill and a valley. At which of those places does it exert the greatest and the least forces on the track?

Multiple-Choice Problems

16. The magnitude of a 15.0-kg dog's weight is (a) 1.53 N; (b) 15.0 N; (c) 147 N; (d) 225 N.
17. A force $\vec{F}_1 = 105$ N $\hat{\imath}$ $- 87$ N $\hat{\jmath}$ is applied to a 1.4-kg box. The resulting acceleration is (a) 75 m/s² $\hat{\imath}$ $-$ 62 m/s² $\hat{\jmath}$; (b) 54 m/s² $\hat{\imath}$ $-$ 39 m/s² $\hat{\jmath}$; (c) $-$75 m/s² $\hat{\imath}$ $+$ 62 m/s² $\hat{\jmath}$; (d) 97 m/s².
18. A force $\vec{F}_1$ acts on a 2-kg mass, giving it acceleration $\vec{a}$. When the same force acts on a 3-kg mass, its acceleration is (a) $\vec{a}$; (b) $\frac{3}{2}\vec{a}$; (c) $2\vec{a}$; (d) $\frac{2}{3}\vec{a}$.
19. A force with magnitude $F_1 = 32$ N and another with $F_2 = 51$ N act on an object. The magnitude F_{net} of the net force is best described by the range (a) 32 N $\leq F_{net} \leq$ 51 N; (b) 0 N $\leq F_{net} \leq$ 83 N; (c) 19 N $\leq F_{net} \leq$ 83 N; (d) 32 N $\leq F_{net} \leq$ 51 N.
20. A 1.0-kg cannonball and a 5.0-kg cannonball are both in free fall. Compared with the 1-kg ball, the gravitational force on the 5-kg ball is (a) one-fifth as great; (b) the same; (c) five times as great; (d) 25 times as great.
21. A block of ice sliding without friction down a 20° slope has acceleration (a) 9.8 m/s²; (b) 4.9 m/s²; (c) 3.4 m/s²; (d) 2.2 m/s².
22. A block sliding down a 20° slope with kinetic friction $\mu_k = 0.25$ has acceleration (a) 4.9 m/s²; (b) 3.4 m/s²; (c) 1.0 m/s²; (d) 0 m/s².
23. What frictional coefficient is required for a car on level ground to round a curve of radius 275 m at 21 m/s? (a) 0.16; (b) 0.24; (c) 0.32; (d) 0.48.
24. A 210-kg rocket lifts off from its launch pad. The force required from the rocket's engine to produce an upward acceleration of 2.5 m/s² is (a) 525 N; (b) 1530 N; (c) 2580 N; (d) 2790 N.
25. A hockey puck slides over ice with $\mu_k = 0.015$. What initial speed does the puck need so it just travels the 61-m length of the rink? (a) 16.2 m/s; (b) 5.7 m/s; (c) 4.2 m/s; (d) 2.8 m/s.
26. An 810-kg compact car rounds a level curve at 25 m/s. A 2430-kg SUV rounds the same curve at 12.5 m/s. Compared with the car, the SUV's centripetal acceleration is (a) the same; (b) three times; (c) one-third; (d) one-half; (e) one-fourth.
27. For the situation of the preceding problem, the ratio of the centripetal force on the SUV to that on the car is (a) 3.00; (b) 1.33; (c) 0.75; (d) 0.67.
28. Consider a tennis ball thrown straight up. After the ball leaves your hand but before its peak, the drag force and the gravitational force are in the _____ direction. After the peak, the drag force and the gravitational force are in the _____ direction. (a) same/same; (b) same/opposite; (c) opposite/same; (d) opposite/opposite.

Problems

Section 4.1 Force and Mass

29. ■ Two forces are applied simultaneously to a crate sliding in a horizontal plane: 13.7 N to the right and 11.5 N to the left. What's the net force on the crate?
30. ■ ■ Two forces acting on a box being dragged across a floor are $\vec{F}_1 = -407$ N $\hat{\imath}$ $- 650$ N $\hat{\jmath}$ and $\vec{F}_2 = 257$ N $\hat{\imath}$ $- 419$ N $\hat{\jmath}$. Find the third force required to give zero net force on the box.

Section 4.2 Newton's Laws of Motion

31. ■■ A constant force of 95.3 N, directed upward, is applied to a 3.5-kg toy rocket by the rocket's engine. What is the rocket's acceleration?

32. **BIO ■ Proteins in motion.** Among the smallest forces biophysicists measure result from motor proteins that move molecules within cells. The protein kinesin exerts a force of 6.0 pN $(6.0 \times 10^{-12}\,\text{N})$. What magnitude of acceleration could it give a molecular complex with mass $3.0 \times 10^{-18}\,\text{kg}$? (Drag forces dominate within the cell, so accelerations last only briefly before terminal speed is reached.)

33. ■■ At time $t = 0$, a 0.230-kg air track glider is moving rightward at 2.0 m/s. At $t = 4.0$ s, it's going leftward at 1.2 m/s. Determine the magnitude and direction of the constant force that acted on the glider during this interval.

34. ■ A constant force applied to a 2.4-kg book produces acceleration $3.4\,\text{m/s}^2\,\hat{\imath} - 2.8\,\text{m/s}^2\,\hat{\jmath}$. What acceleration would result with a 3.6-kg book subject to the same force?

35. ■ Find the magnitude and direction of the force needed to accelerate a 100-g mass with $\vec{a} = -0.255\,\text{m/s}^2\,\hat{\imath} + 0.650\,\text{m/s}^2\,\hat{\jmath}$.

36. ■ Find the weight of a 7.2-kg bowling ball.

37. ■ You throw a tennis ball straight up, and it peaks at height H above the launch point. Ignore air resistance. (a) For each of the following points, draw vectors indicating the net force acting on the ball: (i) just after release; (ii) going up through $H/2$; (iii) at H, the peak; (iv) coming down through $H/2$; (v) coming down through the launch point. (b) For each of the five points, draw the ball's velocity vector.

38. ■■ The cgs unit of force is the *dyne*, equal to 1 g cm/s². Find the conversion factor between dynes and newtons.

39. ■■ Find the (upward) acceleration of Earth due to a 1000-kg object in free fall above Earth. *Hint:* Use Newton's third law.

40. ■■ A 60-kg astronaut floats outside a 3200-kg spacecraft. She's initially stationary with respect to the spacecraft. Then she pushes against the spacecraft, and moves away at 0.350 m/s to the left. Find the velocity of the recoiling spacecraft.

41. ■■ A 920-kg cannon fires a 3.5-kg shell at 95 m/s. What's the cannon's recoil speed?

42. ■■ A 0.075-kg arrow hits the target at 21 m/s and penetrates 3.8 cm before stopping. (a) What average force did the target exert on the arrow? (b) What average force did the arrow exert on the target? (c) An identical arrow strikes the target at 42 m/s. If the target exerts the same average force as before, what's the penetration depth?

Section 4.3 Applications of Newton's Laws

43. ■ A Boeing 777 has takeoff mass of 247,000 kg. Find the force required to give the aircraft a 3.2-m/s² acceleration.

44. ■ A crane lifts a 185-kg steel beam, applying a 1960-N vertical force. Find (a) the net force on the beam and (b) the beam's acceleration.

45. ■■ A net force of 150 N acts on a 5.0-kg block initially at rest. Find the block's speed after 2.5 s.

46. ■ Your 2.25-kg physics book rests on a horizontal tabletop. (a) Find the normal force on the book. (b) Suppose you pull up on the book with a 15.0-N force. Now what's the normal force? (b) Repeat for the case where you push down with a 15.0-N force.

47. ■■ A constant force $35.2\,\text{N}\,\hat{\imath}$ changes a ball's velocity from $-3.25\,\text{m/s}\,\hat{\imath}$ to $+4.56\,\text{m/s}\,\hat{\imath}$ in 3.50 s. Find the ball's mass.

48. ■ A 0.17-kg cue ball rests on the pool table. It's struck by a cue stick applying force $\vec{F_1} = 15\,\text{N}\,\hat{\imath} + 36\,\text{N}\,\hat{\jmath}$. (a) Determine the magnitude and direction of the ball's acceleration. (b) If the force was applied for 0.015 s, what's the ball's final speed?

49. ■■ Two forces act on a 24-kg cart, resulting in acceleration $\vec{a} = -5.17\,\text{m/s}^2\,\hat{\imath} + 2.5\,\text{m/s}^2\,\hat{\jmath}$. One force is $\vec{F_1} = 32\,\text{N}\,\hat{\imath} - 48\,\text{N}\,\hat{\jmath}$. Find the second force.

50. ■ A 31-kg golden retriever spots a squirrel and gives chase. The dog's legs provide an average force of 170 N. (a) What's the dog's acceleration? (b) How much time does it take the dog to go from rest to 10.8 m/s?

51. ■■ An Olympic bobsled starts from rest and slides down a frictionless 7.5° slope. What's its final speed after 25 s?

52. ■ A hot-air balloon experiences an upward *buoyant force* that overcomes gravity. (You'll learn more about buoyant forces in Chapter 10.) (a) Find the buoyant force needed to lift a 480-kg balloon straight up with acceleration 0.50 m/s². (b) What's the buoyant force when you drop back down with the same acceleration? Start each part with a force diagram.

53. **BIO ■■ Escape!** A 65.0-kg woman is rescued from a flooded house by a rope from a helicopter accelerating upward at $0.50g$. She grips the rope equally with both hands. Typically, a person's head accounts for 6.0% of their weight and both legs and feet together account for 34.5%. Find the force (a) due to the rope on each of her hands, (b) on her head, and (c) on each leg at the hip joint. Begin each part with a force diagram.

54. **BIO ■■ Whiplash injuries.** Struck from behind, a 950-kg stopped car accelerates from rest to 32 km/h in 75 ms. Typically, a person's head makes up 6.0% of their body weight. (a) Draw a force diagram for a 65.0-kg person's head during the collision if there's no headrest in the car. (b) What horizontal force on the head would accelerate it along with the rest of the body? Express your answer in newtons and as a multiple of the head's weight. What exerts this force on the head? (c) Would the head in fact accelerate along with the rest of the body? Why? What would it actually do? Explain why this can lead to neck injuries.

55. ■■ A 230-g air track glider is connected to a string hanging over a frictionless pulley (Figure P4.55). A 100-g mass hangs from the other end of the string. (a) Draw force diagrams for the glider and the hanging mass. (b) Find the acceleration of each.

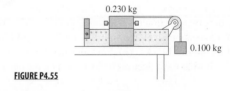

FIGURE P4.55

56. ■■■ Repeat the preceding problem for a track inclined at 10° as in Figure P4.56.

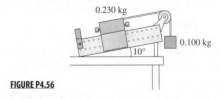

FIGURE P4.56

57. **BIO ■■ Crashing to the floor.** A 78.0-kg person falls straight down from a 1.60-m height (measured from his feet) and lands with weight distributed equally on both feet. To soften the blow, he bends his knees so that it takes 0.200 s for him to stop once his feet touch the ground. (a) What constant force does the floor exert on each foot while he's stopping? (b) Suppose instead that he lands stiff-legged and stops in only 0.100 s. What force does

the floor now exert on each foot? (c) In which case is he more likely to sustain injury? Why?

58. ■ ■ ■ An elevator undergoes constant upward acceleration, taking it from rest to 13.4 m/s in 4.0 s. An 80-kg man stands on a scale in the elevator. (a) Make a force diagram for the man. (b) What does the scale read during the upward acceleration? (c) Nearing the top floor, the elevator slows from 13.4 m/s to rest in the last 22.4 m of upward travel. Assuming constant acceleration, what's the scale reading during this period?

59. ■ ■ Olympic skeleton sled racers can reach speeds of 40 m/s. Find the slope needed to reach that speed after a 30-s run, neglecting friction.

60. ■ ■ A block of ice slides up a frictionless incline, starting at 0.365 m/s. It stops, momentarily, 1.10 s later. Find the inclination angle.

61. BIO ■ ■ **Neck brace.** A patient with a neck injury needs to sit upright with a constant vertically upward force applied to his neck brace, using the system of wires and pulleys shown in Figure P4.61. If $w = 100$ N, what net upward force does this system exert on the neck brace?

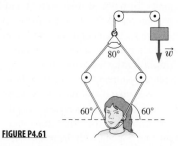

FIGURE P4.61

62. ■ ■ ■ In *Atwood's machine*, two masses m_1 and m_2 hang vertically, connected over a frictionless pulley, as in Figure P4.62. Assume $m_2 > m_1$. (a) Draw force diagrams for both masses. (b) Find the magnitude of the masses' acceleration, in terms of m_1, m_2, and g. (c) Evaluate the acceleration numerically if $m_1 = 0.150$ kg and $m_2 = 0.200$ kg.

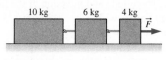

FIGURE P4.62

63. ■ ■ ■ Three blocks of mass m_1, m_2, and m_3 are touching on a frictionless horizontal surface, and a 36-N horizontal force is applied, as shown in Figure P4.63. (a) Find the acceleration of the blocks. (b) Find the net force on each block. (c) With what force does each block push on the one in front of it?

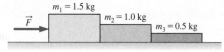

FIGURE P4.63

64. ■ ■ In a sudden stop, a car's seatbelt provides the force that stops the passenger. For a 90-kg passenger, what's the seatbelt force if a car goes from highway speed of 100 km/h to rest in 6.8 s?

65. ■ ■ Galileo drops a 2.5-kg cannon ball from the 58.4-m Tower of Pisa. If the ball makes a 0.130-m-deep hole in the ground, find the average force exerted by the ground on the ball.

66. ■ ■ A man pushes a 32-kg lawn mower using a handle inclined 40° from the horizontal. He pushes with 65 N directed along the handle. (a) What's the mower's weight? (b) What's the normal force on the mower? (c) What's the mower's acceleration (ignoring friction)?

67. ■ ■ BIO **Force from crutches.** A patient with a broken leg stands using a pair of crutches. The crutches support 75% of the 78-kg patient's weight. (a) Find the force each crutch applies to the patient, assuming they're held vertically. (b) Repeat with the crutches pointed slightly outward from the person's sides, each making a 15° angle with the vertical.

68. ■ ■ A 63-kg tightrope walker stands in the middle of a rope. Her weight makes the rope sag, with each half at a 9.5° angle to the horizontal. (a) Make a force diagram for the walker. (b) Find the rope tension.

69. ■ ■ A steel cable lifts a 350-kg concrete block vertically. The maximum safe cable tension is 4200 N. What's the maximum upward acceleration of the block?

70. ■ ■ Three blocks are connected by light strings as shown in Figure P4.70. Force $\vec{F}$ has magnitude 10 N, and the surface is frictionless. (a) Determine the acceleration of the entire system. Find the tensions in the strings between (b) the 10-kg and 6.0-kg blocks and (c) the 6.0-kg and 4.0-kg blocks.

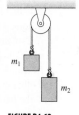

FIGURE P4.70

Section 4.4 Friction and Drag

71. ■ A golfer putts the ball at 2.45 m/s. If the coefficient of rolling friction is 0.045, find (a) the ball's acceleration and (b) how far it travels before stopping.

72. ■ ■ A hockey puck slides over ice with $\mu_k = 0.013$. What initial speed should the puck have so it travels the 61-m length of the rink?

73. ■ ■ In curling, a 19-kg granite stone is released on ice and slides to rest on a target 28.4 m away. A stone released at 1.50 m/s comes to rest due to a constant force of kinetic friction. (a) Make a force diagram and use it to find the stone's acceleration. (b) How much time does it take the stone to stop? (c) Find the coefficient of kinetic friction.

74. ■ ■ A baseball player stealing second base runs at 8.0 m/s. If he slides the last 3.5 m, slowing to a stop at the base, what's the coefficient of kinetic friction between player and ground?

75. ■ ■ A wooden block slides down a 28° incline with acceleration 3.85 m/s². (a) Make a force diagram for the block. (b) Find the coefficient of kinetic friction.

76. ■ ■ A car skids at constant speed down an icy hill inclined at 1.4°. Find the coefficient of friction between the tires and the icy road.

77. ■ ■ A 1.75-kg book is at rest on a wooden board inclined at 15°. (a) Identify each of the forces acting on the book and draw the force diagram. (b) Find the magnitude of each force.

78. ■ ■ A piece of steel rests on a wooden board, with frictional coefficient 0.35 between steel and wood. Find the maximum inclination angle for the board before the steel begins sliding.

79. ■ ■ Your car rolls along a level street at 50 km/h, in neutral and without braking. If the coefficient of rolling friction is 0.023, how far will the car coast before stopping?

80. BIO ■ ■ **Synovial joints.** As a person walks, the femur (upper leg bone) slides in a socket in the hip. This socket contains cartilage and synovial fluid, giving it a low coefficient of friction. Consider a 68-kg person walking on a horizontal sidewalk. Typically, both legs together comprise 34.5% of a person's weight. (a) Find the frictional force on the bone in the hip joint, assuming that hip supports the entire upper body's weight while the other leg swings freely. Is this static or kinetic friction? Use the appropriate value of μ from

Table 4.1. (b) If this person were walking on the Moon, where $g = 1.67 \text{ m/s}^2$, what would be the frictional force at the hip joint? (c) With age or with osteoarthritis, the synovial fluid can dry up, resulting in a much higher coefficient of friction. If this coefficient increases 100-fold (which can happen), what would be the frictional force (on Earth)?

81. ■■ On a wet road, the coefficient of kinetic friction is $\mu_k = 0.25$ for both a 1000-kg car and a 2000-kg truck. (a) Find the stopping distance for a skidding car and skidding truck each with initial speed 50 km/h. (b) Compare the stopping distances for two 1000-kg cars, one going initially at 50 km/h and the other at 100 km/h.

82. **BIO** ■■■ **Hospital patient.** A 68.0-kg patient is suspended in a raised hospital bed as shown in Figure P4.82. The wire is attached to a brace on the patient's neck and pulls parallel to the bed, and the coefficients of kinetic and static friction between the patient and the bed are 0.500 and 0.800, respectively. (a) What is the maximum mass m can be without the patient sliding up the bed? (b) If the wire suddenly breaks, what's the patient's acceleration?

FIGURE P4.82

83. ■■ A student pulls her 22-kg suitcase through the airport at constant velocity. The pull strap makes an angle of 50° above the horizontal. (a) If the frictional force between suitcase and floor is 75 N, what force is the student exerting? (b) What's the coefficient of friction?

84. ■■ A wooden block with mass 0.300 kg rests on a horizontal table, connected to a string that hangs vertically over a frictionless pulley on the table's edge. From the other end of the string hangs a 0.100-kg mass. (a) What minimum coefficient of static friction μ_s between the block and table will keep the system at rest? (b) Find the block's acceleration if $\mu_k = 0.150$.

85. ■■ Repeat the preceding problem if the block is on a 15° incline, upward toward the pulley.

86. ■■ You're using a rope to pull a sled at constant velocity across level snow, with coefficient of kinetic friction 0.050 between sled and the snow. The sled is loaded with physics books, giving a total mass of 48 kg. (a) With what force should you pull with the rope horizontal? (b) Repeat for the rope at 30° above horizontal.

87. ■■ A child starts sledding from rest, going down a 40-m-long 7.5° incline, then coasting across a horizontal stretch. The mass of sled + child is 35 kg, and the coefficient of kinetic friction is 0.060. (a) Make force diagrams for the sled + child on both the hill and the horizontal stretch. (b) Find the sled's speed at the bottom of the incline. (c) How far along the horizontal stretch does the sled travel before stopping? (d) What is the total time for the ride?

88. ■■■ A block of mass $m_1 = 0.560$ kg is placed on top of another block of mass $m_2 = 0.950$ kg, which rests on a frictionless horizontal surface. A horizontal force of 3.46 N is applied to the lower block. (a) Draw force diagrams for the upper block and for the two-block system. (b) What minimum coefficient of static friction between the two blocks will prevent the upper one from slipping?

89. ■■ You're pulling a 173-kg crate across a floor with a rope. The coefficient of static friction is 0.57, and you can exert a maximum force of 900 N. (a) Show that you cannot move the crate by pulling horizontally. (b) If you slowly increase the rope angle while continuing to pull with a 900-N force, at what angle does the crate begin to move?

90. **BIO** ■■ **Walking.** To begin walking horizontally, a person must accelerate her body forward using static friction. A person can typically reach 2.0 m/s over a distance of 30 cm, starting from rest. Assuming constant acceleration, (a) draw a force diagram for the start of the walk. (b) Apply Newton's second law to find the minimum coefficient of static friction for the walk described. (c) Would power walking require more or less friction? Why? (d) Suppose the floor is slippery and has a smaller coefficient of friction than found above. Explain why walking is still possible, but is tiring and awkward.

91. ■■ A truck carries a 3.0-kg box of Washington apples on its flat bed. The coefficient of static friction between box and truck bed is 0.38. (a) What is the maximum acceleration for the truck on a level road if the box isn't to slide? (b) Repeat part (a) assuming the truck is driving toward Snoqualmie Pass on a 4.5° upward incline.

92. ■■ **BIO** **Supporting your weight.** A physical therapy patient supports his entire weight on crutches inclined outward at 25° to the vertical. (a) Find the minimum coefficient of static friction between crutches and floor so the crutches won't slip. (b) If the patient reduces the crutch angle, does that make the required (minimum) frictional coefficient larger, smaller, or the same?

93. ■■ Repeat Problem 70 if there's a coefficient of kinetic friction $\mu_k = 0.10$ between the blocks and the surface.

Section 4.5 Uniform Circular Motion

94. ■ A car rounds a 210-m-radius curve at 11.5 m/s. (a) Draw a force diagram for the car. (b) If the horizontal force acting on the car is 790 N, what's the car's mass?

95. ■■ (a) Use astronomical data in Appendix E to find the gravitational force holding the Moon in its orbit around Earth. (b) Compare your answer with the force holding Earth in its orbit around the Sun.

96. ■■ A child ties a rock to a string and whirls it around in a horizontal circle. (a) The string cannot be perfectly horizontal. Why not? Explain using a force diagram. (b) Assuming a 1.22-m-long string making 25° angle below the horizontal, find (c) the speed of the rock and (d) the period of its uniform circular motion.

97. ■■ A car rounds a flat circular track with radius 225 m. The coefficient of static friction between the tires and track is 0.65. (a) Draw a force diagram for the car. (b) What's the maximum speed for the car if it's to stay on the track?

98. ■■ (a) Repeat the preceding problem, now for a road that's banked at 9.5°. (b) Find the maximum speed for a car on this track in the absence of friction. Compare your result with that of part (a).

99. ■■ The Sun exerts a gravitational force of 5.56×10^{22} N on Venus. Assuming that Venus's orbit is circular with radius 1.08×10^{11} m, determine Venus's orbital period. Check your answer against the observed period, about 225 days.

100. ■■ **BIO** **Centrifuge.** Medical labs often use centrifuges to separate the different components of blood or tissue. In a centrifuge, materials are whirled rapidly so they experience a large centripetal force. (a) Find the acceleration of a blood sample 0.14 m from the center of a centrifuge rotating at 250 revolutions per minute. (b) What is the force acting on a 0.10-g particle at that radius?

101. ■■■ An airplane flies in a vertical circle at a constant 90 m/s. Start each of the following two parts with a force diagram for the pilot, whose mass is m. (a) Find the circle's radius so that the pilot

is "weightless" at the top of the circle. (b) What is the pilot's apparent weight at the bottom of the circle under these conditions?

102. ■■ A block with mass m_1 slides on a frictionless horizontal table. It's connected to a block with mass m_2 by a light cord that passes through a hole in the table, with m_2 hanging vertically. (a) What should m_2 be so that m_1 slides in uniform circular motion in a circle with radius R and speed v? Start with force diagrams for each block. (b) Evaluate your result for $R = 0.50$ m, $v = 0.85$ m/s, and $m_1 = 0.25$ kg.

103. ■■ A 63-kg fighter pilot sits on a scale in the cockpit. (a) What's the scale reading when the plane flies horizontally with constant velocity? Now the plane goes into a vertical loop of radius 1.85 km, flying at a constant 235 m/s. What's the scale reading when the plane is at (b) the bottom and (c) the top of the loop?

104. ■■■ A car rounds a circle of radius 150 m, going counterclockwise at 10.5 m/s. Consider the moment the car is moving in the $+y$-direction. (a) What's the car's acceleration (vector) if its speed is constant? (b) Explain how the acceleration vector differs at that moment if the car's speed is increasing. (c) Compute the new acceleration vector if the speed is increasing at a rate of 2.4 m/s^2.

105. BIO ■■ **Maximum walking speed.** Experiments show that a walking person's hips describe circular arcs centered on the point of contact with the ground, and having radii equal to the leg's length L (see Figure P4.105). Since the person's center of mass (more in Chapter 6) is near the hip, we can model the walker as a mass M moving in a circular arc of radius L. In this case, M is the mass above the hip, which is roughly the person's total mass. At maximum speed, gravity alone is sufficient to provide the centripetal force. (a) Apply Newton's second law and show that the maximum speed at which a person can walk, according to this model, is $v_{max} = \sqrt{Lg}$. (To move faster, one must run.) (b) What's the fastest walking speed for a typical 75-kg adult male? Use measurements on yourself or a friend to determine L.

FIGURE P4.105

106. BIO ■■ **Astronaut locomotion.** An astronaut walks with maximum speed 2.50 m/s on Earth. Use the results of the preceding problem to find her maximum walking speed (a) on the Moon, with $g_{Moon} = 1.67$ m/s^2, and (b) on Mars, with $g_{Mars} = 0.379\,g_{Earth}$.

107. ■■■ A *conical pendulum* consists of a string of length L with one end attached to the ceiling and the other to a mass (Figure P4.107). With the string making a constant angle θ to the vertical, the ball moves in a horizontal circle. (a) Draw a force diagram for the mass. (b) Find the period of the circular motion in terms of L, θ, and the gravitational acceleration g. (c) Find the limiting value of your answer to part (a), in the limit as θ approaches 0.

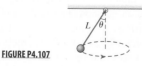

FIGURE P4.107

General Problems

108. ■■ A horizontal force acts on a 0.250-kg puck as it moves over ice. A graph of F_x versus t is shown in Figure GP4.108. Graph the puck's acceleration a_x versus t.

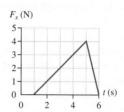

FIGURE GP4.108

109. ■■ You give a 2.90-kg book an initial shove at 2.10 m/s and it comes to rest after sliding 3.25 m across the floor. Find the coefficient of friction between book and floor.

110. ■■ Figure GP4.110 shows velocity versus time for a 140-g puck on a frictionless surface. Graph the net force on the puck as a function of time.

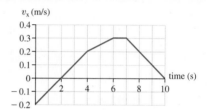

FIGURE GP4.110

111. BIO ■■ **Astronaut training.** Astronauts and test pilots experience extreme accelerations lasting for very short periods. NASA has been studying the effects of such accelerations. In 1954, one test subject was brought to rest in a water trough from a horizontal velocity of 286 m/s over a distance of 112.7 m. Assume constant acceleration and a mass of 77.0 kg. (a) Draw a force diagram for this subject while slowing. (b) What net force was acting during the slow-down? Express your answer in newtons and as a multiple of the subject's weight. (c) For how much time did this force act?

112. ■■ A 670-kg helicopter rises straight up with acceleration 1.20 m/s^2. (a) What upward force must the helicopter's rotor provide? (b) The helicopter then begins its descent with downward acceleration 1.20 m/s^2. Now what force does the rotor provide? Explain why your answers differ.

113. BIO ■■ **Insect locomotion.** According to a 2003 article in *Nature*, froghoppers are typically 6.1 mm long, have mass 12.3 mg, and leave the ground from rest with takeoff speed 2.8 m/s at 58° above the horizontal. In jumping, the insect typically pushes its legs against the ground for 1.0 ms. (a) Draw a force diagram for a froghopper during the takeoff push. (b) Find the horizontal and vertical components of its acceleration (assumed constant) during takeoff. (c) Find the horizontal and vertical components of the force that the ground exerts on the froghopper during takeoff. Express your answer in newtons and as a multiple of the insect's weight.

114. ■■■ (a) Assume that all surfaces in Figure GP4.114 are frictionless. What value should m have so the masses won't accelerate? (b) Now take $\mu_k = 0.15$ on all surfaces and $m = 1.1$ kg. What will be the acceleration of both masses? Start with a force diagram for each.

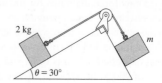

FIGURE GP4.114

115. ■ ■ A frictionless air track is inclined at 5.0° to the horizontal. A glider is given initial speed 1.25 m/s directed up the incline. (a) Draw force diagrams of the glider as it is going both up and down the incline. (b) How far does the glider slide up the incline before stopping? (c) How much time does it take to stop? (d) After stopping, the glider slides back down to its initial position. Find its speed when it returns to its launch point. (e) Draw graphs of velocity and position versus time for the round trip.

116. ■ ■ ■ Repeat the preceding problem, now assuming a coefficient of friction $\mu_k = 0.040$ between glider and track.

117. ■ ■ A 365-kg bobsled starts down an icy slope inclined at 3.4°. (a) If the sled's acceleration along the incline is 0.51 m/s², what is the coefficient of kinetic friction? (b) A velocity-dependent air drag force from the air is directed opposite the sled's velocity. Eventually, the drag reduces the acceleration to zero. What's the magnitude of the drag force at that time?

118. ■ ■ ■ In a popular amusement park ride, people stand on a platform around the inside of a large cylinder. The cylinder spins, the floor drops away, and the people are left pressed with their backs against the cylinder wall. (a) Why don't they fall? Draw a force diagram for a person inside the spinning cylinder to aid your explanation. (b) Take a cylinder with inner radius 3.0 m, rotating with period 3.3 s. Find the minimum coefficient of static friction required between the person and the cylinder wall.

119. ■ ■ You're riding a Ferris wheel while sitting on a scale. (a) Why does the scale reading when you're at the top differ from when you're at the bottom of the wheel? Draw a force diagram for each case. (b) A Ferris wheel with radius 9.6 m has period 36 s. Find the scale reading for a 60-kg person at the bottom and top of the Ferris wheel, assuming a constant rotation rate.

120. BIO ■ ■ ■ **Hypergravity.** The "20-G" centrifuge at NASA's Ames Research Center is used to study the effects of large acceleration ("hypergravity") on astronauts and test pilots. Its 8.84-m-long arm rotates horizontally about one end. The test subject is strapped onto a cot at the other end and rotates with it. A 70.0-kg astronaut who is 1.70 m tall is strapped as shown in Figure GP4.120. The maximum rotation speed of the arm for human study is 35.55 rpm (rev/min). Typically, a person's head comprises 6.0% of his weight, and his two feet together comprise 3.4%. (a) Draw force diagrams for the astronaut's head and for one foot during rotation. (b) Calculate the net force acting on the astronaut's head and on each of his feet when the arm is rotating at its maximum speed. Express your answer in newtons and as a multiple of the weight of the head and foot. Assuming negligible friction from the cot, what exerts the accelerating force on the head and feet? (c) Suppose instead that the arm were rotated at a constant rate in a vertical plane with the astronaut still strapped as shown. Find the maximum and minimum force on the astronaut's head.

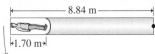

FIGURE GP4.120

Answers to Chapter Questions

Answer to Chapter-Opening Question
No, air resistance reduces the skydiver's acceleration as he falls, even without opening the chute. If he falls long enough, he'll reach terminal speed, at which point his acceleration is zero.

Answers to GOT IT? Questions
Section 4.2　(a)
Section 4.3　(b) The weight
Section 4.4　(c) Horizontal
Section 4.5　(c) < (d) < (b) < (a)

■ While climbing a mountain, these cyclists do work against gravity. Does that work depend on the route chosen?

In this chapter we'll introduce the concept of energy and its close relative, work. Energy is a fundamental idea in physics, and it provides shortcuts to the solution of problems involving force and motion.

Kinetic energy is the energy of motion. Other forms of energy include thermal energy, electrical energy, and nuclear energy. When energy is transformed from one form to another, the total amount of energy is unchanged. This principle—known as conservation of energy—is central throughout physics. We'll introduce conservation of energy in this chapter and will use it frequently throughout the book. You'll come to appreciate that the use and transformation of energy are the keys to modern civilization, and to life itself.

We'll then introduce potential energy, associated with the physical configuration of systems. In many situations the total mechanical energy—the sum of kinetic and potential energy—is conserved (its value remains constant). The interplay between potential and kinetic energy will give new insights into mechanics. Finally, we'll consider power—the rate at which work is done or energy is used.

5.1 Work Done by a Constant Force

Work has several meanings in everyday life. But in physics, the **work** done on an object is a precise quantity that depends on the *forces* applied to the object and the object's *displacement*. We'll first view this with a simple case of an object pulled with constant force.

To Learn

By the end of this chapter you should be able to

■ Understand work and how it's computed.

■ Relate work and kinetic energy.

■ Explain potential and kinetic energy.

■ Apply the principle of conservation of mechanical energy.

■ Explain power, and distinguish it from energy.

Work Done by Constant Forces

Figure 5.1 shows a boy pulling his sled across level snow. We've chosen the $+x$-axis to be in the direction of the sled's motion. Consider the sled moving from x_0 to x, so its displacement is $x = x - x_0$. Figure 5.1 shows that there are four forces acting on the sled, including kinetic friction. We'll take all the forces to be constant.

The *work W done by a constant force* $\vec{F}$ on an object moving along the x-axis is equal to the x-component of the force (F_x) multiplied by the displacement:

$$W = F_x \Delta x \qquad \text{(Work done by a constant force in one-dimensional motion; SI unit: J)} \qquad (5.1)$$

Loosely, work is *force times displacement*. More precisely, Equation 5.1 shows that work involves only the component of force in the direction of displacement. Work is a *scalar* quantity. Multiplying SI units of force (N) and displacement (m) gives the units for work: $\text{N} \cdot \text{m}$. This combination defines a new SI unit, the **joule** (J), with

$$1\,\text{J} = 1\,\text{N} \cdot \text{m}$$

The joule is named for the English physicist James Joule (1818–1889), who helped develop the concepts of work and energy.

The quantities F_x and Δx in Equation 5.1 are both scalars that can be positive, negative, or zero. Therefore, work W can be positive, negative, or zero. Figure 5.2 and Examples 5.1 and 5.2 explore this situation quantitatively.

Net Work

Equation 5.1 defines the work done by any of the *individual forces* acting on an object, for example, each of the four forces on our sled. It's often useful to know the **net work**—the sum of the work done by the individual forces. Symbolically, if there are n forces acting on an object, the net work is

$$W_{\text{net}} = W_1 + W_2 + \cdots + W_n \qquad \text{(Net work done by multiple forces; SI unit: J)} \quad (5.2)$$

Each individual value of work (W_1, W_2, etc.) is defined by Equation 5.1, which leads to another expression for net work:

$$W_{\text{net}} = F_{1x}\Delta x + F_{2x}\Delta x + \cdots + F_{nx}\Delta x$$
$$= (F_{1x} + F_{2x} + \cdots + F_{nx})\Delta x$$

The quantity in parentheses is just the x-component of the net force acting on the object, so

$$W_{\text{net}} = F_{\text{net},x}\Delta x \qquad \text{(Net work done by multiple forces; SI unit: J)} \qquad (5.3)$$

Equations 5.2 and 5.3 give you two equivalent ways to understand and compute net work: as the sum of the work done by individual forces, or as the x-component of the net force times the object's displacement.

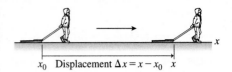

x_0 Displacement $\Delta x = x - x_0$ x

(a) Displacement of sled

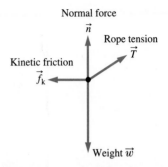

(b) Force diagram for sled

FIGURE 5.1 A sled moving in one dimension.

Because $W = F_x\Delta x$, the sign of work depends on the signs of Δx and F_x:

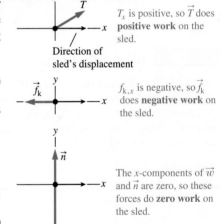

T_x is positive, so $\vec{T}$ does **positive work** on the sled.

$f_{k,x}$ is negative, so $\vec{f}_k$ does **negative work** on the sled.

The x-components of $\vec{w}$ and $\vec{n}$ are zero, so these forces do **zero work** on the sled.

FIGURE 5.2 The work done on the sled by each of the forces in Figure 5.1.

EXAMPLE 5.1 **Pulling the Sled**

The sled in Figure 5.1a has mass 6.35 kg, and it's pulled with constant velocity for 5.00 m. The rope tension is 10.6 N, and the rope makes a 30° angle with the horizontal. Draw a force diagram for the sled. Find the work done on the sled by each of the four forces, and the net work.

ORGANIZE AND PLAN Our diagram (Figure 5.3, next page) shows the four forces on the sled. Work is the x-component of each force multiplied

by displacement. The x-components of the normal force $\vec{n}$ and gravity (the weight $\vec{w}$) are zero. The x-component of the tension follows from trigonometry: $T_x = T\cos\theta$. Thus the sum of the x-components is

$$F_{\text{net},x} = T_x + f_{kx}$$

We're not given the frictional force. But the sled's velocity is constant, so the net force on it is zero. Thus $T\cos\theta + f_{kx} = ma_x = 0$, The

cont'd.

minus sign shows that friction balances the horizontal component of the rope tension.

Known: $m = 6.35$ kg; $T = 10.6$ N, $\theta = 30°$, $\Delta x = 5.00$ m.

SOLVE With no x-components, the normal force and gravity do no work: $W_n = 0$, $W_g = 0$. (We'll consistently use W_g for the work done by gravity.)

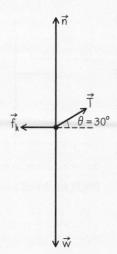

FIGURE 5.3 Force diagram for the sled.

The tension does work

$$W_T = T_x \Delta x = T \cos \theta \, \Delta x$$

or,

$$W_T = (10.6 \text{ N})(\cos 30°)(5.00 \text{ m}) = 45.9 \text{ N} \cdot \text{m} = 45.9 \text{ J}$$

Finally, for the frictional force

$$W_f = f_{kx} \Delta x = -T \cos \theta \, \Delta x = -45.9 \text{ J}$$

The net work is then

$$W_{net} = W_n + W_g + W_T + W_f$$

or

$$W_{net} = 0 \text{ J} + 0 \text{ J} + 45.9 \text{ J} - 45.9 \text{ J} = 0 \text{ J}$$

REFLECT The final result should be obvious: Since the sled moves with constant velocity, there's no net force. So the net work done is zero, regardless of its displacement.

MAKING THE CONNECTION Could the data in this example be used to find the coefficient of kinetic friction between sled and snow?

ANSWER Yes: Recall that $f_k = \mu_k n$. Analyzing the equations $F_{net,x} = 0$ and $F_{net,y} = 0$ leads to $n = 56.9$ N and $\mu_k = 0.16$.

EXAMPLE 5.2 **The Sled Accelerates**

The boy pulls his 6.35-kg sled another 5.00 m on level snow, using the same rope orientation and tension. Find the work done by each force and the net work in two cases: (a) an icy patch, with the sled accelerating at 0.390 m/s²; (b) a slushy patch, on which the sled slows with acceleration −0.390 m/s².

ORGANIZE AND PLAN Figure 5.4 shows force diagrams for each case. Again, the normal force and gravity do no work. In both cases the rope

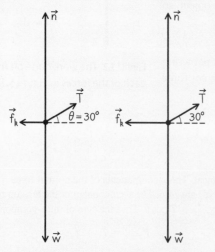

(a) With less friction
$a_x = 0.390$ m/s²

(b) With more friction
$a_x = -0.390$ m/s²

FIGURE 5.4 Two cases: lower friction and higher friction.

tension is the same as in the preceding example—and so, therefore, is the work done by tension. But the frictional force is different, so the work done by friction and the net work will differ.

We'll focus on kinetic friction, since it's the only force that differs. The x-component of Newton's law is as before, except that now the acceleration isn't zero:

$$F_{net,x} = T_x + f_{kx} = T \cos \theta + f_{kx} = ma_x$$

Solving for f_{kx} gives

$$f_{kx} = ma_x - T \cos \theta$$

Known: $T = 10.6$ N; $m = 6.35$ kg, $\theta = 30°$.

SOLVE (a) With $a_x = 0.390$ m/s², the friction component is

$$f_{kx} = ma_x - T \cos \theta = (6.35 \text{ kg})(0.390 \text{ m/s}^2) - (10.6 \text{ N})(\cos 30°)$$
$$= -6.70 \text{ N}$$

where the minus sign shows that friction acts opposite the sled's motion.

Then the work done by friction is

$$W_f = f_{kx} \Delta x = (-6.70 \text{ N})(5.00 \text{ m}) = -33.5 \text{ J}$$

The work done by the other forces is the same as in the preceding example: $W_T = 45.9$ J and $W_n = W_g = 0$ J. So the net work is

$$W_{net} = W_n + W_g + W_T + W_f = 0 \text{ J} + 0 \text{ J} + 45.9 \text{ J} - 33.5 \text{ J}$$
$$= 12.4 \text{ J}$$

cont'd.

(b) With $a_x = -0.390 \text{ m/s}^2$, similar calculations give

$$f_{kx} = ma_x - T\cos\theta$$
$$= (6.35 \text{ kg})(-0.390 \text{ m/s}^2) - (10.6 \text{ N})(\cos 30°)$$
$$= -11.66 \text{ N}$$

and

$$W_f = f_{kx}\Delta x = (-11.66 \text{ N})(5.00 \text{ m}) = -58.3 \text{ J}$$

The other Ws remain unchanged, so the net work is now

$$W_{net} = W_n + W_g + W_T + W_f = 0 \text{ J} + 0 \text{ J} + 45.9 \text{ J} - 58.3 \text{ J}$$
$$= -12.4 \text{ J}$$

REFLECT The net work done on an object is a function of its acceleration, positive if it's speeding up and negative if it's slowing. Here accelerations of the same magnitude but opposite signs give net works that are, correspondingly, of the same magnitude but opposite sign.

MAKING THE CONNECTION Can μ_k be computed in this example? How should its values in parts (a) and (b) compare?

ANSWER The icy patch in part (a) results in positive acceleration and implies a smaller friction coefficient than in part (b), where friction is great enough to slow the sled. Calculations confirm this: with $f_k = \mu_k n$, the results are (a) $n = 56.9 \text{ N}$ and $\mu_k = 0.12$; (b) $n = 56.9 \text{ N}$ and $\mu_k = 0.20$.

Reviewing New Concepts

The last two examples show a pattern that illustrates an important relation between net work done on an object and the change in its motion:

- Positive net work $(W_{net} > 0) \rightarrow$ increasing speed
- Zero net work $(W_{net} = 0) \rightarrow$ constant speed
- Negative net work $(W_{net} < 0) \rightarrow$ decreasing speed

Although we've seen this pattern only for specific examples, it's true in general. In Section 5.3, we'll prove the **work-energy theorem**, which expresses the precise mathematical relationship between net work and change in speed.

Computing Work: General Rules

You may have noticed another pattern, relating work to the relative *directions* of force and displacement. Refer again to Figure 5.2. If the angle θ between force and displacement is less than 90°, then work is positive. That's true in general, because when $\theta < 90°$, F_x and Δx have the same sign, and so their product (W) is positive. Similarly, when $\theta > 90°$, F_x and Δx have opposite signs, so work is negative. The intermediate case, with force and displacement perpendicular $(\theta = 90°)$, gives zero work because the force has no component in the direction of the displacement.

This pattern suggests an alternative way to compute the work done by a force. In general, the x-component of force is

$$F_x = F\cos\theta$$

where F is the magnitude and θ the angle between $\vec{F}$ and the $+x$-axis. Therefore, by the definition of work (Equation 5.1),

$$W = (F\cos\theta)\Delta x \quad \text{(Work done by a constant force in one-dimensional motion, geometric view; SI unit: J)} \quad (5.4)$$

Equation 5.4 gives you a way to think about work geometrically, in terms of the magnitude of the force, the displacement, and the angle θ between force and displacement.

✓ **TIP**

You can compute the work done by a force using components (Equation 5.1) or using the force's magnitude and direction (Equation 5.4).

CONCEPTUAL EXAMPLE 5.3 **Uniform Circular Motion**

What's the work done by the Sun's gravity on a planet orbiting in a circle at constant speed?

SOLVE The Sun's gravity does no work on the planet. The planet's velocity and hence its displacement over small intervals are always tangent to the circle, and hence perpendicular to the centrally directed gravitational force (Figure 5.5). Whenever a force is perpendicular to displacement, that force does zero work (recall Figure 5.2 and Equation 5.4).

REFLECT This example shows that displacement over a very small interval is in the direction of velocity. Therefore, a force that's always perpendicular to an object's velocity does no work on the object. That's the case with the centripetal force in uniform circular motion. No matter what provides that force—gravity, tension, friction,

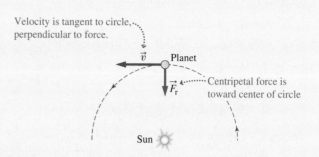

FIGURE 5.5 The Sun's gravity does no work on a planet in a *circular* orbit.

magnetism, whatever—**the centripetal force in circular motion does no work**.

Another case where a force does no work is when an object's displacement is zero. Suppose you're pushing a stuck car, but it doesn't budge. You're pushing hard and wearing yourself out, but the work you do *on the car* is $W = 0$, because $\Delta x = 0$. This is one instance where the everyday meaning of "work" differs from its physics definition.

One more general rule regarding work follows from Equation 5.4 and Figure 5.2: **The work done by kinetic friction or drag forces is always negative.** These forces are always directed opposite the motion. Thus, the angle θ in Equation 5.4 is 180°. With $\cos(180°) = -1$, the work is negative. In Chapter 13 you'll see how frictional work raises the temperature of the interacting surfaces or, with ice, causes melting.

Work Done by Gravity

You saw in Chapter 2 that near Earth, the gravitational force on an object with mass m is $\vec{w} = m\vec{g}$, where $\vec{g}$ points down, with magnitude $g = 9.80 \text{ m/s}^2$. (Chapter 9 treats the case of large distances from Earth and other bodies, where g isn't constant.)

We'll adopt the usual coordinate system with x horizontal and y vertically upward; then the gravitational force $\vec{w}$ is in the $-y$-direction. Equation 5.1 defines work done by a constant force in one dimension. Using Δy rather than Δx in that equation, the work done by gravity is $W_g = w_y \Delta y$. Because $\vec{w}$ points in the $-y$-direction and has magnitude mg, $w_y = -mg$, and the work becomes

$$W_g = -mg\Delta y \quad \text{(Work done by gravity; SI unit: J)} \quad (5.5)$$

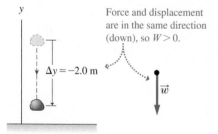

FIGURE 5.6 Work done by gravity on a falling rock.

Drop a 4.5-kg rock and let it fall 2.0 m (Figure 5.6); then $y = -2.0$ m, and the work done by gravity is

$$W_g = -mg\Delta y = -(4.5 \text{ kg})(9.80 \text{ m/s}^2)(-2.0 \text{ m}) = +88.2 \text{ kg} \cdot \text{m}^2/\text{s}^2 = 88.2 \text{ J}$$

Does it make sense that this work is positive? Yes: Force and displacement are in the same direction (Figure 5.6), hence positive work. In contrast, a projectile heading upward has negative work done by gravity, because the force (downward) and displacement (upward) are opposite. Computationally, an upward-moving object has displacement $\Delta y > 0$, but $w_y < 0$, so the work W_g in Equation 5.5 is negative.

✓**TIP**

The work done by gravity is positive when an object is going down and negative when it's going up.

| EXAMPLE 5.4 | **Baseball Work** |

A 0.145-kg baseball pops straight up from the ground at 21.4 m/s. (a) What maximum height does it reach? (b) Find the work done by gravity on the baseball during its upward flight. (c) Find the work done by gravity for the entire round trip, from launch to when the ball hits the ground.

ORGANIZE AND PLAN The work done by gravity is $W_g = -mg\Delta y$, so you'll need Δy for the two cases. For part (b), the upward flight, Δy is the maximum height (Figure 5.7a). Finding that is a one-dimensional kinematics problem, using Equation 2.13:

$$v_y^2 = v_{0y}^2 - 2g\Delta y$$

In part (c), the ball ends up back on the ground (Figure 5.7b), so $\Delta y = 0$.

Known: $m = 0.145$ kg, $v_{0y} = 21.4$ m/s.

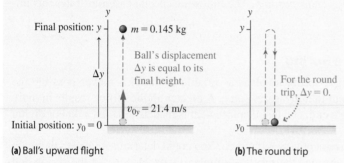

(a) Ball's upward flight   **(b)** The round trip

FIGURE 5.7 Work done by gravity on a baseball.

SOLVE (a) Solving Equation 2.13 for Δy and using $v_y = 0$ at the top of its flight gives the maximum height:

$$\Delta y = \frac{v_{0y}^2 - v_y^2}{2g} = \frac{(21.4 \text{ m/s})^2 - (0 \text{ m/s})^2}{2(9.80 \text{ m/s}^2)} = 23.4 \text{ m}$$

(b) Then the work done by gravity during the upward flight is

$$W_g = -mg\Delta y = -(0.145 \text{ kg})(9.80 \text{ m/s}^2)(23.4 \text{ m})$$
$$= -33.3 \text{ kg} \cdot \text{m}^2/\text{s}^2 = -33.3 \text{ J}$$

(c) After reaching the ground again, $\Delta y = 0$. Then

$$W_g = -mg\Delta y = 0 \text{ J}$$

The total work is zero.

REFLECT Zero work for the round trip means that gravity does positive work on the way down, equal in magnitude to the negative work done going up. A ball would have to fall below its launch point for the total gravitational work to be positive.

MAKING THE CONNECTION What's the connection between the sign of the net work done on the baseball and changes in its speed?

ANSWER You've seen that negative net work implies a reduction in speed. While the ball is going up, the work is negative, and indeed it slows. Going down, the work is positive, and speed increases. When the ball reaches the ground, the total work done on it is zero, and its speed is the same as at launch. We'll soon make this connection between work and speed change more explicit.

Work in Two-Dimensional Motion

How would the preceding example change if the baseball were launched at an angle as in Figure 5.8? For the work done by gravity, there's no difference: It's still given by $W_g = -mg\Delta y$, as in Equation 5.5, where Δy is the y-component of the baseball's displacement. Any horizontal displacement is perpendicular to the gravitational force, so there's no work involved.

Until now we've considered only motion in a single direction, and we've seen that only the component of force in that direction contributes to work. More generally, both force and displacement can have components in different directions, and displacement components vary for motion on curved paths. We now expand our definition of work done by a constant force $\vec{F}$ to two dimensions. In general, the force $\vec{F}$ has components F_x and F_y, while the displacement components are Δx and Δy. Then the work is

$$W = F_x \Delta x + F_y \Delta y \quad \text{(Work in two-dimensional motion; SI unit: J)} \quad (5.6)$$

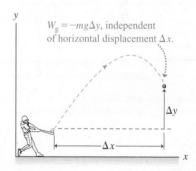

FIGURE 5.8 Work in two-dimensional motion.

GOT IT? Section 5.1 A piano hangs from a steel cable. As it's lowered from a third-floor apartment to the ground, the work done on the piano by the cable is (a) positive; (b) negative; (c) zero; or (d) it cannot be determined without more information.

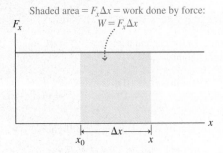

Shaded area = $F_x \Delta x$ = work done by force:
$$W = F_x \Delta x$$

(a) Work done by a constant force

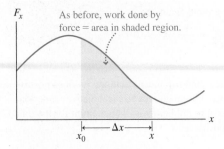

As before, work done by force = area in shaded region.

(b) Work done by a variable force

FIGURE 5.9 Work done equals area under the force-versus-position graph.

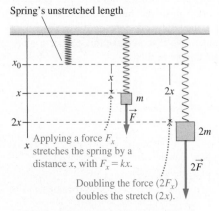

Spring's unstretched length

Applying a force F_x stretches the spring by a distance x, with $F_x = kx$.

Doubling the force $(2F_x)$ doubles the stretch $(2x)$.

(a) A force applied by a hanging weight stretches a spring according to Hooke's law

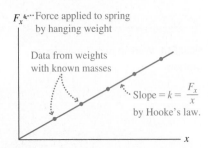

Force applied to spring by hanging weight

Data from weights with known masses

Slope = $k = \dfrac{F_x}{x}$ by Hooke's law.

(b) Determining the spring constant k

FIGURE 5.10 Hooke's law relates the force applied to a spring to the amount by which the spring stretches (or is compressed).

5.2 Work Done by a Variable Force

Section 5.1 considered only constant forces. But many forces vary with position. Even in a simple task like lifting a box, the force you apply probably varies throughout the motion. This section presents a general way to think about work done by variable forces, and then focuses on the important case of a spring.

Work from the Force-versus-Position Graph

Figure 5.9a graphs force versus position for a constant force with component F_x in the direction of an object's motion. You know that the work done by this force as the object undergoes displacement Δx is

$$W = F_x \Delta x$$

In Figure 5.9a, this work corresponds to the rectangular area under the force-versus-distance graph.

As Figure 5.9b suggests, that remains true for *any* force resulting in one-dimensional motion, whether it's constant or not: *The work done by a force with x-component F_x is the area under the graph of F_x versus position.* The following special case illustrates this important point.

A Variable Force: The Spring

Extending or compressing a spring involves a variable force. This makes possible the spring scale—an instrument that measures force. A greater applied force results in more displacement of the spring's end, and that displacement is read as force on a dial or digital readout.

How does spring displacement vary with force? You could determine this with the setup in Figure 5.10. For most springs, the result is simple: displacement x of the spring end from its unstretched (equilibrium) position is *directly proportional* to the applied force F_x:

$$F_x = kx \qquad \text{(Hooke's law; SI unit: N)} \qquad (5.7)$$

This is **Hooke's law**, named for the English physicist Robert Hooke (1635–1703), who worked with Newton on force and motion. We stress that Hooke's law isn't a fundamental law, but rather is *approximately* true for many springs. Even when Hooke's law works, a spring loses its elasticity if you stretch it too far.

The proportionality constant k in Equation 5.7 is the **spring constant** and has SI units of N/m. You could measure k by hanging different masses from a spring and measuring its stretch (Figure 5.10a); a graph of weight versus stretch should be a straight line whose slope is k (Figure 5.10b). Suppose a 0.250-kg mass stretches the spring by 0.120 m. Then the spring constant is

$$k = \frac{F_x}{x} = \frac{mg}{x} = \frac{(0.250 \text{ kg})(9.80 \text{ m/s}^2)}{0.120 \text{ m}} = 20.4 \text{ N/m}$$

The spring constant measures the spring's stiffness; this result is a fairly typical value for a spring you might use in physics labs. Stiffer springs, such as those in your car's suspension, have much higher k values. For example, suppose a 1040-kg car is supported by four identical springs, which compress 3.5 cm (0.035 m) under the car's weight. Then the spring constant for each spring, supporting one-fourth of the car's weight $(mg/4)$, is

$$k = \frac{F_x}{x} = \frac{mg/4}{x} = \frac{(1040 \text{ kg})(9.80 \text{ m/s}^2)}{4(0.035 \text{ m})} = 7.28 \times 10^4 \text{ N/m}$$

That's over 70 kN/m—a very stiff spring.

Work Done on a Spring

How much work is done by the force applied to a spring? You can find out using the method of Figure 5.9, which gives work as the area under the force-versus-position graph. For a spring that obeys Hooke's law, that area is a triangle. Figure 5.11 shows that the work is

$$W = \tfrac{1}{2}(x)(kx)$$

or

$$W = \tfrac{1}{2}kx^2 \qquad \text{(Work done stretching a spring; SI unit: J)} \qquad (5.8)$$

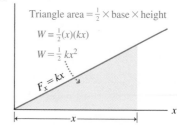

F_x (applied force)

Triangle area $= \tfrac{1}{2} \times$ base $\times$ height

$W = \tfrac{1}{2}(x)(kx)$

$W = \tfrac{1}{2} kx^2$

$F_x = kx$

FIGURE 5.11 Work done in stretching a spring.

Equation 5.8 gives the work done to stretch a spring from its equilibrium position (taken as $x = 0$) through a distance x. If the spring has already been stretched to a position x_A, then the additional work required to stretch from x_A to x_B is $W = \tfrac{1}{2}kx_B^2 - \tfrac{1}{2}kx_A^2$.

Consider the spring in Figure 5.10, with $k = 20.4$ N/m. Hang a 500-g mass from the spring and wait until it comes to rest. From Hooke's law, $F_x = kx$, the end of the spring undergoes displacement x given by

$$x = \frac{F_x}{k} = \frac{mg}{k} = \frac{(0.500 \text{ kg})(9.80 \text{ m/s}^2)}{20.4 \text{ N/m}} = 0.240 \text{ m}$$

Then the work done in stretching the spring follows from Equation 5.8:

$$W = \tfrac{1}{2}kx^2 = \tfrac{1}{2}(20.4 \text{ N/m})(0.240 \text{ m})^2 = 0.588 \text{ J}$$

Add another 500-g mass and it stretches another 0.240 m, to $x_B = 0.480$ m. The additional work is

$$W = \tfrac{1}{2}kx_B^2 - \tfrac{1}{2}kx_A^2 = \tfrac{1}{2}(20.4 \text{ N/m})(0.480 \text{ m})^2 - 0.588 \text{ J} = 1.76 \text{ J}$$

where $\tfrac{1}{2}kx_A^2$ is the 0.588 J we just calculated. Note that the work required for the second 0.240-m stretch is greater than for the first one. That's because the already-stretched spring exerts a greater force.

Many springs have the same spring constant whether they're stretched or compressed (Figure 5.12). We'll take that to be a property of an "ideal spring," one that follows Hooke's law with the same force constant for stretch and compression. Hooke's law still works; with compression, x is negative and the force direction is reversed.

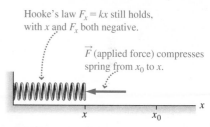

Hooke's law $F_x = kx$ still holds, with x and F_x both negative.

$\vec{F}$ (applied force) compresses spring from x_0 to x.

FIGURE 5.12 Compressing a spring.

✓ **TIP**

An "ideal spring" also has no mass. All real springs have mass, so the Hooke's law ideal is a good approximation for vertical springs only when the forces acting on the spring are much larger than its weight.

Springs and Newton's Third Law

Stretch a spring by pulling with your hand, and the spring pulls back with a force of equal magnitude and opposite direction. The force you apply and the spring force constitute a Newton's third-law pair. If the applied force is $F_{\text{applied},x} = kx$, then the spring force is $F_{\text{spring},x} = -kx$. The spring force is a **restoring force**, so called because it tends to restore the spring to equilibrium. The restoring force works equally well whether the spring is stretched or compressed. If you compress a spring by pushing to the left, as in Figure 5.12, the restoring force $-kx$ pushes back to the right.

What's the work done by the spring's restoring force? Because the restoring force opposes the applied force with equal magnitude, the work done by the restoring force is just the negative of the work done by the applied force. Thus, using Equation 5.8, the work done by the spring is $W = -\tfrac{1}{2}kx^2$. The work done *by* the spring is just the negative of the work done *on* the spring.

- The force applied to stretch a spring a distance x from equilibrium is $F_{applied,x} = kx$.
- The spring's restoring force is $F_{spring,x} = -kx$.
- The work done by an external force to stretch the spring from equilibrium is $W_{applied} = \frac{1}{2}kx^2$.
- The work done by the spring's restoring force is $W_{spring} = -\frac{1}{2}kx^2$.

GOT IT? Section 5.2 Rank in order the work done in stretching each of these springs from equilibrium through the displacement as shown.

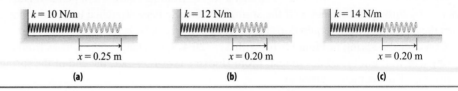

$k = 10$ N/m	$k = 12$ N/m	$k = 14$ N/m
$x = 0.25$ m	$x = 0.20$ m	$x = 0.20$ m
(a)	(b)	(c)

5.3 Kinetic Energy and the Work-Energy Theorem

Relating Net Work to Changing Speed

In Section 5.1 we made the connection between net work and changing speed. We'll now develop a relationship that makes that connection quantitatively useful.

Suppose a constant net force $F_{net,x}$ acts on an object undergoing displacement Δx. Equation 5.3 gives the net work done on the object:

$$W_{net} = F_{net,x}\,\Delta x$$

By Newton's second law, $F_{net,x} = ma_x$. Substituting this for the net force,

$$W_{net} = ma_x \Delta x$$

You've seen the product $a_x \Delta x$ before in kinematic Equation 2.10:

$$v_x^2 = v_{0x}^2 + 2a_x \Delta x$$

So $a_x \Delta x = \frac{1}{2}(v_x^2 - v_{0x}^2)$. Substituting this in our equation for net work,

$$W_{net} = \frac{1}{2}m(v_x^2 - v_{0x}^2)$$

For one-dimensional motion, the square of v_x is equal to the square of the speed v, so we can replace the velocity components v_x and v_{0x} with the speeds v and v_0:

$$W_{net} = \frac{1}{2}m(v^2 - v_0^2)$$

or

$$W_{net} = \frac{1}{2}mv^2 - \frac{1}{2}mv_0^2 \quad \text{(Work-energy theorem; SI unit: J)} \quad (5.9)$$

Equation 5.9 is the **work-energy theorem**. The quantity $\frac{1}{2}mv^2$ is the **kinetic energy (K)** of an object of mass m moving with speed v. Symbolically,

$$K = \frac{1}{2}mv^2 \quad \text{(Kinetic energy; SI unit: J)} \quad (5.10)$$

Kinetic energy is a scalar and has SI units of joules, just like work. Using our definition of kinetic energy, the work-energy theorem can be rewritten in terms of the initial and final kinetic energies, K_0 and K:

$$W_{net} = K - K_0 = \Delta K \quad \text{(Work-energy theorem revisited; SI unit: J)} \quad (5.11)$$

In words, the work-energy theorem states that *the net work done on an object equals its change in kinetic energy*. Equation 5.11 tells you that positive net work results in an increase in kinetic energy, and negative net work results in a decrease in kinetic energy.

The work-energy theorem holds in one, two, or three dimensions, even though our derivation was in one dimension. The work-energy theorem is a powerful tool that relates the net work done on an object to the object's changing speed. It provides an alternative to the detailed description of motion provided by Newton's second law, an alternative that makes solving some problems much easier. Upcoming examples show how this works.

Kinetic energy $K = \frac{1}{2}mv^2$ increases as the *square* of an object's speed. Double your driving speed, and your kinetic energy quadruples! Even a modest increase from 70 km/h to 100 km/h more than doubles your kinetic energy. This rapid increase in kinetic energy with speed is one reason excessive speed is dangerous. Stopping your car requires that the brakes do negative work, to rid the car of its kinetic energy. So even a modest speed increase makes it much harder on your brakes.

 TIP

Kinetic energy is the energy of motion. Any moving object has kinetic energy, which depends on its mass and the square of its speed. From its definition, you can see that kinetic energy is *always* positive. Some other forms of energy can be positive or negative.

CONCEPTUAL EXAMPLE 5.5 **Changing Kinetic Energy**

Give an example of a net force that does positive work on an object, and show that the resulting motion is consistent with the work-energy theorem. Repeat for a net force that does zero net work and for one that does negative net work.

SOLVE Positive net work: Drop a ball (Figure 5.13a), and the gravitational force provides the net work on the ball. Gravity and displacement have the same direction, so the net work is positive, and it increases as the ball falls. The ball's speed and therefore its kinetic energy also increase—as the work-energy theorem requires.

Zero net work: A good example is uniform circular motion (Figure 5.13b). You've seen that the net force—the centripetal force—is perpendicular to displacement, so it does no work. And the speed in uniform circular motion is constant, consistent with the work-energy theorem when there's no net work.

Negative net work (Figure 5.13c): A stopping car is a good example. The net work done by friction is negative. The change in kinetic energy is also negative, because speed decreases.

REFLECT None of these examples required constant acceleration. That's especially obvious with circular motion, where you know that

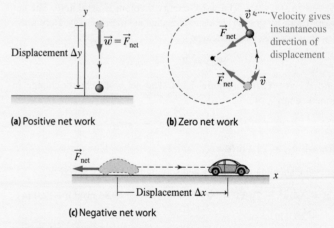

(a) Positive net work   (b) Zero net work

(c) Negative net work

FIGURE 5.13 Work done in three situations.

the acceleration continually changes direction. So the work-energy theorem frees you from the assumption of constant acceleration.

Using the Work-Energy Theorem

The work-energy theorem involves only initial and final speeds. It skips the details of how motion varies with time, often providing an easier solution than applying Newton's second law—at least when those details aren't needed. Using the work-energy theorem is also easier mathematically, because Newton's second law is a vector equation, while work and energy are scalars.

The Work-Energy Theorem

Some of the steps in applying the work-energy theorem are similar to those of other problem-solving strategies, but others—italicized below—are unique to work-energy problems.

ORGANIZE AND PLAN

- Visualize situation and make schematic diagram.
- Understand what forces are present (gravity, springs, friction, etc.).
- *Relate net force to work done.*
- *Set net work done equal to change in kinetic energy.*
- Review information you have; plan how to use it to solve for unknown(s).

SOLVE

- Gather information.
- *Combine and solve equation(s) resulting from work-energy theorem for unknown quantity.*
- Insert numerical values and solve.

REFLECT

- Check dimensions and units of answer. Are they reasonable?
- If problem relates to something familiar, assess whether answer makes sense.

EXAMPLE 5.6 **Cliff Launch**

You throw a ball at 23.4 m/s from a cliff. When you release the ball, it's 12.0 m above the ground below. Ignoring air resistance, what is the ball's speed when it hits the ground?

ORGANIZE AND PLAN The work-energy theorem is ideal here. The net work (done by gravity) leads to a change in kinetic energy. Here gravity does positive work, because

$$W_{net} = W_g = -mg\Delta y$$

and $\Delta y = -12.0$ m (Figure 5.14). The net work equals change in kinetic energy, or

$$-mg\Delta y = \tfrac{1}{2}mv^2 - \tfrac{1}{2}mv_0^2$$

Known: $v_0 = 23.4$ m/s.

SOLVE Canceling the masses and solving for the final speed v,

$$v = \sqrt{v_0^2 - 2g\Delta y} = \sqrt{(23.4 \text{ m/s})^2 - 2(9.80 \text{ m/s}^2)(-12.0 \text{ m})}$$
$$= 28.0 \text{ m/s}$$

REFLECT As expected, the ball moves faster at the ground than at launch. Note that our result is independent of launch angle.

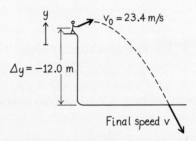

FIGURE 5.14 Gravity does work on a projectile.

MAKING THE CONNECTION The projectile's final *speed* is independent of launch angle, but is its final *velocity*?

ANSWER No. Throw the ball straight down and it hits the ground with its velocity vertically down. Toss it horizontally and it hits with the same horizontal velocity it started at, and thus with a reduced vertical component. In both cases its *speed*—a scalar—is the same, but its *velocity*—a vector—isn't.

CONCEPTUAL EXAMPLE 5.7 **The Drag Force in Projectile Flight**

A baseball is hit from level ground at an upward angle. Use work and energy to describe and compare its trajectories with and without the drag force of air resistance.

SOLVE Without drag, the flight is a parabola (Chapter 3), symmetric about the peak (Figure 5.15). Gravity does negative work on the way up, positive on the way down, for zero gravitational work. But the drag force opposes the ball's velocity, so it does negative work throughout the flight, decreasing both components of the ball's velocity. Therefore, the ball doesn't get as high as in the drag-free case, and the peak height occurs after a smaller horizontal displacement.

Air resistance continues to do negative work on the way down. When the ball reaches the ground, the net work done on it (gravity plus drag) is negative, so it's moving slower than at launch. The net result is that the ball doesn't go as high or as far.

REFLECT The work-energy theorem is a powerful tool for this kind of qualitative analysis. A more quantitative analysis requires details of the drag force.

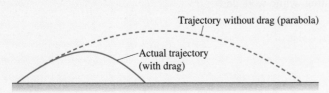

Trajectory without drag (parabola)

Actual trajectory (with drag)

FIGURE 5.15 Projectile motion with and without drag.

Thinking about Energy

"Energy" is another of those terms in everyday conversation, as in "I have a lot of energy today." But what does *energy* mean in physics?

Kinetic energy—the energy an object has by virtue of its motion—is one of many forms of energy you'll encounter. In Sections 5.4 and 5.5, we'll define potential energy and total mechanical energy. In later chapters, we'll examine thermal energy and electrical energy, as well as the energy associated with gravitational, electric, and magnetic fields. In relativity there is rest energy, the energy equivalent of a mass at rest.

Energy is central to physics, joining mass as a fundamental quantity throughout the universe. Think of energy as what makes things happen; without energy, there would be no motion, no activity, no change. One overriding principle is the **principle of conservation of energy**, which states that energy can change from one form to another, but the total amount of energy remains constant. You'll see in Section 5.5 how this important principle operates.

GOT IT? Section 5.3 A 0.20-kg ball is dropped from rest. Assume air resistance is negligible. After falling through a distance of 2.5 m, the ball's kinetic energy is (a) 2.5 J; (b) 4.9 J; (c) 7.7 J; (d) 12.3 J.

5.4 Potential Energy

Conservative and Nonconservative Forces

Throw a rock off a hill, targeting a tin can at the bottom. There are different trajectories for the rock, depending on launch velocity (Figure 5.16). However, the work done by gravity on the rock is the same for all of them, because it depends only on the vertical displacement Δy. By the same reasoning, you can convince yourself that the work done by gravity on any object moving between any two points is independent of the path taken. Forces for which the work is independent of path are **conservative** forces.

> **Conservative forces:** If the work done by a force on an object moving between two points does not depend on the path taken, the force is conservative.

Not all forces are conservative. In our rock example, the drag force of air also acts on the rock. The drag force depends on velocity, so the work it does won't be the same for different paths. Therefore, drag is a **nonconservative** force.

> **Nonconservative forces:** If the work done by a force on an object moving between two points depends on the path taken between those points, the force is nonconservative.

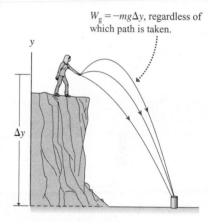

$W_g = -mg\Delta y$, regardless of which path is taken.

FIGURE 5.16 Work done by gravity is independent of path.

CONCEPTUAL EXAMPLE 5.8 **Conservative or Nonconservative?**

Categorize the following forces as conservative or nonconservative: (a) kinetic friction; (b) the force of a spring that obeys Hooke's law.

SOLVE (a) The work done by kinetic friction depends on the path (Figure 5.17a, next page). Slide a heavy box across the floor from A to B, and if friction is the same everywhere you'll do less work on the shortest path—a straight line. If you have to go around an obstacle, you'll do more work. The work is path dependent, so the force is nonconservative.

(b) Attach a mass to the spring, compress the spring, and let go (Figure 5.17b, next page). The mass oscillates back and forth, passing

cont'd.

:nergy

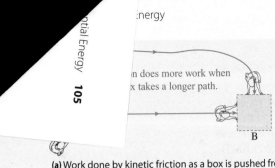

n does more work when
x takes a longer path.

B

(a) Work done by kinetic friction as a box is pushed from A to B along two routes

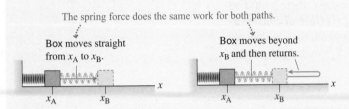

The spring force does the same work for both paths.

Box moves straight from x_A to x_B.

Box moves beyond x_B and then returns.

(b) Work done by a spring as a box moves from x_A to x_B along two routes

FIGURE 5.17 (a) Work done by kinetic friction; (b) Work done by a spring.

through the same point many times. From Section 5.2, the work done on the spring as it goes from position x_A to position x_B is

$$W_{\text{on spring}} = \tfrac{1}{2}kx_B^2 - \tfrac{1}{2}kx_A^2$$

As discussed in Section 5.2, the work done *by* the spring is the negative of the work done *on* the spring, or

$$W_{\text{by spring}} = -W_{\text{on spring}} = \tfrac{1}{2}kx_A^2 - \tfrac{1}{2}kx_B^2$$

There's no mention here of how many times the mass has oscillated back and forth—that is, of the path it's taken. So here the work is path independent, and this force is conservative.

REFLECT In part (b), the fact that the work is the same tells you that for a mass on a spring, absent friction, speed is the same at a given point whether it's going right or left. So the motion is symmetric, just as in free fall, when an object's speed at a given point is the same going up and down.

Defining Potential Energy

Throw a ball straight up. Its kinetic energy decreases as it rises, then returns as the ball falls—as if energy were stored and then returned. The stored energy is called **potential energy**. Potential energy (symbol U) is energy a system has due to the relative positions of objects—in this example, the ball relative to Earth.

Suppose a conservative force does work on an object. We define the resulting change in potential energy ΔU as the negative of the work done by that force. Symbolically,

$$\Delta U = -W_{\text{conservative}} \quad \text{(Definition of potential energy; SI unit: J)} \quad (5.12)$$

Gravitational Potential Energy

Gravity provides an example of potential energy. Toss a baseball upward, and its height changes by Δy (Figure 5.18). Then gravity does work $W_g = -mg\Delta y$ on the ball. So, by the definition in Equation 5.12, the ball's potential energy changes by

$$\Delta U = -W_g = mg\Delta y \quad \text{(Gravitational potential energy; SI unit: J)} \quad (5.13)$$

Like work and kinetic energy, potential energy is a scalar, with SI units of joules. For a standard 145-g baseball undergoing vertical displacement of 10.0 m, the change in potential energy is

$$\Delta U = mg\Delta y = (0.145\,\text{kg})(9.80\,\text{m/s}^2)(10.0\,\text{m}) = 14.2\,\text{J}$$

It's worth noting that the same baseball coming *down* through a height of 10.0 m has a change in potential energy equal to

$$\Delta U = mg\Delta y = (0.145\,\text{kg})(9.80\,\text{m/s}^2)(-10.0\,\text{m}) = -14.2\,\text{J}$$

so the overall change in potential energy when the ball returns to its starting height is $14.2\,\text{J} - 14.2\,\text{J} = 0$.

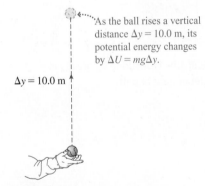

As the ball rises a vertical distance $\Delta y = 10.0$ m, its potential energy changes by $\Delta U = mg\Delta y$.

$\Delta y = 10.0$ m

FIGURE 5.18 The baseball's potential energy changes as a function of its height.

✓ **TIP**

Work and energy (in any form) are always scalar quantities.

Elastic Potential Energy

An ideal spring provides another example of a conservative force and its associated potential energy. Figure 5.19a shows a spring with $k = 55$ N/m extended from its equilibrium position x_A a distance of 0.10 m to x_B. Section 5.2 showed that the work done by a spring as it's stretched from x_A to x_B is

$$W_{by\ spring} = \tfrac{1}{2}kx_A^2 - \tfrac{1}{2}kx_B^2$$

The corresponding change in the spring's potential energy is

$$\Delta U = -W_{net} = -W_{by\ spring} = \tfrac{1}{2}kx_B^2 - \tfrac{1}{2}kx_A^2 \quad \text{(Elastic potential energy; SI unit: J)} \quad (5.14)$$

Equation 5.14 shows that the potential energy of a spring increases as it's stretched or compressed (Figure 5.19a) and decreases as it's relaxed toward its equilibrium configuration ($x = 0$, Figure 5.19b). For the same final situation, the potential-energy decrease balances the increase, resulting in zero overall change—just as in the gravitational case when the ball returned to its starting height.

The Zero of Potential Energy

Equation 5.12 defines potential energy in terms of a *change* and not as an *absolute* amount. (In contrast, kinetic energy $\tfrac{1}{2}mv^2$ is an unambiguous quantity that's always positive.) It's possible to define potential energy as a function of position, provided you first assign a position where the potential energy is zero. That zero point is arbitrary, but once you've assigned it, all other potential energy values are defined as changes from that zero.

For example, in problems involving gravity, you may want to assign the ground ($y = 0$) to be the place where potential energy is zero. Then with $\Delta U = mg\Delta y$ for gravity, the potential energy at any height y is $U = mg(y - 0)$, or

$$U = mgy \quad \text{(Gravitational potential energy; SI unit: J)} \quad (5.15)$$

With the assignment of $U = 0$ at $y = 0$, an 18.8-kg concrete block 12.5 m above the ground has potential energy

$$U = mgy = (18.8\ \text{kg})(9.80\ \text{m/s}^2)(12.5\ \text{m}) = 2300\ \text{J}$$

Note that Equation 5.15 is only valid near Earth's surface, where g is essentially constant. In Chapter 9 you'll deal more generally with gravitational potential energy.

In problems involving springs, it's best to assign $U = 0$ at $x = 0$, the spring's equilibrium position. Then at any other position x,

$$U = \tfrac{1}{2}kx^2 - \tfrac{1}{2}k(0)^2$$

or

$$U = \tfrac{1}{2}kx^2 \quad \text{(Potential energy for a spring; SI unit: J)} \quad (5.16)$$

For example, a spring with $k = 1250$ N/m stretched 0.15 m from equilibrium has potential energy

$$U = \tfrac{1}{2}kx^2 = \tfrac{1}{2}(1250\ \text{N/m})(0.15\ \text{m})^2 = 14\ \text{J}$$

We stress that the assignment of a zero point for potential energy is truly arbitrary. For a given problem, you might want to assign a different zero point (say above the ground if that's the launch point of a projectile). You've grown accustomed to hard and fast definitions in physics, so maybe this seems like too much freedom! But it's ultimately *changes* in potential energy that are related to work, and thus to motion. Changes in potential energy are independent of where you choose the zero point. You'll see this repeatedly in the examples of the next section.

① When spring is extended from x_A to x_B,
$$\Delta U = \tfrac{1}{2}kx_B^2 - \tfrac{1}{2}kx_A^2$$
$$= \tfrac{1}{2}(55\ \text{N/m})(0.10\ \text{m})^2 - \tfrac{1}{2}(55\ \text{N/m})(0\ \text{m})^2$$
$$= 0.275\ \text{J}$$

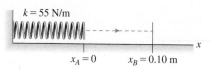

$k = 55$ N/m

$x_A = 0$ $x_B = 0.10$ m

(a)

② When spring relaxes from x_B to x_A,
$$\Delta U = \tfrac{1}{2}kx_A^2 - \tfrac{1}{2}kx_B^2$$
$$= -0.275\ \text{J}$$

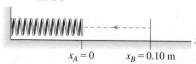

$x_A = 0$ $x_B = 0.10$ m

③ The net change for the round trip is
$$\Delta U = 0.275\ \text{J} + (-0.275\ \text{J}) = 0$$

(b)

FIGURE 5.19 (a) Potential energy increases when the spring is stretched. (b) Potential energy decreases when the spring returns to equilibrium.

Potential Energy Functions for Conservative Forces

Potential energy depends only on position, so for the concept to be meaningful, it must be the case that the potential energy difference between your zero point and any other point is independent of the path taken. That's true only for conservative forces, so it's only for conservative forces that we can define potential energy. Physically, that's because conservative forces store the work done against them as potential energy and can give it back as kinetic energy. Nonconservative forces dissipate energy into random thermal motions, and that energy becomes unavailable.

GOT IT? Section 5.4 The four balls shown are identical. Rank them in order of decreasing potential energy.

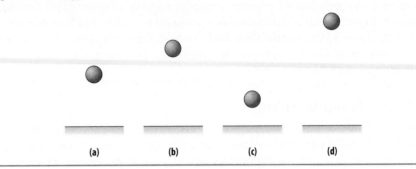

    (a)   **(b)**   **(c)**   **(d)**

5.5 Conservation of Mechanical Energy

So far we've related changes in both kinetic and potential energy to work. Now we'll use those relationships to connect kinetic and potential energy directly.

From Section 5.3 (Equation 5.11), the work-energy theorem shows how net work results in a change in kinetic energy:

$$W_{net} = K - K_0 = \Delta K$$

From Section 5.4 (Equation 5.12), the definition of potential energy also relates net work done by conservative forces to change in potential energy:

$$\Delta U = -W_{net}$$

Combining these relationships,

$$W_{net} = \Delta K = -\Delta U$$

The second equality $\Delta K = -\Delta U$ can be rearranged:

$$\Delta K + \Delta U = 0 \quad \text{(Kinetic and potential energy changes; SI unit: J)} \quad (5.17)$$

What does this equation say? *The sum of the changes in kinetic energy and potential energy is zero for an object subject to only conservative forces.* Put another way, *the sum of the kinetic and potential energy is constant.* This follows directly from Equation 5.17, because any change in kinetic energy must be offset by an opposite change in potential energy. The sum of kinetic and potential energy is the **total mechanical energy**, E. Thus,

$$E = K + U = \text{constant} \quad \text{(Total mechanical energy; SI unit: J)} \quad (5.18)$$

This statement is known as the principle of **conservation of mechanical energy**. It's a powerful tool for solving motion problems whenever we're dealing with conservative forces for which a potential energy function is known. The following examples should help you appreciate this important idea.

APPLICATION **Pole Vault**

Pole vaulters undergo several energy conversions. The initial run gives the vaulter kinetic energy. The vaulter plants the pole, transforming kinetic energy into potential energy of the deformed pole. Then the pole straightens and lifts the vaulter over the bar, transforming its elastic potential energy into gravitational potential energy. The athlete then falls toward the pit, exchanging gravitational potential energy for kinetic energy. Finally, kinetic energy is dissipated in deforming the landing cushion.

PROBLEM-SOLVING STRATEGY 5.2 **Conservation of Mechanical Energy**

Here's a strategy for problems involving conservation of mechanical energy. We've italicized the steps that differ from earlier strategies.

ORGANIZE AND PLAN

- Visualize situation and make schematic diagram.
- Understand what forces are present (gravity, a spring, friction, etc.). *Be sure they're conservative forces.*
- *Gather information on kinetic energy and potential energy of all objects in system.*
- *Equate total mechanical energy (kinetic plus potential) at two different points in the motion.*

SOLVE

- *Combine and solve equation(s) expressing conservation of energy for unknown quantity.*
- Insert numerical values and solve.

REFLECT

- Check dimensions and units of answer. Are they reasonable?
- If problem relates to something familiar, assess whether answer makes sense.

EXAMPLE 5.9 **Roller Coaster**

A roller coaster car starts from rest at the top of the track, 20.0 m above the ground. Ignoring friction, how fast is the car moving (a) at a height of 10.0 m and (b) when it reaches ground level?

ORGANIZE AND PLAN Gravity is a conservative force, and we're neglecting friction, so the car's total mechanical energy E is conserved. We can find E at the top of the track, and the potential energy U at any height. Conservation of energy then gives the kinetic energy K, and hence the speed of the car.

The car starts from rest at the top of the track, so here $K = 0$ (Figure 5.20). The gravitational potential energy (Equation 5.15) is $U = mgy$, where we've chosen to measure y from the ground. Starting at $h = 20.0$ m, the car's total mechanical energy is

$$E = K + U = 0 + mgy = mgh$$

At any other height y, the car's kinetic energy is $K = \frac{1}{2}mv^2$, and the total energy is

$$E = K + U = \frac{1}{2}mv^2 + mgy$$

Because total mechanical energy is conserved, E is always mgh, as it was at the top. Thus

$$mgh = \frac{1}{2}mv^2 + mgy$$

Known: Starting height $h = 20.0$ m.

SOLVE Solving for the speed v gives

$$v^2 = 2g(h - y)$$

Evaluating at $y = 10$ m and at ground level,

(a) $v_{10\,m} = \sqrt{2g(h - y)}$
$= \sqrt{2(9.80 \text{ m/s}^2)(20.0 \text{ m} - 10.0 \text{ m})} = 14.0$ m/s

(b) At ground level ($y = 0$), the speed is

$v_{\text{ground}} = \sqrt{2g(h - y)}$
$= \sqrt{2(9.80 \text{ m/s}^2)(20.0 \text{ m} - 0.0 \text{ m})} = 19.8$ m/s

REFLECT The varying slope of the track in Figure 5.20 means that the car's acceleration along the track is anything but constant—so you couldn't have solved this problem with the kinematic equations for constant acceleration. But note how the powerful conservation-of-energy principle cuts through all those details, allowing easy calculation of the speed at any point on the track.

MAKING THE CONNECTION If a later section of track goes back up to a height of 20.0 m, how fast would the car be moving there?

ANSWER At that height, just as at the start, potential energy equals total mechanical energy. Thus, kinetic energy is zero, and so is speed. In reality, friction "robs" the car of energy, and it wouldn't make it back to its starting height.

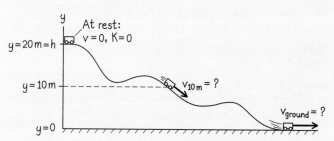

FIGURE 5.20 Conservation of energy for a roller coaster car.

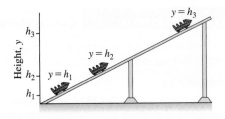
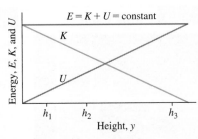

FIGURE 5.21 In a conservative system, total mechanical energy E is constant.

The roller coaster provides a good example of mechanical energy conservation. The coaster goes up and down many times. Whenever it goes up, it gains potential energy and loses kinetic energy. When it goes down, it gains kinetic energy and loses potential energy. Figure 5.21 shows a graph of the roller coaster's kinetic and potential energy as functions of height. At any height, the sum of kinetic and potential energies is constant, illustrating conservation of total mechanical energy.

EXAMPLE 5.10 **Tranquilizing a Rhinoceros**

A biologist sits in a tree 5.64 m above the ground, using a spring-loaded gun to shoot tranquilizer darts into a rhinoceros. The spring is compressed with the dart resting against it. Pulling the trigger releases the spring and launches the dart. Given spring constant $k = 740$ N/m, and spring compression $d = 12.5$ cm, find the speed of the 38.0-gram dart (a) as it leaves the gun, and (b) when it strikes the rhinoceros, at 1.31 m above the ground.

ORGANIZE AND PLAN (a) Both the spring force and gravity are conservative, so conservation of mechanical energy applies. The spring's potential energy gets converted into the dart's kinetic energy. Thus

$$\tfrac{1}{2}kd^2 = \tfrac{1}{2}mv_0^2$$

where v_0 is the dart's speed leaving the gun. You can solve this equation for the unknown v_0.

(b) After launch, you can apply conservation of total mechanical energy to find the final kinetic energy and speed. At launch, the dart has speed v_0 at height $y_0 = 5.64$ m (Figure 5.22). When the dart reaches the rhino at height $y = 1.31$ m, its speed is v. Equating the initial and final energy,

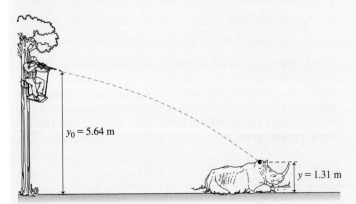

$y_0 = 5.64$ m

$y = 1.31$ m

FIGURE 5.22 Tranquilizing a rhinoceros.

$$E = K + U = \tfrac{1}{2}mv_0^2 + mgy_0 = \tfrac{1}{2}mv^2 + mgy$$

With all other quantities known, you can solve for the dart's final speed v.

Known: $k = 740$ N/m; $m = 0.0380$ kg; $d = 0.125$ m; $y_0 = 5.64$ m; $y = 1.31$ m.

SOLVE (a) Solving $\tfrac{1}{2}kd^2 = \tfrac{1}{2}mv_0^2$ for v_0,

$$v_0 = \sqrt{\frac{kd^2}{m}} = \sqrt{\frac{(740 \text{ N/m})(0.125 \text{ m})^2}{0.0380 \text{ kg}}} = 17.4 \text{ m/s}$$

(b) Note that in the expression $\tfrac{1}{2}mv_0^2 + mgy_0 = \tfrac{1}{2}mv^2 + mgy$, the mass m cancels, leaving

$$\tfrac{1}{2}v_0^2 + gy_0 = \tfrac{1}{2}v^2 + gy$$

Solving for the final speed v,

$$v = \sqrt{v_0^2 + 2g(y_0 - y)}$$
$$= \sqrt{(17.4 \text{ m/s})^2 + 2(9.8 \text{ m/s}^2)(5.64 \text{ m} - 1.31 \text{ m})}$$

or $v = 19.7$ m/s.

REFLECT The speed doesn't increase a whole lot as the dart drops, because the spring's potential energy is considerably greater than the change in gravitational potential energy. Note that the launch angle doesn't matter here, although strictly speaking a nonhorizontal launch would require you to consider the change in gravitational energy as the spring decompresses—a quantity that would be negligible in this case.

MAKING THE CONNECTION What's the total mechanical energy of this system?

ANSWER It can be found at any stage—as the spring's initial energy, or as the dart's total mechanical energy at launch or when it hits the rhinoceros. Try doing it all three ways. The answer is about 5.78 J.

EXAMPLE 5.11 Golf Ball Trajectory

A golfer hits a 45.7-gram ball with her 9-iron, launching it at $v_0 = 30.9$ m/s and $\theta = 42°$. Take the zero of potential energy at the ground, and ignore air resistance. Find (a) the ball's total mechanical energy, (b) its kinetic energy at the peak of its flight, and (c) its maximum height.

ORGANIZE AND PLAN Since we're ignoring air resistance, gravity is the only force acting after the ball leaves the club, so total mechanical energy is conserved throughout the flight. At the top the ball is moving horizontally, so its speed is $v = v_x$ at that point (Figure 5.23). This fact can be used to find the kinetic energy and height.

Total mechanical energy is $E = K + U$, and gravitational potential energy is $U = mgy$. The x-component of velocity (from the launch conditions) is $v_x = v_0 \cos \theta$, which is also its speed at the peak. From this you can get kinetic energy at the peak. The potential energy is then $U = E - K$, and since $U = mgh$, you can solve for h.

Known: $m = 0.0457$ kg; $v_0 = 30.9$ m/s; $\theta = 42°$.

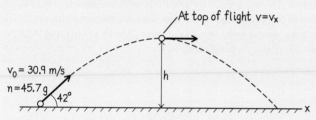

FIGURE 5.23 Conservation of energy for a golf ball.

SOLVE (a) Using launch conditions, total mechanical energy is

$$E = K + U = \tfrac{1}{2} mv^2 + mgy$$
$$= \tfrac{1}{2}(0.0457 \text{ kg})(30.9 \text{ m/s})^2 + (0.0457 \text{ kg})(9.80 \text{ m/s}^2)(0.0 \text{ m})$$
$$= 21.8 \text{ J}$$

(b) At the peak $(y = h)$, total mechanical energy is still 21.8 J, because it's conserved. The ball's speed is

$$v = v_x = v_0 \cos(42°) = (30.9 \text{ m/s})(\cos 42°) = 23.0 \text{ m/s}$$

Therefore, the kinetic energy at the peak is

$$K = \tfrac{1}{2} mv^2 = \tfrac{1}{2}(0.0457 \text{ kg})(23.0 \text{ m/s})^2 = 12.1 \text{ J}$$

(c) Potential energy is $U = E - K = mgh$. Solving for h,

$$h = \frac{E - K}{mg} = \frac{21.8 \text{ J} - 12.1 \text{ J}}{(0.0457 \text{ kg})(9.80 \text{ m/s}^2)} = 21.7 \text{ m}$$

REFLECT You could have answered (c) using kinematics. The principle of conservation of total mechanical energy provides an alternative.

MAKING THE CONNECTION Suppose, more realistically, that the ball loses 15% of its mechanical energy to drag over its entire flight. What's its speed when it hits the ground?

ANSWER The total mechanical energy is now $(0.85)(21.8 \text{ J}) = 18.5 \text{ J}$. This is all kinetic energy when the ball hits the ground, because $U = 0$ there. Using $E = 18.5 \text{ J} = K = \tfrac{1}{2}mv^2$ gives $v = 28.5$ m/s, well below the launch speed.

Nonconservative Forces

In Section 5.4 we stressed that potential energy can only be defined for conservative forces. Here we've applied the concept of potential energy in problems using conservation of mechanical energy $E = K + U$. You might think potential energy useless in any problem involving nonconservative forces. Fortunately, that's not the case, as we'll illustrate with an example.

Consider the golf ball in Example 5.11. In the "Making the Connection," we included the nonconservative drag force. This force does negative work on the ball, lowering its kinetic energy without increasing potential energy. So *frictional or drag forces reduce the total mechanical energy* of a system. We'll state this result in the form of an equation:

$$E_{\text{final}} = E_{\text{initial}} + W_f \qquad \text{(Energy and non-conservative forces; SI unit: J)} \quad (5.19)$$

where W_f is the work done by frictional or drag forces. Because W_f is *negative*, E_{final} is *less* than E_{initial} in the presence of drag or friction.

For the golf ball, potential energy was the same at the beginning and end of its flight. But you can apply Equation 5.19 to problems that involve changing potential energy. That's because the term W_f in Equation 5.19 depends only on the work done by nonconservative forces, while the work done by conservative forces is accounted for by changes in potential energy. The next example illustrates this.

EXAMPLE 5.12 **Friction and Skiing**

A 65.0-kg downhill skier starts from rest at the top of a slope that drops 120 m vertically. At the bottom, she's moving at 32.5 m/s. Find the work done by frictional forces.

ORGANIZE AND PLAN This problem involves the change in mechanical energy due to frictional forces (Equation 5.19). Total mechanical energy is still given by $E = K + U$. The skier starts from rest, so her initial kinetic energy is zero. You can set the bottom of the slope at $y = 0$ and the top at $y = h = 120$ m (Figure 5.24) to find the potential energy at the top and bottom.

Known: $m = 65.0$ kg; $h = 120$ m; final speed $v = 32.5$ m/s.

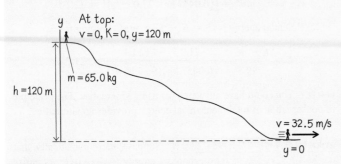

FIGURE 5.24 Energy losses in skiing.

SOLVE Initial energy is all potential:

$$E_{initial} = K + U = 0 + mgh = mgh$$

Final energy is all kinetic:

$$E_{final} = K + U = \tfrac{1}{2}mv^2 + 0 = \tfrac{1}{2}mv^2$$

Using these energies in Equation 5.19 gives

$$W_f = E_{final} - E_{initial} = \tfrac{1}{2}mv^2 - mgh$$
$$= \tfrac{1}{2}(65.0 \text{ kg})(32.5 \text{ m/s})^2 - (65.0 \text{ kg})(9.8 \text{ m/s}^2)(120 \text{ m})$$
$$= -42.1 \text{ kJ}$$

REFLECT As expected, the work done by friction is negative. Here "friction" includes both surface friction and air resistance, and the computation doesn't reveal how much each contributes.

MAKING THE CONNECTION Compare the work done by friction with the work done by gravity in this example.

ANSWER The work done by gravity is $mgh = 76.4$ kJ. The absolute value of the work done by friction is a little more than half this. The net work, which is the sum of the work done by gravity and the work done by friction, is positive, consistent with the work-energy theorem.

APPLICATION **Shock Absorbers**

A car's shock absorber system puts nonconservative forces to good use. Shock absorbers use springs that help change kinetic energy imparted by road bumps in the road into elastic potential energy. In addition, the spring or a piston inside a separate cylinder is immersed in heavy oil, which dissipates the system's energy as the spring relaxes. The result is a smoother ride.

Reviewing New Concepts

- Kinetic energy $K = \tfrac{1}{2}mv^2$.
- The work-energy theorem states that the net work done on an object equals the change in that object's kinetic energy.
- Total mechanical energy is $E = K + U$.
- When only conservative forces are present, $E = K + U$ is constant.
- When frictional or drag forces are present, $E_{final} = E_{initial} + W_f$.
- The work done by nonconservative forces is always negative.

GOT IT? Section 5.5 Rank the speed of the roller coaster car, in order of increasing speed, at the four positions shown. Neglect friction.

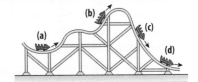

5.6 Power

A sprinter expends a lot of energy in a short time. A long-distance runner sustains a more modest energy output over a long time. Your car's engine does work at a greater rate climbing a hill than cruising on level ground. In all these cases we're talking not about the total *amount* of work or energy, but about the *rate* at which energy is expended or work done. That rate is called **power** (symbol P), and it's defined as

$$\text{Power } P = \frac{\text{work}}{\text{time}} = \frac{\text{energy delivered}}{\text{time}} \qquad \text{(Definition of power; SI unit: W)} \quad (5.20)$$

The alternate expressions in Equation 5.20, involving work or energy, follow from the close relation between work and energy, as expressed in the work-energy theorem and conservation of mechanical energy.

Whether it's work or energy being considered, the SI units for power are J/s, which defines the **watt** (W):

$$1 \text{ W} = 1 \text{ J/s}$$

The watt is named in honor of Scottish engineer and inventor James Watt (1736–1819), whose work with engines advanced the understanding of mechanical work and energy. Watt himself defined the **horsepower** (hp), estimated to be a typical sustained work rate for a horse. Although we use SI units consistently in physics, you'll see power of automobiles and other machines listed in horsepower. The conversion is 1 hp = 745.7 W.

For example, suppose a construction crane lifts a 13,200-kg steel girder 35.0 m straight up to frame a building. The crane lifts with constant speed for 14.7 s. Lifting with constant speed requires an upward pull equal to the girder's weight mg. With the direction of pull and displacement ($\Delta y = 35.0$ m) in the same direction, the work done is $W = F_y \Delta y = mg \Delta y$, and the power required is

$$P = \frac{\text{work}}{\text{time}} = \frac{mg \Delta y}{t} = \frac{(13{,}200 \text{ kg})(9.80 \text{ m/s}^2)(35.0 \text{ m})}{14.7 \text{ s}} = 3.08 \times 10^5 \text{ J/s} = 308 \text{ kW}$$

Note that the dimensions of work or energy are power multiplied by time. (In SI, 1 J = 1 W·s.) You pay for electrical energy by the kilowatt-hour (kWh), which is energy consumed at a rate of 1 kW (1000 W) for one hour; typical cost is around 10¢/kWh, but there's considerable variation with geographical region, season, and even time of day. Although primarily used for electrical energy, the kWh is a perfectly good (though non-SI) unit for any form of energy. Given that 1 h = 3600 s, the conversion between kWh and J is

$$1 \text{ kWh} = (1000 \text{ W})(3600 \text{ s}) = 3.6 \times 10^6 \text{ W·s} = 3.6 \times 10^6 \text{ J}$$

Average Power and Instantaneous Power

You saw in Section 5.1 that the work done by a constant force as an object moves in one dimension is $W = F_x \Delta x$. If this work is done over a time interval Δt, then the power is

$$P = \frac{\text{work}}{\text{time}} = \frac{F_x \Delta x}{\Delta t} \tag{5.21}$$

There are two ways to interpret this expression. The quantity $\Delta x / \Delta t$ is the average velocity $\bar{v}_x$. Then the corresponding quantity on the left side of Equation 5.21 is the average power $\bar{P}$. That is,

$$\bar{P} = F_x \bar{v}_x \quad \text{(Average power; SI unit: W)} \tag{5.22}$$

If the force varies with time, so does the power. Then we need the **instantaneous power**. Like other instantaneous quantities, this comes from taking a limit as the time interval Δt approaches zero. From kinematics, the instantaneous velocity is

$$v_x = \lim_{\Delta t \to 0} \frac{\Delta x}{\Delta t}$$

so the instantaneous power is

$$P = F_x v_x \quad \text{(Instantaneous power; SI unit: W)} \tag{5.23}$$

Knowing the power in terms of velocity can simplify problem solving. In the example with the construction crane, the girder was lifted at a constant velocity

$$v_y = \frac{\Delta y}{\Delta t} = \frac{35.0 \text{ m}}{14.7 \text{ s}} = 2.38 \text{ m/s}$$

Equation 5.23 (using y for the vertical motion) gives the instantaneous power:

$$P = F_y v_y = (13{,}200 \text{ kg})(9.80 \text{ m/s}^2)(2.38 \text{ m/s}) = 3.08 \times 10^5 \text{ W}$$

That's the same answer we found in the example.

EXAMPLE 5.13   **Hilly San Francisco**

Near the corner of Filbert and Leavenworth Streets in San Francisco, the slope on Filbert is about 17°. An 1120-kg car climbs this hill at a constant 50 km/h (13.9 m/s) while working against a combined friction and drag force of 890 N. Find the power required under these conditions.

ORGANIZE AND PLAN Figure 5.25 shows the forces acting on the car. The drive wheels supply the applied force that propels the car upslope (which we take as the +x-direction). The four forces sum to zero,

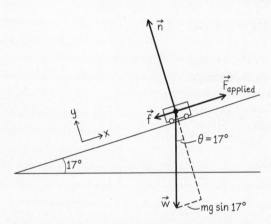

FIGURE 5.25 Force diagram for the car on a steep hill.

giving the car its constant velocity. Then the sum of the x-components of all the forces is

$$F_{\text{applied}} - mg \sin\theta - f = 0$$

where the x-components of gravity and friction are negative because they point down the slope. Solving for the applied force gives $F_{\text{applied}} = mg \sin\theta + f$. Then the power is $P = F_{\text{applied}} v_x$.

Known: $m = 1120$ kg; $v = 13.9$ m/s; $f = 890$ N.

SOLVE Compute the applied force:

$$\begin{aligned}F_{\text{applied}} &= mg \sin(17°) + f \\ &= (1120 \text{ kg})(9.80 \text{ m/s}^2)\sin(17°) + 890 \text{ N} = 4.1 \text{ kN}\end{aligned}$$

Thus the required power is

$$P = F_{\text{applied}} v_x = (4.1 \text{ kN})(13.9 \text{ m/s}) = 57 \text{ kW}$$

REFLECT Here most of the power is required to overcome gravity, much less to overcome friction. Drag becomes increasingly important at highway speeds, where a larger power output is needed just to maintain steady speed.

MAKING THE CONNECTION   If this car is rated at 150 hp maximum, what fraction is being used?

ANSWER 57 kW = 76 hp, or just about half the available power.

EXAMPLE 5.14   **High-Energy Society**

The annual energy consumption of entire nations is often expressed in quads (Q, for quadrillion British thermal units (Btu), where 1 Q = 10^{15} Btu, with 1 Btu = 1054 J). The United States's annual energy consumption is approximately 100 Q, roughly one-fourth of the world total. Given the U.S. population of about 300 million, find the per capita energy consumption rate in watts.

ORGANIZE AND PLAN We'll convert the annual energy consumption to joules, then divide by the number of seconds in a year to get watts, and then divide by the population.

Known: annual energy consumption = 100 Q = 100×10^{15} Btu, 1 Btu = 1054 J.

SOLVE The U.S. total energy consumption rate is

$$P = \frac{\text{energy}}{\text{time}} = \frac{(100 \text{ Q})(10^{15} \text{ Btu/Q})(1054 \text{ J/Btu})}{(365.25 \text{ d/y})(24 \text{ h/d})(3600 \text{ s/h})}$$

$$= 3.3 \times 10^{12} \text{ J/s} = 3.3 \times 10^{12} \text{ W}$$

Then the per capita energy consumption rate is

$$\frac{\text{power}}{\text{population}} = \frac{3.3 \times 10^{12} \text{ W}}{300 \times 10^6 \text{ persons}}$$

$$= 11 \times 10^3 \text{ W/person} = 11 \text{ kW/person}$$

REFLECT Our answer is more than 100 times the average human body's power output of 100 W or 0.1 kW. That's what it means to live in a high-energy society!

MAKING THE CONNECTION   Do we put all that energy to good use? You'll see in Chapter 14 that physics imposes definite limitations on the efficiency with which we can convert some forms of energy. Partly for that reason, but also because of avoidable inefficiencies, more than half our energy consumption is wasted.

GOT IT? Section 5.6   In these diagrams, the same car takes the times indicated to travel up the four hills shown. Each car drives at constant speed, not necessarily the same speed as the others. Neglecting friction, rank in increasing order of the power required.

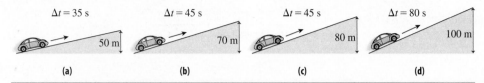

Chapter 5 in Context

This chapter has introduced the concepts of *work* and *energy* and shown how they're related through the *work-energy theorem*. *Kinetic energy* is energy of motion, and *potential energy* is stored energy. Their sum, *total mechanical energy*, is constant when only *conservative forces* act. A system's mechanical energy decreases when *nonconservative forces are present*. Using all these principles, we've addressed problems involving force and motion that go beyond what you can readily solve with Newton's laws alone.

Looking Ahead The kinematics, dynamics, and energy concepts you've studied to this point apply to single particles, or to objects that can be treated as such. In the chapters that follow we'll expand energy concepts you've learned here to multiparticle systems.

CHAPTER 5 SUMMARY

Work Done by a Constant Force

(Section 5.1) The **work** done on an object depends on the applied forces and the object's displacement.

Work done by a constant force in one dimension: $W = F_x \Delta x$
in two dimensions: $W = F_x \Delta x + F_y \Delta y$

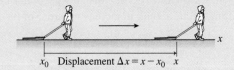

x_0 Displacement $\Delta x = x - x_0$ x

Work Done by a Variable Force

(Section 5.2) The work done by a variable force in one dimension is the area under the force-versus-position graph.

Hooke's law for a spring: $F_x = kx$

Work done on a spring: $W = \frac{1}{2}kx^2$

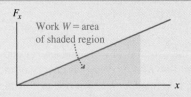

Work W = area of shaded region

Kinetic Energy and the Work-Energy Theorem

(Section 5.3) An object's **kinetic energy (K)** depends on the object's mass m and the speed v.

The **work-energy theorem** states that the net work done on an object equals its change in kinetic energy.

Kinetic energy: $K = \frac{1}{2}mv^2$

Work-energy theorem: $W_{net} = \Delta K$

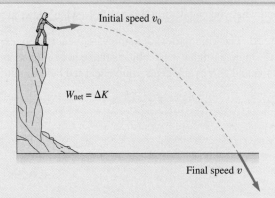

Initial speed v_0

$W_{net} = \Delta K$

Final speed v

Potential Energy

(Section 5.4) **Potential energy (U)** is stored energy in a system due to the relative positions of objects in the system.

Potential energy defined: $\Delta U = -W_{net}$

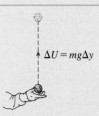

$\Delta U = mg\Delta y$

Conservation of Mechanical Energy

(Section 5.5) The sum of kinetic and potential energy is called the **total mechanical energy, E.**

Conservation of mechanical energy states that the total mechanical energy is constant for an object subject to conservative forces.

Total mechanical energy: $E = K + U$ = constant (for conservative forces)

Mechanical energy and nonconservative forces:
$E_{final} = E_{initial} + W_f$, where W_f is the work done by frictional forces or drag.

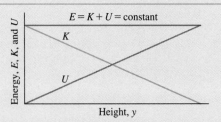

$E = K + U$ = constant

K

U

Energy, E, K, and U

Height, y

Power

(Section 5.6) **Power** is work per unit time.

Power: $P = \dfrac{\text{work}}{\text{time}} = \dfrac{\text{energy delivered}}{\text{time}}$

NOTE: Problem difficulty is labeled as ■ straightforward to ■ ■ ■ challenging. Problems labeled BIO are of biological or medical interest.

Conceptual Questions

1. How much total work does gravity do on you as you climb a mountain and descend to your starting point? Given this result, why do you feel so tired when you return from such a hike?
2. A car rounds a circular curve while its speed decreases. Is the net work done on the car positive, zero, or negative? Explain.
3. Give an example of how the word "work" used in casual conversation differs from its meaning in physics.
4. A factory worker pushes hard against a heavy toolbox to keep it at rest on a ramp. Is he doing work?
5. A nonzero net force is applied to an object, but its kinetic energy doesn't change. Explain why the force must be perpendicular to the object's velocity.
6. If an object's speed triples, by what factor does its kinetic energy increase?
7. You drag a box 2 m across the floor at constant speed. Next, you drag the same box 2 m across the same floor, giving it constant acceleration. Compare the work done by kinetic friction in the two cases.
8. Lift a hammer a fixed distance at constant velocity. Next, lift the same hammer the same distance with a constant upward acceleration. Compare the work you do in the two cases.
9. Is rolling friction a conservative or nonconservative force?
10. You launch two projectiles off a cliff at the same speed, one 30° above the horizontal, the other 30° below. Ignoring air resistance, compare their speeds when they hit the ground. Repeat, accounting for air resistance.
11. Discuss whether each of these quantities can ever be negative: (a) kinetic energy; (b) gravitational potential energy; (c) potential energy of a spring; (d) total mechanical energy; (e) work done by the air on a projectile.
12. Hiking trails on steep hillsides often follow zigzag paths ("switchbacks"). Use energy and power to explain their usefulness.
13. Describe the energy transformations throughout a pole vault, from the initial run until the athlete has come to rest on the cushion beneath the bar.

Multiple-Choice Problems

14. A book moves 2.15 m in the $+x$-direction under the influence of a 45.0-N force, also in the $+x$-direction. The work done on the book is (a) 20.9 J; (b) 45.0 J; (c) 48.4 J; (d) 96.8 J.
15. The work done by gravity on a 0.50-kg projectile that falls from $y = 12.5$ m to $y = 1.5$ m is (a) 5.5 J; (b) 27 J; (c) 54 J; (d) 81 J.
16. A 0.168-kg hockey puck slides at 11.4 m/s. The work needed to stop the puck is (a) -21.1 J (b) -12.4 J; (c) -10.9 J; (d) -8.3 J.
17. In a "dead lift," a weight lifter grabs a 185-kg barbell and lifts it 0.550 m from the floor. If the barbell started and ended at rest, how much work did the weight lifter do? (a) 997 J; (b) 498 J; (c) 249 J; (d) 102 J.
18. A Hooke's law spring with $k = 135$ N/m is compressed 9.50 cm from equilibrium. The work required to do this is (a) 12.8 J; (b) 1.22 J; (c) 0.61 J; (d) 0.35 J.
19. A Hooke's law spring has $k = 500$ N/m. The work done in extending the spring from $x = 0.30$ m to $x = 0.40$ m is (a) 17.5 J; (b) 20.0 J; (c) 25.0 J; (d) 40.0 J.

20. A 24.5-kg boulder that falls from a 13.4-m cliff strikes the ground with kinetic energy (a) 3220 J; (b) 1610 J; (c) 1450 J; (d) 328 J.
21. At one moment an electron is moving right with speed v and kinetic energy K. Later, the same electron is moving left with a speed $2v$. Now what's its kinetic energy? (a) $2K$; (b) $-2K$; (c) $4K$; (d) $-4K$.
22. A 2.15-kg rock has kinetic energy 346 J. After you do -211 J of work on the rock, its speed is (a) 11.2 m/s; (b) 17.9 m/s; (c) 22.8 m/s; (d) 322 m/s.
23. What's the change in potential energy of a 70-kg mountaineer going from sea level to the 8850-m summit of Mt. Everest? (a) 8850 J; (b) 6.2×10^5 J; (c) 3.0×10^6 J; (d) 6.1×10^6 J.
24. A Hooke's law spring stores 18 J of energy when compressed 0.14 m. What's its spring constant? (a) 1840 N/m; (b) 920 N/m; (c) 460 N/m; 120 N/m.
25. The change in kinetic energy of a 1.25-kg projectile that rises 12.8 m is (a) -16 J; (b) -102 J; (c) -157 J; (d) $+102$ J.
26. The power required to lift a 2.85-kg brick 10.0 m in 2.50 s is (a) 11.4 W; (b) 55.9 W; (c) 112 W; (d) 147 W.
27. A box slides across a horizontal floor to the right, with the net force on it toward the left. Which of the following is *not* true? (a) The box is slowing. (b) The net work done on the box is negative. (c) The work done by gravity is negative. (d) The box will not continue to move indefinitely.

Problems

Section 5.1 Work Done by a Constant Force
28. ■ If Galileo dropped a 2.50-kg cannon ball from the 58.4-m Tower of Pisa, how much work did gravity do on the ball?
29. ■ You push a heavy box, applying a 540-N horizontal force in the direction of motion while the box slides 3.5 m across the floor. How much work do you do?
30. ■ An object moves 2.50 m in the $+x$-direction under the influence of a 125-N force directed 50° above the x-axis. Find the work done on the object.
31. ■ ■ A 1320-kg car moves in the $+x$-direction with speed 21.5 m/s. Assuming constant braking and drag forces, find (a) the force and (b) the work needed to stop the car in a distance of 145 m.
32. ■ ■ Arranged on the floor are five concrete blocks, each 25.0 kg and 0.305 m tall. What is the minimum work required to stack all five vertically?
33. ■ ■ A 1.52-kg book slides 1.24 m along a level surface. The coefficient of kinetic friction between book and surface is 0.140. Find the work done by friction.
34. ■ ■ The book in the preceding problem is initially moving at 1.81 m/s. Find (a) the distance traveled before it stops and (b) the work done by friction in bringing the book to rest.
35. ■ ■ A force $\vec{F} = 2.34$ N $\hat{\imath} + 1.06$ N $\hat{\jmath}$ is applied to a cement block on a level floor. Find the work done by this force if the block's displacement is (a) 2.50 m $\hat{\imath}$; (b) -2.50 m $\hat{\imath}$; (c) 2.50 m $\hat{\imath}$ +2.50 m $\hat{\jmath}$.
36. ■ ■ A force $\vec{F} = 13$ N $\hat{\imath} + 13$ N $\hat{\jmath}$ acts on a hockey puck. Determine the work done if the force results in the puck's displacement by 4.2 m in the $+x$-direction and 2.1 m in the $-y$-direction.
37. ■ ■ A model rocket with mass 1.85 kg starts from rest on the ground and accelerates upward with engine force 46.2 N. From launch until the rocket reaches a height of 100 m, find (a) the

work done by the rocket engine, (b) the work done by gravity, and (c) the net work.

38. ■■ A 6.1-kg cannon ball is launched at a 45° angle on level ground. The cannon muzzle is 1.8 m above the ground. (a) Find the work done on the ball by gravity from launch until the time it reaches the ground. (b) Repeat part (a) if the ball is launched from the edge of a 19-m-high cliff.

39. ■■ A 1.25-kg block is pulled at constant speed up a frictionless 15° incline with constant force $\vec{F}$ directed up the incline. (a) Identify all forces acting on the block, and use Newton's first law to find $\vec{F}$. (b) Find the work done by $\vec{F}$ in moving the block 0.60 m up the incline. (c) Find the work done by gravity over the same path. (d) Combine your results to find the net work done on the block.

40. ■■ A 45.0-kg crate is dragged at constant velocity 8.20 m across a horizontal floor with a rope making a 30° angle above the horizontal. The coefficient of kinetic friction is 0.250. Find the work done (a) by friction and (b) by the rope.

41. ■■■ A glider (mass $m_1 = 0.15$ kg) on a frictionless horizontal air track is connected by a light string over a pulley to a metal block (mass $m_2 = 0.10$ kg) hanging vertically (Figure P5.41). The objects are released from rest and move 0.50 m. (a) Find the objects' acceleration. (b) Find the net work done on each. (c) Find the work done by the string on each. (d) Find the work done by gravity on the hanging mass.

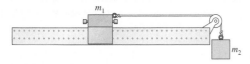

FIGURE P5.41

42. ■■ BIO **Weight lifting.** In a bench press, a weight lifter presses a 105-kg barbell 0.485 m straight up. If the barbell starts and ends at rest, how much work did the weight lifter do?

Section 5.2 Work Done by a Variable Force

43. ■ A Hooke's law spring hangs vertically with the top end fixed. Attaching a 0.150-kg mass to the bottom end stretches the spring 0.125 m. (a) Find the spring constant. (b) What will be the total stretch if a 1.00-kg mass is hung from the spring?

44. ■■ BIO **Stretching DNA.** With its double-helix structure, DNA is coiled like a spring. A biophysicist grabs the ends of a DNA strand with optical tweezers and stretches it 26 μm, producing 1.2-pN tension in the strand. What's the DNA's spring constant?

45. ■ If 13.4 J of work compresses a spring 2.37 cm, what's the spring constant?

46. ■ How much work does it take to compress a spring with $k = 25.0$ N/m by 0.450 m?

47. ■ Find the work done in extending a spring with $k = 150$ N/m from $x = 0.10$ m to $x = 0.30$ m.

48. BIO ■ **Tendons.** Muscles are attached to bones by elastic bundles called *tendons*. For small stretches, tendons can be modeled as springs obeying Hooke's law. Experiments on the Achilles tendon found that it stretched 2.66 mm with a 125-kg mass hung from it. (a) What is the spring constant of the Achilles tendon? (b) By how much would it have to stretch to store 50.0 J of energy?

49. ■■ Refer to the force-versus-position graph in Figure P5.49. How much work is done by the force for a displacement from (a) 0 to

10 cm; (b) 5 cm to 10 cm; (c) 0 to 15 cm? (d) How much work is done by the force for a displacement from 10 cm to 0 cm?

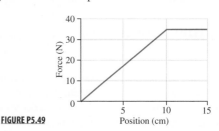

FIGURE P5.49

50. ■■ BIO **Spider silk.** Spider silk is one of the most remarkable elastic materials known. Consider a silk strand suspended vertically with a 0.35-g fly stuck on the end. With the fly attached, the silk measures 28.0 cm in length. The resident spider, of mass 0.66 g, senses the fly and climbs down the silk to investigate. With both spider and fly at the bottom, the silk measures 37.5 cm. Find (a) the spring constant and (b) the equilibrium length of the silk.

51. ■■ Refer to the force-versus-position graph in Figure P5.51. The force is in the x-direction (positive or negative as indicated), and position is measured along the $+x$-axis. How much work is done by the force for a displacement from (a) 0 to 2 m; (b) 2 m to 3 m; (c) 3 m to 5 m; (d) 0 m to 5 m? (e) How much work is done by the force for a displacement from 2 m to 0?

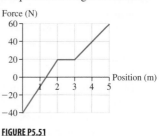

FIGURE P5.51

52. ■ Four identical springs with $k = 63.4$ kN/m support a car, with the car's weight distributed equally among them. Find the maximum weight for the car if the springs should be compressed no more than 4.0 cm when the car is at rest.

53. ■■ How much lower does the car in the preceding problem ride with four 90-kg passengers?

54. ■■■ A force $F_x = 4x + 12$ (in N, with x in m) acts on an object in one-dimensional motion. (a) Graph the force as a function of position. (b) Find the work done by that force in moving an object from $x = 0$ to $x = 5.0$ m.

55. ■■■ A spring with $k = 25.0$ N/m is oriented vertically with one end fixed to the ground. A 0.100-kg mass on top of the spring compresses it. Find the spring's maximum compression in each of these cases: (a) You hold the mass while you gently compress the spring, and when you release the mass it sits at rest atop the spring. (b) You place the mass on the uncompressed spring and release it. (c) You drop the mass from 10.0 cm above the spring.

Section 5.3 Kinetic Energy and the Work-Energy Theorem

56. ■ BIO **Typical animal kinetic energies.** For each case below, calculate the kinetic energy of the animal described. In each case, express your answer in joules and in joules per kilogram of body mass. (a) A 62-kg person walking at 1.0 m/s; (b) a 62-kg athlete running a 4-minute mile at constant speed; (c) a 72-kg cheetah running at its top speed of 72 mph (32 m/s); (d) a 12.3-mg froghopper that leaves the ground with initial speed 2.8 m/s.

57. ■ A boulder flies through the air at 12.4 m/s with kinetic energy 305 J. (a) What's its mass? (b) What's the boulder's kinetic energy if its speed (b) doubles or (c) is halved?

58. ■ At room temperature, a nitrogen molecule (mass = 4.65×10^{-26} kg) in air has kinetic energy 6.07×10^{-21} J. Find its speed.

59. ■■ A fully loaded 737 airliner has mass 68,000 kg. (a) Ignoring drag, how much work do the engines need to do to achieve takeoff speed of 250 km/h? (b) What minimum force should the engines supply to achieve takeoff in a distance of 1.20 km? (c) The 737 is powered by two engines, each of which can produce 117 kN of force. Are they powerful enough for the takeoff of part (b)?

60. ■■ How much work is required to lift the aircraft of the preceding problem to its 10.5-km cruising altitude? Compare with the work required to achieve takeoff speed.

61. ■■ A baseball with mass 0.145 kg is pitched at 39.0 m/s. Upon reaching home plate, 18.4 m away, its speed is 36.2 m/s. If the decrease is due entirely to drag, find (a) the work done by the drag force and (b) the magnitude of the (assumed constant) drag force.

62. ■ The Moon's mass is 7.36×10^{22} kg, and its (assumed circular) orbit has radius 3.84×10^8 m and a period 27.3 days. Find the Moon's kinetic energy.

63. ■■ A 0.145-kg baseball is struck by a bat 1.20 m above the ground, popping straight up at 21.8 m/s. (a) What's the ball's kinetic energy when it leaves the bat? (b) How much work is done by gravity once the ball reaches maximum height? (c) Use your answer in part (b) to find that maximum height. (d) Find the work gravity does on the ball from when it's batted until it hits the ground. (e) Ignoring air resistance, use your answer in part (d) to find the ball's speed at the ground.

64. ■■■ A projectile is fired horizontally from a 35-m cliff at 26 m/s. Find the speed and velocity of the projectile when it strikes the ground.

65. ■■ A rock is dropped from a 10-m-high ledge. (a) What's its speed when it hits the ground? (b) What's its height when its speed is half the value found in part (a)?

66. ■■ An archer fires a 0.175-kg arrow at 27 m/s at a 45° angle. What's the arrow's kinetic energy at the moment it's fired? (b) What's its kinetic energy at the peak of its flight? (c) What's its peak height?

67. ■■ If a 25-gram bullet with a speed of 310 m/s travels 15 cm into a tree before stopping, what's the average force exerted to stop the bullet?

68. ■■ A 75-g toy rocket is launched straight up from the ground at 19 m/s. (a) What's its kinetic energy? (b) Find the work done by gravity and the rocket's new kinetic energy after it has risen 10 m. (c) Use your answer to part (b) to find the rocket's speed at 10 m.

69. ■■ A crane lifts a 750-kg girder 8.85 m. How much work does the crane do lifting (a) at constant speed and (b) with upward acceleration 1.20 m/s^2?

70. ■■ The force graphed in Figure P5.49 is applied to a 1.8-kg box initially at rest at $x = 0$ on a frictionless, horizontal surface. Find the box's speed at (a) $x = 5$ cm; (b) $x = 10$ cm; (c) $x = 15$ cm.

71. ■■ Repeat the preceding problem if the box was moving in the $+x$-direction at 1.0 m/s when it was at $x = 0$.

72. ■■■ A 1250-kg car going 21 m/s has to stop suddenly. The driver locks the brakes, and the car skids to a halt in a distance of 65 m. (a) What was the car's acceleration while stopping? (b) How much work was done by friction to stop the car? (c) What is the coefficient of kinetic friction between tires and road?

Section 5.4 Potential Energy

73. ■ Find the change in gravitational potential energy of a 60-kg woman climbing from sea level to the 4390-m summit of Mt. Rainier.

74. ■ How far must you compress a spring with $k = 650$ N/m in order to store 450 J of energy?

75. ■ A spring with $k = 125$ N/m is initially compressed a distance $d = 0.125$ m from equilibrium, then it's extended the same distance from equilibrium. What's the change in potential energy?

76. ■■ You throw a 0.13-kg rock from a 15-m cliff. (a) Taking zero of potential energy at the cliff top, find the rock's potential energy when first released and when it hits the ground. Then find the change in potential energy. (b) Repeat part (a), this time taking $U = 0$ at the ground. (c) Compare and discuss the results of parts (a) and (b).

77. BIO ■■ **Food energy.** Energy is stored in food as the potential energy of the electrical bonds in molecules. Your body converts food energy to mechanical energy and heat. Food energy is expressed in "calories," which are actually kilocalories (kcal, with 1 kcal = 4.186 kJ). (a) How many joules are in a 120-kcal serving of breakfast cereal? (b) A glass of 1% milk contains 130 kcal. How many glasses would a 62-kg person have to drink to get the energy needed to climb a hill 125 m high, assuming all the milk's energy is converted to her potential energy?

78. BIO ■■ **Use of food energy.** (See the preceding problem.) When the body "burns" food, only about 20% of the food energy is available as mechanical energy. Suppose that a 75-kg person consumes ice cream containing 280 kcal. (a) How high a hill would he have to climb to "work off" those calories? (b) If he wanted to do a series of chin-ups in which he lifted his body by 50.0 cm, how many of these would he have to do to work off the ice cream?

79. BIO ■■ **Exercise program.** You're at the gym, using a weight machine to do arm raises. Each raise lifts a 20.0-N weight 45 cm. How many raises must you do to work off 100 kcal? Is this a reasonable workout session? Assume 20% conversion of food energy to mechanical energy.

Section 5.5 Conservation of Mechanical Energy

80. ■ The total mechanical energy of an object moving at 29.2 m/s is 563 J, and its potential energy is 175 J. What's its mass?

81. ■ Take the ground as the zero of potential energy. (a) Find the total mechanical energy of a 45.9-gram golf ball 23.4 m above the ground and moving at 31.2 m/s. (b) Ignoring drag forces, what's the ball's speed when it hits the ground?

82. ■ At the Zero Gravity Thrill Amusement Park near Dallas, Texas, people drop from a 30-m tower into a net below. At what speed do they reach the net?

83. ■■ Two men pass a 5.0-kg "medicine ball" back and forth. (a) If one man launches the ball by pushing it from rest with a 138-N horizontal force over 0.50 m, how fast is the ball going when it leaves his hands? (b) How much work must the other man do to stop the ball?

84. ■■ (a) A horizontal spring with $k = 35$ N/m is compressed 0.085 m and used to launch a 0.075-kg marble. (a) Find the marble's launch speed. (b) Repeat for a vertical launch.

85. ■ A roller coaster going 19.2 m/s starts up a hill. Ignoring friction, what's its speed after it has risen 12.2 m vertically?

86. ■ A horizontal spring with $k = 75$ N/m has one end attached to a wall and the other end free. An 85-g wad of putty is thrown horizontally at 3.4 m/s directly toward the free end. Find the maximum spring compression.

87. ■■ A spring with $k = 1340$ N/m is oriented vertically with one end attached to the ground. A 7.27-kg bowling ball is dropped from 1.75 m above the top of the spring. Find the maximum spring compression.

88. ■■ A horizontal spring with $k = 120$ N/m has one end attached to the wall. A 250-g block is pushed onto the free end, compressing the spring by 0.150 m. The block is then released, and the

spring launches it outward. (a) Neglecting friction, what's its speed when it leaves the spring? (b) Repeat part (a) if the coefficient of kinetic friction is 0.220.

89. ■■ A rubber ball is dropped from rest 2.4 m onto level ground. (a) What's the ball's speed when it hits the ground? (b) Bouncing back, the ball loses 25% of its mechanical energy. To what height does it rebound?

90. ■■ A 4.75-kg radio-controlled model airplane is flying 23.5 m above the ground with velocity 12.9 m/s $\hat{\imath}$ + 3.48 m/s $\hat{\jmath}$, with the x-axis horizontal and the y-axis vertical. (a) Taking $y = 0$ at the ground, what's the plane's total mechanical energy? (b) If the engine fails and the plane plummets, what's its speed when it crashes? Neglect air resistance.

91. ■■■ A 980-kg car's parking brake fails on a 3.6° incline. The coefficient of rolling friction is 0.030, and the car rolls 35 m down the incline. Find the work done by (a) friction and (b) gravity. (c) Find the car's final speed.

92. ■■ A snowboarder reaches the bottom of a frictionless "half-pipe" with speed 15.9 m/s. The half-pipe is a half-cylinder with curvature radius 11.0 m. How high above the edge of the half-pipe will the snowboarder fly?

93. ■■■ A spring with $k = 42.0$ N/m is mounted horizontally at the edge of a 1.20-m-high table (Figure P5.93). The spring is compressed 5.00 cm, and a 25.0-g pellet is placed at its end. When the spring is released, how far (horizontally) from the edge of the table does the pellet strike the floor?

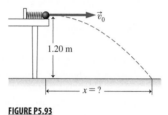

FIGURE P5.93

94. ■■■ A frictionless roller coaster starts from rest 25 m above the ground. (a) What's its speed when it reaches the ground? (b) Upon reaching the ground, the track goes into a vertical, circular loop. Find the maximum loop radius such that the car will maintain contact with the track at the top.

95. ■■ A cat jumps to a 1.15-m-high dresser, leaving the floor at 75° above the horizontal. What minimum speed must it have?

96. ■■■■ A simple pendulum consists of a ball of mass m attached to a light string of length L. The other end of the string is attached to the ceiling, so the ball swings freely in a vertical plane. The ball is pulled aside until the string makes an angle θ with the vertical, at which point the ball is released from rest. Use conservation of energy to find the ball's speed when it reaches the bottom of its arc, as a function of L and θ. Evaluate numerically for $\theta = 45°$ and $L = 1.20$ m.

97. ■■■ A large spring is placed at the bottom of an elevator shaft to minimize the impact in case the elevator cable breaks. A loaded car has mass 480 kg, and its maximum height above the spring is 11.8 m. In order to minimize the shock, the maximum acceleration of the car after hitting the spring is $4g$. What should be the spring constant k?

Section 5.6 Power

98. ■ What power is needed to lift a 350-kg crate of bricks from the ground to the top of a 23.8-m-high building in 1 minute?

99. ■ Find the work done by a motor operating at a constant 8.5 kW for 30 s.

100. ■ A woman takes 1.2 s to lift a 65-kg barbell 0.45 m straight up in a bench press. What's her average power output?

101. ■■ Victoria Falls in Africa drops about 100 m, and in the rainy season as much as 550 million m³ of water per minute rush over the falls. What's the total power in the waterfall? *Hint:* The density of water is 1000 kg/m³.

102. ■■ Your sofa won't fit through the door of your new sixth-floor apartment, so you use a 1.12-kW motor to lift the 86.1-kg sofa 17.2 m from the street. How much time does the lift take?

103. ■■ A motorized lift runs along a stairway inclined at 30°. (a) Find the work done in lifting a 75-kg person and 22-kg chair if the track's length is 5.6 m. (b) What power must the motor deliver if the person is to make it from bottom to top in 12 s?

104. ■■ A 58-kg skier is being pulled up a 12° frictionless slope. What power is required for the skier to cover the entire 1.20-km slope in 4.5 min?

105. ■■ Suppose your 1320-kg sports car has a 280-hp engine that's 40% efficient. (That is, 40% of the 280 hp can be converted into the car's motion.) Find the car's maximum speed after accelerating from rest for 4.0 s.

106. ■■ A constant force F_x acts along the x-axis on an object of mass m initially at rest. Find the instantaneous power delivered by that force, as a function of time.

107. ■■■ A man normally consumes 8.4 MJ of food energy per day. He then begins running a distance of 8 km four times per week. If he expends energy at the rate of 450 W while running at 12 km/h, how much more food energy should he consume daily in order to maintain constant weight?

108. ■■ A 62-kg student jogs upstairs from the first floor to the sixth, a vertical distance of 19.2 m, in 55 s. (a) Find the power the student expends working against gravity and compare with her average power expenditure of 100 W. (b) She then jogs back down to the first floor and notes that the total work she's done against gravity is zero for the round trip. Why does she still feel tired?

109. ■■ A 0.150-kg apple falls 2.60 m to the ground. (a) Find the work done by gravity. (b) Make a graph of the power supplied by gravity as a function of time over the entire fall. (c) Show that the work done by gravity is equal to the average multiplied by the fall time.

110. **BIO** ■■■ **Kleiber's law.** The basal metabolic rate (BMR) measures an animal's typical resting power use. For mammals, BMR approximately obeys the equation BMR $\approx A\,m^{3/4}$ (Kleiber's law), where m is the mass of the animal and A is a constant whose value depends on the species. (a) What are the SI units of A? (b) According to Kleiber's law, what's the BMR of a 75-kg person if $A = 3.4$ in SI units? (c) What's the value of A for a polar bear, which has a mass of 700 kg and BMR = 460 W? (d) A 180-kg gorilla has a BMR of 170 W. Use Kleiber's law to predict the BMR of King Kong, a 1000-kg gorilla, assuming A is the same for all gorillas.

111. **BIO** ■■ **The heart.** A person typically contains 5.0 L of blood of density 1.05 g/mL. When at rest, it normally takes 1.0 min to pump all this blood through the body. (a) How much work does the heart do to lift all that blood from feet to brain, a distance of 1.85 m? (b) What average power does the heart expend in the process? (c) The heart's actual power consumption, for a resting person, is typically 6.0 W. Why is this greater than the power found in part (b)? Besides the potential energy to lift the blood, where else does this power go?

112. ■■ According to the U.S. Department of Energy, the United States consumed about 1.03×10^{20} J of energy in 2003. Find the energy consumed in kWh and the cost assuming a rate of $0.12 per kWh.

General Problems

113. BIO ■■ **Brisk walking.** A 175-lb person walking briskly on level ground at 4.5 mph consumes 7.0 kcal per minute. What *distance* should this person walk to "burn off" 125 kcal (1 kcal = 4.186 kJ)?

114. BIO ■■ **Power walking.** Power walking on level ground for 20 min consumes 175 kcal. For a 70.0-kg person walking at 1.5 m/s, how much food energy would be consumed in 20 min walking up a 10° incline? (Assume 20% conversion of food energy to mechanical energy.)

115. ■■ Figure GP5.115 shows the force that acts on an object moving along the *x*-axis. Determine the work done as the object moves from (a) $x = 0$ to $x = 7.0$ m and (b) $x = 0$ to $x = 12.0$ m.

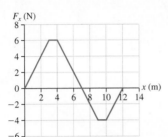

FIGURE GP5.115

116. BIO ■■■ **Elasticity of human hair.** Hair is somewhat elastic, so you can model it as an ideal spring. Experimental tests on a single strand of hair show that it stretches 2.55 cm when a 0.100-kg mass is hung from it. (a) What's the spring constant of this strand? (b) How much energy is stored in it if it stretches by 15.0 mm? (c) Suppose you combine 200 identical, parallel strands. What will be the spring constant of this bundle, and how much potential energy will it store if it's stretched 15.0 mm?

117. BIO ■■ **Insect energy.** The froghopper is the insect world's champion jumper. These insects are typically 6.1 mm long, have mass 12.3 mg, and leave the ground at 2.8 m/s at 58° above the horizontal. (a) How high does a froghopper go in such a leap? (b) The energy for the leap is stored in the muscles of the insect's legs, which you can model as ideal springs. If the initial compression of *each* of the two legs is one-third of the body length, what is their spring constant?

118. ■■ A mass hanging from a vertical spring has gravitational potential energy, and the spring has elastic potential energy. (a) Determine how far the spring ($k = 16$ N/m) stretches when a 100-gram mass is hung from it and allowed to come to rest. (b) If the mass is pulled down 3 cm further, determine the change in each type of potential energy.

119. ■■■ The force graphed in Figure P5.51 is applied to a 2.0-kg block that was sliding to the right (the $+x$-direction) over a frictionless surface with speed 5.0 m/s at $x = 0$. (a) Is the block ever at rest? If so, where? (b) Find a position (other than $x = 0$) when the block is again moving to the right at 5.0 m/s.

120. ■■■ Consider again Atwood's machine, described in Problem 4.62, in which two masses m_1 and m_2 are connected over a pulley. Assume that $m_2 > m_1$. The masses are released from rest. The potential energy of m_2 decreases by 7.2 J, while its kinetic energy increases by 3.6 J. The potential energy of m_1 increases by 2.4 J, as its kinetic energy increases by 1.2 J. (a) Determine the net work done on the system (the two masses) by external forces. (b) What force does this work? (c) What is the ratio of the two masses? (d) Is total mechanical energy conserved?

121. ■■■ A 1500-kg roller coaster (including passengers) passes point A at 3 m/s (Figure GP5.121). Due to safety concerns, you must design the track so that at point B the passengers do not experience an upward force that exceeds 4*g*. If the arc at B is circular with radius 15 m, (a) determine the minimum value of *h* that satisfies this requirement, and (b) determine the speed of the coaster at C.

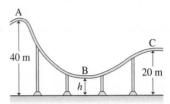

FIGURE GP5.121

122. ■■ While driving on a straight level highway in your 1450-kg car, you take your foot off the gas and find that your speed drops from 65 mi/h to 55 mi/h over one-tenth of a mile. Assuming your average speed during this interval was 60 mi/h, find the power (in watts and horsepower) needed to keep your car moving at a constant 60 mi/h.

123. ■■ A golf ball with mass 45.9 g leaves the ground at 42.6 m/s. It subsequently hits the ground at 31.9 m/s. How much work was done by air resistance (drag)?

124. ■■■ Consider the air-track experiment shown in Figure P5.41, with $m_1 = 0.250$ kg and $m_2 = 0.125$ kg. The system is released from rest, and the hanging mass drops 0.40 m. (a) How much work is done by gravity? (b) Use the work-energy theorem to find the speeds of both blocks. (c) Use the blocks' speeds to find their acceleration.

125. ■■■ A spring-loaded gun has $k = 72.0$ N/m. The spring is compressed 3.20 cm and shoots a 15-g pellet horizontally from 1.20 m above the ground. (a) What's the pellet's speed when it leaves the gun? (b) What's its speed when it hits the ground? (c) How far does the pellet travel horizontally? (See Figure P5.93.)

126. ■■ The 36-kg wheel of an airplane flying at 245 m/s at an altitude of 7300 m falls off during a flight. (a) If the wheel hits the ground at 372 m/s, how much work was done on the wheel by air resistance (drag) during its fall? (b) If there had been no drag, what would have been the wheel's speed when it hit the ground?

Answers to Chapter Questions

Answer to Chapter-Opening Question
The work done against gravity depends on the net elevation change, but not the exact route. A rider on a bicycle with a combined mass of 80 kg does about 400 kJ (100 kcal) of work against gravity, regardless of the path up a 500-m-tall hill. To climb such a hill in 20 minutes, the rider's power output must exceed 300 W.

Answers to GOT IT? Questions
Section 5.1   (b) negative
Section 5.2   (a) > (c) > (b)
Section 5.3   (b) 4.9 J
Section 5.4   (d) > (b) > (a) > (c)
Section 5.5   (b) < (c) < (a) < (d)
Section 5.6   (d) < (a) < (b) < (c)

■ As the diver flies through the air, most parts of his body follow complex trajectories. But one special point follows a parabola. What's that point, and why is it special?

This chapter introduces the concept of momentum, which arises from Newton's second law. We'll prove that momentum is conserved in the absence of external forces, and then we'll use this fact to analyze collisions between objects. We'll consider both elastic collisions, which conserve kinetic energy, and inelastic collisions, in which kinetic energy is lost. This chapter also includes a brief discussion of the center of mass.

6.1 Introduction to Momentum

As introduced in Chapter 4, Newton's second law, $\vec{F}_{net} = m\vec{a}$, is the backbone of dynamics, because it relates force and acceleration. In Chapter 5, Newton's second law was crucial in developing the relation between work and kinetic energy.

There's another way to consider Newton's law, one that leads to new concepts and problem-solving tools. Central to this approach is **momentum**—another term in everyday use but with a more precise physics meaning. Momentum applies both to single objects and to entire systems. We'll use the momentum of systems to explore how objects interact.

Net Force Revisited

Newton's second law relates the net force on an object to its mass and acceleration. With constant acceleration (Section 3.3), the acceleration over *any* interval Δt is $\vec{a} = \Delta\vec{v}/\Delta t$. Then Newton's second law can be expressed

$$\vec{F}_{net} = m\vec{a} = m\frac{\Delta\vec{v}}{\Delta t}$$

With mass m constant, we can bring the factor m inside the Δ symbol, giving

$$\vec{F}_{net} = \frac{\Delta(m\vec{v})}{\Delta t} \qquad (6.1)$$

The quantity $m\vec{v}$ is the object's **momentum** $\vec{p}$. *The momentum of an object is the product of its mass and velocity:*

$$\vec{p} = m\vec{v} \qquad \text{(Definition of momentum; SI unit: kg·m/s)} \qquad (6.2)$$

With this definition, Newton's second law (Equation 6.1) becomes

$$\vec{F}_{net} = \frac{\Delta\vec{p}}{\Delta t} \qquad \text{(Newton's second law, expressed in terms of momentum)} \qquad (6.3)$$

for constant acceleration, hence constant force. If force isn't constant, then the instantaneous net force is the limiting value of $\Delta\vec{p}/\Delta t$ as Δt approaches zero.

 TIP

Momentum is a vector quantity, and it points in the same direction as the velocity vector.

Interpreting the Results

What makes momentum important is evident from Equation 6.3: *Momentum is what changes when a net force is applied to an object.* In fact, Equation 6.3 is a more general form of Newton's second law than $\vec{F}_{net} = m\vec{a}$, which is valid only for objects that don't gain or lose mass. More generally, an object's momentum $\vec{p} = m\vec{v}$ can change by changing velocity, mass, or both. A classic example is a rocket (Figure 6.1), which is propelled forward by ejecting some of its mass—exhaust gases—in the opposite direction.

Some Basics of Momentum

Momentum $\vec{p} = m\vec{v}$ is a vector. Because mass m is a positive scalar, the momentum $\vec{p}$ always points in the same direction as the velocity $\vec{v}$ (Figure 6.2). Like any vector, momentum has magnitude and direction. The magnitude p is

$$p = mv \qquad (6.4)$$

Thus the magnitude of an object's momentum equals its mass multiplied by its speed. The SI units for momentum are kg·m/s. There's no specific name for this combination.

You've seen the essential relationship between momentum and force (Equation 6.3). Momentum is also related to kinetic energy. Both kinetic energy $K = \frac{1}{2}mv^2$ and momentum magnitude $p = mv$ depend on mass and speed. Taking $v = p/m$ and substituting in the equation for K gives

$$K = \tfrac{1}{2}mv^2 = \tfrac{1}{2}m\left(\frac{p}{m}\right)^2$$

or

$$K = \frac{p^2}{2m} \qquad (6.5)$$

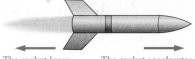

The rocket's momentum depends on both its mass and its velocity.

The rocket loses mass as it propels exhaust gases backward.

The rocket accelerates, so its velocity also changes.

FIGURE 6.1 Changing mass and velocity in a rocket.

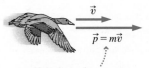

Velocity and momentum vectors point in same direction.

FIGURE 6.2 Direction of the momentum vector.

Kinetic energy is a *scalar*, so this expression depends only on the momentum's magnitude p, not the direction of the vector $\vec{p}$.

EXAMPLE 6.1 **Vector Momentum**

In pool, a 162-g cue ball rolls at 3.24 m/s at a 25° angle to the $+x$-axis, as in Figure 6.3a (table edges define the coordinate axes). Find the components of the ball's momentum.

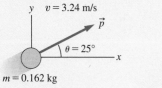

(a) Momentum vector

(b) Momentum vector resolved into its components

FIGURE 6.3 Momentum is a vector with components.

ORGANIZE AND PLAN Like any vector in two dimensions, the components of the momentum vector, shown in Figure 6.3b, follow from trigonometry. Here the right-triangle geometry shows that

$$p_x = p \cos \theta$$
$$p_y = p \sin \theta$$

where $p = mv$ is the magnitude of the momentum.

Known: mass $m = 0.162$ kg; speed $v = 3.24$ m/s; angle $\theta = 25°$.

SOLVE Inserting the numerical values,

$$p_x = p \cos \theta = (0.162 \text{ kg})(3.24 \text{ m/s})(\cos 25°) = 0.476 \text{ kg} \cdot \text{m/s}$$
$$p_y = p \sin \theta = (0.162 \text{ kg})(3.24 \text{ m/s})(\sin 25°) = 0.222 \text{ kg} \cdot \text{m/s}.$$

REFLECT A look at Figure 6.3b shows that the momentum's x-component should indeed be larger than the y-component, consistent with our results. In a real pool game, the ball's spin is important. A spinning ball has *angular momentum*, which we'll discuss in Chapter 8.

MAKING THE CONNECTION Given the momentum components p_x and p_y, how do you find the magnitude and direction of $\vec{p}$?

ANSWER As with any vector, the magnitude comes from the Pythagorean relationship $p^2 = p_x{}^2 + p_y{}^2$, so $p = \sqrt{p_x{}^2 + p_y{}^2}$. Figure 6.3b shows that the direction angle θ is $\theta = \tan^{-1}(p_y/p_x)$.

Impulse

Imagine yourself at bat, a baseball hurtling toward you (Figure 6.4). In physics terms, your goal is to *change the baseball's momentum*. You'll do that by applying a force with your bat. Initially the ball's momentum is directed from pitcher toward home plate. You'd like to change the ball's momentum so it moves rapidly away from you. That change depends on the net force and the length of time it acts. By Equation 6.3,

$$\vec{\overline{F}}_{\text{net}} = \frac{\Delta \vec{p}}{\Delta t} \qquad (6.6)$$

Here we've used the *average* net force (indicated by the bar over the vector $\vec{F}$), because forces usually vary throughout the time they're acting. The net force and time interval during which it acts define the **impulse** (symbol $\vec{J}$):

$$\vec{J} = \vec{\overline{F}}_{\text{net}} \Delta t \qquad \text{(Definition of impulse; SI unit: kg} \cdot \text{m/s)} \qquad (6.7)$$

Comparing Equations 6.6 and 6.7, you can see that

$$\vec{J} = \Delta \vec{p} \qquad \text{(Impulse and momentum; SI unit: kg} \cdot \text{m/s)} \qquad (6.8)$$

That is, the impulse from a net force equals the change in momentum. This is the **impulse-momentum theorem**. It's useful, for example, to estimate the contact force in a brief encounter between two objects, like ball and bat. Suppose our baseball (mass 0.15 kg) approaches the bat horizontally at 32 m/s and leaves in the opposite direction at 38 m/s. Let the direction on approach be $-x$, so the batted ball moves in the $+x$-direction. Then the change in the ball's x-momentum (here the only nonzero component) is $(0.15 \text{ kg})(38 \text{ m/s} - (-32 \text{ m/s})) = 10.5 \text{ kg} \cdot \text{m/s}$. By the impulse-momentum theorem, this equals the impulse $J = \overline{F}_{\text{net}} \Delta t$. If the contact time is 0.50 ms, then the average force the bat exerts on the ball is

$$\overline{F}_{\text{net}} = \frac{J}{\Delta t} = \frac{\Delta p}{\Delta t} = \frac{10.5 \text{ kg} \cdot \text{m/s}}{0.50 \times 10^{-3} \text{ s}} = 21 \text{ kN}$$

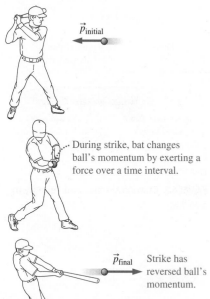

... During strike, bat changes ball's momentum by exerting a force over a time interval.

Strike has reversed ball's momentum.

FIGURE 6.4 The bat changes the baseball's momentum.

This large but brief force would be difficult to measure directly. Note that we've only computed the *average* force. The actual force varies significantly during the contact time, increasing as the ball gets squashed (Figure 6.5). Figure 6.6a graphs this variation, while Figure 6.6b shows that the area under the force-versus-time graph gives the impulse in one-dimensional motion. So you can find the impulse for a varying force, provided you have a force-versus-time graph.

Reviewing New Concepts

- Momentum is a vector defined by $\vec{p} = m\vec{v}$
- Momentum is related to net force by Newton's second law, $\vec{F}_{net} = \dfrac{\Delta \vec{p}}{\Delta t}$, for a constant net force
- Impulse equals change in momentum: $\vec{J} = \Delta \vec{p} = \overline{\vec{F}}_{net}\,\Delta t$

FIGURE 6.5 The bat exerts a force on the ball.

GOT IT? Section 6.1   Rank the momentum of the following animals in order, from lowest to highest: (a) a 50-kg cheetah running at 27 m/s; (b) a 75-kg human sprinting at 11 m/s; (c) a 180-kg lion running at 14 m/s; (d) a 120-kg dolphin swimming at 16 m/s.

6.2 Conservation of Momentum

The Principle of Conservation of Momentum

Figure 6.7 shows an astronaut outside her spacecraft. Both are in free fall and so, as we'll see in Chapter 9, we can ignore gravity in their reference frame. The astronaut gives the spacecraft a slight push, applying force $\vec{F}_{12}$ as shown. By Newton's third law, the spacecraft pushes back on the astronaut with force $\vec{F}_{21} = -\vec{F}_{12}$.

✓**TIP**

You might review Newton's third law in Chapter 4; it's closely tied to momentum conservation.

Equation 6.3 relates these two forces to the corresponding rates of change of momentum:

$$\frac{\Delta \vec{p}_1}{\Delta t} = -\frac{\Delta \vec{p}_2}{\Delta t}$$

The contact time Δt is the same for both astronaut and spacecraft, so

$$\Delta \vec{p}_1 = -\Delta \vec{p}_2 \tag{6.9}$$

Equation 6.9 says that *in an interaction between two objects, the change in one object's momentum is of the same magnitude and opposite direction as the other object's momentum change.* Later, this will help you understand what happens when two objects collide.

Rearranging Equation 6.9 gives another important result:

$$\Delta \vec{p}_1 + \Delta \vec{p}_2 = 0 \tag{6.10}$$

That is, the sum of the momentum changes is zero. That means the total momentum $\vec{p}_1 + \vec{p}_2$ of the two objects doesn't change when they interact:

$$\vec{p}_1 + \vec{p}_2 = \text{constant} \quad \begin{array}{l}\text{(Momentum conservation, in a system with} \\ \text{zero net external force; SI unit: kg} \cdot \text{m/s)}\end{array} \tag{6.11}$$

We'll state this result as a general principle, known as the principle of **conservation of momentum**: *Whenever two objects interact, in the absence of external forces, the total momentum is conserved.*

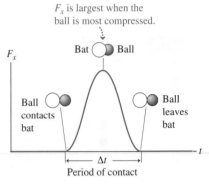

(a) Force versus time for a bat striking a ball

(b) Impulse equals area under curve

FIGURE 6.6 Finding the impulse graphically.

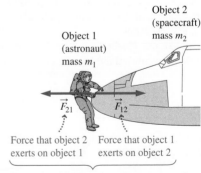

By Newton's third law, these forces are equal in magnitude and opposite in direction.

FIGURE 6.7 Astronaut pushing off the spacecraft.

A Wider View of Momentum Conservation

The conservation of momentum principle extends to many-particle systems. In such systems, momentum conservation holds, as we've established, between any pair of interacting particles. If the momentum of each interacting pair is conserved, so is the system's total momentum.

Another way to understand this is to consider forces that act *within* a system of particles. For every such force, Newton's third law requires a second force of equal magnitude but opposite direction, such as the colliding billiard balls in Figure 6.8. The net force *on the entire system* includes the sum of all such pairs—and that sum is always zero, by the third law. Therefore, as long as the only forces acting are *internal* to the system, the system's total momentum doesn't change.

The general principle of momentum conservation therefore states: *For any system of particles not subject to a net external force, the total momentum of the system of particles is conserved.*

The condition *no net external force* is crucial. Although forces acting from outside the system are subject to the third law, the forces they're paired with *don't act on the system being considered*, so they don't cancel the external forces. Thus external forces *can* change a system's momentum. To apply momentum conservation, you need to identify a system of particles that interact with each other but whose interaction with the outside world is negligible.

The principle of momentum conservation works for all kinds of forces, whether or not there's direct contact. Imagine a skydiver starting to fall toward Earth with initially zero momentum (Figure 6.9). Then the total momentum of the skydiver + Earth system is initially zero. Gravity accelerates the skydiver, increasing his momentum as he falls. But gravity acts between the two components of the system, skydiver and Earth—so it's an internal force, and the system's total momentum is conserved. Therefore, Earth must gain upward momentum, to keep the total momentum constant. You don't notice Earth rushing upward to meet a falling body, because Earth is so massive that it can have large momentum with very little speed ($p = mv$). As with gravity, conservation of momentum holds for other action-at-a-distance forces including electric and magnetic forces.

The forces come in force pairs that follow Newton's third law. For example, $\vec{F}_{31}$ is equal in magnitude and opposite in direction to $\vec{F}_{13}$.

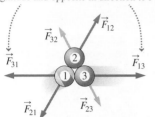

FIGURE 6.8 Three billiard balls colliding, illustrating the forces that act between each pair of balls.

Skydiver's momentum increases as he falls.

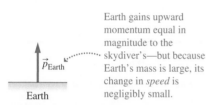

Earth gains upward momentum equal in magnitude to the skydiver's—but because Earth's mass is large, its change in *speed* is negligibly small.

FIGURE 6.9 Momentum is conserved in the Earth + skydiver system.

EXAMPLE 6.2   **Spacewalk**

The astronaut of Figure 6.7 (135 kg, including spacesuit) and spacecraft (4620 kg) are initially at rest. After pushing away, the astronaut is moving at 1.42 m/s. What's the spacecraft's velocity?

ORGANIZE AND PLAN Before the push, both spacecraft and astronaut are at rest (with respect to our orbiting reference frame), so the system's total momentum is zero. The only force acting is the push, which is internal to the system. So momentum is conserved, and the total momentum after the push must still be zero. This fact, along with the momentum of the astronaut, can be used to find the unknown velocity of the spacecraft.

Take the $+x$-direction to point to the right. Then the astronaut's velocity is toward the left (Figure 6.10), in the $-x$-direction, and so

$v_{1x} = -1.42$ m/s. Then the spacecraft must move in the $+x$-direction, so the two momentum vectors can add to zero, the initial momentum.

Conservation of momentum gives $p_{1x} + p_{2x} = 0$. With $\vec{p} = m\vec{v}$,

$$m_1 v_{1x} + m_2 v_{2x} = 0$$

Known: $m_1 = 135$ kg; $v_{1x} = -1.42$ m/s; $m_2 = 4620$ kg.

SOLVE Solving for the unknown v_{2x},

$$v_{2x} = -\frac{m_1 v_{1x}}{m_2} = -\frac{(135\ \text{kg})(-1.42\ \text{m/s})}{4620\ \text{kg}} = 0.0415\ \text{m/s}$$

or about 4.15 cm/s in the $+x$-direction.

REFLECT As expected, the heavier object recoils more slowly. In fact, you can show that the speed ratio is the inverse of the mass ratio whenever two objects start from rest.

MAKING THE CONNECTION   How can the astronaut get herself back to the spacecraft?

ANSWER She needs some other change of momentum. She could carry a rocket pack, or the spacecraft could fire its engines to approach her. In desperation, she could throw a tool opposite the direction to the spacecraft; if she gives it enough momentum, she'll reverse her motion.

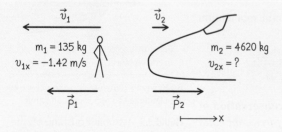

$\vec{v}_1$

$\vec{v}_2$

$m_1 = 135$ kg
$v_{1x} = -1.42$ m/s

$m_2 = 4620$ kg
$v_{2x} = ?$

$\vec{p}_1$

$\vec{p}_2$

x

FIGURE 6.10 Momentum conservation in space.

Some Applications of Momentum Conservation

Rocket propulsion uses momentum conservation. Figure 6.11 shows a rocket initially at rest in a reference frame where outside forces such as gravity can be ignored.

Before its engine fires, the system of rocket + fuel has zero momentum. Combustion in the rocket engine then expels hot gases with backward-pointing momentum. To conserve momentum, the rest of the rocket accelerates forward. A launch from Earth, analyzed in Earth's reference frame, would be more complicated because now gravity plays a role. But in either case detailed analysis shows the rocket pushing on gas molecules to expel them, with the gases then pushing back on the rocket to accelerate it.

Momentum transfer plays an essential role in most ball games. The club strikes a stationary golf ball, giving it momentum. In tennis and baseball, a player strikes an incoming ball to reverse its direction and maybe increase its speed (Figure 6.4). Although external forces including gravity and muscles may be acting, the contact forces of the impact are so much larger that momentum conservation is approximately valid. Therefore, club, bat, and racquet all lose momentum that the ball gains.

Before ignition

Rocket + fuel at rest

After ignition

Engine burns fuel to form hot gases, which exit at high speed.

To balance high momentum of exhaust, rocket gains forward momentum (accelerates).

FIGURE 6.11 The rocket uses conservation of momentum to accelerate.

PROBLEM-SOLVING STRATEGY 6.1 **Conservation of Momentum**

Applying momentum conservation requires modifying your familiar problem-solving strategy. Here we've italicized instructions specific to momentum conservation.

ORGANIZE AND PLAN

- Visualize situation. *Make sure you've identified a system with no external forces, so momentum is conserved.*
- *Pick appropriate coordinate system for the motions you're considering.*
- *Draw schematic diagram. Consider "before" and "after" diagrams to show how individual momenta change throughout an interaction, while system momentum remains constant.*
- *Equate initial ("before") momentum and final ("after") momentum.*

SOLVE

- Gather information, especially masses and velocities.
- *Solve momentum conservation equation(s) for unknown quantity (quantities).*
- Insert numerical values and solve.

REFLECT

- Check dimensions and units of answer. Are they reasonable?
- If problem relates to something familiar, assess whether answer makes sense.

EXAMPLE 6.3 **Squid Propulsion**

A 13.1-kg squid fires a burst of water, giving it average acceleration $2.0g$ (19.6 m/s^2) over a 100-ms time interval. (a) Find the associated impulse. (b) Assuming the squid starts from rest, what are its momentum change and final speed?

ORGANIZE AND PLAN Given the squid's mass and acceleration, we can use Newton's law to find the force; multiplying by Δt gives the impulse. By the impulse-momentum theorem, that's the momentum change. Momentum is mass times velocity, which we can use to find the speed.

Take the x-direction to be that of the squid's motion (Figure 6.12), so the x-components of acceleration, velocity, and impulse are positive. The impulse is

$$J_x = F_x \Delta t = ma_x \Delta t$$

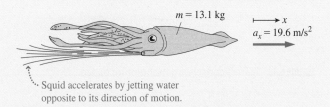

$m = 13.1 \text{ kg}$
$a_x = 19.6 \text{ m/s}^2$

Squid accelerates by jetting water opposite to its direction of motion.

FIGURE 6.12 The squid accelerates.

Then by the impulse-momentum theorem

$$J_x = \Delta p_x, = p_x$$

because the initial momentum was zero. Finally, $p_x = mv_x$, which allows you to solve for v_x which, in this one-dimensional situation, is the speed.

cont'd.

Known: $m = 13.1$ kg; $a_x = 19.6$ m/s²; $\Delta t = 0.100$ s.

SOLVE The impulse is

$$J_x = F_x \Delta t = ma_x \Delta t = (13.1 \text{ kg})(19.6 \text{ m/s}^2)(0.10 \text{ s})$$
$$= 25.7 \text{ kg} \cdot \text{m/s}$$

By the impulse-momentum theorem, the momentum change is also 25.7 kg · m/s. Because the squid started from rest, its final momentum $p_x = mv_x$. Thus the x-component of velocity is

$$v_x = \frac{p_x}{m} = \frac{25.7 \text{ kg} \cdot \text{m/s}}{13.1 \text{ kg}} = 1.96 \text{ m/s}$$

which in this case is also the speed.

REFLECT The squid's final speed, found using conservation of momentum and the impulse-momentum theorem, is consistent with the rules of kinematics—in this case, $v_x = a_x \Delta t$.

MAKING THE CONNECTION How fast is that, compared with the squid's top speed?

ANSWER Squids are the fastest marine invertebrates, topping out at over 10 m/s. They accomplish this by repeated water propulsion, along with the motion of their tentacles. The water's drag force keeps the squid from going faster.

EXAMPLE 6.4　**Newton on Ice, Revisited**

Consider again the skaters of Example 4.3, with masses $m_A = 50$ kg and $m_B = 80$ kg. They stand together at rest and then push off each other with a 200-N force maintained for 0.40 s. Use momentum concepts to determine the skaters' velocities when they separate.

ORGANIZE AND PLAN As in Example 4.3, let the lighter skater, A, move in the +x-direction, and B in the −x-direction (Figure 6.13). If we neglect friction for the short time of their push, the skaters constitute an isolated system with no net external force, so momentum conservation applies.

Before the push, both skaters are at rest, so the total momentum of the system is zero. Therefore, the skaters' total momentum after separation is

$$\vec{p}_A + \vec{p}_B = 0$$

or

$$\vec{p}_B = -\vec{p}_A$$

That is, the skaters have momenta with equal magnitude but opposite direction. Each skater's momentum comes from an impulse with magnitude

$$\Delta p = F \Delta t$$

Knowing momentum and mass then gives the velocities.

Known: $m_A = 50$ kg; $m_B = 80$ kg; $F = 200$ N; $\Delta t = 0.40$ s.

SOLVE The magnitude of the impulse for each skater is

$$\Delta p = F \Delta t = (200 \text{ N})(0.40 \text{ s}) = 80 \text{ kg} \cdot \text{m/s}$$

Since each started at rest, this is also the magnitude of the skaters' momenta after the push. Therefore, the two velocities are

Skater A: $v = \dfrac{p}{m} = \dfrac{80 \text{ kg} \cdot \text{m/s}}{50 \text{ kg}} = 1.6$ m/s in the +x-direction

Skater B: $v = \dfrac{p}{m} = \dfrac{80 \text{ kg} \cdot \text{m/s}}{80 \text{ kg}} = 1.0$ m/s in the −x-direction

REFLECT The answers agree with Example 4.3. Newton's third law and the principle of momentum conservation express equivalent information.

MAKING THE CONNECTION In the expression $\Delta p = F \Delta t$, are each of the three quantities (Δp, F, and Δt) required to be the same for both skaters? Why?

ANSWER Yes. The Δp values are the same because of momentum conservation. The forces have the same magnitude by Newton's third law. And the Δt quantities are also the same, because the contact time is the same for both skaters.

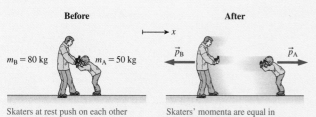

Before	After
$m_B = 80$ kg　$m_A = 50$ kg	$\vec{p}_B$　　$\vec{p}_A$
Skaters at rest push on each other with force $F_{AB} = F_{BA} = 200$ N for time $\Delta t = 0.40$ s.	Skaters' momenta are equal in magnitude and opposite in direction; total momentum remains zero: $\vec{p}_A + \vec{p}_B = 0$

FIGURE 6.13 Skaters conserving momentum on ice.

GOT IT? Section 6.2 A 0.5-kg ball bounces off the wall as shown. The ball's change in momentum is (a) 0; (b) 2.0 kg · m/s; (c) 6.0 kg · m/s; (d) 10.0 kg · m/s.

Before　12 m/s　　After　8 m/s

6.3 Collisions and Explosions in One Dimension

A **collision** is a brief encounter between objects that involves strong interaction forces and results in sudden, dramatic changes in the motions of one or both objects. Collisions in nature range from subatomic particle collisions to colliding galaxies. Everyday collisions occur in games where objects are struck, kicked, batted, or clubbed. Accidental collisions occur on the highway. The principle of momentum conservation is the key to understanding collisions.

Some collisions occur in one dimension. We'll start with these in this section, because they're the simplest collisions. Then in Section 6.4, we'll consider two-dimensional collisions.

Types of Collisions

The forces between colliding objects are so strong that any external forces are negligible in comparison, and thus momentum conservation becomes an excellent approximation. However, total mechanical energy may or may not be conserved. The distinction leads to three types of collision:

In an **elastic collision**, the total mechanical energy of the colliding objects is conserved.

In an **inelastic collision**, the total mechanical energy of the system of objects is not conserved.

In a **perfectly inelastic collision**, the colliding objects stick together. The total mechanical energy of the system of objects is not conserved.

Perfectly Inelastic Collisions in One Dimension

Perfectly inelastic collisions in one dimension are the simplest, so we'll analyze them first. This type of collision happens when one car rear-ends another and the two stick together. More benignly, you might study perfectly inelastic collisions by attaching Velcro to air-track gliders' bumpers.

Suppose an air-track glider with mass m_1 approaches a second with mass m_2, initially at rest (Figure 6.14a). The moving glider has a velocity v_{1xi} (i for "initial"), so its momentum is $p_{1xi} = m_1 v_{1xi}$. Since only m_1 is moving, that's the total momentum. The gliders stick together and move with velocity v_{xf} (f for "final" in Figure 6.14b). With combined mass $m_1 + m_2$, the momentum is $p_{xf} = (m_1 + m_2)v_{xf}$. Conservation of momentum states that the total momentum before the collision equals the total momentum after the collision. In this case, that means

$$m_1 v_{1xi} = (m_1 + m_2)v_{xf}$$

Knowing the masses and initial velocity, we can solve for v_{xf},

$$v_{xf} = \frac{m_1 v_{1xi}}{m_1 + m_2}$$

Before collision:
Moving glider approaches glider at rest

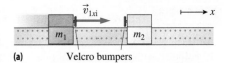

(a)   Velcro bumpers

The gliders experience zero net external force, so their momentum is conserved throughout.

After collision

(b)   Gliders stuck together

FIGURE 6.14 Perfectly inelastic collision on an air track.

EXAMPLE 6.5   **Truck-Car Collision**

A 3470-kg truck going 11.0 m/s runs into a 975-kg stopped car, and they stick together. How fast is the pair moving just after colliding?

ORGANIZE AND PLAN We start with "before" and "after" pictures (Figure 6.15). This is the same physical situation as our air-track gliders. We'll solve the same way, using conservation of momentum. As shown in the text, the final velocity, as a function of the masses and initial velocity of the oncoming truck, is

$$v_{xf} = \frac{m_1 v_{1xi}}{m_1 + m_2}$$

Known: Truck mass $m_1 = 3470$ kg; car mass $m_2 = 975$ kg; truck initial velocity $v_{1xi} = 11.0$ m/s.

Before

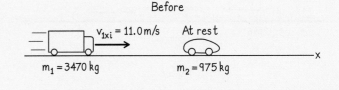

After

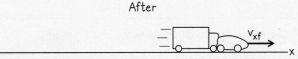

FIGURE 6.15 Conserving momentum in a truck-car collision.

cont'd.

SOLVE Inserting values,

$$v_{xf} = \frac{m_1 v_{1xi}}{m_1 + m_2} = \frac{(3470 \text{ kg})(11.0 \text{ m/s})}{3470 \text{ kg} + 975 \text{ kg}} = 8.59 \text{ m/s}$$

REFLECT Consistent with the discussion in Conceptual Example 6.6, the final velocity of the combined vehicles isn't much less than the truck's initial velocity.

MAKING THE CONNECTION How long would the truck-car combination travel at 8.59 m/s?

ANSWER Note that the problem statement used the phrase "just after colliding." Friction would slow the wrecked pair quickly unless the collision occurred on an icy road.

CONCEPTUAL EXAMPLE 6.6 Final Velocity?

Consider the perfectly inelastic collision described in Figure 6.14. What can you say about the final velocity of the combined objects, relative to the initial velocity, if the objects have equal mass ($m_1 = m_2$)? What if $m_1 > m_2$? What if $m_1 < m_2$?

SOLVE Your intuition may be a good guide, but to be safe, consider the equation for v_{xf} derived in the text:

$$v_{xf} = \frac{m_1 v_{1xi}}{m_1 + m_2}$$

When $m_1 = m_2$, the fraction $m_1/(m_1 + m_2) = \frac{1}{2}$. Therefore, the final velocity is exactly half the velocity of the initially moving glider. When $m_1 > m_2$, the fraction $m_1/(m_1 + m_2) > \frac{1}{2}$, and the final velocity is more than half the initial velocity. When $m_1 < m_2$, $m_1/(m_1 + m_2) < \frac{1}{2}$, and the final velocity is less than half the initial velocity.

REFLECT Note that in a perfectly inelastic collision with one object initially stationary, the objects can't rebound backward. The combined objects have to keep moving forward to conserve momentum.

There's nothing that says one object has to be at rest prior to a perfectly inelastic collision. But the principle of momentum conservation still applies, as the following example shows.

EXAMPLE 6.7 Different Masses, Different Velocities

Air-track glider 1, mass $m_1 = 0.150$ kg, is moving right at 35.2 cm/s. It collides with glider 2, mass $m_2 = 0.100$ kg, moving left at 44.7 cm/s, and the two stick together. What's the velocity of the joined gliders after they collide?

ORGANIZE AND PLAN Figure 6.16 includes our "before" and "after" diagrams. With external forces negligible, momentum conservation governs this collision. The total momentum of the two gliders before the collision equals the momentum of the joined gliders after the collision.

Let the $+x$-axis point rightward. Equating the x-components of momentum before and after the collision,

$$p_{1xi} + p_{2xi} = p_{xf}$$

After the collision there's only one object, with mass $m_1 + m_2$, so there's only one "final" term in the equation. With $p_x = mv_x$, we have

$$m_1 v_{1xi} + m_2 v_{2xi} = (m_1 + m_2)v_{xf}$$

Known: $m_1 = 0.150$ kg; $v_{1xi} = 0.352$ m/s; $m_2 = 0.100$ kg, $v_{2xi} = -0.447$ m/s (negative because it's moving to the left).

SOLVE Solving for v_{xf},

$$v_{xf} = \frac{m_1 v_{1xi} + m_2 v_{2xi}}{m_1 + m_2}$$
$$= \frac{(0.150 \text{ kg})(0.352 \text{ m/s}) + (0.100 \text{ kg})(-0.447 \text{ m/s})}{0.150 \text{ kg} + 0.100 \text{ kg}}$$
$$= 0.0324 \text{ m/s}$$

REFLECT The final velocity is positive, meaning the joined gliders move to the right. This might have turned out differently: If glider 2 were moving somewhat faster, the total momentum would have been leftward. Or, with just the right combination of velocities, the colliding gliders could end up at rest.

MAKING THE CONNECTION What velocity for glider 2 would make the pair stop completely?

ANSWER The system's total momentum must be zero. That is, $m_1 v_{1xi} + m_2 v_{2xi} = 0$. Solving for v_{2xi}, you'll find $v_{2xi} = -0.528$ m/s, just a bit faster than the velocity in the original problem.

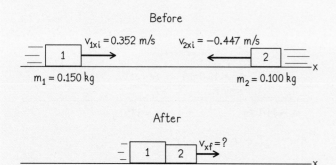

FIGURE 6.16 Air-track collision between moving gliders.

You can show that kinetic energy isn't conserved in the inelastic collisions we've considered. For the gliders in Example 6.7, the total kinetic energy before the collision was

$$K = \tfrac{1}{2}mv_{1xi}^2 + \tfrac{1}{2}mv_{2xi}^2 = \tfrac{1}{2}(0.150\ \text{kg})(0.352\ \text{m/s})^2 + \tfrac{1}{2}(0.100\ \text{kg})(0.447\ \text{m/s})^2$$
$$= 0.019\ \text{J}$$

After colliding, the joined gliders have kinetic energy

$$K = \tfrac{1}{2}mv_{1xi}^2 + \tfrac{1}{2}mv_{2xi}^2 = \tfrac{1}{2}(0.150\ \text{kg})(0.0324\ \text{m/s})^2 + \tfrac{1}{2}(0.100\ \text{kg})(0.0324\ \text{m/s})^2$$
$$= 1.3 \times 10^{-4}\ \text{J}$$

Clearly, lots of kinetic energy is lost in this collision. The fraction lost depends on the masses and velocities of the colliding objects.

Elastic Collisions in One Dimension

Consider a collision between two amusement park bumper cars. Ideally, there's no energy lost in the springy bumpers, so the collision is *elastic*. In a perfectly elastic collision, the total kinetic energy remains the same before and after the collision.

Elastic collisions are more complicated than the perfectly inelastic collisions we've just analyzed. That's because in elastic collisions, two final velocities need to be determined. Fortunately, there's also more information, because total kinetic energy is the same before and after the collision. (*During* the collision, kinetic energy becomes, briefly, potential energy—but we don't need the details to compare "before" and "after" situations.)

Consider first a mass m_1 that collides elastically with mass m_2 initially at rest, as shown in Figure 6.17. If the collision is head-on, then both masses end up moving along the line defined by the m_1's initial velocity. Then we have a *one-dimensional elastic collision*, and we'll take the initial velocity to define the x-axis.

The incoming object m_1 has initial velocity component v_{1xi}. If we know this velocity and the two masses, then we can predict the final velocities of both objects. In a one-dimensional collision we're then after two unknowns, v_{1xf} and v_{2xf}. As usual, external forces are negligible during the collision, so the system's total momentum is conserved. Equating the initial and final momentum components,

$$m_1 v_{1xi} = m_1 v_{1xf} + m_2 v_{2xf}$$

In an *elastic* collision, mechanical energy is also conserved, so we equate initial and final kinetic energies:

$$\tfrac{1}{2}m_1 v_{1xi}^2 = \tfrac{1}{2}m_1 v_{1xf}^2 + \tfrac{1}{2}m_2 v_{2xf}^2$$

Solving these two equations for v_{1xf} and v_{2xf} requires some algebra. We'll leave this to the end-of-chapter problems, and present the results:

$$v_{1xf} = \frac{m_1 - m_2}{m_1 + m_2} v_{1xi} \tag{6.12}$$

and

$$v_{2xf} = \frac{2m_1}{m_1 + m_2} v_{1xi} \tag{6.13}$$

Note the special case $m_1 = m_2$. (Think of this as the "pool table" case, because it's what happens when a cue ball strikes a stationary ball of the same mass head-on.) With $m_1 = m_2$, Equations 6.12 and 6.13 give $v_{1xf} = 0$ and $v_{2xf} = v_{1xi}$. Thus the cue ball stops, and the target ball acquires the velocity the cue ball had before the collision.

Before collision:
Moving glider approaches glider at rest

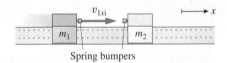

Spring bumpers

Because the gliders collide elastically, no kinetic energy is lost during the collision.

After collision

FIGURE 6.17 One-dimensional elastic collision on an air track.

Exploring Cases of Elastic Collisions

For a head-on elastic collision with m_2 initially at rest, explore what happens when $m_1 > m_2$ and when $m_1 < m_2$. Consider also the extreme cases, $m_1 \gg m_2$ and $m_1 \ll m_2$.

SOLVE The big difference between the cases $m_1 > m_2$ and $m_1 < m_2$ is in the final velocity of the incoming object. If $m_1 > m_2$, then Equation 6.12 gives $v_{1xf} > 0$, showing that the incoming mass m_1 continues moving to the right after the collision (Figure 6.18a). However, if $m_1 < m_2$, then $v_{1xf} < 0$, so m_1 rebounds to the left (Figure 6.18b). In

either case v_{2xf} is positive—it's impossible for an object struck from the left to move any direction but right.

When $m_1 \gg m_2$, the ratio $(m_1 - m_2)/(m_1 + m_2)$ in Equation 6.12 is nearly 1, so the velocity of the incoming mass hardly changes. That's reasonable, because the velocity of a heavy object shouldn't be affected much by its striking something much lighter. A good example is a massive golf club striking a ball. And with $m_1 \gg m_2$, we can neglect m_2 in Equation 6.13, showing that v_{2xf} is approximately $2v_{1xi}$.

When $m_1 \ll m_2$, $v_{1xf} < 0$ and is approximately $-v_{1xi}$. Thus the light object rebounds with nearly its initial speed. (Think of a ball bouncing off a brick wall.) Meanwhile v_{2xf} is positive but negligibly small—it's hard for a light object to move a much heavier one.

REFLECT Try to think of other examples—from sports, games, or everyday life—where heavy and light objects collide. Do they follow the patterns described here?

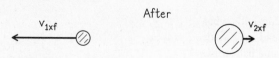

(a) $m_1 > m_2$ **(b)** $m_1 < m_2$

FIGURE 6.18 Results for the two cases with different masses.

An elastic collision is an idealization in the macroscopic world, because energy is always lost in deforming the colliding bodies. But elastic collisions are common in the subatomic world, as the next example shows.

EXAMPLE 6.9 **Inside a Nuclear Reactor**

One type of nuclear reactor uses graphite (solid carbon) to slow neutrons, enhancing nuclear fission. Suppose a neutron (mass 1.67×10^{-27} kg) moving at 2.88×10^5 m/s collides elastically with a stationary carbon nucleus (mass 1.99×10^{-26} kg). Find the velocities of the two particles after colliding.

ORGANIZE AND PLAN As usual, we neglect external forces during the brief collision, so momentum is conserved. Because this collision is elastic, mechanical energy is also conserved. For convenience, let the neutron's original motion define the $+x$-direction (Figure 6.19).

With momentum and energy both conserved, and m_2 initially at rest, the velocities after the collision follow from Equations 6.12 and 6.13.

Known: Neutron mass $m_1 = 1.67 \times 10^{-27}$ kg; carbon mass $m_2 = 1.99 \times 10^{-26}$ kg; neutron's initial velocity $v_{1xi} = 2.88 \times 10^5$ m/s.

SOLVE Using the given values,

$$v_{1xf} = \frac{m_1 - m_2}{m_1 + m_2} v_{1xi}$$
$$= \left(\frac{1.67 \times 10^{-27} \text{ kg} - 1.99 \times 10^{-26} \text{ kg}}{1.67 \times 10^{-27} \text{ kg} + 1.99 \times 10^{-26} \text{ kg}} \right) (2.88 \times 10^5 \text{ m/s})$$
$$= -2.43 \times 10^5 \text{ m/s}$$

$$v_{2xf} = \frac{2m_1}{m_1 + m_2} v_{1xi}$$
$$= \left(\frac{2(1.67 \times 10^{-27} \text{ kg})}{1.67 \times 10^{-27} \text{ kg} + 1.99 \times 10^{-26} \text{ kg}} \right) (2.88 \times 10^5 \text{ m/s})$$
$$= 4.46 \times 10^4 \text{ m/s}$$

As expected, the lighter neutron rebounds, while the carbon acquires a modest speed.

REFLECT The neutron's speed reduction is about 16%. Although carbon has some advantages as a neutron-slowing material, its large mass relative to the neutron means that many collisions are required for sufficient slowing.

MAKING THE CONNECTION The neutron-slowing substance in a nuclear reactor is the *moderator*. What might be a more appropriate moderator than graphite?

ANSWER A head-on collision with $m_1 = m_2$ completely stops the incident particle. Therefore, we want a particle with mass closer to the neutron's. One such particle is the proton comprising the nucleus of hydrogen, and indeed many reactors use ordinary water (H_2O) as the moderator.

Before

$v_{1xi} = 2.88 \times 10^5$ m/s   At rest

Neutron
$m = 1.67 \times 10^{-27}$ kg

Carbon nucleus
$m = 1.99 \times 10^{-26}$ kg

After

v_{1xf} v_{2xf}

FIGURE 6.19 Slowing a neutron.

The elastic collisions in this section all involved the special case where one object is initially at rest. In the end-of-chapter problems, you'll have the chance to explore the case where both are initially moving.

Explosions

In Example 6.7, we suggested that it's possible for two colliding objects to come to rest after the collision. This must happen in a perfectly inelastic collision, whenever the total momentum is zero.

Now consider a time-reversed video of that zero-momentum collision. You'd start with two objects stuck together, and then they'd suddenly fly apart. That's an *explosion*, of which Figure 6.20 shows a simple example.

You can think of an explosion as a perfectly inelastic collision in reverse. The big difference isn't momentum (which is zero in both collision and explosion), but energy. In a collision with zero total momentum, all the kinetic energy of the two objects gets converted to other forms (such as heat and sound). The reverse process, the explosion, requires an energy source to give the objects their kinetic energy. In Figure 6.20 that's provided by a spring. A more typical explosion involves many particles and is driven by chemical energy.

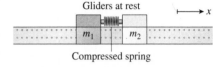

Before explosion: zero momentum

An energy source (here, the spring) gives the gliders their kinetic energy.

After explosion: momentum conserved

FIGURE 6.20 Two gliders explode apart.

TIP

Think of an explosion as a perfectly inelastic collision in reverse.

An important example of a two-particle explosion is *alpha decay*, in which an atomic nucleus splits spontaneously into a smaller nucleus and an alpha particle (a helium nucleus). Many heavy radioactive nuclei undergo this process. We'll discuss alpha decay further in Chapter 25.

EXAMPLE 6.10   **Alpha Decay of Uranium**

A uranium-238 nucleus at rest undergoes radioactive decay, splitting into an alpha particle (helium nucleus) with mass 6.64×10^{-27} kg and a thorium nucleus with mass 3.89×10^{-25} kg. The measured kinetic energy of the alpha particle is 6.73×10^{-13} J. Find the momentum and kinetic energy of the thorium.

ORGANIZE AND PLAN As in the neutron-slowing example, momentum is conserved, but in this reverse-inelastic process, kinetic energy isn't. Knowing the alpha particle's kinetic energy lets you find its momentum. This, along with momentum conservation, gives the momentum of the thorium nucleus, and that lets you find its kinetic energy too. Equation 6.5 relates momentum and kinetic energy: $K = p^2/2m$.

We'll take the $+x$-direction to be that of the alpha particle (Figure 6.21). The total momentum of the system is zero, because that's what it was before the decay. Therefore, the thorium's momentum has the same magnitude and opposite direction of the alpha particle's momentum:

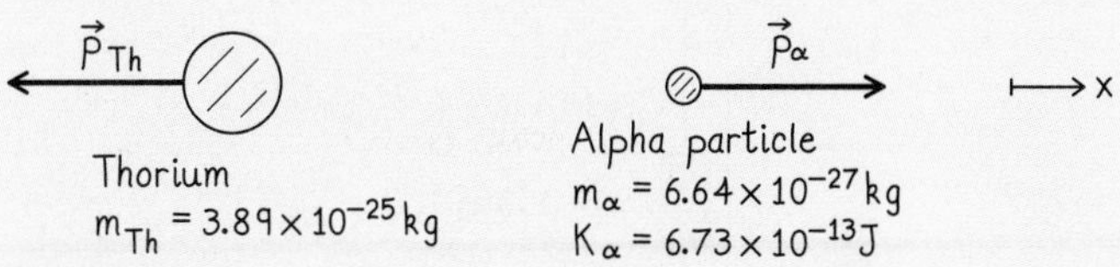

Known: $m_\alpha = 6.64 \times 10^{-27}$ kg; $m_{Th} = 3.89 \times 10^{-25}$ kg; $K_\alpha = 6.73 \times 10^{-13}$ J.

SOLVE Solving Equation 6.5 for momentum p gives

$$p_\alpha = \sqrt{2m_\alpha K_\alpha} = \sqrt{2(6.64 \times 10^{-27} \text{ kg})(6.73 \times 10^{-13} \text{ J})}$$
$$= 9.45 \times 10^{-20} \text{ kg} \cdot \text{m/s}$$

in the $+x$-direction. Therefore, the thorium nucleus has the momentum of magnitude $p_{Th} = 9.45 \times 10^{-20}$ kg·m/s but in the $-x$-direction. The thorium's kinetic energy is

$$K_{Th} = \frac{p_{Th}^2}{2m_{Th}} = \frac{(9.45 \times 10^{-20} \text{ kg} \cdot \text{m/s})^2}{2(3.89 \times 10^{-25} \text{ kg})} = 1.15 \times 10^{-14} \text{ J}$$

REFLECT Note that the thorium's kinetic energy is much smaller than the alpha's. This is a consequence of momentum conservation: Because the particles started from rest, the heavier particle is moving much slower.

MAKING THE CONNECTION Should you worry about being hit by an alpha particle with this kinetic energy?

ANSWER That's enough energy to do significant damage to the DNA in your cells, possibly inducing mutations or cancer. That's unlikely with a single alpha particle. But exposure to a radioactive alpha-emitter like U-238 poses serious health risks.

$\vec{p}_{Th}$ Thorium $m_{Th} = 3.89 \times 10^{-25}$ kg

$\vec{p}_\alpha$ Alpha particle $m_\alpha = 6.64 \times 10^{-27}$ kg $K_\alpha = 6.73 \times 10^{-13}$ J

$\longmapsto x$

FIGURE 6.21 Alpha decay of uranium-238.

GOT IT? Section 6.3 On a train track, a moving train car strikes a heavier car that's initially at rest. Compare the two cases where the collision is elastic or inelastic. Which collision transfers more momentum to the heavier car? (a) the elastic collision; (b) the inelastic collision; (c) neither.

Reviewing New Concepts

In a collision:
- Momentum of an isolated system is conserved.
- Total mechanical energy—usually kinetic—is conserved in an elastic collision.
- Total mechanical energy is not conserved in an inelastic collision.

Think of an explosion as a perfectly inelastic collision in reverse.

6.4 Collisions and Explosions in Two Dimensions

The same principles we've applied to one-dimensional collisions also apply to two-dimensional collisions, because momentum is still conserved. The fact that momentum is a vector quantity, however, makes two-dimensional collisions more complex.

Kinetic energy may or may not be conserved in a two-dimensional collision, and the classifications of Section 6.3 still apply: collisions are elastic, inelastic, or perfectly inelastic. Once again, we'll start with perfectly inelastic collisions, which are easier to understand because there's only one object after the collision.

Perfectly Inelastic Collisions in Two Dimensions

Figure 6.22 shows a perfectly inelastic collision in two dimensions. It's two-dimensional because the initial velocities aren't along the same line.

Kinetic energy isn't conserved in a perfectly inelastic collision, so the only principle to apply here is conservation of momentum—as usual, assuming external forces are negligible during the collision. Equating the total momentum before and after the collision,

$$m_1 \vec{v}_{1i} + m_2 \vec{v}_{2i} = (m_1 + m_2)\vec{v}_f \tag{6.14}$$

Because we're in two dimensions, the vector notation is essential here. There's only one final velocity $\vec{v}_f$, because the objects stick together in a perfectly inelastic collision. We can solve for that velocity:

$$\vec{v}_f = \frac{m_1 \vec{v}_{1i} + m_2 \vec{v}_{2i}}{m_1 + m_2} \tag{6.15}$$

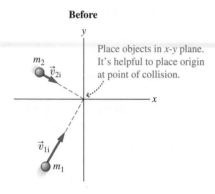

Before

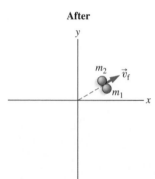

After

FIGURE 6.22 Two-dimensional perfectly inelastic collision.

The following example shows how this works.

EXAMPLE 6.11 **Accident at an Intersection**

Car 1, mass 925 kg, is traveling eastward at 25 mph (11.2 m/s), and Car 2 (1150 kg) is traveling northward at 30 mph (13.4 m/s). They collide at an intersection and stick together. Find the velocity of the wreckage immediately after the collision.

ORGANIZE AND PLAN This is a perfectly inelastic collision in which momentum is conserved, neglecting any external forces acting over the short collision time. Conservation of momentum is all that's needed to solve the problem. In any two-dimensional problem, it helps to draw a before/after diagram with appropriate coordinate system. Let the +x-direction be east and the +y-direction be north (Figure 6.23).

The final velocity in this perfectly inelastic collision is given by Equation 6.15:

$$\vec{v}_f = \frac{m_1 \vec{v}_{1i} + m_2 \vec{v}_{2i}}{m_1 + m_2}$$

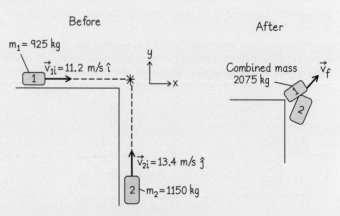

FIGURE 6.23 Cars colliding at intersection.

cont'd.

Because you'll be adding vectors, it's convenient to express the velocities in unit vector notation.

Known: Car 1: $m_1 = 925$ kg, $\vec{v}_{1i} = 11.2$ m/s $\hat{\imath}$

Car 2: $m_2 = 1150$ kg, $\vec{v}_{2i} = 13.4$ m/s $\hat{\jmath}$

SOLVE Using the given values,

$$\vec{v}_f = \frac{m_1 \vec{v}_{1i} + m_2 \vec{v}_{2i}}{m_1 + m_2}$$

$$= \frac{(925 \text{ kg})(11.2 \text{ m/s } \hat{\imath}) + (1150 \text{ kg})(13.4 \text{ m/s } \hat{\jmath})}{925 \text{ kg} + 1150 \text{ kg}}$$

$$= 4.99 \text{ m/s } \hat{\imath} + 7.43 \text{ m/s } \hat{\jmath}$$

REFLECT The final velocity has a larger y-component than x-component. This makes sense, because Car 2—initially traveling in the y-direction—carried more momentum than Car 1, by virtue of both its greater mass and greater speed.

MAKING THE CONNECTION In what direction is the wreckage headed just after the collision?

ANSWER The angle of the final velocity is $\theta = \tan^{-1} 7.43/4.99 = 56.1°$ north of east.

✓ **TIP**

Keep track of both components of momentum in two-dimensional collisions.

Explosions in Two Dimensions

For an explosion to be two-dimensional, there must be multiple pieces moving off in different directions. As in one dimension, think of the explosion as a time-reversed perfectly inelastic collision. Momentum is conserved, but not kinetic energy. If the exploding object is initially at rest, the total momentum is zero. If the object is initially moving, its momentum before the explosion equals the net momentum of all the fragments.

 EXAMPLE 6.12 **Bomb**

A bomb of mass M explodes into three pieces. One (mass $M/4$) flies off in the $-y$-direction at 16.8 m/s. A second (mass $M/4$) flies off in the $-x$-direction at 11.4 m/s. What's the velocity of the third piece?

ORGANIZE AND PLAN Figure 6.24 is our diagram. With no external forces, momentum is conserved, but kinetic energy isn't, because the blast supplies kinetic energy to the exploding pieces. Applying conservation of momentum is all that's needed to find the unknown velocity.

Mass is conserved, so the third piece has mass $M - M/4 - M/4 = M/2$. The total momentum of the system is zero, because the

device was at rest prior to the explosion. Because momentum is conserved, the net momentum of the three pieces is zero.

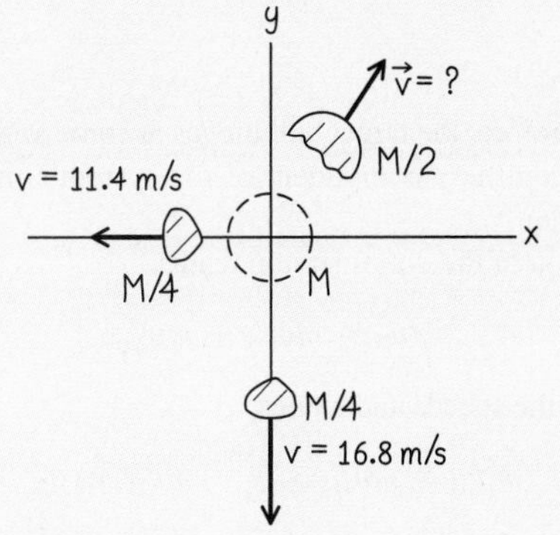

Known: Initial velocities of two $M/4$ pieces: -11.4 m/s $\hat{\imath}$ and -16.8 m/s $\hat{\jmath}$.

SOLVE Here conservation of momentum reads

$$\left(\frac{M}{4}\right)(-11.4 \text{ m/s } \hat{\imath}) + \left(\frac{M}{4}\right)(-16.8 \text{ m/s } \hat{\jmath}) + \left(\frac{M}{2}\right)\vec{v} = 0$$

where $\vec{v}$ is the velocity of the third piece. Solving for $\vec{v}$,

$$\vec{v} = \left(\tfrac{1}{2}\right)(11.4 \text{ m/s } \hat{\imath}) + \left(\tfrac{1}{2}\right)(16.8 \text{ m/s } \hat{\jmath}) = 5.7 \text{ m/s } \hat{\imath} + 8.4 \text{ m/s } \hat{\jmath}$$

REFLECT As a check, you can verify that the net momentum of the three pieces is in fact zero. This makes sense because the three vectors in Figure 6.24—in the $-y$-direction, $-x$-direction, and first quadrant—clearly add to zero.

MAKING THE CONNECTION If $M = 2.8$ kg, how much energy was released in the explosion?

ANSWER The energy released shows up as kinetic energy of the three pieces. Adding the kinetic energies $\tfrac{1}{2}mv^2$ gives 216 J.

FIGURE 6.24 Motion of the shell fragments.

Elastic Collisions in Two Dimensions

A two-dimensional elastic collision conserves both momentum and kinetic energy. You can study this situation experimentally with colliding pucks on an air table.

Figure 6.25 shows two colliding billiard balls with masses m_1 and m_2. For reasons that will soon be clear, we'll write the initial and final velocity vectors of the two objects in terms of their x- and y-components. Table 6.1 summarizes the objects' velocities before and after the collision.

Before

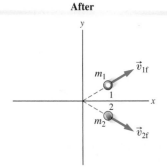

After

FIGURE 6.25 An elastic collision in two dimensions.

TABLE 6.1 Summary of the Velocities Before and After the Collision

Object	Mass	Velocity before collision	Velocity after collision
1	m_1	$\vec{v}_{1i} = v_{1xi}\,\hat{i} + v_{1yi}\,\hat{j}$	$\vec{v}_{1f} = v_{1xf}\,\hat{i} + v_{1yf}\,\hat{j}$
2	m_2	$\vec{v}_{2i} = v_{2xi}\,\hat{i} + v_{2yi}\,\hat{j}$	$\vec{v}_{2f} = v_{2xf}\,\hat{i} + v_{2yf}\,\hat{j}$

Conservation of momentum applies to each component, so we can equate separately the x- and y-components before and after the collision:

$$m_1 v_{1xi} + m_2 v_{2xi} = m_1 v_{1xf} + m_2 v_{2xf} \tag{6.16}$$

$$m_1 v_{1yi} + m_2 v_{2yi} = m_1 v_{1yf} + m_2 v_{2yf} \tag{6.17}$$

In this elastic collision, kinetic energy is also conserved:

$$\tfrac{1}{2}m_1 v_{1i}^2 + \tfrac{1}{2}m_2 v_{2i}^2 = \tfrac{1}{2}m_1 v_{1f}^2 + \tfrac{1}{2}m_2 v_{2f}^2 \tag{6.18}$$

Note that the speeds in Equation 6.18 aren't the velocity components of Equations 6.16 and 6.17. That's because kinetic energy is a *scalar*, which depends on the *magnitude* of the velocity, given by

$$v_{1i}^2 = v_{1xi}^2 + v_{2yi}^2$$

and similarly for the other three.

Equations 6.16, 6.17, and 6.18 express conservation of momentum and energy for elastic collisions in two dimensions. There are four final velocity components, but only three equations. So we need to know something else—a final speed, one of the final velocity components, or the angle between the final velocity vectors.

It's sometimes helpful to express the velocities not in terms of components but as a magnitude and direction. That's handy because the magnitudes—speeds—appear in Equation 6.18. The following example illustrates this.

EXAMPLE 6.13   **Billiard Ball Collision**

A 120-g cue ball moving at 0.42 m/s strikes a stationary target ball with the same mass. After the collision, the cue ball moves at 0.21 m/s at 60° from its original line of motion. Assuming an elastic collision, what's the final velocity of the target ball? Express your answer in magnitude (speed) and direction.

ORGANIZE AND PLAN This is an elastic collision, so both momentum and kinetic energy are conserved; for momentum, x- and y-components are individually conserved.

We'll choose a coordinate system with the incoming cue ball moving in the +x-direction at $v_{1i} = 0.42$ m/s (Figure 6.26), so its momentum (the system's total momentum) is in that direction. Note that the cue ball must hit the target ball off-center in order to move off at a 60° angle, and we've chosen that angle to be θ_1 below the x-axis

as shown. Therefore, the target ball moves at some angle θ_2 above the +x-axis, to keep the y-component of the system's momentum zero after the collision.

Conservation of the x-momentum reads:

$$mv_{1i} = mv_{1xf} + mv_{2xf}$$

or, in terms of the speeds and angles,

$$mv_{1i} = mv_{1f}\cos\theta_1 + mv_{2f}\cos\theta_2$$

Similarly, conserving the y-momentum gives

$$0 = -mv_{1yf}\sin\theta_1 + mv_{2yf}\sin\theta_2$$

cont'd.

Conservation of kinetic energy gives a third equation,

$$\tfrac{1}{2}mv_{1i}^2 = \tfrac{1}{2}mv_{1f}^2 + \tfrac{1}{2}mv_{2f}^2$$

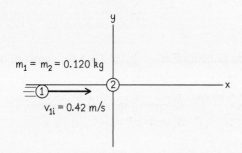

Before

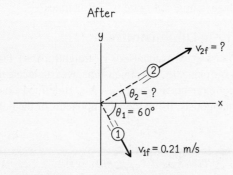

After

FIGURE 6.26 Elastic billiard ball collision.

Known: $m = 0.120$ kg; $v_{1i} = 0.42$ m/s; $v_{1f} = 0.21$ m/s; $\theta_1 = 60°$.

SOLVE Note that m cancels in all three equations, so you don't really need the mass. With known values $v_{1i} = 0.42$ m/s and $v_{1f} = 0.21$ m/s, the conservation-of-energy equation can be solved for the speed v_{2f} of the target ball:

$$v_{2f} = \sqrt{v_{1i}^2 - v_{1f}^2} = \sqrt{(0.42\ \text{m/s})^2 - (0.21\ \text{m/s})^2} = 0.364\ \text{m/s}$$

Now that v_{2f} is known, either momentum equation can be solved for θ_2. The y equation is easier:

$$\sin\theta_2 = \frac{v_{1f}\sin\theta_1}{v_{2f}} = \frac{(0.21\ \text{m/s})[\sin(60°)]}{0.364\ \text{m/s}} = 0.500$$

so $\theta_2 = 30°$.

REFLECT Note that the total angle between the balls' velocities after the collision is 90°. This is *not* a coincidence, but a consequence of their equal mass. When objects of equal mass collide, with one initially at rest, their final velocities are *always* perpendicular. Now you know the pool player's secret!

MAKING THE CONNECTION If the speeds of the two balls were the same after colliding, what would be their directions (relative to the x-axis)?

ANSWER The 90° rule above, along with the fact that the y-momentum is zero, means that the balls move at 45° above and below the x-axis.

GOT IT? Section 6.4 Three identical ball bearings are stuck together. A small explosive charge blows them apart (see figure). Two move perpendicularly as shown, with the same speed v_0. The third ball's speed (a) is 0; (b) is less than v_0; equal to v_0; (c) is greater than v_0; (d) cannot be determined without more information.

6.5 Center of Mass

Try balancing a dinner plate on one finger. You instinctively place your finger as close as possible to the center of the symmetric plate. Now try balancing a nonsymmetric object, such as a baseball bat or cordless phone. You'll discover that the object balances on just one point, its **center of mass**. Here we'll define center of mass for a system of particles, first in one dimension and then in higher dimensions. We'll show some applications and explain how center of mass relates to momentum and collisions.

Center of Mass in One Dimension

We'll first consider the center of mass of several objects (small enough to be considered point particles) on the x-axis (Figure 6.27). The center of mass position is defined as a *weighted average*, in which objects with greater mass are given proportionally more weight:

$$\text{Center of mass} = X_{cm} = \frac{m_1x_1 + m_2x_2 + m_3x_3 + \cdots + m_nx_n}{m_1 + m_2 + m_3 + \cdots + m_n} \quad (6.19)$$

$m_1 = 2$ kg $m_2 = 2$ kg $m_3 = 4$ kg

$x_1 = -2$ m $x_2 = 0$ X_{cm} $x_3 = 3$ m

Center of mass for system of particles is the average of the particles' positions weighted by their masses:

$$X_{cm} = \frac{m_1x_1 + m_2x_2 + m_3x_3}{m_1 + m_2 + m_3}$$

$$= \frac{(2\ \text{kg})(-2\ \text{m}) + (2\ \text{kg})(0) + (4\ \text{kg})(3\ \text{m})}{2\ \text{kg} + 2\ \text{kg} + 4\ \text{kg}}$$

$$= 1\ \text{m}$$

FIGURE 6.27 Center of mass in one dimension.

This is expressed more compactly using summation notation:

$$\text{Center of mass} = X_{cm} = \frac{\displaystyle\sum_{i=1}^{n} m_i x_i}{\displaystyle\sum_{i=1}^{n} m_i}$$

The denominator is just the total mass M; that is, $M = \displaystyle\sum_{i=1}^{n} m_i$. Then the center of mass in one dimension is

$$\text{Center of mass } X_{cm} = \frac{1}{M}\sum_{i=1}^{n} m_i x_i \qquad \text{(Center of mass, one dimension;}\atop \text{SI unit: m)} \qquad (6.20)$$

It's clear that the center of mass is a *position*, measured in meters in SI. There may or may not be a particle at that position, as the example in Figure 6.27 illustrates.

Center of Mass in Two and Three Dimensions

Extending our definition of center of mass to two dimensions is straightforward. Figure 6.28 illustrates some point particles in the x-y plane. In two dimensions, we can locate the center of mass by its position vector, with components X_{cm} and Y_{cm}. X_{cm} is defined exactly as in one dimension, and Y_{cm} is defined analogously, using the y-components of the positions of the individual particles. That is,

$$X_{cm} = \frac{1}{M}\sum_{i=1}^{n} m_i x_i, \quad Y_{cm} = \frac{1}{M}\sum_{i=1}^{n} m_i y_i \qquad (6.21a, b)$$

In three dimensions, there's a third component:

$$Z_{cm} = \frac{1}{M}\sum_{i=1}^{n} m_i z_i \qquad (6.22)$$

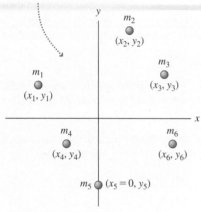

Center of mass for this system of particles will fall somewhere within the system.

FIGURE 6.28 Center of mass in two dimensions.

EXAMPLE 6.14 **Space Station**

A space station consists of three modules forming an equilateral triangle, connected by struts of length L and negligible mass. Two modules have mass m, the other $2m$. Find the station's center of mass.

ORGANIZE AND PLAN The station is essentially two-dimensional, so Equations 6.21 give the center-of-mass position:

$$X_{cm} = \frac{1}{M}\sum_{i=1}^{n} m_i x_i, \quad Y_{cm} = \frac{1}{M}\sum_{i=1}^{n} m_i y_i$$

We've sketched the space station and chosen a coordinate system as shown in Figure 6.29, with the most massive module on the y-axis.

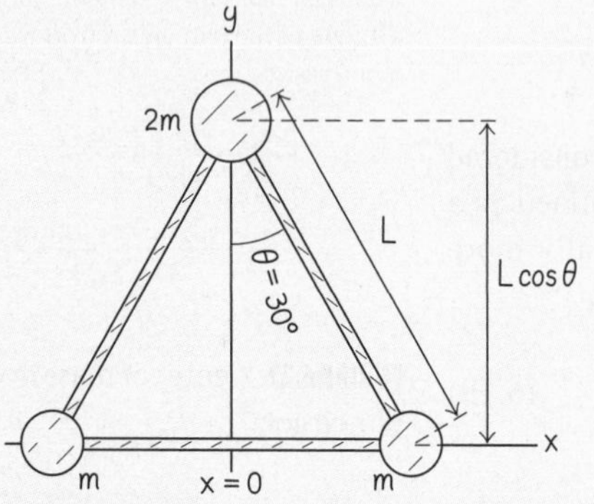

FIGURE 6.29 Geometry of the space station.

Symmetry makes it obvious that $X_{cm} = 0$ in our coordinates, so we'll only need to calculate Y_{cm}.

Known: $m_1 = m_2 = m$; $m_3 = 2m$; $y_1 = y_2 = 0$; $y_3 = L\cos\theta$; $\theta = 30°$ (half-angle of equilateral triangle).

SOLVE Using the given values,

$$\begin{aligned} Y_{cm} &= \frac{1}{M}\sum_{i=1}^{n} m_i y_i \\ &= \frac{1}{4m}\{(m)(0\text{ m}) + (m)(0\text{ m}) + (2m)[L\cos(\theta)]\} \\ &= \frac{\sqrt{3}}{4}L \approx 0.43L \end{aligned}$$

REFLECT Since $\cos(30°) = \sqrt{3}/2$, that's exactly halfway between the lower two modules and the upper one in Figure 6.29. That's not surprising; the mass distribution has $2m$ at $y = 0$ and $2m$ at $y = \sqrt{3}L/2$.

MAKING THE CONNECTION What if two of the modules had mass $2m$ and one of them m?

ANSWER Then the center of mass would have been closer to the line joining the $2m$ modules. A calculation similar to the above gives $Y_{cm} = \sqrt{3}L/8$, where now $y = 0$ is on the line joining the heavier modules.

Center of Mass for Extended Objects

Objects in your everyday world aren't point particles. How can you find the center of mass of an extended object, like a soccer ball or a soccer player?

Symmetry helps: That soccer ball is spherical, with all the particles making it up equidistant from the center. So its center of mass must be at its center. The soccer player is more complicated. When she's standing, the near-symmetry of the human body puts the horizontal position of her center of mass near the vertical plane bisecting her body. The vertical component of the center of mass is harder to find and depends on the body shape and mass distribution. The following rule helps with such a calculation:

> **Rule for extended objects:** In any example in which you need to find the center of mass of a collection of extended objects, you can treat each individual object as a point particle located at that object's center of mass.

Keep this rule in mind when you work a problem involving extended objects.

Many objects are symmetric, or nearly so. Examples include the dinner plate that opened this section and the soccer ball mentioned above. Given that center of mass is a weighted average position of the masses making up an object, this rule follows:

> **Rule for symmetric objects:** The center of mass of any perfectly symmetric object is located at the object's geometric center.

No macroscopic object is perfectly symmetric, so in practice this rule gives an approximate center-of-mass position. Nature provides some nearly symmetric bodies, such as the planet you're standing on. Earth has a slight bulge at its equator (see Chapter 8), but to a fairly good approximation, it's a sphere. Mountains and valleys are small compared with Earth's overall size. Also, although Earth has different onion-like layers that vary significantly in composition and density, the layers are in *reasonably* symmetric spherical shells (Figure 6.30). The approximation as a symmetric sphere is also good for our Moon, as well as many other planets and stars. For all of them, the center of mass is essentially the geometric center.

Some objects have limited symmetry, about a single line or plane. An example is the space station in Example 6.14. Figure 6.29 shows symmetry about a vertical line, here the y-axis, so the center of mass must be centered in the x-direction—giving $X_{cm} = 0$. But there's no horizontal line of symmetry, and therefore we had to calculate Y_{cm}.

Center of Mass and Collisions

Center of mass is closely related to momentum and its application to collisions. To see why, imagine a one-dimensional collision between masses m_1 and m_2. At any moment, the center of mass is given by the usual relationship

$$X_{cm} = \frac{1}{M} \sum_{i=1}^{n} m_i x_i = \frac{1}{M} \left(m_1 x_1 + m_2 x_2 \right)$$

If the particles move, undergoing displacements Δx_1 and Δx_2, then the center-of-mass position may change:

$$\Delta X_{cm} = \frac{1}{M} \left(m_1 \Delta x_1 + m_2 \Delta x_2 \right)$$

Dividing by the time interval Δt in which these changes occur,

$$\frac{\Delta X_{cm}}{\Delta t} = \frac{1}{M} \left(m_1 \frac{\Delta x_1}{\Delta t} + m_2 \frac{\Delta x_2}{\Delta t} \right)$$

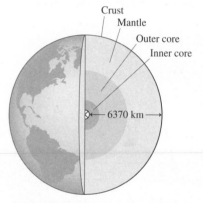

FIGURE 6.30 Earth's interior layers are nearly symmetric.

APPLICATION

Center of Mass and Balance

In order to stand balanced on one foot, your center of mass has to be above that foot. The ballerina stands perched on one leg, her arms and other leg extended. She's learned to make small adjustments in arm and leg positions to keep her center of mass above her supporting toes.

In the limit as Δt approaches zero, $\Delta x / \Delta t$ becomes the velocity, giving

$$V_{cm,x} = \frac{1}{M}(m_1 v_{1x} + m_2 v_{2x})$$

where $V_{cm,x}$ is the velocity of the center of mass. Notice that the terms in parentheses, $m_1 v_{1x}$ and $m_2 v_{2x}$, are just the momenta of the two objects. Therefore, the sum in parentheses is the total momentum of the system, which you know is conserved in a collision during which external forces are negligible. With the total mass M also constant, the left side of the equation (the center-of-mass velocity) is constant. We'll state this as a general rule, which is also valid in two and three dimensions:

> In a system with zero net external force, the velocity of the center of mass is constant.

What happens to the center of mass when a net external force acts on the system? As an example, imagine a toy rocket following the parabolic trajectory of a projectile. It explodes in flight, but the center of mass of all the fragments continues along the same parabolic trajectory (Figure 6.31). This is an example of another general principle, which we'll state as follows.

> In a system with total mass M subject to a net external force $\vec{F}_{net}$, the center of mass undergoes acceleration $\vec{A}_{cm}$. This acceleration is consistent with Newton's second law, such that $\vec{F}_{net} = M\vec{A}_{cm}$.

Note that this principle doesn't distinguish the case of a single object from a swarm of objects with their own individual motions; all that matters to the center of mass is the net force on the entire system. Our fireworks rocket, for example, has total mass M and therefore experiences a net gravitational force $M\vec{g}$, where $\vec{g}$ is the downward acceleration due to gravity. Then $\vec{F}_{net} = M\vec{A}_{cm} = M\vec{g}$, so the center-of-mass acceleration is $\vec{A}_{cm} = \vec{g}$, both before and after the explosion. So the center of mass continues to follow the original parabolic trajectory.

The same principle allows you to understand the motion of the diver shown at the beginning of the chapter. The diver may complete several somersaults and twists as he executes his dive. Internal muscle forces help his body assume different configurations, but the only external force is gravity. So he's a projectile, and his center of mass follows a parabolic path.

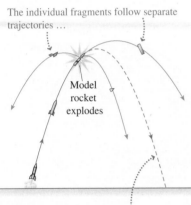

The individual fragments follow separate trajectories ...

Model rocket explodes

... but the center of mass of the set of fragments follows the rocket's original trajectory.

FIGURE 6.31 Center-of-mass motion of the rocket under the influence of gravity.

Chapter 6 in Context

This chapter continued our study of forces and motion, building on Chapter 4's introduction of Newton's laws and Chapter 5's concepts of work and energy. Chapter 6 has been all about *momentum*, a concept that's especially useful in studying the behavior of systems of interacting particles. Applications include *collisions*, *explosions*, and *center-of-mass motion*.

Looking Ahead Now that you've seen the basics of force, energy, and momentum, you're ready to move on and consider some special but important types of motion: oscillations (Chapter 7) and rotational motion (Chapter 8). You'll be able to apply many of the tools you've learned by this point to motions ranging from swinging pendulums to swaying suspension bridges, swirling dancers, and spinning bicycle wheels.

CHAPTER 6 SUMMARY

Introduction to Momentum

(Section 6.1) An object's **momentum** is the product of its mass and velocity.

Momentum is what changes when a net force is applied to an object, as described by Newton's law.

Momentum: $\vec{p} = m\vec{v}$

Newton's second law in terms of momentum: $\vec{F}_{net} = \dfrac{\Delta \vec{p}}{\Delta t}$ for a constant force

Momentum and kinetic energy: $K = \dfrac{p^2}{2m}$

Conservation of Momentum

(Section 6.2) **Conservation of momentum:** Whenever two objects interact, isolated from outside forces, the total momentum of the system is conserved.

In an interaction between two objects, the momentum changes of the two are equal in magnitude but opposite in direction.

Conservation of momentum (absent external forces):

$\Delta \vec{p}_1 = -\Delta \vec{p}_2$, for an interaction between two objects

$\vec{p}_1 + \vec{p}_2 =$ constant. The total momentum of two interacting objects is conserved.

Collisions and Explosions in One Dimension

(Section 6.3) **Elastic collisions** conserve mechanical energy and momentum.

Inelastic collisions do not conserve mechanical energy but do conserve momentum.

In a **perfectly inelastic collision**, the colliding objects stick together.

Kinetic energy is conserved in an elastic collision but not in an inelastic collision.

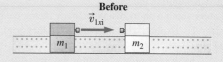

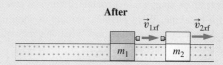

Momentum and energy conservation applied to elastic collisions in one dimension: $v_{1xf} = \dfrac{m_1 - m_2}{m_1 + m_2} v_{1xi}$ $v_{2xf} = \dfrac{2m_1}{m_1 + m_2} v_{1xi}$

Collisions and Explosions in Two Dimensions

(Section 6.4) When momentum is conserved in a collision in two dimensions, both components must be conserved separately.

Kinetic energy may or may not be conserved in a two-dimensional collision, just as in one-dimensional collisions.

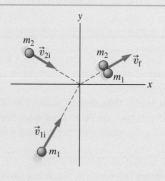

In two-dimensional elastic collisions with no outside forces, the kinetic energy and both components of momentum are conserved.

Perfectly inelastic collision in two dimensions: $m_1\vec{v}_{1i} + m_2\vec{v}_{2i} = (m_1 + m_2)\vec{v}_f$

Elastic collision in two dimensions: $m_1 v_{1xi} + m_2 v_{2xi} = m_1 v_{1xf} + m_2 v_{2xf}$
$$m_1 v_{1yi} + m_2 v_{2yi} = m_1 v_{1yf} + m_2 v_{2yf}$$

Center of Mass

(Section 6.5) An object's **center of mass** is a weighted average of positions of all the mass present.

Center of mass in one dimension: $X_{cm} = \dfrac{1}{M} \sum\limits_{i=1}^{n} m_i x_i$

Center of mass in two dimensions: $X_{cm} = \dfrac{1}{M} \sum\limits_{i=1}^{n} m_i x_i$ and $Y_{cm} = \dfrac{1}{M} \sum\limits_{i=1}^{n} m_i y_i.$

NOTE: Problem difficulty is labeled as ■ straightforward to ■■■ challenging. Problems labeled BIO are of biological or medical interest.

Conceptual Questions

1. A ball is thrown against a wall. Is the average force exerted on the wall by the ball greater if the ball sticks to the wall or if it rebounds?

2. If an adult human and a cat are running with the same momentum, which has greater kinetic energy?

3. If a system of particles has zero momentum, is the kinetic energy of the system necessarily zero? If a system of particles has zero kinetic energy, does it necessarily have zero momentum?

4. Soldiers are trained to fire their rifles with the end of the weapon tight against their shoulder. Why?

5. A skeptic claims that a rocket should not work in outer space, because there's no air to push off. How would you respond?

6. Automobile engineers design cars with "crumple zones" that collapse in a collision. Why is this better than making the outer surfaces of the car elastic, like bumper cars?

7. As you drive with gradually increasing speed, which increases at a faster rate, your car's momentum or its kinetic energy? Explain.

8. How does conservation of momentum explain "whiplash" injuries in rear-end auto accidents?

9. Why do golf and tennis coaches recommend a strong "follow-through" when hitting the ball?

10. Two skaters stand a short distance apart, facing each other on frictionless ice. Explain what happens when they begin tossing a ball back and forth.

11. A classic "Physics Olympics" event is the "egg drop," in which the winner drops an egg into a container from the greatest height without it breaking. What physical principles should you consider in designing such a container? What characteristics should you engineer into your container?

12. Why is a car with airbags and seatbelts safer than one with seatbelts only?

13. Is it necessary for a high jumper's center of mass to clear the bar? Explain.

14. Give an example of an object whose center of mass is not within the object.

15. Why must the center of mass of two objects lie on a line between them? Does the center of mass of three objects have to lie in the plane determined by the three?

16. Estimate where you expect the center of mass of an 85-cm-long baseball bat to be. If you have access to a bat, how could you determine the center of mass position experimentally without destroying the bat?

17. In the physics toy known as Newton's cradle, one steel ball swings in to hit the others (Figure CQ6.17). A ball on the opposite end swings out in response. If you pull two balls to the side and let go, you'll find that two opposite balls swing out. But conservation of momentum would also permit one ball to swing out with twice the speed of the incident pair. Why doesn't this happen?

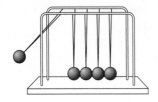

FIGURE CQ6.17

Multiple-Choice Problems

18. The momentum of a 1240-kg car going 100 mph (44.7 m/s) is (a) 27.7 kg·m/s; (b) 4470 kg·m/s; (c) 5.54×10^4 kg·m/s; (d) 1.24×10^6 kg·m/s.

19. To stop a 960-kg car traveling at 25 m/s in a time of 15 s requires an average force of magnitude (a) 1600 N; (b) 3200 N; (c) 6400 N; (d) 20,000 N.

20. A constant net force of 28.0 N applied for 12.5 s to a 0.168-kg hockey puck initially at rest results in momentum (a) 58.8 kg·m/s; (b) 350 kg·m/s; (c) 750 kg·m/s; (d) 2080 kg·m/s.

21. A 5.00-kg boulder dropped from a 12.0-m-high tower strikes the ground with momentum (a) 60.0 kg·m/s; (b) 76.7 kg·m/s; (c) 120 kg·m/s; (d) 588 kg·m/s.

22. A 0.145-kg baseball approaches the bat at 29.1 m/s. The bat gives it an impulse of 9.75 kg·m/s in the opposite direction. What's the ball's new speed? (a) 0 m/s; (b) 19.0 m/s; (c) 29.1 m/s; (d) 38.1 m/s.

23. A 75-g ball strikes a wall horizontally at 12 m/s and rebounds in the opposite direction at 10 m/s. The ball's change of momentum has magnitude (a) 0.150 kg·m/s; (b) 0.75 kg·m/s; (c) 0.90 kg·m/s; (d) 1.65 kg·m/s.

24. A 177-g air-track glider moving at 0.350 m/s collides elastically with a 133-g glider initially at rest. The final speed of the lighter glider is (a) 0.200 m/s; (b) 0.263 m/s; (c) 0.350 m/s; (d) 1.13 m/s.

25. If the collision of the preceding problem is perfectly inelastic, the final speed of the combined gliders is (a) 0.200 m/s; (b) 0.300 m/s; (c) 0.400 m/s; (d) 0.500 m/s.

26. An alpha particle (mass = 6.64×10^{-27} kg) moving at 4.65 Mm/s undergoes a head-on elastic collision with a stationary sodium nucleus (mass = 3.82×10^{-26} kg) at rest. At what speed does the alpha particle rebound? (a) 3.27 Mm/s; (b) 4.65 Mm/s; (c) 6.50 Mm/s; (d) 9.30 Mm/s.

27. Two identical wads of putty are traveling perpendicular to one another, both at 2.50 m/s, when they undergo a perfectly inelastic collision. What's the speed of the combined wad after the collision? (a) 5.00 m/s; (b) 3.54 m/s; (c) 2.10 m/s; (d) 1.77 m/s.

28. A billiard ball moving at 2.38 m/s collides elastically with an identical ball initially at rest. After the collision, the speed of one ball is 1.19 m/s. What's the speed of the other? (a) 1.19 m/s; (b) 2.06 m/s; (c) 2.38 m/s; (d) 4.25 m/s

29. Three masses are at different points along a 2.00-m stick: 0.45 kg at 0.80 m, 0.60 kg at 1.10 m, and 1.15 kg at 1.60 m. Where's the center of mass? (a) 1.1 m; (b) 1.2 m; (c) 1.3 m; (d) 1.4 m.

30. Three objects lie in the x-y plane: 1.40 kg at (3.9 m, 9.56 m); 1.90 kg at (−6.58 m, −15.6 m); and 2.40 kg at (0 m, −14.4 m). What are the center of mass coordinates? (a) (2.12 m, −6.90 m); (b) (0 m, −7.67 m); (c) (1.75 m, −12.9 m); (d) (1.22 m, −8.90 m).

31. A 0.20-kg ball and a 0.40-kg ball approach from opposite directions. Initially the 0.20-kg ball is moving at 3.0 m/s to the right, and the 0.40-kg ball at 2.0 m/s to the left. After a collision (not necessarily elastic) the 0.20-kg ball is moving 3.0 m/s to the left and the 0.40-kg ball is (a) at rest; (b) moving 1.0 m/s to the right; (c) moving 1.5 m/s to the right; (d) moving 2.0 m/s to the right.

32. A ball is thrown against a wall; it rebounds with speed unchanged. Which of the following is true? (a) The ball's kinetic energy is also unchanged. (b) The ball's momentum is also

unchanged. (c) Both kinetic energy and momentum are unchanged. (d) Both kinetic energy and momentum change.

Problems

Section 6.1 Introduction to Momentum

33. ■ What minimum force is needed to accelerate a bicycle and rider (combined mass 105 kg) to 15 m/s in 8.0 s? What's the system's momentum at that time?

34. ■ (a) Find the change in momentum of a 1120-kg car that increases its speed from 5 m/s to 15 m/s in a time of 12 s while traveling in the same direction? (b) What minimum force, applied in the direction of motion, is needed to do this?

35. ■ What's the momentum (magnitude) of a 64-kg person running at a speed of 7.3 m/s?

36. ■ A 135-g ball has a kinetic energy 29.8 J. (a) What's the magnitude of its momentum? (b) Use your answer to find its speed.

37. ■ A 145-g baseball moves horizontally toward the batter at 34.5 m/s. If the ball leaves the bat at 39.2 m/s in the opposite direction, find the impulse imparted to the ball.

38. ■ In a crash test, a 1240-kg car slams into a concrete wall at 17.9 m/s and comes to rest in 0.275 s. (a) Find the magnitude of the impulse delivered to the car. (b) Find the average stopping force.

39. ■ Model rocket engines are characterized by the total impulse they deliver, measured in newton-seconds. (a) Show that 1 N · s is the same as 1 kg · m/s, the unit we've used for impulse. (b) What speed can a 7.5-N · s engine give a rocket whose mass is 140 g at the end of the engine firing?

40. ■■ (a) A 145-g baseball approaches the bat at 32.3 m/s. The batter imparts a 12.0-kg · m/s impulse in the opposite direction. What are the ball's new speed and direction? (b) If contact time was 1.10 ms, what was the average force applied by the bat?

41. BIO ■■ **Insect locomotion.** High-speed photos of a 220-μg flea jumping vertically indicate that the jump lasts 1.2 ms and involves an average vertical acceleration of 100g. What average (a) force and (b) impulse does the ground exert on the flea during its jump? (c) What's the flea's change in momentum during its jump?

42. ■■ A 73.5-kg-long jumper leaves the ground at 20° above the horizontal with speed 8.25 m/s. What are the magnitude and direction of the jumper's momentum at that instant?

43. ■■ A net force $0.340\,\text{N}\,\hat{\imath} + 0.240\,N\,\hat{\jmath}$ is applied to a 170-g hockey puck for 4.50 s. (a) If the puck was initially at rest, find its final momentum. (b) If the puck was initially moving with velocity $2.90\,\text{m/s}\,\hat{\imath} + 1.35\,\text{m/s}\,\hat{\jmath}$, find its final momentum.

44. ■■ A 150-g ball moving at 0.45 m/s in the +x-direction is subject to a 0.15-N force in the −y-direction for 1.5 s. Find the ball's (a) momentum change and (b) final velocity.

45. ■■ Two cannon balls, with masses of 1.50 kg and 4.50 kg, are dropped from a 19.0-m-high tower. (a) Find the momentum of each ball as it strikes the ground. (b) Use your answer to part (a) to find each ball's kinetic energy just before it strikes the ground.

46. ■■ A 230-g ball is thrown horizontally at 19.8 m/s. (a) As it falls under the influence of gravity, find its change in momentum and final momentum after 1.00 s. (b) Show that the change in momentum in part (a) satisfies the relation $\Delta \vec{p} = \vec{F}\Delta t$.

47. ■■ A 64-kg runner rounds a circular track of radius 63.7 m at a constant 5.30 m/s. Find the magnitude her momentum changes after she's gone (a) one-half of the circle; (b) one-fourth of the circle; (c) the full circle.

48. ■■ (a) Use the impulse-momentum theorem to find the time it takes a ball falling from a tall building to double its speed from 8 m/s to 16 m/s. (Ignore air resistance.) (b) Use a kinematic equation to verify your answer.

49. ■■ A 160-g billiard ball traveling at 1.67 m/s approaches the table's side cushion at a 30° angle (measured from the cushion) and rebounds at the same angle. Find the change in the ball's momentum if its speed after striking the cushion is (a) unchanged and (b) reduced to 1.42 m/s. (c) For both cases, find the average force the cushion exerts on the ball if they're in contact for 25 ms.

50. BIO ■■ **Tissue biopsy.** Physicians use the *needle biopsy* technique to take tissue samples from internal organs. A spring-loaded gun shoots a hollow needle into the tissue; extracting the needle brings out the tissue core. A particular device uses 8.3-mg needles that take 90 ms to stop in the tissue, which exerts a stopping force of 41 μN. (a) Find the impulse imparted by the spring. (b) How far into the tissue does the needle penetrate?

51. ■■■ A120-g ice chunk is ejected from a comet, subject to the force shown in Curve 1 in Figure P6.51. (a) Find the total impulse imparted to the chunk over the time intervals 0 to 0.5 s and 0.5 s to 1.0 s. (b) If the chunk was initially at rest, find its velocity after 1.0 s.

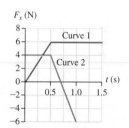

FIGURE P6.51

52. ■■■ Repeat the preceding problem for the force graphed in Curve 2 of Figure P6.51.

53. BIO ■■ **Charging rhinoceros.** Wildlife biologists fire 20-g rubber bullets to stop a rhinoceros charging at 0.81 m/s. The bullets strike the rhino and drop vertically to the ground. The biologists' gun fires 15 bullets each second, at 73 m/s, and it takes 34 s to stop the rhino. (a) What impulse does each bullet deliver? (b) What's the rhino's mass? Neglect forces between rhino and ground.

Section 6.2 Conservation of Momentum

54. ■ A 79-kg skater and a 57-kg skater stand at rest on frictionless ice. They push off, and the 57-kg skater moves at 2.4 m/s in the +x-direction. What's the other's velocity?

55. ■ A 130-g meteoroid (a space rock) is at rest at the origin when internal pressure causes it to burst in two. One piece, with mass 55 g, moves off at 0.65 m/s in the +x-direction. Find the velocity of the other piece.

56. BIO ■ **Squid motion.** An 18.5-kg squid hovering at rest in water fires a burst of water with impulse 32.0 kg · m/s. What's its new speed? (Neglect the water's drag force.)

57. ■■ A 250-g golf club head moving horizontally at 24.2 m/s strikes a stationary 45.7-g ball. The ball comes off the club moving in the club's initial direction at 37.6 m/s. Find the club head's speed immediately after contact. (Assume there's no additional force from the golfer.)

58. ■■ A child playing tee ball hits a stationary 75-g ball with a 240-g bat swinging initially at 4.5 m/s. Bat and ball move in the same direction after the impact, the ball at 6.2 m/s. What's the bat's post-impact speed?

Section 6.3 Collisions and Explosions in One Dimension

59. ■ A 1030-kg car going 3.4 m/s through a parking lot hits and sticks to the bumper of a stationary 1140-kg car. Find the speed of the joined cars immediately after the collision.

60. ■ A 95-tonne (1 t = 1000 kg) spacecraft moving in the $+x$-direction at 0.34 m/s docks with a 75-tonne craft moving in the $-x$-direction at 0.58 m/s. Find the velocity of the joined spacecraft.

61. ■ A 60.0-kg ice skater moving at 1.85 m/s in the $+x$-direction collides elastically with an 87.5-kg skater initially at rest. Find the velocities of the two skaters after the collision.

62. ■ A proton (mass 1.67×10^{-27} kg) is moving at $1.25 \times 10^{+6}$ m/s directly toward a stationary helium nucleus (mass 6.64×10^{-27} kg). After a head-on elastic collision, what are the particles' velocities?

63. ■■■ Consider an elastic head-on collision between two masses m_1 and m_2. Their velocities before colliding are $v_{1xi}\,\hat{\imath}$ and $v_{2xi}\,\hat{\imath}$; afterward, they're $v_{1xf}\,\hat{\imath}$ and $v_{2xf}\,\hat{\imath}$. (a) Write equations expressing conservation of momentum and conservation of energy. (b) Use your equations to show that $v_{1xi} - v_{2xi} = -(v_{1xf} - v_{2xf})$. (c) Interpret the results of part (b).

64. ■■■ A billiard ball moving right at 1.67 m/s undergoes a head-on elastic collision with an identical ball moving left at 2.25 m/s. (a) What are the balls' velocities after the collision? (b) Show that your results are consistent with the expression in part (b) of the preceding problem.

65. ■■ A moving car hits an identical stopped car and the two stick together. Show explicitly that kinetic energy is not conserved. (*Hint:* Momentum *is* conserved.)

66. ■■ A 2.64-g bullet moving at 280 m/s enters and stops in an initially stationary 2.10-kg wooden block on a horizontal frictionless surface. (a) What's the speed of the bullet/block combination? (b) What fraction of the bullet's kinetic energy was lost in this perfectly inelastic collision? (c) How much work was done in stopping the bullet? (d) If the bullet penetrated 5.00 cm into the wood, what was the average stopping force?

The following three problems concern the *ballistic pendulum*, used to determine the speed of bullets. A bullet is fired into a wooden block, which hangs at rest but is free to swing like a pendulum (Figure P6.67). The block's vertical rise is used to deduce the bullet's original speed.

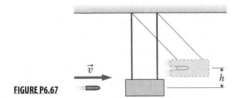

FIGURE P6.67

67. ■■ Assess the ballistic pendulum from the standpoint of energy conservation. During which part of the process is mechanical energy conserved? During which part is mechanical energy not conserved?

68. ■■■ A 9.72-g bullet is fired into a 4.60-kg block suspended as in Figure P6.67. The bullet stops in the block, which rises 16.8 cm above its initial position. Find the bullet's speed.

69. ■■■ Answer the following questions for a bullet of mass m that enters the block with a speed v and stops in the block, which originally had mass M. (a) Find the speed V of the bullet/block combination just after the bullet stops. (b) Find the bullet's original speed v as a function of m, M, and h, the height to which the block swings.

70. ■■ Two 0.85-m-long pendulums hang side by side. The masses of the pendulum bobs are 75 g and 95 g. The lighter bob is pulled aside until its string is horizontal and is then released from rest. It swings down and collides elastically with the other bob at the bottom of its arc. To what height does each bob rebound?

71. BIO ■■ **An eagle's dive.** When diving for prey, bald eagles can reach speeds of 200 mph (322 km/h). Suppose a 6.0-kg eagle flying horizontally at 322 km/h suddenly grasps and holds a stationary 5.0-kg jackrabbit in her talons. (a) What's the eagle's speed just after grasping the rabbit? (b) What percent of her kinetic energy is lost during the process? (c) What would be the eagle's final speed if the rabbit were initially running toward her at 2.0 m/s?

72. BIO ■■ **Animal aggression.** During mating season, bull elk engage in fierce fights by running head to head. Suppose a 450-kg elk running northward at 35 km/h crashes into a 500-kg elk charging directly toward him at the same speed. The impact lasts 0.50 s, and the animals lock horns. (a) Find the velocity (magnitude and direction) of the elk just after they lock horns. (b) What average force does each elk exert on the other during this collision?

73. ■■■ Assume a perfectly elastic collision between a 916-g bat and a 145-g baseball. Initially the ball was moving at 32 m/s toward the bat and afterward it's heading away at 40 m/s. Find the bat's speed before and after the collision.

74. ■■ The *coefficient of restitution* (COR) is the fraction of the initial kinetic energy remaining after a collision. COR = 1 for elastic collisions, and COR < 1 for inelastic collisions. What's the COR when a ball moving initially at 26.0 m/s rebounds from a wall at 21.0 m/s?

75. ■■ The COR (see preceding problem) of a ball bouncing on the ground is 0.82. If it's dropped from rest, to what fraction of its original height does the ball rebound after three bounces?

76. ■ In *beta decay*, a neutron at rest decays into a proton and an electron, ejected in opposite directions. (More on beta decay in Chapter 25.) If the electron's speed is 7.25 Mm/s, what's the proton's speed?

77. ■■ A croquet ball moving at 0.95 m/s strikes an identical ball initially at rest. This is a one-dimensional inelastic collision, with 10% of the original kinetic energy lost. Find the final velocities of the balls.

78. ■■ Two railroad freight cars with masses 100 Mg and 140 Mg approach with equal speeds of 0.34 m/s. They collide, the lighter car rebounding opposite its original direction at 0.32 m/s. (a) Find the velocity of the heavier car after the collision. (b) What fraction of the original kinetic energy was lost in this inelastic collision?

Section 6.4 Collisions and Explosions in Two Dimensions

79. ■■ Two identical wads of putty are moving perpendicular to one another with the same speed, 1.45 m/s, when they undergo a perfectly inelastic collision. What's the velocity of the putty after the collision?

80. ■■ A 1070-kg car and 3420-kg truck undergo a perfectly inelastic collision. Before the collision, the car was traveling southward at 1.45 m/s, and the truck westward at 9.20 m/s. Find the velocity of the wreckage immediately after the collision.

81. ■■ Two identical cars suffer a perfectly inelastic collision at an intersection. It's known that one car was originally traveling east at 25 mi/h (11.2 m/s). The other car was traveling north. Skid marks show that the wreckage moved 54° north of east just after the collision. Was the second car exceeding the 30-mi/h speed limit?

82. ■■ A bomb explodes into three identical pieces. One flies in the $+y$-direction at 13.4 m/s, a second in the $-x$-direction at 16.1 m/s. What's the velocity of the third piece?

83. ■■ A billiard ball moving at 1.65 m/s in the $+x$-direction collides with an identical ball initially at rest. After the collision, one

ball moves off at 45° above the x-axis. Find both balls' post-collision velocities.

84. **BIO** ▪▪ **A hunting cheetah.** Cheetahs can sustain speeds of 25 m/s over short distances. One of their prey is Thomson's gazelle, with top speed about 22 m/s. If a 50.0-kg cheetah running southward at top speed grabs and holds onto a 20.0-kg gazelle running westward at 22 m/s, find the velocity of these animals just after the attack.

85. ▪▪ In *Compton scattering*, a photon scatters elastically off an electron initially at rest. Suppose the incoming photon has momentum of 1.0×10^{-21} kg·m/s $\hat{\imath}$. The scattered photon's momentum is 1.8×10^{-22} kg·m/s $\hat{\imath} - 3.1 \times 10^{-28}$ kg·m/s $\hat{\jmath}$. Find the momentum of the recoiling electron.

86. ▪▪▪ In *Rutherford scattering*, a helium nucleus (mass 6.64×10^{-27} kg) scatters elastically off a stationary gold nucleus $(3.27 \times 10^{-25}$ kg). If a helium nucleus with initial speed 2.50×10^6 m/s is scattered at 22.5° with respect to its initial velocity, what are the velocities of the recoiling gold and scattered helium nuclei?

87. ▪▪▪ A 25.0-kg mortar shell is launched from level ground at 35.0 m/s at 60° above the horizontal. At the top of its trajectory, the shell explodes into three pieces. A 10.0-kg piece flies forward at 38.0 m/s. Another 10.0-kg piece flies straight up at 11.5 m/s. What's the velocity of the remaining piece?

88. ▪▪▪ Figure P6.88 shows a collision between two identical pucks on an air table. (The incoming pucks approach the origin from below the x-axis.) Given the data shown, determine (a) the velocity marked "?" and (b) whether the collision is elastic.

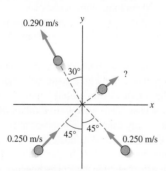

FIGURE P6.88

Section 6.5 Center of Mass

89. ▪ Two masses are placed at different points along a meterstick of negligible mass: 0.250 kg at 0.200 m and 0.500 kg at 0.500 m. Where's the center of mass of this system?

90. ▪ (a) Treating Sun and Earth as symmetric spheres, find the center of mass of the Earth-Sun system. Where is this point in relation to the Sun's surface? (b) Repeat for the Earth-Moon system. (See Appendix E for astrophysical data.)

91. ▪ Treating Sun and Jupiter as symmetric spheres, find the center of mass of the Jupiter-Sun system. Where is this point in relation to the Sun's surface?

92. ▪ In a carbon monoxide (CO) molecule, the carbon and oxygen atoms are separated by 0.112 nm. Using atomic masses in Appendix D, find the molecule's center of mass.

93. ▪▪ On a seesaw, a 28-kg child sits 2.8 m from the pivot point. Where should a 38-kg child sit to put the center of mass at the pivot?

94. ▪ You have a 120-g meterstick. (a) Where should you place the meterstick atop your finger so it balances? (b) Where's the balance point if you place an additional 100-g mass at the 20-cm mark and a 200-g mass at the 80-cm mark?

95. ▪▪ Two identical air-track gliders are moving to the right, the leftmost glider at 0.350 m/s and the rightmost one at 0.250 m/s. (a) What's the velocity of their center of mass? (b) The faster glider strikes the slower one and they undergo an elastic collision. Find their velocities after the collision. (c) Use your answer in part (b) to find the center-of-mass velocity after the collision, and compare with your answer in part (a).

96. **BIO** ▪▪ **Leg center of mass.** A person sits on a chair with her thigh horizontal and lower leg vertical. The entire thigh has mass 12.9 kg and length 44.6 cm. The lower leg and foot have a combined mass of 8.40 kg and length 48.0 cm. Assume the center of mass of each part is at its center. Find the center of mass of the entire leg in this configuration.

97. ▪ With the origin at one corner of a soccer field, there's a 59.0-kg player standing at coordinates (24.3 m, 35.9 m) and a 71.5-kg player at (78.8 m, 21.5 m). Where's the center of mass of these two players?

98. ▪▪ Figure P6.98 shows the water molecule, with its bond angle 104.5° and bond length 95.7 pm. The oxygen atom's mass is 16 times that of each hydrogen. Find the molecule's center of mass.

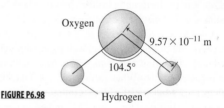

FIGURE P6.98

99. **BIO** ▪▪▪ **Leg raises.** A person of mass M lies on the floor doing leg raises. His leg is 95 cm long and pivots at the hip. Treat his legs (including feet) as uniform cylinders, with both legs comprising 34.5% of body mass, and the rest of his body as a uniform cylinder comprising the rest of his mass. He raises both legs 50.0° above the horizontal. (a) How far does the center of mass of each leg rise? (b) How far does the entire body's center of mass rise? (c) Since the center of mass of his body rises, there must be an external force acting. Identify this force.

100. **BIO** ▪▪ **Arm raises.** A weight lifter does a series of arm raises while holding a 2.50-kg weight. Her forearm (including the hand) starts out horizontal and pivots to a vertical position about her elbow. The mass of the lower arm plus hand is 5.00 kg, uniformly distributed along its 35.0-cm length. With each arm raise, (a) by how much does the center of mass of her forearm (including her hand and the weight) rise, and (b) what is the change in its gravitational potential energy?

101. ▪▪▪ Prove this rule stated in the text: If a net force $\vec{F}_{net}$ is applied to a system of particles with total mass M, then the center of mass of the system accelerates according to Newton's second law, $\vec{F}_{net} = M\vec{A}_{cm}$. (*Hint:* Start with the equation for the center-of-mass velocity and divide by t.)

102. ▪▪▪ For the billiard balls in Example 6.13, find the system's center-of-mass velocity before and after the collision. Are they the same? Should they be?

General Problems

103. ▪▪ A constant force $\vec{F}$ acts on an electron for 3.0 s, changing its velocity from 6.2×10^6 m/s $\hat{\imath} - 5.8 \times 10^5$ m/s $\hat{\jmath}$ to -3.7×10^6 m/s $\hat{\imath} - 15.8 \times 10^6$ m/s $\hat{\jmath}$. Find $\vec{F}$.

104. ▪▪ A locomotive pulls a train of 10,000-kg cars under a grain elevator. Each car remains under the hopper for 8.0 s while 4500 kg of wheat drops in. Determine the force the engine must exert to keep the train's speed constant.

105. BIO ■■ **Biomechanics in athletics.** A 68.0-kg tennis player (including racket) leaps vertically to return a serve. The 57.0-g ball was initially moving horizontally at 50.0 m/s. Due to the ball's height, the player can't swing her racket, but merely allows the ball to bounce off of it, which reverses the ball's direction and reduces its speed by 75.0%. If the ball/racket contact lasts for 35.0 ms, find (a) the average force the racket exerts on the ball and (b) the racket's recoil speed.

106. ■ Marie Curie first identified the element radium in 1898. She found that radium-226 (mass 3.77×10^{-25} kg) decays by emitting an alpha particle (mass 6.64×10^{-27} kg). If the alpha particle's speed is 2.4×10^6 m/s, what's the speed of the recoiling nucleus?

107. ■■ A popcorn kernel in a hot pan bursts into two pieces, with masses 71 mg and 92 mg. The more massive piece moves horizontally at 48 cm/s. Describe the motion of the other piece.

108. ■■■ Two 0.16-kg billiard balls undergo a glancing collision. One was initially moving at 1.24 m/s at 40° below the +x-axis, the other at 3.15 m/s at 80° above the +x-axis. After the collision one ball moves at 80° above the +x-axis and the other in the +y-direction. (a) Determine (a) the speeds of both balls after the collision and (b) whether the collision was elastic.

109. ■■■ A 4.0-kg wooden block is at rest on a tabletop. A 20-g bullet enters the block going at 800 m/s. It passes through the 20-cm-thick block, emerging at 425 m/s. (a) How much time was the bullet in the block? (b) What average force did the bullet exert on the block? (c) What was the block's initial velocity when the bullet emerged? (d) If the block slides 81 cm before stopping, what's the coefficient of friction between block and table?

110. ■■ A billiard ball moving at 1.50 m/s in the +x-direction strikes an identical ball at rest. After the collision, one ball moves at 35° above the +x-axis, and the other ball at 55° below the +x-axis. Find the speeds of both balls.

111. ■■ A runaway toboggan of mass 8.6 kg is moving horizontally at 23 km/h. As it passes under a tree, 12 kg of snow drops on it. What is its subsequent speed?

112. ■■ Gliders with masses 0.120 kg and 0.180 kg are at rest on an air track with a spring between them. The spring is compressed 1.25 cm and then released. The lighter glider moves off at 0.730 m/s. (a) Find the speed of the heavier glider. (b) Find the spring constant k, assuming all the spring's potential energy was converted into the gliders' kinetic energy.

113. ■■■ A 950-kg car is at the top of a 36-m-long, 2.5° incline. Its parking brake fails and it starts rolling down the hill. Halfway down, it strikes and sticks to a 1240-kg parked car. (a) Ignoring friction, what's the speed of the joined cars at the bottom of the incline? (b) Compare your answer in part (a) with what the first car's speed would have been at the bottom had it not struck the second car.

114. ■■ A 2.0-kg ball and a 3.0-kg ball, each moving at 0.90 m/s, undergo a head-on collision. The lighter ball rebounds opposite its initial direction, with speed 0.90 m/s. (a) Find the post-collision velocity of the heavier ball. (b) How much mechanical energy was lost in this collision? Express your answer in J and as a fraction of the system's initial mechanical energy.

115. ■■ An astronaut (mass 128 kg, with equipment) floats in space, 15.0 m from his spacecraft and at rest relative to it. (a) In order to get back, he throws a 1.10-kg wrench at 5.40 m/s in the direction opposite the spacecraft. How much time does it take him to reach the spacecraft? (b) Repeat part (a) if the astronaut was initially moving at 2.85 cm/s away from the spacecraft.

116. ■■ A student drops a hard rubber ball from his dorm roof, 23.6 m above the sidewalk. The ball rebounds to a height of 18.1 m. Assuming negligible air resistance, what momentum change did the ball undergo on colliding with the sidewalk?

117. BIO ■■ **Human center of mass.** A man with mass M stands upright with both legs straight and arms vertically downward at his sides. Biometric measurements suggest that you can model his body as follows, treating each segment as having uniform density. Assume his shoulders are at the base of his head (that is, ignore his neck).

Head: 6.90% of body mass, sphere of diameter 25.0 cm
Trunk: 46.0% of body mass, cylinder 60.0 cm long
Legs and feet: 34.6% of body mass, cylinders 95.0 cm long
Arms and hands: 12.5% of body mass, cylinders 65.0 cm long
(a) Find the height of his center of mass. (b) Find the new center of mass if he raises his arms vertically.

118. BIO ■■■ **Bending the back.** An 80-kg person bends forward by pivoting his upper body 10.0° from the vertical. His legs remain straight and his arms hang vertically. Model his body using the same assumptions as in the preceding problem. Find this person's center of mass, relative to the foot/floor contact point.

119. BIO ■■■ **Exercise program.** The person in the preceding problem does jumping jacks. He starts with arms raised vertically upright and legs 60.0° apart, and then jumps into a position with his legs together and his arms vertically downward. (a) Find the location of his center of mass in each of the two positions. (b) By how much vertical distance does his center of mass move with each cycle of this exercise?

120. ■■ Four thin uniform metal rods are attached to form a square. Each is 30 cm long, with masses 1 kg, 2 kg, 3 kg, and 4 kg, in order around the square. Locate the system's center of mass.

121. ■■■ Use conservation of momentum and energy to derive Equations 6.12 and 6.13.

122. ■■■ An air-track glider with mass m_1 and speed v_{1xi} collides perfectly inelastically with a stationary glider of mass m_2. (a) Show that the fraction of the initial kinetic energy remaining after the collision is $m_1/(m_1 + m_2)$. (b) Explain why this result means that kinetic energy cannot be conserved in such a collision.

123. ■■■ Canada's CANDU nuclear reactors use heavy water to slow neutrons (see Example 6.9). The neutrons (mass 1.67×10^{-27} kg) collide with deuterium nuclei (hydrogen-2, mass 3.34×10^{-27} kg). Suppose a neutron with initial speed v_i collides head-on and elastically with a stationary deuterium nucleus. (a) Find the neutron's speed after the collision, in terms of v_i. (b) How many such collisions will it take before the neutron slows to less than 1% of its initial speed?

Answers to Chapter Questions

Answer to Chapter-Opening Question

The diver's center of mass follows the simple parabolic path of a projectile, because as Newton's laws show, the diver's mass acts like it's all concentrated there.

Answers to GOT IT? Questions

Section 6.1   (b) < (a) < (d) < (c).
Section 6.2   (d) 10.0 kg · m/s
Section 6.3   (a) the elastic collision
Section 6.4   (c) greater than v_0

■ After the initial plunge, the bungee jumper oscillates up and down, enjoying the view. What determines the frequency of those oscillations and how quickly they die out?

This chapter introduces oscillatory motion. You'll consider in detail the important special case of simple harmonic motion, exemplified by a mass oscillating on an ideal spring. For this model, you'll learn how the oscillation period depends on mass and the spring constant. You'll also see how the position, velocity, and acceleration vary with time. These concepts follow naturally from a discussion of energy conservation for the simple harmonic oscillator.

The simple pendulum comes next. For small-angle swings it, too, undergoes simple harmonic motion. Frictional forces often damp oscillations, so we'll explore such damping. Finally, we'll consider driven oscillations, which lead to the phenomenon of resonance when the driving frequency matches the oscillator's natural frequency.

7.1 Periodic Motion

Nature and technology provide many examples of **periodic motion**—motion that repeatedly follows the same trajectory. Uniform circular motion (Chapter 3), such as the Moon in its orbit, is a good example; so is a comet's repeating elliptical orbit. A clock's pendulum swings back and forth in periodic motion. A more contemporary timepiece, your watch, contains a tiny crystal undergoing periodic vibrations 32,768 times each second.

Period and Frequency

The **period** of any periodic motion is the time to complete one cycle through the full trajectory. The period of Earth's orbital motion is 1 year; the period of your beating heart is about 1 s. Consistent with Section 3.5 (Uniform Circular Motion), we'll use T for period, to distinguish it from variable times designated t. The **frequency** f of periodic motion is the reciprocal of the period, so

$$f = \frac{1}{T}$$

Although this equation shows that the units of frequency are s^{-1}, scientists and engineers usually measure frequency f in **hertz** (Hz, named for the German physicist Heinrich Hertz, who first generated radio waves in 1887). Mathematically, 1 Hz is the same as $1\,s^{-1}$, but in using Hz it's understood that we're describing the number of full cycles each second. You'll soon see a different measure of frequency that's measured explicitly in s^{-1}.

You can think of frequency as the *rate* at which periodic motion repeats—the number of cycles per second. Bounce a basketball with period $T = 0.25$ s, and $f = 1/T = 4.0$ Hz—meaning you're bouncing the ball four times each second. Before the unit Hz became commonplace, frequency was given in "cycles per second"—which means the same as the hertz.

✓ **TIP**

You can think of frequency either as (a) the reciprocal of the period, or (b) the number of times per second the motion is repeated.

EXAMPLE 7.1 **Period and Frequency**

(a) Earth orbits the Sun every 365 days. Find the frequency of this motion, in hertz. (b) A tuning fork set to "concert A" (A above middle C) vibrates at 440 Hz. What's the period of its motion?

ORGANIZE AND PLAN The reciprocal relationship between period and frequency is what's needed here. Given frequency or period, the other is its reciprocal. Thus

$$f = \frac{1}{T}$$

To get frequency in hertz, the period must be in seconds. Earth's 365-day period converts to 3.15×10^7 s.

SOLVE (a) Earth's orbital frequency is the reciprocal of the period $T = 3.15 \times 10^7$ s:

$$f = \frac{1}{T} = \frac{1}{3.15 \times 10^7\ s} = 3.17 \times 10^{-8}\ s^{-1} = 3.17 \times 10^{-8}\ Hz$$

(b) The tuning fork's period is the reciprocal of its frequency:

$$T = \frac{1}{f} = \frac{1}{440\ Hz} = \frac{1}{440\ s^{-1}} = 2.27 \times 10^{-3}\ s$$

REFLECT In part (b), the frequency 440 Hz and period on the order of 10^{-3} s are typical for audible sound, because this is near the middle of the musical scale. In Chapter 11 you'll learn about sound and the range of human hearing.

MAKING THE CONNECTION Which planets' orbits have frequencies higher than Earth's, and which have lower?

ANSWER The key is the reciprocal relationship between frequency and period. Mercury and Venus have shorter orbital periods than Earth, so their frequencies are higher. The outer planets (Mars, Jupiter, and beyond), have longer periods than Earth, hence lower frequencies.

Oscillations

An **oscillation** is any motion that proceeds *back and forth* over the same path. A vibrating prong of that tuning fork is one example; others are a pendulum's swing, a car bouncing on its springs, and the movement of your heart's muscle wall as it beats.

Not all periodic motion is oscillatory. A car continuously circling a closed track undergoes periodic motion, but not oscillatory motion. However, if the car drives clockwise around the track, then reverses and returns counterclockwise to its starting point, that's an oscillation.

GOT IT? Section 7.1 A hummingbird beats its wings at a frequency of 71 Hz. The period of one up-and-down wing oscillation is about (a) 71 s; (b) 0.71 s; (c) 0.14 s; (d) 0.014 s.

7.2 Simple Harmonic Motion

Figure 7.1 graphs three oscillatory motions. Each satisfies our oscillation criteria, because each cycles regularly back and forth between the same maximum and minimum positions. The **sinusoidal oscillation** is particularly important. Its position-versus-time curve has the smooth, undulating shape of the sine or cosine function. Recall that sine and cosine have similar graphs, just shifted by one-fourth of a cycle. Here it's a cosine function, but either way we use the term *sinusoidal*.

A system whose motion describes a sinusoidal function of time is a **simple harmonic oscillator**, and its motion is **simple harmonic motion (SHM)**. SHM is nature's fundamental oscillation, and it's also widespread in technology.

Amplitude and Angular Frequency in SHM

The paradigm of a simple harmonic oscillator is an object with mass m on an ideal spring (Figure 7.2). Recall from Section 5.2 that an ideal spring obeys Hooke's law, with its force directly proportional to displacement from equilibrium: $F = -kx$. The symmetry of the Hooke's law force dictates that the mass travels equal distances either side of equilibrium. That distance is the **amplitude** A. What's less obvious is that the *linear* relationship between displacement and force is what makes the motion sinusoidal, that is, what ensures that the mass-spring system undergoes SHM.

Proof that the mass-spring system undergoes SHM requires calculus, but this can be verified experimentally by tracking the motion using video capture or a motion detector. Oscillations between $x = +A$ and $x = -A$ are described using a cosine function, which varies from $+1$ and -1. The cosine applies when you release the mass at position $x = A$, taking the time to be $t = 0$. Then the mass's simple harmonic motion is described by

$$x = A \cos(\omega t) \quad \text{(Position of an object in simple harmonic motion; SI unit: m)} \quad (7.1)$$

where ω (lowercase Greek omega) is a constant related to m and k. Figure 7.3 graphs this motion. Had we chosen to start our clock as the mass moved past $x = 0$, we would have had sine instead of cosine.

The constant ω is the **angular frequency**, and it's closely related to the period T. That's because the cosine function completes a full cycle as ωt goes from 0 to 2π. Our oscillator completes a full cycle in one period, as time goes from $t = 0$ to $t = T$. Therefore, for an oscillator following Equation 7.1, $\omega T = 2\pi$, or

$$\omega = \frac{2\pi}{T} \quad (7.2)$$

Because the oscillator frequency f is the reciprocal of the period ($f = 1/T$), the angular frequency is

$$\omega = 2\pi f$$

Later you'll see physically why ω is called *angular* frequency; for now, note that we usually consider arguments to sine and cosine as angles, and that there are 2π radians of angle in a full circle—hints that ω has something to do with angles.

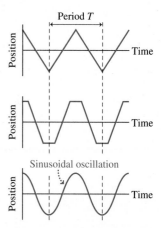

FIGURE 7.1 These curves all represent oscillations, which cycle repeatedly back and forth between minimum and maximum positions.

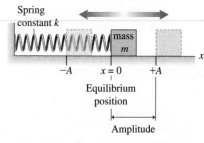

FIGURE 7.2 Mass on an ideal spring, used to illustrate simple harmonic motion.

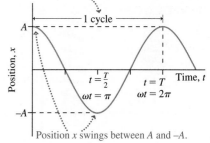

FIGURE 7.3 Position versus time for the simple harmonic oscillator.

CONCEPTUAL EXAMPLE 7.2   **Simple Harmonic Motion**

A block on a spring undergoes simple harmonic motion with amplitude A, its position given by $x = A \cos(\omega t)$. Where is the block (a) when the force on it is zero, (b) when the force has the greatest magnitude, (c) when its speed is zero, and (d) when its speed is greatest?

SOLVE Figure 7.4 shows the solutions.

(a) The spring force on the block is $F_x = -kx$. Therefore, the force is zero when $x = 0$.

(b) The force $-kx$ has maximum magnitude when the magnitude of x is largest, which occurs at $x = A$ and $x = -A$.

(c) The block's speed is zero when $x = \pm A$, where it's momentarily stopped while changing direction. This is consistent with the position-versus-time graph in Figure 7.3. Remember that velocity is the slope of the tangent to the position-versus-time graph. The positions $x = \pm A$ occur at the top and the bottom of the cosine curve, where the tangent slope is zero. This implies zero velocity and zero speed at those points.

(d) The speed is maximum at $x = 0$. Physically, a block released at $x = A$ experiences a force $F_x = -kx$ that's in the $-x$-direction as long as $x > 0$. The block accelerates to the left, gaining speed. But once it's in the region $x < 0$, the force is to the right, slowing the block. The same process happens in reverse on the return part of the cycle, beginning at $x = -A$.

REFLECT What a wealth of information comes from the equation of motion for the oscillator and from Hooke's law! We'll return to the question of velocity and speed later in this section.

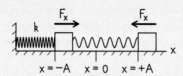

At $x = -A$ and $x = +A$, the block is momentarily motionless ($v = 0$), and it feels the maximum force (push or pull) from the spring.

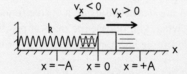

At $x = 0$, the block feels no force (because the spring is momentarily relaxed), and it is moving the fastest.

FIGURE 7.4 Solutions for the four cases discussed.

Period of the Simple Harmonic Oscillator

If you want to build a simple harmonic oscillator, it's easier to avoid friction if you hang the spring vertically, as in Figure 7.5. Frictional forces (including drag from the air) will gradually diminish the amplitude of the oscillations, as we'll discuss in Section 7.5, but for now we'll neglect the drag force.

Does gravity affect the oscillator? Not significantly. As Figure 7.5 shows, gravity only changes the equilibrium position by a distance d. Oscillations about the new equilibrium occur exactly as they do for the horizontal oscillator. That's because the force of gravity is *constant*, while oscillations result from the *varying* force of the spring. So the period T for the vertical oscillator is the same as it would have been for a horizontal one.

Experiment with different springs and varying masses; you'll soon see that the period depends on the spring constant k and mass m. A stiffer spring has a larger k, which increases the force and hence acceleration, thus shortening the period. Higher mass means more inertia, lengthening the period. Do enough experiments and you can verify that the period is

$$T = 2\pi\sqrt{\frac{m}{k}} \qquad \text{(Period of simple harmonic motion; SI unit: s)} \qquad (7.3)$$

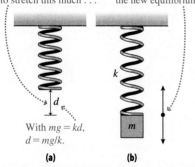

When a block is added, weight causes the spring to stretch this much . . .   . . . so the block oscillates about the new equilibrium.

With $mg = kd$, $d = mg/k$.

(a)   (b)

FIGURE 7.5 A vertical simple harmonic oscillator.

Equation 7.3 is exact only for an ideal spring, one that has no mass and obeys Hooke's law, but it's an excellent approximation when the spring mass is much less than the oscillating mass, and when the spring doesn't stretch beyond the range where the force and displacement are proportional. By using Newton's second law and calculus, Equation 7.3 can be proven to be exact for an ideal spring. The calculus derivation also highlights the mathematically central role of the angular frequency, which for us follows by Equation 7.2:

$$\omega = \frac{2\pi}{T} = \frac{2\pi}{2\pi\sqrt{m/k}}$$

or

$$\omega = \sqrt{\frac{k}{m}} \quad \text{(Angular frequency of a harmonic oscillator; SI unit: s}^{-1}\text{)} \quad (7.4)$$

Reviewing New Concepts

Important ideas about simple harmonic motion:
- A mass attached to an ideal Hooke's law spring oscillates in simple harmonic motion.
- Position in simple harmonic motion is a sinusoidal function of time.
- The period of a simple harmonic oscillator with mass m and spring constant k is $T = 2\pi\sqrt{m/k}$.

EXAMPLE 7.3   **Bad Shocks!**

A car with failed shock absorbers is effectively a mass-spring system that starts oscillating when the car hits a bump. A 1240-kg car has one-fourth of its mass supported by each shock absorber with a spring constant 15.0 kN/m. Find the period, frequency, and angular frequency of the oscillation.

ORGANIZE AND PLAN The period is related to the known parameters k and m. The effective mass is one-fourth the car's mass, which is $m = 1240 \text{ kg}/4 = 310 \text{ kg}$. After you know the period, the basic relations between period, frequency, and angular frequency let you complete the problem.

The period is

$$T = 2\pi\sqrt{\frac{m}{k}}$$

and frequency is its reciprocal: $f = 1/T$. Angular frequency ω follows from $\omega = 2\pi f$.

Known: $k = 15.0$ kN/m and effective mass $m = 310$ kg.

SOLVE The values given yield the period:

$$T = 2\pi\sqrt{\frac{m}{k}} = 2\pi\sqrt{\frac{310 \text{ kg}}{15.0 \times 10^3 \text{ N/m}}} = 0.903 \text{ s}$$

then the frequency

$$f = \frac{1}{T} = \frac{1}{0.903 \text{ s}} = 1.11 \text{ s}^{-1} = 1.11 \text{ Hz}$$

and angular frequency

$$\omega = 2\pi f = 2\pi(1.11 \text{ s}^{-1}) = 6.97 \text{ s}^{-1}$$

REFLECT A period around 1 s seems about right for a car bouncing on a bumpy road. You could also compute the angular frequency directly, using Equation 7.4.

MAKING THE CONNECTION Should you add or remove passengers from the car to get a period greater than 1 s?

ANSWER Period is proportional to the square root of mass, so *more* mass is needed to increase the period.

CONCEPTUAL EXAMPLE 7.4   **Atomic Oscillations**

In Chapter 5 we noted that systems ranging from molecules to stars exhibit springlike behavior. Therefore, these systems can undergo simple harmonic motion. A useful model for a solid consists of atoms arranged in a regular pattern and connected by springs (Figure 7.6). The interatomic

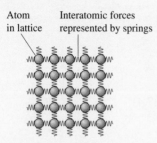

Atom in lattice   Interatomic forces represented by springs

FIGURE 7.6 Atoms in a solid in simple harmonic motion.

forces are actually electrical, but for small oscillations these "springs" obey Hooke's law fairly well. Electrical forces are strong over this short range, so the effective spring constants are large. The atomic masses are very small. What does this imply about oscillation periods and frequencies?

SOLVE With large spring constant k and small mass m, Equation 7.3 for SHM implies a short period T and correspondingly high frequency $f = 1/T$. In real solids, oscillation frequencies are typically on the order of terahertz (THz = 10^{12} Hz), with periods therefore on the order of picoseconds (ps = 10^{-12} s). Atomic vibrations in solids are extremely rapid!

REFLECT High frequencies are reasonable only at the atomic scale, with masses on the order of 10^{-27} kg to 10^{-25} kg. Note that the mass appears in the denominator of the frequency formula and in the numerator of the period.

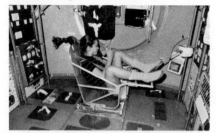
Period and Amplitude

Equation 7.3 doesn't contain A, showing that the period of simple harmonic motion doesn't depend on its amplitude. This might seem surprising, because with greater amplitude, an oscillator has to cover greater distance. But greater amplitude results in greater spring forces and larger accelerations—exactly compensating for the greater distance and giving the same period regardless of amplitude. This is a special feature of simple harmonic motion, and it results only when the force tending to restore the system to equilibrium is *linear* in the displacement.

✓TIP

The period of a harmonic oscillator (a) increases with increased mass, (b) decreases with increased spring constant, and (c) is independent of the amplitude.

GOT IT? Section 7.2   Rank in order, from lowest to highest, the periods of the five oscillators.

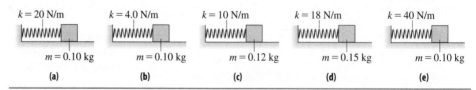

$k = 20$ N/m	$k = 4.0$ N/m	$k = 10$ N/m	$k = 18$ N/m	$k = 40$ N/m
$m = 0.10$ kg	$m = 0.10$ kg	$m = 0.12$ kg	$m = 0.15$ kg	$m = 0.10$ kg
(a)	(b)	(c)	(d)	(e)

7.3 Energy in Simple Harmonic Motion

In Chapter 5 you saw that mechanical energy is conserved in systems subject to conservative forces. The force of an ideal spring is conservative, so the simple harmonic oscillator is such a system. Neglecting friction, the total mechanical energy of a simple harmonic oscillator remains constant.

Total Mechanical Energy

Oscillator has kinetic energy $K = \frac{1}{2}mv^2$ associated with mass's motion.

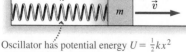

Oscillator has potential energy $U = \frac{1}{2}kx^2$ associated with spring.

FIGURE 7.7 Energy in simple harmonic motion.

Recall from Chapter 5 that a system's total mechanical energy E is the sum of its kinetic energy K and potential energy U: $E = K + U$. In the harmonic oscillator, kinetic energy is in the mass m moving with speed v (Figure 7.7):

$$K = \tfrac{1}{2}mv^2$$

Potential energy is in the spring, and depends on displacement x:

$$U = \tfrac{1}{2}kx^2$$

Therefore, the oscillator's total mechanical energy is:

$$E = K + U = \tfrac{1}{2}mv^2 + \tfrac{1}{2}kx^2 \tag{7.5}$$

The spring force is conservative, so E is constant. What's its value? An easy answer comes from considering the endpoints of the motion, $x = \pm A$. At these points, the mass stops momentarily to reverse direction, so speed and kinetic energy are both zero. Then the total energy is the spring's potential energy:

$$E = \tfrac{1}{2}kA^2$$

The total mechanical energy depends only on the spring constant k and oscillation amplitude A.

Speed and Velocity

Knowing the oscillator's total mechanical energy lets you find the speed v as a function of position. Substituting $E = \tfrac{1}{2}kA^2$ in Equation 7.5 gives

$$\tfrac{1}{2}kA^2 = \tfrac{1}{2}mv^2 + \tfrac{1}{2}kx^2$$

Solving for v^2:

$$v^2 = \frac{k}{m}(A^2 - x^2)$$

Taking the square root gives

$$v = \sqrt{\frac{k}{m}(A^2 - x^2)} \quad \text{(Speed of a harmonic oscillator; SI unit: m/s)} \tag{7.6}$$

We've taken the positive root because speed is always positive (or zero).

Figure 7.8 shows speed v as a function of position x, obtained from Equation 7.6. The maximum speed, which occurs when the oscillator passes through its equilibrium position $x = 0$, follows by setting $x = 0$ in Equation 7.6:

$$v_{\text{max}} = \sqrt{\frac{k}{m}}\,A \tag{7.7}$$

Using Section 7.2's result that $\sqrt{k/m} = \omega$ gives an alternate expression for the maximum speed in terms of the angular frequency ω:

$$v_{\text{max}} = \omega A$$

In one-dimensional motion, speed is just the absolute value of velocity v_x. Therefore,

$$v_x = \pm\sqrt{\frac{k}{m}(A^2 - x^2)} \quad \begin{array}{l}\text{(Velocity versus position for a}\\ \text{harmonic oscillator; SI unit: m/s)}\end{array} \tag{7.8}$$

where the positive sign applies when the mass is moving to the right (the $+x$-direction), and the negative root when it's moving to the left. Given the maximum speed $v_{\text{max}} = \sqrt{k/m}A$, the velocity takes the extreme values

$$v_{x,\text{max}} = \sqrt{\frac{k}{m}}A, \quad v_{x,\text{min}} = -\sqrt{\frac{k}{m}}A$$

These occur as the mass passes through $x = 0$, going, respectively, right and left.

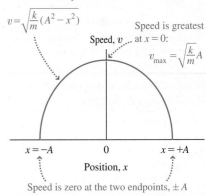

Speed of a simple harmonic oscillator is a function of position x:

$$v = \sqrt{\frac{k}{m}(A^2 - x^2)}$$

Speed, v

Speed is greatest at $x = 0$:

$$v_{\text{max}} = \sqrt{\frac{k}{m}}A$$

$x = -A$ 0 $x = +A$

Position, x

Speed is zero at the two endpoints, $\pm A$

FIGURE 7.8 Speed of the simple harmonic oscillator as a function of position.

Consider a simple harmonic oscillator with $k = 10.0$ N/m and $m = 250$ g. If the oscillation amplitude is 3.5 cm, the maximum velocity is

$$v_{x,\text{max}} = \sqrt{\frac{k}{m}}A = \sqrt{\frac{10.0 \text{ N/m}}{0.250 \text{ kg}}}(0.035 \text{ m}) = 0.22 \text{ m/s}$$

Note that doubling the amplitude doubles the maximum velocity—in this case, to 0.44 m/s.

✓TIP

The harmonic oscillator's maximum speed is reached when it passes through its equilibrium position $(x = 0)$.

Velocity versus Time

Equation 7.8 gives velocity as a function of *position*. But in simple harmonic motion, Equation 7.1 gives position as a function of time: $x = A \cos(\omega t)$. Using this expression for x in Equation 7.7 gives us velocity as a function of *time*.

$$v_x = \pm\sqrt{\frac{k}{m}(A^2 - A^2 \cos^2(\omega t))} = \pm\sqrt{\frac{k}{m}A^2(1 - \cos^2(\omega t))}$$

Factoring A^2 out of the square root and using $\sqrt{k/m} = \omega$ reduces this to

$$v_x = \pm\omega A\sqrt{1 - \cos^2(\omega t)}$$

The trig identity $\sin^2\theta + \cos^2\theta = 1$ gives $\sin\theta = \pm\sqrt{1 - \cos^2\theta}$. Applying this result to our velocity equation,

$$v_x = \pm\omega A \sin(\omega t)$$

Which sign? Here it's *negative*. Our choice of cosine in $x = A\cos(\omega t)$ implies time $t = 0$ when the mass is stopped at $x = A$. From there it moves *left*, in the $-x$-direction, showing that v_x goes negative as time t increases from zero (Figure 7.9). So the velocity-versus-time function is

$$v_x = -\omega A \sin(\omega t) \qquad \text{(Velocity versus time for a harmonic oscillator; SI unit: m/s)} \qquad (7.9)$$

Because velocity follows a sine function, the oscillator moves to the left for one half of each cycle, then to the right for the other half. Figure 7.9 graphs this sinusoidal dependence of velocity on time.

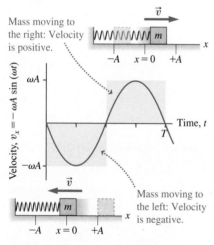

Mass moving to the right: Velocity is positive.

Mass moving to the left: Velocity is negative.

FIGURE 7.9 Velocity versus time for the simple harmonic oscillator.

CONCEPTUAL EXAMPLE 7.5 **Velocity and Maximum Speed**

Show that the equation for velocity $v_x = -\omega A \sin(\omega t)$ (Equation 7.9) gives the correct value for maximum speed, as deduced from conservation of energy.

SOLVE The velocity function is

$$v_x = -\omega A \sin(\omega t)$$

In one-dimensional motion, speed v is the absolute value of the velocity v_x. Since sine varies between +1 and −1, v_x varies between $+\omega A$ and $-\omega A$, so the maximum speed (absolute value of velocity) is

$$v_{\text{max}} = \omega A$$

as we found using energy conservation.

REFLECT These equations have the right units for speed or velocity. Angular frequency has units s^{-1}, and amplitude is in m, so the product of the two has units m/s.

Another Look at Energy

Knowing the simple harmonic oscillator's velocity and position as functions of time lets you see what happens to the oscillator's energy throughout a complete cycle. As usual, total mechanical energy is the sum of kinetic and potential energy:

$$E = K + U = \tfrac{1}{2}mv^2 + \tfrac{1}{2}kx^2$$

Here $v^2 = v_x^2$, because $v = |v_x|$, and squaring eliminates any negative signs. Using Equation 7.9 for v_x and $x = A\cos(\omega t)$ for position, the total energy becomes

$$E = \tfrac{1}{2}m\omega^2 A^2 \sin^2(\omega t) + \tfrac{1}{2}kA^2 \cos^2(\omega t)$$

Recalling that $\omega = \sqrt{k/m}$, so $m\omega^2 = k$, the energy simplifies to

$$E = \tfrac{1}{2}kA^2 \sin^2(\omega t) + \tfrac{1}{2}kA^2 \cos^2(\omega t) = \tfrac{1}{2}kA^2[\sin^2(\omega t) + \cos^2(\omega t)] \quad (7.10)$$

That's it—the harmonic oscillator's total mechanical energy E as a function of time. You should recognize the trig identity $\sin^2\theta + \cos^2\theta = 1$, confirming that the total mechanical energy is indeed constant, given by $E = \tfrac{1}{2}kA^2$. Moreover, the intermediate equality in Equation 7.10 shows how kinetic energy (the left-hand term) and potential energy (the right-hand term) each vary with time. Figure 7.10 graphs the kinetic, potential, and total energy for the simple harmonic oscillator. There's a beautiful symmetry here, with the continual exchange of energy between kinetic and potential!

✓ TIP

Absent frictional forces, the total mechanical energy of a simple harmonic oscillator is constant. The kinetic and potential energy each vary sinusoidally with time, but their sum is constant.

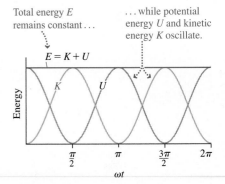

FIGURE 7.10 Energy (kinetic and potential) as a function of time.

EXAMPLE 7.6 Exchange of Kinetic and Potential Energy

Given a simple harmonic oscillator with period T, find the first time after $t = 0$ when the energy is (a) all kinetic and (b) all potential. Express your answer in terms of T.

ORGANIZE AND PLAN Both kinetic energy and potential energy are known functions of time (Figure 7.10). Because they involve squares of the sine and cosine functions, each has its maximum when the sine or cosine function reaches −1 or +1.

From Equation 7.10, the oscillator has kinetic energy $K = \tfrac{1}{2}kA^2 \sin^2(\omega t)$ and potential energy $U = \tfrac{1}{2}kA^2 \cos^2(\omega t)$. To solve the problem, we need the first time after $t = 0$ when (a) $\sin^2(\omega t)$ and (b) $\cos^2(\omega t)$ first reaches 1.

SOLVE (a) The kinetic energy is maximum when $\sin^2(\omega t) = 1$, and that happens when $\omega t = \pi/2$ (refer again to Figure 7.10). Since $\omega = 2\pi/T$, this gives

$$t = \frac{\pi}{2\omega} = \frac{\pi}{2(2\pi/T)} = \frac{T}{4}$$

That is, kinetic energy peaks after one-fourth of a period.

(b) The function $\cos^2(\omega t)$ has its maximum value 1 when $t = 0$ and again when $\omega t = \pi$. So the potential energy's first maximum after $t = 0$ occurs when

$$t = \frac{\pi}{\omega} = \frac{\pi}{(2\pi/T)} = \frac{T}{2}$$

or after half a period.

REFLECT Notice that these results, which were determined analytically, agree with the rise and fall of kinetic and potential energy in the graph in Figure 7.10.

MAKING THE CONNECTION Do the equations for kinetic and potential energy still work after one period has elapsed ($t > T$)?

ANSWER Yes. The trig functions are defined for all values of t, and each cycle is identical, so there's no problem with $t > T$. Without friction, simple harmonic motion would continue forever.

Position, Velocity, and Acceleration

So far we've explored the time dependence of position and velocity in SHM. What about acceleration? That follows from Newton's second law, which in one dimension is

$$F_{net,x} = ma_x$$

For the simple harmonic oscillator, the net force is the spring force $-kx$, where again the minus sign shows that the spring pulls or pushes the system back toward equilibrium. Using the spring force in Newton's law gives $-kx = ma_x$, or

$$a_x = -\frac{k}{m}x$$

Because position is $x = A\cos(\omega t)$, acceleration becomes

$$a_x = -\frac{k}{m}A\cos(\omega t)$$

Finally, because $k/m = \omega^2$, we have

$$a_x = -\omega^2 A\cos(\omega t) \qquad \begin{array}{l}\text{(Acceleration versus time for a}\\ \text{harmonic oscillator; SI unit: m/s}^2\text{)}\end{array} \qquad (7.11)$$

Amazingly, position, velocity, and acceleration are all sinusoidal functions of time (Figure 7.11). You might compare the graphs in Figure 7.11 with Figures 2.10 and 2.14, where we showed that, in general, velocity is the slope of the position-versus-time curve and acceleration is the slope of the velocity-versus-time curve.

✓ **TIP**

The position, velocity, and acceleration of the simple harmonic oscillator all vary sinusoidally in time.

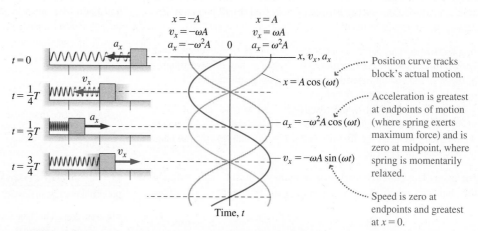

FIGURE 7.11 Graphs of position, velocity, and acceleration versus time for a simple harmonic oscillator. The graphs are rotated to show their relation to the motion of an oscillating mass-spring system.

EXAMPLE 7.7 **Swaying Skyscraper!**

Tall buildings are flexible and undergo simple harmonic motion when excited by wind. Suppose the top floor of a skyscraper is in SHM, swaying back and forth with frequency $f = 0.15$ Hz and amplitude $A = 1.7$ m. Find the maximum speed and acceleration experienced by top-floor occupants.

ORGANIZE AND PLAN Equation 7.7 gives the maximum speed in SHM: $v_{max} = \omega A$. The maximum acceleration follows from the general expression for acceleration (Equation 7.11), which reached a maximum when $\cos(\omega t) = -1$, giving $a_{x,max} = \omega^2 A$. Note that both these results correspond to the extreme values in Table 7.1. To find the angular frequency ω, recall that $\omega = 2\pi f$.

Known: $f = 0.15$ Hz and $A = 1.7$ m.

TABLE 7.1 Position, Velocity, Speed, and Acceleration at Key Points in the Cycle (T = period, ω = angular frequency, A = amplitude)

Time t	Position x	Velocity v_x	Speed v	Acceleration a_x
0	A	0	0	$-\omega^2 A$
$T/4$	0	$-\omega A$	ωA	0
$T/2$	$-A$	0	0	$\omega^2 A$
$3T/4$	0	ωA	ωA	0
T	A	0	0	$-\omega^2 A$

SOLVE First, the angular frequency ω is

$$\omega = 2\pi f = 2\pi(0.15 \text{ Hz}) = 0.9425 \text{ s}^{-1}$$

Then

$$v_{max} = \omega A = (0.9425 \text{ s}^{-1})(1.7 \text{ m}) = 1.6 \text{ m/s}$$

and

$$a_{x,max} = \omega^2 A = (0.9425 \text{ s}^{-1})^2(1.7 \text{ m}) = 1.5 \text{ m/s}^2$$

REFLECT That maximum acceleration is about $0.15g$, and because the acceleration is continually changing it can make building occupants feel seasick!

MAKING THE CONNECTION How might you minimize sway in a tall building?

ANSWER You could make the building stiffer—but that means more massive construction and greater expense—and may actually make the building less safe in an earthquake. The following application describes a more intelligent approach.

APPLICATION **Tuned Mass Dampers**

Modern skyscrapers use *tuned mass dampers*, which are simple harmonic oscillators with huge masses and periods tuned to the period of the building's natural sway. The damper oscillates out of phase with the building, so when the building sways left, the damper goes right. That moves the center of mass of the building so it tends to stay more upright. Tuned mass dampers not only make the building's occupants more comfortable, but also minimize earthquake damage. The photos show the world's largest tuned mass damper and the building that houses it, Taiwan's Taipei 101 skyscraper.

GOT IT? Section 7.3 Each oscillator shown has an amplitude of 10 cm. Rank in order, from lowest to highest, the total energy of the five oscillators.

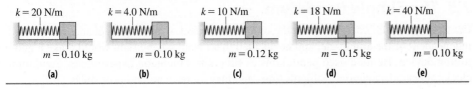

$k = 20$ N/m	$k = 4.0$ N/m	$k = 10$ N/m	$k = 18$ N/m	$k = 40$ N/m
$m = 0.10$ kg	$m = 0.10$ kg	$m = 0.12$ kg	$m = 0.15$ kg	$m = 0.10$ kg
(a)	(b)	(c)	(d)	(e)

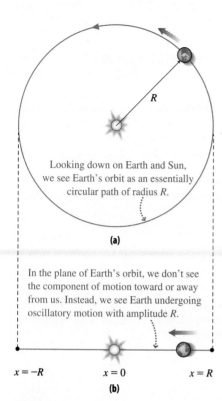

Looking down on Earth and Sun, we see Earth's orbit as an essentially circular path of radius R.

(a)

In the plane of Earth's orbit, we don't see the component of motion toward or away from us. Instead, we see Earth undergoing oscillatory motion with amplitude R.

$x = -R$ $x = 0$ $x = R$

(b)

FIGURE 7.12 Two views of Earth's orbital motion.

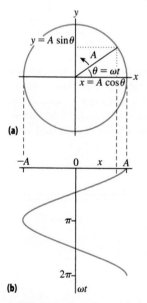

(a)

(b)

FIGURE 7.13 Uniform circular motion with radius A and angle $\theta = \omega t$.

7.4 SHM and Uniform Circular Motion

Figure 7.12a shows Earth orbiting the Sun—essentially the uniform circular motion you studied in Chapter 3. Figure 7.12b shows what you would see viewing Earth in the plane of its orbit: The planet would appear to move back and forth in oscillatory motion. What you're seeing is a single component of Earth's circular motion—the *projection* of the full circular motion onto a line, which we'll take as the *x*-axis. We'll now show that this projected motion is simple harmonic.

In one period T, Earth or any other object in uniform circular motion sweeps out a full circle—360° or 2π radians. Because the object's speed is constant, the time to move through an arbitrary angle θ is related to the period as θ is to 2π:

$$\frac{t}{T} = \frac{\theta}{2\pi}$$

or

$$\theta = \frac{2\pi}{T} t$$

You've seen that $2\pi/T$ before: in simple harmonic motion, it's the angular frequency ω. Using the same terminology, the angle the object's position makes with the *x*-axis becomes

$$\theta = \omega t$$

Figure 7.13 illustrates this: In uniform circular motion, an object's position on the circle is described by an angle θ that increases continually with time. The *x*-component of the object's position is $x = A \cos \theta$, where A is the circle's radius. But you've just seen that $\theta = \omega t$, so the *x*-component is

$$x = A \cos (\omega t)$$

That's precisely our description of simple harmonic motion! What we've shown here is that **the projection of uniform circular motion onto a diameter of the circle gives simple harmonic motion.**

Now you can see why ω is called *angular frequency*. Although there aren't any real angles in one-dimensional SHM, you can always imagine a related circular motion and think of the quantity ωt as the angular position of an object on the circular path; ω itself is the speed of the circular motion, expressed in radians per second or, since radians are dimensionless, just s^{-1}. Similarly, we speak of a quarter-cycle of SHM as being 90° or $\pi/2$ radians, even though there's no actual angle involved.

There's a practical side to the relationship between circular motion and one-dimensional SHM. Many mechanical devices turn one kind of motion into the other. In your car's engine, and more obviously in an old steam locomotive, the back-and-forth motion of a piston is converted into wheel rotation. The opposite occurs in sewing machines and oil wells, where circular motion gets converted into the up-and-down oscillatory motion of the sewing needle or oil pump mechanism.

7.5 The Simple Pendulum

Centuries ago, Galileo noticed that the oscillation period of a swinging chandelier seemed independent of the amplitude, and he decided that meant a pendulum would make a good timing device. He then used pendulums as timers in kinematic experiments. Pendulum clocks appeared soon after Galileo's time, and they've been around ever since.

What makes the pendulum such a good timer? How does it work? Is its period really independent of amplitude, like the harmonic oscillator? We'll answer these questions by exploring the simple pendulum.

The Simple Pendulum

A **simple pendulum** is defined as a point particle of mass m (the pendulum bob) suspended by a cord or rod of length L. A real pendulum approximates this ideal if (1) the bob is small compared with the length L, (2) the cord or rod mass is much less than the bob's mass, and (3) the cord or rod remains straight and doesn't stretch. Pull the pendulum bob aside and let go; the pendulum then swings back and forth. Neglecting air drag and friction at the pendulum's pivot, these oscillations are periodic.

Figure 7.14a shows that the pendulum bob swings in a circular arc with radius equal to the cord length L. When the cord makes an angle θ with the vertical, the bob is displaced a distance $s = L\theta$, with θ in radians, from its vertical equilibrium. Figure 7.14b shows the two forces acting on the bob at that point—gravity and the cord tension. We'll break those forces into components tangent and perpendicular to the arc. The perpendicular components sum to provide the centripetal acceleration v^2/L that keeps the pendulum in its circular path. We're more interested in the force component tangent to the arc—the one that causes acceleration *along* the arc—which Figure 7.14b shows is $-mg\sin\theta$. That minus sign is important; it shows that the tangent force acts *opposite* the pendulum's displacement. So the tangential component of Newton's second law reads

$$F_s = ma_s = -mg\sin\theta$$

where the s subscripts designate components along the arc, since s is the bob's position measured along the arc.

This expression of Newton's second law is the exact equation governing the motion of a simple pendulum. Because of the sine term, it's not a harmonic oscillator equation, and without some advanced math, there's nothing more we can do with it as is. But there's an approximation that's quite accurate in many situations. It's called the "small-angle approximation," because it's good as long as the pendulum swings only through small angles. This makes the simple pendulum behave like a simple harmonic oscillator and lets us apply the results of our SHM analysis.

The Small-Angle Approximation

Put your calculator into radian mode and explore the sine function. You'll find that for small angles ($\theta \ll 1$ radian), θ and $\sin\theta$ are nearly equal. For example, at $10°$ ($\theta = 0.1745$ radians), $\sin\theta = 0.1736$—only half a percent difference. For angles less than $10°$, the difference is even smaller. Figure 7.15 shows why: for small values of its argument, the function $f(\theta) = \sin\theta$ is nearly indistinguishable from the straight line $f(\theta) = \theta$. So if a pendulum's maximum angular displacement is small, we can approximate θ by $\sin\theta$ in our expression of Newton's law for the pendulum, giving

$$ma_s = -mg\theta$$

✓ **TIP**

The angle θ is measured in radians, because it relates arc length to radius.

The mass m cancels, and recalling that $s = L\theta$,

$$a_s = -\frac{g}{L}s$$

Compare this with the harmonic oscillator's acceleration (Section 7.3):

$$a_x = -\frac{k}{m}x$$

The two equations have exactly the same form, with acceleration proportional to displacement—so they must describe analogous physical behavior. Therefore:

In the small-angle approximation, the simple pendulum behaves like a simple harmonic oscillator.

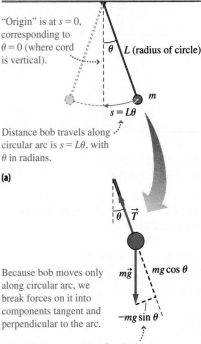

"Origin" is at $s = 0$, corresponding to $\theta = 0$ (where cord is vertical).

θ L (radius of circle)

m

$s = L\theta$

Distance bob travels along circular arc is $s = L\theta$, with θ in radians.

(a)

θ $\vec{T}$

Because bob moves only along circular arc, we break forces on it into components tangent and perpendicular to the arc.

$m\vec{g}$ $mg\cos\theta$

$-mg\sin\theta$

With θ taken as positive for displacement to the right, tangential force component is $-mg\sin\theta$.

(b)

FIGURE 7.14 (a) The pendulum swings in a circular arc. (b) Force diagram for the simple pendulum.

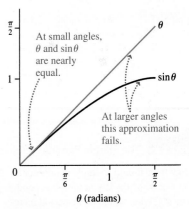

$\frac{\pi}{2}$

At small angles, θ and $\sin\theta$ are nearly equal.

θ

1

$\sin\theta$

At larger angles this approximation fails.

0 $\frac{\pi}{6}$ 1 $\frac{\pi}{2}$

θ (radians)

FIGURE 7.15 For an angle θ much less than 1 radian, $\sin\theta$ and θ are nearly equal.

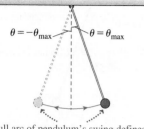

$\theta = -\theta_{max}$ $\theta = \theta_{max}$

Full arc of pendulum's swing defines angle θ_{max}.

FIGURE 7.16 End points of the pendulum's motion.

TABLE 7.2 Period versus Amplitude for a Real Pendulum

Amplitude θ_{max} (radians)	Amplitude θ_{max} (degrees)	Period T (s)
0.1	5.73	1.80
0.2	11.5	1.81
0.4	22.9	1.82
0.6	34.4	1.84
0.8	45.8	1.88
1.0	57.3	1.92
1.2	68.8	1.98
1.4	80.2	2.05
$\pi/2 (\approx 1.57)$	90	2.13

In analogy with the simple harmonic oscillator's period $T = 2\pi\sqrt{m/k}$, the pendulum's period is

$$T = 2\pi\sqrt{\frac{L}{g}} \qquad \text{(Period of a simple pendulum, small-angle approximation; SI unit: s)} \qquad (7.12)$$

Also by analogy, the equation of motion for the simple pendulum is

$$\theta = \theta_{max} \cos(\omega t) \qquad (7.13)$$

where the amplitude θ_{max} is the maximum angle the cord makes with the vertical, and $\omega = 2\pi/T = \sqrt{g/L}$ is the angular frequency. So the pendulum swings back and forth, with angle θ ranging from $-\theta_{max}$ to θ_{max} as shown in Figure 7.16.

The mathematical results presented here are valid *only* in the small-angle approximation. Table 7.2 shows some data obtained for the period of a real pendulum with length 80.0 cm, for which Equation 7.12 gives

$$T = 2\pi\sqrt{\frac{L}{g}} = 2\pi\sqrt{\frac{0.800 \text{ m}}{9.80 \text{ m/s}^2}} = 1.80 \text{ s}$$

The first few measured periods are close to this value, but the period increases significantly for larger amplitudes. Thus a real pendulum does not behave like a harmonic oscillator, except when the maximum angular displacement is small.

✓ **TIP**

The simple pendulum behaves like a simple harmonic oscillator only for small-angle oscillations.

Reviewing New Concepts: The Simple Pendulum

- A simple pendulum exhibits oscillations between $-\theta_{max}$ and θ_{max}.
- For small oscillations, the simple pendulum's period is approximately $T = 2\pi\sqrt{\frac{L}{g}}$.
- The simple pendulum's motion approximates simple harmonic motion only for small amplitudes.

EXAMPLE 7.8 **The Simple Pendulum Takes Off**

(a) How long should a simple pendulum be so that its small-oscillation period is 2.00 s? (b) If you took this pendulum to the Moon ($g_{Moon} = 1.60 \text{ m/s}^2$), what would be its period there?

ORGANIZE AND PLAN Equation 7.12 relates length and period of a simple pendulum: $T = 2\pi\sqrt{L/g}$. We'll need the correct g for each astronomical body.

Known: $T_{Earth} = 2.00$ s; $g_{Moon} = 1.60 \text{ m/s}^2$.

SOLVE (a) Solve for length, using g_{Earth}:

$$L = \frac{gT^2}{4\pi^2} = \frac{(9.80 \text{ m/s}^2)(2.00 \text{ s})^2}{4\pi^2} = 0.993 \text{ m}$$

(b) Use this result to find T_{Moon}:

$$T_{Moon} = 2\pi\sqrt{\frac{L}{g_{Moon}}} = 2\pi\sqrt{\frac{0.993 \text{ m}}{1.60 \text{ m/s}^2}} = 4.95 \text{ s}$$

REFLECT It's an interesting coincidence that a 1-m-long pendulum has a period around 2 s on Earth. On the Moon, its period increases to nearly 5 s. The pendulum is gravity driven, so its period depends on the strength of gravity.

MAKING THE CONNECTION How would a pendulum work in the apparent weightlessness of free fall, as on an orbiting spacecraft?

ANSWER It wouldn't work at all, unless the spacecraft were rotating to simulate gravity.

GOT IT? Section 7.5 Rank in order, from lowest to highest, the periods of the four pendulums.

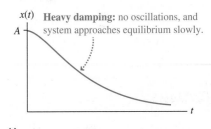

(a) $L = 0.75$ m, $m = 0.65$ kg

(b) $L = 0.75$ m, $m = 0.68$ kg

(c) $L = 0.91$ m, $m = 0.54$ kg

(d) $L = 0.67$ m, $m = 0.71$ kg

TABLE 7.3 Conditions for Light Damping, Critical Damping, and Heavy Damping

Damping type	Damping parameter b
Light	$b^2 < 4mk$
Critical	$b^2 = 4mk$
Heavy	$b^2 > 4mk$

7.6 Damped and Driven Oscillations

Start a mass-spring system oscillating, or a pendulum swinging, and the oscillations eventually die out. All oscillating systems are subject to friction, drag, or other energy losses, which damp the oscillations. Engineers often need to design systems so that energy is supplied to overcome such losses. For example, a pendulum clock is driven by the potential energy of a raised weight. The force associated with an energy source can even increase the oscillation amplitude. Here we'll discuss these phenomena of damped and driven oscillators, then consider some practical applications.

Damped Harmonic Motion

You saw in Chapter 4 that the drag force on an object is directed opposite its motion, and in some cases is proportional to the object's velocity:

$$\vec{F}_{\text{drag}} = -b\vec{v}$$

where b is a constant that measures the importance of drag.

If the damping force isn't too great, oscillations occur, but their amplitude declines exponentially. This is **light damping** (Table 7.3 and Figure 7.17a). Increase the damping parameter b and eventually you reach **critical damping**, where the system no longer oscillates. Displace a critically damped mass-spring system, and it goes smoothly back toward equilibrium (Figure 7.17b). Increase b further, and the increased damping force still prohibits oscillations but also delays the object's return to equilibrium. This is **heavy damping** (Figure 7.17c).

The Light-Damping Solution

Light damping is the case most often encountered. Using calculus to solve Newton's law for this case gives position versus time for the lightly damped harmonic oscillator:

$$x = Ae^{-bt/2m} \cos(\omega_{\text{damped}} t) \tag{7.14}$$

with the oscillation frequency

$$\omega_{\text{damped}} = \sqrt{\frac{k}{m} - \frac{b^2}{4m^2}} \tag{7.15}$$

Equation 7.14 makes sense if you think of it as the product of two factors: an exponentially decaying amplitude $Ae^{-bt/2m}$ and a sinusoidal oscillation $\cos(\omega_{\text{damped}} t)$. Figure 7.18 graphs the position function, with amplitude $Ae^{-bt/2m}$ superimposed as an "envelope" on the oscillation, which continues with constant period even as its amplitude decays.

Note that the frequency ω_{damped} in Equation 7.15 is smaller than the undamped frequency $\omega = \sqrt{k/m}$. The period of damped oscillations is then $T_{\text{damped}} = 2\pi/\omega_{\text{damped}}$, which is *larger* than the undamped period. That makes sense: The drag force is slowing the oscillations as well as damping their amplitude.

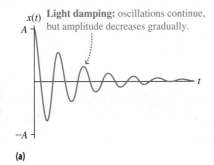

(a) Light damping: oscillations continue, but amplitude decreases gradually.

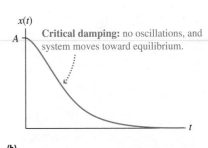

(b) Critical damping: no oscillations, and system moves toward equilibrium.

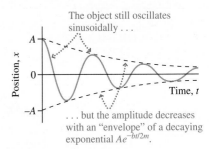

(c) Heavy damping: no oscillations, and system approaches equilibrium slowly.

FIGURE 7.17 Position versus time for the three types of damping.

The object still oscillates sinusoidally ... but the amplitude decreases with an "envelope" of a decaying exponential $Ae^{-bt/2m}$.

FIGURE 7.18 Position versus time with light damping.

Light Damping and Critical Damping

What happens to frequency and period as critical damping is approached? Explain why your answers are consistent with our description of critical damping.

SOLVE Table 7.3 shows that critical damping occurs when $b^2 = 4mk$. As b approaches this value, the light-damping frequency (Equation 7.15) approaches

$$\omega_{\text{damped}} = \sqrt{\frac{k}{m} - \frac{b^2}{4m^2}} \rightarrow \sqrt{\frac{k}{m} - \frac{4mk}{4m^2}} = \sqrt{\frac{k}{m} - \frac{k}{m}} = 0$$

A frequency approaching zero corresponds to a period $(T = 2\pi/\omega)$ approaching infinity. That's consistent with nonperiodic motion, as described earlier for critical damping.

REFLECT For heavy damping, the expression under the square root is negative. The square root of a negative number is imaginary, which you can take to mean that there's no oscillation frequency with heavy damping. That's also consistent with what's observed.

✓ **TIP**

When a simple harmonic oscillator is damped, the resulting motion is oscillatory only if the damping is light.

Applications of Damping

Damping isn't necessarily bad, and in some applications it's quite useful. Your car's shock absorbers are one example. Each shock contains a piston immersed in a heavy fluid, typically mounted inside the car's suspension springs. The result is rapid damping of spring oscillations. (Try this yourself: Push down your car's hood, let go, and watch the oscillation stop quickly.) Without shock absorbers, the car would bounce up and down long after passing over a bump! Ideally, shocks should provide critical damping, to reduce the oscillation as quickly as possible. This isn't exactly possible in practice, because the critical damping condition depends on the car's mass, which varies with passenger and cargo loads. Huge but conceptually similar shock absorbers are used in buildings and bridges to damp earthquake-induced oscillations.

The bungee jumper in the photo at the beginning of this chapter leaps, bottoms out, and then oscillates vertically on the springy bungee cord. The motion is lightly damped, due to air drag and nonconservative forces within the bungee cord. The light damping causes a gradual decrease in the oscillation's amplitude. Once the amplitude is small enough, the excitement is over!

Driven Oscillations

You're pushing a child on a playground swing. If you push in the direction she's already moving, you add energy. The right push just compensates for energy losses, maintaining a constant-amplitude swing. Push harder, and you're increasing the system's energy. The oscillation amplitude increases—and higher she goes! This is an example of a **driven oscillation**.

Driven oscillations are widespread in nature and in technology. They can be beneficial or harmful. Your stereo speakers are driven electromagnetically to oscillate at frequencies that produce sound waves. (You'll learn more about sound in Chapter 11.) Wind and earthquakes drive potentially dangerous oscillations in buildings, and you've seen two approaches to countering this effect. Bridges, airplane wings, and other engineered structures are also subject to driven oscillations. Driven oscillations in individual atoms are responsible for myriad natural phenomena, including reflection of light and the sky's blue color.

Resonance

Consider again that child on the swing. There's some "natural" frequency of the swing's pendulumlike oscillation (see Section 7.4). If you push with that same frequency, the amplitude grows rapidly. This is **resonance**, which occurs when the driving frequency matches the natural frequency of an oscillator. Push at some other frequency, and you'll have more trouble building up the oscillation amplitude.

Figure 7.19 quantifies this phenomenon. The figure shows a typical **resonance curve**, a graph of oscillation amplitude as a function of driving frequency. For most driving frequencies, the amplitude is low. That's because the driving force isn't always in sync with the way the system "wants" to move. But when the driving frequency matches the natural frequency, every push is in the right direction to make the amplitude larger.

 TIP

Resonance occurs when an oscillator is driven at or near its natural oscillation frequency.

FIGURE 7.19 Resonance curve, with ω_0 equal to the natural frequency $\sqrt{k/m}$.

Resonance can be disastrous. A famous example is the Tacoma Narrows Bridge (Figure 7.20), which was destroyed by resonant oscillations during a windstorm in November, 1940, just months after it opened. A new bridge was built in 1950 with much stiffer supporting materials, to prevent such resonant oscillations from developing. That bridge has stood the test of time.

Resonance also occurs in microscopic systems. The fundamental physics behind global warming involves resonant oscillations in carbon dioxide molecules, whose natural frequencies are in the range of infrared radiation by which Earth sheds heat. More atmospheric CO_2 means more infrared absorption, and that leads to a higher surface temperature.

Reviewing New Concepts: Damped and Driven Oscillations

- A damping force applied to a simple harmonic motion results in light damping, critical damping, or heavy damping, depending on the strength of the damping force.
- Driven oscillations approach resonance when the driving frequency is equal to the system's natural frequency.

FIGURE 7.20 Collapse of the Tacoma Narrows Bridge in November, 1940—only four months after its opening—followed the resonant growth of large-amplitude oscillations.

Chapter 7 in Context

Chapters 2 and 3 gave you the essentials of kinematics. Chapters 4 and 5 introduced the important concepts of force and energy and gave numerous applications.

Here you've seen how both force and energy help us understand *oscillations*. We've focused on two specific types of oscillations: *simple harmonic motion* and the *simple pendulum*. Forces drive oscillations, and you've seen how the spring force $-kx$ leads to simple harmonic motion. The simple pendulum behaves like a simple harmonic oscillator as long as its oscillation amplitude is small. Finally, you saw how *damping* and *driving* forces affect oscillations.

Looking Ahead: Rotational motion and waves In Chapter 8 we'll apply force and energy to study rotational motion. In Chapter 7 you saw how simple harmonic motion is related to uniform circular motion, and in Chapter 8 you'll see more connections between circular motion, oscillatory motion, and rotating bodies. The concepts of period, frequency, and angular frequency will again prove relevant. Finally, in Chapters 11 and 20 you'll study waves, which are closely related to simple harmonic motion.

CHAPTER 7 SUMMARY

Periodic Motion

(Section 7.1) **Periodic motion** repetitively covers the same path. The **period** is the time to traverse the path. **Frequency** is the reciprocal of the period and is the rate at which periodic motion is repeated. **Oscillatory motion** is periodic motion that repeats a back-and-forth path.

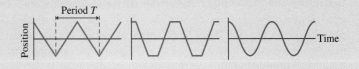

Frequency of periodic motion: $f = \dfrac{1}{T}$

Simple Harmonic Motion

(Section 7.2) **Simple harmonic motion** depends sinusoidally on time; its paradigm is the mass-spring system. The mass oscillates about its equilibrium position $x = 0$, undergoing a maximum displacement A—the **amplitude**—in either direction. The **angular frequency** ω is an alternate measure of frequency in simple harmonic motion. The period of a **simple harmonic oscillator** depends on the spring constant and mass but not the amplitude.

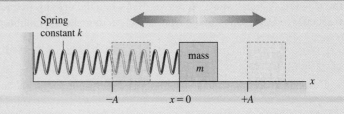

Simple harmonic motion: $x = A \cos(\omega t)$

Period of simple harmonic motion: $T = 2\pi\sqrt{\dfrac{m}{k}}$

Angular frequency: $\omega = \sqrt{\dfrac{k}{m}} = \dfrac{2\pi}{T} = 2\pi f$

Energy in Simple Harmonic Motion/SHM and Uniform Circular Motion

(Sections 7.3 and 7.4) Absent friction, the **total mechanical energy** E (kinetic plus potential) of a simple harmonic oscillator is conserved. The maximum velocity of a simple harmonic oscillator occurs as the object passes through $x = 0$. The velocity, acceleration, and position of a simple harmonic oscillator all vary sinusoidally with time. The projection of **uniform circular motion** onto the circle's diameter is simple harmonic motion.

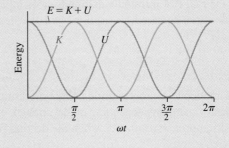

Total mechanical energy: $E = K + U = \frac{1}{2}mv^2 + \frac{1}{2}kx^2$

Oscillator's speed as a function of position: $v = \sqrt{\dfrac{k}{m}(A^2 - x^2)}$

Oscillator's velocity as a function of time: $v_x = -\omega A \sin(\omega t)$

Oscillator's acceleration: $a_x = -\omega^2 A \cos(\omega t)$

The Simple Pendulum

(Section 7.5) A **simple pendulum** acts like a simple harmonic oscillator for small oscillations. The period of a small-amplitude simple pendulum depends on gravity and the pendulum's length but not on amplitude or mass.

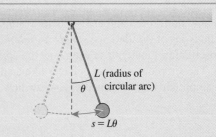

Simple pendulum period: $T = 2\pi\sqrt{\dfrac{L}{g}}$

Damped and Driven Oscillations

(Section 7.6) In a damped harmonic oscillator, friction or drag forces retard the oscillator's motion. A **lightly damped** system oscillates with decreasing amplitude. **Critically** and **heavily damped** systems cease to oscillate. Driving a harmonic oscillator at its natural frequency results in **resonance**.

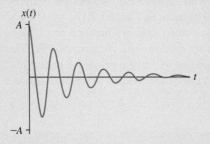

Position versus time for lightly damped oscillator:
$x = Ae^{-bt/2m} \cos(\omega_{\text{damped}} t)$

NOTE: Problem difficulty is labeled as ■straightforward to ■■■challenging. Problems labeled BIO are of biological or medical interest.

Conceptual Questions

1. Give an example of motion that's periodic but not oscillatory.
2. Which has the longer period, an oscillator with a frequency of 200 Hz or one with $f = 300$ Hz?
3. At what point(s) in the oscillation cycle is a simple harmonic oscillator's acceleration zero? At what point(s) is its velocity zero?
4. A mass-spring system undergoes simple harmonic motion. If mass is doubled, what happens to the period?
5. A mass-spring system undergoes simple harmonic motion. If the spring constant is doubled, what happens to the period?
6. A mass-spring system undergoes simple harmonic motion with amplitude A. What fraction of the system's energy is kinetic energy when the object's position is $x = A/2$?
7. Explain how simple harmonic motion could be used to determine the mass of an astronaut in the apparent weightlessness of an orbiting spacecraft.
8. Your car oscillates at a certain frequency after hitting a bump. How does that frequency change as passengers are added?
9. What is the net work done by the spring in a simple harmonic oscillator over one complete oscillation cycle? When during the cycle is the spring doing positive work, and when is it doing negative work?
10. If you double the length of a simple pendulum, what happens to the period?
11. For a pendulum of fixed length, compare what happens to the period if you double the amplitude (a) from 2° to 4° and (b) from 20° to 40°?
12. Suppose you have a pendulum clock that runs accurately at sea level. Will it still be accurate when taken to a mountaintop where g is slightly less? If not, will it run fast or slow?
13. If you want to cut the period of a simple pendulum in half, how should you change its length?
14. What two forces act on a pendulum bob? Draw a force diagram showing each of the forces and the net force on the bob when it's at maximum displacement, $\theta = \theta_{max}$. Repeat for $\theta = 0$.
15. A bungee jumper leaps from a bridge and then oscillates on the bungee cord. Discuss the forms of energy present and how they change throughout the leap and the oscillation.
16. Give some examples of resonance in everyday life.

Multiple-Choice Problems

17. The period of a tuning fork vibrating at 200 Hz is (a) 200 s; (b) 0.5 s; (c) 0.005 s; (d) 0.002 s.
18. A hummingbird flaps its wings once every 12.5 ms. The corresponding frequency is (a) 12.5 Hz; (b) 20 Hz; (c) 40 Hz; (d) 80 Hz.
19. The position of a simple harmonic oscillator with period T is $x = A \cos\left(\dfrac{2\pi}{T} t\right)$. The time it takes the oscillator to go from $x = A$ to $x = 0$ is (a) $T/8$; (b) $T/4$; (c) $T/2$; (d) T.
20. A simple harmonic oscillator has $k = 15.0 \text{ N/m}$ and $m = 0.100$ kg. Its period is (a) 0.0067 s; (b) 0.150 s; (c) 0.333 s; (d) 0.513 s.
21. A simple harmonic oscillator has $m = 1.50$ kg, $k = 20.0 \text{ N/m}$, and $A = 2.00$ m. Its maximum speed is (a) 2.00 m/s; (b) 7.30 m/s; (c) 13.3 m/s; (d) 26.7 m.

22. A simple harmonic oscillator with amplitude A_0 has total mechanical energy E_0. If the amplitude changes to $2A_0$ but the system is otherwise unchanged, the total mechanical energy becomes (a) $\sqrt{2}E_0$; (b) $2E_0$; (c) $4E_0$; (d) $8E_0$.
23. A mass-spring system has a period of 5.00 s. If the mass is doubled, the new period is (a) 3.54 s; (b) 5.00 s; (c) 7.07 s; (d) 10.0 s.
24. A mass-spring system has a period of 5.00 s. If the spring constant is doubled, the new period is (a) 3.54 s; (b) 5.00 s; (c) 7.07 s; (d) 10.0 s.
25. The period of a mass-spring simple harmonic oscillator increases when which of the following is increased? (a) the spring constant; (b) the oscillation amplitude; (c) the ratio k/m; (d) the mass.
26. You hang a spring vertically from the ceiling and attach a 0.240-kg mass to the bottom. At rest, the mass stretches the spring 12 cm. If you pull the mass down and release, it will oscillate with period (a) 0.08 s; (b) 0.30 s; (c) 0.50 s; (d) 0.70 s.
27. The period of a simple pendulum with length 13.4 m is (a) 7.35 s; (b) 8.59 s; (c) 11.0 s; (d) 13.4 s.
28. A simple pendulum with small-amplitude oscillations has a period of 1.24 s. If its length is tripled, the period becomes (a) 2.15 s; (b) 2.94 s; (c) 3.72 s; (d) 11.2 s.
29. Astronauts exploring a new planet measure the period of a 1.20-m pendulum and get 2.40 s. The planet's gravitational acceleration is (a) 2.00 m/s²; (b) 2.62 m/s²; (c) 5.24 m/s²; (d) 8.22 m/s².
30. How long should you make a pendulum so that its period is 1.0 s? (a) 0.25 m; (b) 0.54 m; (c) 0.79 m; (d) 1.0 m.

Problems

Section 7.1 Periodic Motion

31. ■ (a) Is Earth's orbital motion periodic, oscillatory, or both? (b) What's the frequency of Earth's orbital motion, in Hz?
32. ■ Jupiter takes 11.9 years to orbit the Sun. What's the frequency of its orbital motion?
33. ■ In musical notation, Presto designates a fast tempo with a frequency around 160 beats per minute. What is the period of one beat?
34. BIO ■ **The EEG.** Physicians use electroencephalograms (EEGs) to study electrical oscillations in the human brain. So-called beta oscillations, whose enhancement has been linked to alcoholism, have frequencies from 12 to 30 Hz. What's the corresponding range in periods?
35. ■ The turbines in a jet engine rotate with frequency 16 kHz. Find (a) the period and (b) the angular frequency.
36. BIO ■ **Heart rate.** After running a 1500-m race, an athlete's heart beats 145 times per minute. Find the period and frequency of the heartbeat.

Section 7.2 Simple Harmonic Motion

37. ■■ The period of a simple harmonic oscillator depends on only the spring constant k and the mass m. Using dimensional analysis, show that the only combination of those two parameters that gives units of time is $\sqrt{m/k}$.
38. ■ A mass-spring system has $k = 55.2 \text{ N/m}$ and $m = 0.450$ kg. Find the frequency, angular frequency, and period of its simple harmonic motion.

39. BIO ■ **Oscillating spider.** A 1.4-g spider dangles from the end of her silk thread, undergoing vertical oscillations at 1.1 Hz. What's the spring constant of the spider's thread?

40. ■ ■ A bungee jumper undergoes simple harmonic motion with amplitude 5.0 m and frequency 0.125 Hz. Graph position versus time for two oscillation cycles.

41. ■ ■ Assume the bungee jumper in the preceding problem follows the simple harmonic motion equation $x = A \cos(\omega t)$. Find the jumper's position at the following times: (a) 0.25 s; (b) 0.50 s; (c) 1.0 s.

42. ■ ■ Using the information in the preceding two problems, determine the bungee jumper's velocity at the following times: (a) 0.25 s; (b) 0.50 s; (c) 1.0 s.

43. ■ A mass-spring system has mass 0.975 kg and oscillates with period 0.500 s. Find the spring constant.

44. ■ A simple model of carbon dioxide has two oxygen atoms connected by springs to a carbon atom (Figure P7.44). Oscillations of this molecule are responsible for the infrared absorption that results in global warming. In one oscillation mode, the carbon stays fixed while the oxygen atoms undergo SHM in opposite directions, as shown. If the effective spring constant is 1.7 kN/m, what's the oscillation frequency?

FIGURE P7.44

45. ■ ■ A mass-spring system with $m = 0.200$ kg undergoes simple harmonic motion with period 0.55 seconds. When an additional mass Δm is added, the period increases by 20%. Find Δm.

46. ■ ■ A mass-spring system has $k = 110$ N/m and $m = 1.45$ kg. If it is undergoing simple harmonic motion, how much time does it take the mass to go from $x = A$ to $x = 0$?

47. ■ ■ Suppose the oscillation amplitude in the preceding problem is $A = 0.50$ m. Graph position versus time for the first two periods, assuming the oscillation began with the mass stationary at $x = A$.

48. ■ ■ A spring ($k = 65.0$ N/m) hangs vertically with its top end fixed. A mass attached to the bottom of the spring then displaces it 0.250 m, establishing a new equilibrium. If the mass is further displaced and then released, what's the period of the resulting oscillations?

49. ■ ■ Consider an oscillator with a position given by $x = A \cos(\omega t + \pi/2)$. (a) Graph position versus time for one complete cycle, from $t = 0$ to $t = T$. (b) Is this simple harmonic motion? Explain.

50. ■ ■ Rework Conceptual Example 7.2, parts (c) and (d), this time using energy concepts to explain where a simple harmonic oscillator's speed is the least and greatest.

51. ■ ■ You hang a spring vertically from the ceiling and attach a 660-g mass to the bottom. When that mass is at rest, it stretches the spring a distance d. If you pull the mass down and release so it oscillates with a period of 1.04 s, what is d?

52. ■ ■ Four identical springs support equally the weight of a 1200-kg car. (a) If a 95-kg driver gets in, the car drops 6.5 mm. What is k for each spring? (b) The car driver goes over a speed bump, causing a small vertical oscillation. Find the oscillation period, assuming the springs aren't damped.

53. ■ ■ ■ A spring with $k = 25$ N/m is attached to a 0.23-kg block as shown in Figure P7.53. The block slides without friction along the horizontal surface. On top of the 0.23-kg block rests a 0.11-kg block. The coefficient of static friction between the blocks is 0.14. (a) If the blocks move together as one mass, what

is the oscillation period? (b) Find the maximum amplitude permissible if the upper block isn't to slip.

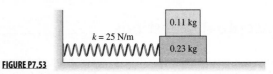

FIGURE P7.53

54. ■ ■ ■ A piston oscillates vertically at 5.0 Hz. Find the maximum oscillation amplitude so a coin resting atop the piston doesn't come off at any point in the cycle.

Section 7.3 Energy in Simple Harmonic Motion

55. ■ ■ Given a simple harmonic oscillator with period T, find the first time after $t = 0$ when the energy is half kinetic and half potential.

56. ■ ■ Find the total mechanical energy of a mass-spring system with $m = 1.24$ kg and maximum speed 0.670 m/s.

57. ■ ■ The maximum speed and acceleration of a simple harmonic oscillator are 0.95 m/s and 1.56 m/s². Find the oscillation amplitude.

58. ■ ■ A simple harmonic oscillator has spring constant $k = 5.0$ N/m, amplitude $A = 10$ cm, and maximum speed 4.2 m/s. What's the oscillator's speed when it's at $x = 5.0$ cm?

59. ■ ■ ■ A simple harmonic oscillator with $k = 94.0$ N/m and $m = 1.06$ kg has amplitude 0.560 m. (a) Find its total mechanical energy. (b) Find the oscillator's speed when $x = -0.210$ m. (c) What fraction of its energy is kinetic energy when $x = 0.500$ m?

60. ■ ■ A spider sits on its web, undergoing simple harmonic motion with amplitude A. What fraction of each cycle does the spider spend at positions with $x > 0.9A$? *Hint*: Use the analog of uniform circular motion.

61. ■ ■ For a simple harmonic oscillator with $k = 4.0$ N/m and $A = 0.75$ m, the time between maximum velocity and maximum acceleration is 2.50 s. (a) What's the oscillation period? (b) What's the mass? (c) Find the maximum velocity and maximum acceleration.

62. ■ ■ A simple harmonic oscillator with $m = 0.750$ kg and total energy $E = 125$ J has amplitude 1.50 m. Find (a) the spring constant, (b) the period, and (c) the maximum speed and acceleration.

63. ■ ■ A 0.60-kg mass undergoes simple harmonic motion on a spring with $k = 14$ N/m. The oscillator's speed is 0.95 m/s when it's at $x = 0.22$ m. Find (a) the oscillation amplitude, (b) the total mechanical energy, and (c) the oscillator's speed when it's at $x = 0.11$ m.

64. ■ ■ ■ A puck with mass 0.16 kg slides along a frictionless horizontal surface at 5.5 m/s. It hits and sticks to the free end of a spring with $k = 15$ N/m (Figure P7.64). Find the amplitude and period of the subsequent simple harmonic motion.

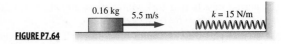

FIGURE P7.64

Section 7.4 SHM and Uniform Circular Motion

65. ■ ■ A steam locomotive uses the back-and-forth motion of a drive rod to turn its 1.42-m-diameter wheel. Each compete oscillation of the rod corresponds to one wheel revolution. What's the rod's oscillation frequency when the train is moving at 26 m/s?

66. ■ ■ Show that the angular frequency ω of simple harmonic motion is equal to the rate at which the corresponding circular motion undergoes angular displacement, measured in radians per second.

67. ■ ■ A wheel rotates at 600 rpm. Viewed from the edge, a point on the wheel appears to undergo SHM. Find (a) the frequency in Hz and (b) the angular frequency for this SHM.

Section 7.5 The Simple Pendulum

68. ■ What's the gravitational acceleration on a planet where the period of a 2.20-m-long pendulum is 2.87 s?

69. ■ ■ On a swing, a 35-kg girl undergoes small-amplitude oscillations with period 4.1 s. (a) How long are the ropes that support the swing? (b) How is the period affected if the girl's 48-kg brother replaces her?

70. ■ ■ For what angle (in radians) is there a 1% difference between the angle and its sine?

71. ■ ■ A pendulum makes 25 oscillations in 32 s. Find (a) the period and (b) the pendulum's length.

72. ■ What are the small-amplitude period and frequency of a 10.0-m-long pendulum?

73. ■ ■ A simple pendulum with length 1.50 m is released from rest at a 5° angle. Graph its angular position θ from $t = 0$ to $t = 10$ s.

74. ■ ■ Consider a pendulum of length L and amplitude θ_{max}. (a) Take the zero of gravitational potential energy at the pendulum's lowest point, $\theta = 0$. In terms of the given parameters, find (a) the pendulum's total mechanical energy and (b) its maximum speed. (c) Do your answers to parts (a) and (b) depend on the small-angle approximation? Explain.

75. ■ ■ ■ Thermal expansion increases the length of a clock's pendulum by a fractional increase of 5.0×10^{-5}. Under these conditions, how much error in timekeeping accumulates over 1 day?

76. ■ ■ ■ A simple pendulum oscillates with amplitude $\theta_{max} = 20°$. (a) Sketch the pendulum's trajectory and (b) indicate the direction of the net force on the pendulum bob at positions $\theta = 0°$ and $\theta = 20°$. (c) Compare the net force on the bob at $\theta = 0°$ when it's swinging and when it's hanging at rest.

77. ■ ■ At sea level, g varies from about 9.78 m/s² near the equator to 9.83 m/s² near the North Pole. Find the difference between the small-amplitude periods of a 2.00-m-long pendulum at these locations.

Section 7.6 Damped and Driven Oscillations

78. ■ Show that the units of the damping parameter b are kg/s.

79. ■ A simple harmonic oscillator has $m = 1.50$ kg, $k = 80.0$ N/m, and damping parameter $b = 2.65$ kg/s. Is the motion lightly damped, critically damped, or heavily damped?

80. ■ A simple harmonic oscillator has $m = 1.10$ kg, $k = 9.25$ N/m, and $b = 12.1$ kg/s. Is the motion lightly damped, critically damped, or heavily damped?

81. ■ ■ A simple harmonic oscillator has $m = 3.15$ kg, $k = 150.0$ N/m, and $b = 8.15$ kg/s. (a) Show that the motion is lightly damped. (b) Find the oscillation period, and compare it with the period of undamped oscillations for the same system. (c) After how many oscillations (approximately) will the amplitude drop to half its initial value?

82. ■ ■ What damping parameter b is needed for the car of Problem 52 if the shock absorbers are to provide critical damping with just the driver in the car?

83. ■ ■ The amplitude of a harmonic oscillator with $m = 0.50$ kg and $k = 12$ N/m drops to half its initial value after 12 oscillation periods. Find the damping parameter b.

84. ■ ■ Find the fraction of mechanical energy lost during each oscillation cycle of a damped oscillator with mass m, spring constant k, and damping parameter b.

85. ■ ■ A 70-kg bungee jumper leaps from a bridge and finds himself oscillating vertically with amplitude 8.50 m and period 3.75 s. (a) Find k for the bungee cord. (b) If the damping parameter is 4.50 kg/s, how much time does it take the oscillation amplitude to drop by half?

General Problems

86. ■ ■ A kitchen scale can measure a maximum mass of 250 g; it uses a vertical spring whose compression is calibrated to the mass in its weighing pan. (a) What spring constant is needed if the spring compresses 0.50 cm when the scale reads its maximum value? (b) What is the oscillation period with this maximum mass on the scale?

87. **BIO** ■ ■ **Oscillating protein.** The protein dynein powers the flagella that propel some unicellular organisms. Biophysicists have found that dynein is intrinsically oscillatory and that it exerts peak forces of about 1.0 pN when it attaches to structures called microtubules. The resulting oscillations have amplitude 15 nm. (a) If this system is to be modeled as a mass-spring system, what's the associated spring constant? (b) If the oscillation frequency is 70 Hz, what's the effective mass?

88. ■ ■ ■ A 0.25-kg block oscillates between two 16-N/m springs, as shown in Figure GP7.88. (a) What's the oscillation period? (b) Compare with the period of the same block oscillating on a single 16-N/m spring.

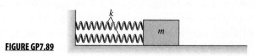

FIGURE GP7.88

89. ■ ■ ■ A 0.25-kg block oscillates under the influence of two 16-N/m springs, as shown in Figure GP7.89. (a) What's the oscillation period? (b) Compare with the period of the same block oscillating on a single 16-N/m spring.

FIGURE GP7.89

90. ■ ■ A 0.25-kg block oscillates on the end of two 16-N/m springs placed end to end, as shown in Figure GP7.90. (a) What's the oscillation period? (b) Compare with the period of the same block oscillating on a single 16-N/m spring.

FIGURE GP7.90

91. ■ ■ Two mass-spring systems with the same mass are undergoing oscillatory motion with the same amplitudes. System 1 has twice the frequency of system 2. How do (a) their frequencies and (b) their maximum accelerations compare?

92. ■ ■ ■ A 1.35-m-long pendulum hangs with its 0.535-kg bob at rest. A 0.110-kg ball of putty moving horizontally at 1.86 m/s strikes the bob and sticks. Find the resulting period and amplitude of the swinging pendulum.

93. ■ ■ A spring with $k = 34.0$ N/m is vertical with one end attached to the floor. You place a 0.50-kg mass on top of the spring and depress it to start it oscillating vertically. The mass simply rests on top of the spring and isn't firmly attached. (a) Find the maximum oscillation amplitude that allows the mass to stay on the spring throughout the cycle. (b) If you exceed this maximum amplitude just slightly, at what point in the cycle would the mass come off the spring?

94. ■ ■ ■ A snowboarder oscillates back and forth along a semicircular "half-pipe" of radius 8.0 m (Figure GP7.94), rising to a maximum height h as shown. (a) Find the snowboarder's maximum speed if

$h = 8.0$ m. (b) Does the oscillation period depend on h? Explain. (c) Compute the period of *small* oscillations ($h \ll 8.0$ m).

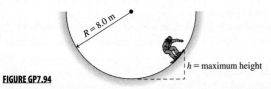

FIGURE GP7.94

95. ■ ■ An atom in a solid vibrates at 12 THz with amplitude 10 pm. (a) Find the atom's maximum speed and maximum acceleration. (b) If it's a carbon atom, what's its energy?

96. ■ ■ ■ A 6.50-g bullet traveling at 495 m/s embeds itself in a 1.76-kg wooden block at rest on a frictionless surface. The block is attached to a spring with $k = 85.0$ N/m (Figure GP7.96). Find (a) the period and (b) the amplitude of the subsequent simple harmonic motion. (c) Find the total energy of the bullet+block+spring system before and after the bullet enters the block.

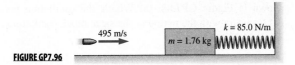

FIGURE GP7.96

97. ■ ■ ■ A spring with $k = 250$ N/m stands vertically with one end attached to the floor. A 2.15-kg brick is dropped from 25.0 cm above the top of the spring and sticks to the spring. (a) How far does the spring compress? (b) What are the amplitude and period of the resulting simple harmonic motion?

98. ■ ■ ■ A car with mass 1220 kg including the driver travels a rough road with bumps spaced 5.5 m apart. At a speed of 12.0 m/s, the car bounces up and down with exceptionally large amplitude. Two more 80-kg passengers now get in the car. (a) What speed will result in maximum oscillation amplitude? (b) How much did the car's suspension sag with the two added passengers?

99. ■ ■ ■ A tightrope walker of mass m stands at rest midway along a cable of length L and negligible mass. The cable is stretched tightly between two supports, giving it a tension F. If this equilibrium is disturbed, the tightrope walker undergoes small-amplitude vertical oscillations. Show that the period of these oscillations is $T = 2\pi\sqrt{\dfrac{mL}{4F}}$. *Hint*: For small oscillations, the cable tension doesn't change. You can set up Newton's second law and show that the force tending to restore equilibrium is directly proportional to the displacement.

100. ■ ■ ■ A 1.2-kg block is attached to a horizontal spring with $k = 23$ N/m, as shown in Figure GP7.100. The block is oscillating with amplitude 10 cm. (a) What is the period of oscillation? (b) A second block with mass 0.80 kg traveling at 1.7 m/s, as shown, hits the oscillating block at the rightmost point of its oscillation. The two blocks stick together. What is the resulting period and amplitude of SHM?

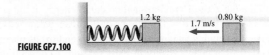

FIGURE GP7.100

Answers to Chapter Questions

Answer to Chapter-Opening Question
The flexible cord acts like an oscillating spring. The oscillation frequency depends on the stiffness of the cord (the "spring constant") and the jumper's mass. The oscillations are damped by frictional forces, particularly the drag force on the person moving through air.

Answers to GOT IT? Questions
Section 7.1   (d) 0.014 s
Section 7.2   (e) < (a) < (d) < (c) < (b)
Section 7.3   (b) < (c) < (d) < (a) < (e)
Section 7.5   (d) < (a) = (b) < (c).

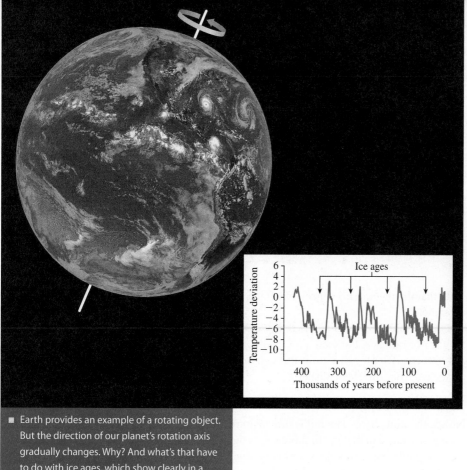

8 Rotational Motion

Earth provides an example of a rotating object. But the direction of our planet's rotation axis gradually changes. Why? And what's that have to do with ice ages, which show clearly in a 420,000-year record of Earth's temperature?

To Learn

By the end of this chapter you should be able to

- Explain angular velocity and angular acceleration.
- Recognize the kinematic equations for rotational motion with constant angular acceleration.
- Understand the analogy between quantities describing rotational motion and the corresponding quantities in one-dimensional motion.
- Distinguish the radial and tangential components of acceleration.
- Determine the kinetic energy of a rotating object.
- Explain rotational inertia and how it depends on an object's shape.
- Apply the work-energy theorem to rolling objects.
- Understand torque as the rotational analog of force.
- Describe the rotational analog of Newton's second law.
- Describe the conditions for an object to be in mechanical equilibrium.
- Understand angular momentum and conditions under which it's conserved.
- Describe angular motion quantities as vectors.
- Explain precession.

In this chapter we'll explore the kinematics and dynamics of rotating objects, first using analogies with quantities you understand from one-dimensional motion. You'll learn about torque, the rotational analog of force, and will develop a rotational analog of Newton's second law. We'll then explore static equilibrium, which requires zero net force *and* zero net torque. Finally, we'll look briefly at the more complicated situation in which rotational quantities are treated as vectors.

8.1 Rotational Kinematics

Rotation and Translation

All around you are things that rotate. A DVD spins as a laser reads the movie information it contains. That's a **pure rotation**, a solid object rotating about a fixed object. Your car's wheels rotate and simultaneously move forward; that's a combination of **rotational motion** and **translational motion**. A baseball hurtles toward home plate in translational motion, but the pitcher's given it spin, so it's also rotating. Earth, too, rotates while simultaneously orbiting the Sun.

In this section we'll consider pure rotation, developing rotational analogs of the position, velocity, and acceleration you're familiar with from one-dimensional translational motion.

Angular Position

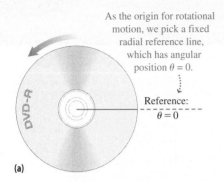

As the origin for rotational motion, we pick a fixed radial reference line, which has angular position $\theta = 0$.

Reference: $\theta = 0$

(a)

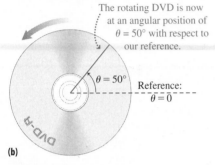

The rotating DVD is now at an angular position of $\theta = 50°$ with respect to our reference.

$\theta = 50°$ Reference: $\theta = 0$

(b)

FIGURE 8.1 Angular position for a rotating object is defined relative to a fixed reference that serves the same function as the origin for a Cartesian coordinate axis.

A spinning compact disk or DVD provides an example of rigid-body rotation. Imagine drawing a line marking the disk's radius and a fixed reference that initially coincides with your marked radius (Figure 8.1a). As the disk rotates (Figure 8.1b), your marked radius makes an increasing angle θ with the reference. This angle θ defines the **angular position**.

We'll take angles to be positive going counterclockwise (CCW) from our $\theta = 0$ reference and negative going clockwise (CW). This is somewhat arbitrary, because a rotation that's counterclockwise viewed from above appears clockwise from below. Choosing a positive direction for angles is like choosing the $+x$-direction in Cartesian coordinates.

In studying rotational motion, it's advantageous to measure angular positions in **radians**, not degrees. The radian measure of an angle is the ratio of arc length to radius:

$$\theta \text{ (in radians)} = \frac{\text{arc length}}{\text{radius}} = \frac{s}{r} \tag{8.1}$$

Defined as a ratio of two distances, angle is a *dimensionless* quantity. The "unit" radian is equivalent to having no units at all. We'll use the radian label (abbreviated *rad*) to remind ourselves that this particular dimensionless number measures an angle. For example, suppose you travel 13.0 m along an arc of radius 2.0 m. Then you've moved through an angle

$$\theta = \frac{s}{r} = \frac{13.0 \text{ m}}{2.0 \text{ m}} = 6.5 \text{ rad}$$

Note that the meters (m) cancel, leaving a dimensionless result labeled "rad" because it's an angle.

One advantage of radians is that you can turn Equation 8.1 around to solve for arc length s or radius r. For example, a point on the outer edge of a CD or DVD is at radius $r = 6.0$ cm. After the disk turns through one complete revolution ($\theta = 2\pi$ rad), the point on the edge has traveled a distance

$$s = r\theta = (0.06 \text{ m})(2\pi \text{ rad}) = 0.38 \text{ m}$$

Notice the units here: The rad is dimensionless, so we dropped it from the final answer, giving a distance in meters. In this case—one full revolution—the arc length is just the circle's circumference. Angular position is sometimes given in revolutions (rev). The conversion factor is 1 rev $= 360° = 2\pi$ rad.

✓**TIP**

Equation 8.1 applies only when θ is in radians.

EXAMPLE 8.1 **Our Rotating Planet**

Earth rotates once every 24 hours. In a reference frame fixed to Earth, how far does a point on the equator move in 1 hour?

ORGANIZE AND PLAN The distance moved along a circular arc (Figure 8.2) is $s = r\theta$. Appendix E gives Earth's radius as $R_E = 6.37 \times 10^6$ m. The angle θ is 1/24 of a revolution, which we'll need to convert to radians.

Known: $R_E = 6.37 \times 10^6$ m.

SOLVE Earth has rotated through $\theta = 1/24$ rev, or:

$$\frac{1}{24} \text{ rev} \times \frac{2\pi \text{ rad}}{1 \text{ rev}} = 0.262 \text{ rad}$$

Then Equation 8.1 gives:

$$s = r\theta = (6.37 \times 10^6 \text{ m})(0.262 \text{ rad}) = 1.67 \times 10^6 \text{ m}$$

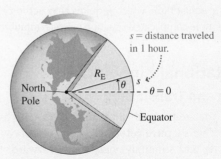

$s =$ distance traveled in 1 hour.

North Pole

R_E

θ s $\theta = 0$

Equator

View from above the North Pole

FIGURE 8.2 Finding the distance traveled along an arc on the Equator.

cont'd.

REFLECT Is our answer reasonable? It's 1670 km, about the distance from Denver to Chicago. That makes sense, because there's a 1-hour time zone difference between Denver and Chicago. But that's only approximate; see "Making the Connection."

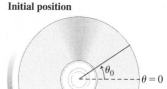

MAKING THE CONNECTION Do points in the United States travel that same distance in each hour?

ANSWER No, because as you move away from the equator, the circumference of the circle on which you move gets smaller. Of all the places in the United States, in Alaska you'd travel along the smallest circle.

Angular Displacement and Velocity

Figure 8.3 shows our rotating DVD at some initial angular position θ_0, and later at a new position θ. We define **angular displacement** as the difference between two angular positions:

$$\Delta\theta = \theta - \theta_0 \quad \text{(Angular displacement; SI unit: rad)} \tag{8.2}$$

This definition is analogous to the definition of displacement in one-dimensional motion (Equation 2.1). In Chapter 2 we defined average velocity as displacement divided by the corresponding time interval Δt, and we define average angular velocity analogously:

$$\overline{\omega} = \frac{\Delta\theta}{\Delta t} \quad \text{(Average angular velocity; SI unit: rad/s)} \tag{8.3}$$

We use the same lowercase Greek omega (ω) as we did for angular frequency in the previous chapter—a reflection of Section 7.4's connection between simple harmonic motion and circular motion. With displacement in radians and time in seconds, the unit for angular velocity is radians per second (rad/s). You'll also see angular velocity measured in revolutions per second (rev/s), revolutions per minute (rev/min or rpm), or degrees per second (deg/s).

Average angular velocity uses only information from the endpoints of the interval, so it can't give you details of the motion. For those details, you need to consider ever shorter time intervals. Because it is analogous to the instantaneous velocity for linear motion (Equation 2.4), we define **instantaneous angular velocity** as the average angular velocity in the limit as the time interval approaches zero. That is,

$$\omega = \lim_{\Delta t \to 0} \frac{\Delta\theta}{\Delta t} \quad \text{(Instantaneous angular velocity; SI unit: rad/s)} \tag{8.4}$$

Instantaneous angular velocity (or just angular velocity) is positive when angular position θ is increasing, negative if θ is decreasing, and zero if the object isn't rotating.

Constant Angular Velocity

Just as in one-dimensional motion with constant velocity, the special case of *constant angular velocity* has the same average and instantaneous values, both given by $\omega = \Delta\theta/\Delta t$. Then each rotation takes the same time, which is defined as the **period** of the rotational motion. In one period T, any point in the object rotates through angle 2π, so the angular velocity is

$$\omega = \frac{2\pi}{T} \quad \text{(Constant angular velocity)} \tag{8.5}$$

Examples of constant angular velocity include a car's wheel when the car moves at constant speed, ancient phonograph records with typical speed $33\frac{1}{3}$ rpm, and motors of all sizes—including a remarkable biological motor described below. CDs and some DVDs don't rotate with constant angular velocity. We'll explain why later.

Rotational motion is common in mechanical systems and astronomical bodies, but less obvious in biology, where the most familiar examples are the limited rotations of limbs

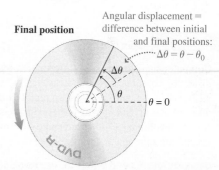

FIGURE 8.3 Defining angular displacement.

about shoulder, knee, hip, and elbow joints. But, remarkably, true rotation also occurs at the cellular level. The mechanism that drives the flagellum—the propulsion organ—in the bacterium *E. coli* spins at some 100 rev/s, or more than 600 rad/s, propelling the bacterium at speeds of around 25 μm/s.

Angular Acceleration

In Chapter 2, we defined acceleration as the rate of change of velocity. **Angular acceleration** is defined analogously. In analogy with Equation 2.6, average angular acceleration is

$$\bar{\alpha} = \frac{\Delta\omega}{\Delta t} \quad \text{(Average angular acceleration; SI unit: rad/s}^2\text{)} \tag{8.6}$$

That's the lowercase Greek alpha (α) for angular acceleration. The usual limiting procedure then defines instantaneous acceleration:

$$\alpha = \lim_{\Delta t \to 0} \frac{\Delta\omega}{\Delta t} \quad \text{(Instantaneous angular acceleration; SI unit: rad/s}^2\text{)} \tag{8.7}$$

Angular acceleration has units of radians per second per second, or rad/s^2. It may be positive, zero, or negative.

EXAMPLE 8.2 **Spin Up!**

Computer hard drives store information magnetically on spinning disks. To conserve battery energy, laptop drives spin down to rest when not in use, and then spin up rapidly when needed. A particular disk operates at 7200 rpm and takes 65 ms to spin up. Find (a) the time for one revolution at full speed and (b) the average angular acceleration during spin-up.

ORGANIZE AND PLAN The time for one revolution is the period T, which Equation 8.5 relates to the angular velocity ω. The average angular acceleration is the change in angular velocity divided by the spin-up time. To get results in standard units, it's necessary to convert the angular velocity from rpm to rad/s.

Known: $\omega = 7200$ rpm; $\Delta t = 65$ ms $= 65 \times 10^{-3}$ s.

SOLVE (a) Converting angular velocity from rpm to rad/s:

$$7200 \text{ rev/min} \times \frac{1 \text{ min}}{60 \text{ s}} \times \frac{2\pi \text{ rad}}{\text{rev}} = 754 \text{ rad/s}$$

Then Equation 8.5 gives

$$T = \frac{2\pi}{\omega} = \frac{2\pi \text{ rad}}{754 \text{ rad/s}} = 8.33 \times 10^{-3} \text{ s}$$

(b) The average angular acceleration is (Equation 8.6)

$$\bar{\alpha} = \frac{\Delta\omega}{\Delta t} = \frac{754 \text{ rad/s}}{65 \times 10^{-3} \text{ s}}$$
$$= 1.16 \times 10^4 \text{ rad/s}^2$$

REFLECT Our calculation shows that it takes a little less then one-hundredth of a second for the disk to make one revolution. The acceleration is huge, but that's necessary to avoid delays in accessing data.

MAKING THE CONNECTION If the disk takes a more leisurely 1.3 s to spin down, what's the average angular acceleration?

ANSWER Now the initial angular velocity is 754 rad/s and the final is zero. That means $\Delta\omega = 0 - 754$ rad/s $= -754$ rad/s, so the average angular acceleration is $\bar{\alpha} = -754$ rad/s/1.3 s $= -580$ rad/s^2. The angular acceleration is negative because the angular velocity decreased, from positive to zero.

CONCEPTUAL EXAMPLE 8.3 **Rigid and Fluid Bodies**

At the beginning of this section, we introduced the idea of a rigid rotating body. What are some of the characteristics of a *rigid* rotator? How might these differ from a rotating *fluid*?

SOLVE In a rigid rotator, such as our spinning CD or a bicycle wheel, every point goes through the same angular displacement in a given time. Thus, every point in the rigid body has the same angular velocity and the same angular acceleration.

However, in a fluid (such as the hurricane in Figure 8.4) there's no solid connection, so different points can have different angular velocities and accelerations. The gaseous Sun is a huge example: In contrast

cont'd.

to Earth, it rotates with greater angular vel___
the poles.

REFLECT The common angular velocity and ac___
tors don't imply that every point has the same ___
ample, the outer end of a bicycle spoke travels ___ ___ inner
end. Similarly, the outer tracks of a DVD move ___ ___ than the inner
ones at a given angular velocity; to "read" information at a constant
rate, the DVD's rotation slows as information is read from tracks pro-
gressively farther out. We'll address the relationship between angular
and linear velocity later in this chapter.

FIGURE 8.4 A hurricane is an example of a nonrigid rotator.

Reviewing New Concepts

- Average angular velocity is angular displacement divided by the corresponding time interval.
- Instantaneous angular velocity is the average angular velocity in the limit as the interval approaches zero.
- Average angular acceleration is the change in angular velocity divided by the corresponding time interval.
- Instantaneous angular acceleration is the average angular acceleration in the limit as the interval approaches zero.

GOT IT? Section 8.1   A wheel rotates clockwise while slowing down. What are the signs of its angular velocity ω and angular acceleration α? (a) Both are positive; (b) ω is positive and α is negative; (c) ω is negative and α is positive; (d) both are negative.

8.2 Kinematic Equations for Rotational Motion

In Section 2.4 you saw the kinematic equations (Equations 2.8, 2.9, and 2.10), which relate position, velocity, and acceleration for the case of constant acceleration. Our definitions of angular velocity and acceleration are exactly analogous to those of translational velocity and acceleration from Chapter 2. Therefore, we can get the kinematic equations for constant angular acceleration just by replacing each translational variable with its rotational counterpart, as outlined in Table 8.1.

Table 8.2 uses these analogies to express the rotational versions of the kinematic equations. The strategies for solving kinematics problems with constant angular acceleration are the same ones you already know for translational motion. Problem-Solving Strategy 8.1 gives some hints and reminders.

TABLE 8.1 Rotational Analogs of Translational Quantities

Translational quantity	Rotational quantity
Position x	Angular position θ
Displacement Δx	Displacement $\Delta\theta$
Velocity v_x	Angular velocity ω
Acceleration a_x	Angular acceleration α
Time t	Time t

TABLE 8.2 Kinematic Equations for Constant Acceleration

Translational equation		Rotational equation	
$v_x = v_{x0} + a_x t$	(2.8)	$\omega = \omega_0 + \alpha t$	(8.8)
$x = x_0 + v_{x0}t + \frac{1}{2}a_x t^2$	(2.9)	$\theta = \theta_0 + \omega_0 t + \frac{1}{2}\alpha t^2$	(8.9)
$v_x^2 = v_{x0}^2 + 2a_x \Delta x$	(2.10)	$\omega^2 = \omega_0^2 + 2\alpha\Delta\theta$	(8.10)

PROBLEM-SOLVING STRATEGY 8.1 — Kinematics Problems with Constant Angular Acceleration

ORGANIZE AND PLAN
- Draw a sketch showing the rotating object.
- If applicable, choose a reference and identify the positive rotation direction.
- Review what quantities you know and what you're trying to find. This will help you identify which kinematic equation(s) to use.

SOLVE
- Gather the given information and select the kinematic equation(s) you need.
- Remember that these equations are valid only for constant angular acceleration α.
- Solve for the unknown(s).
- Insert numerical values and compute the answer(s), giving appropriate units. Note where the radian (rad) appears or disappears as needed.

REFLECT
- Are the dimensions and units of the answer(s) correct?
- If the problem relates to something familiar, think about whether the answer makes sense.

EXAMPLE 8.4 — Tall Grass

A woman mows her lawn with an electric mower, its blade turning at 1500 rpm. She encounters tall grass and increases the speed to 2000 rpm over 3.40 s with constant angular acceleration. (a) What was the angular acceleration? (b) Through how many revolutions did the blade turn while accelerating?

ORGANIZE AND PLAN We'll take the blade's rotation to define the positive direction. With its angular velocity increasing, the blade's angular acceleration will be positive. The known quantities are initial and final angular velocity (ω_0 and ω) and time t. Given these quantities, Equation 8.8 will give angular acceleration. Then either of the other kinematic equations can be used to find the blade's angular displacement. You should also change angular velocity from rpm to rad/s.

Known: $\omega_0 = 1500$ rpm (rev/min), $\omega = 2000$ rpm (rev/min), and $t = 3.40$ s.

SOLVE (a) First, converting from rpm to rad/s:

$$\omega_0 = 1500 \text{ rev/min} \times \frac{1 \text{ min}}{60 \text{ s}} \times \frac{2\pi \text{ rad}}{\text{rev}} = 157 \text{ rad/s}$$

and

$$\omega = 2000 \text{ rev/min} \times \frac{1 \text{ min}}{60 \text{ s}} \times \frac{2\pi \text{ rad}}{\text{rev}} = 209 \text{ rad/s}$$

Then solving Equation 8.8 for angular acceleration and inserting numerical values,

$$\alpha = \frac{\omega - \omega_0}{t} = \frac{209 \text{ rad/s} - 157 \text{ rad/s}}{3.40 \text{ s}} = 15.3 \text{ rad/s}^2$$

(b) At this point, either of the other equations yields the angular displacement $\Delta\theta$. Choosing Equation 8.9,

$$\Delta\theta = \theta - \theta_0$$
$$= \omega_0 t + \tfrac{1}{2}\alpha t^2$$
$$= (157 \text{ rad/s})(3.40 \text{ s}) + \tfrac{1}{2}(15.3 \text{ rad/s}^2)(3.40 \text{ s})^2$$
$$= 622 \text{ rad}$$

The problem asked how many revolutions the blade turned, so we convert 622 rad to revolutions:

$$622 \text{ rad} \times \frac{1 \text{ rev}}{2\pi \text{ rad}} = 99 \text{ rev}$$

REFLECT Is 99 revolutions a reasonable answer? With constant acceleration, the average angular velocity is 1750 rev/min (average of 1500 rev/min and 2000 rev/min). With that average number of revolutions per minute, 99 rev in 3.4 s is just right.

MAKING THE CONNECTION How much time would it take the blade to stop, with the same initial 1500 rpm and an acceleration of -15.3 rad/s^2?

ANSWER With the ω_0 and ω known (the latter is now 0), along with angular acceleration α, Equation 8.8 can be solved for time to get $t = 10.3$ s.

EXAMPLE 8.5 **As the CD Turns**

A CD starts out spinning at 500 rpm and drops to 200 rpm at the end of its playing time; that way, information is "read" from the disk at a constant rate. If the disk undergoes constant angular acceleration -7.08×10^{-3} rad/s², find (a) its angular displacement and (b) the time involved.

ORGANIZE AND PLAN Here you're not given the time, which suggests that Equation 8.10, which doesn't contain time, might give the angular displacement. And it will: You're given initial and final angular velocities and angular acceleration, so you can solve for $\Delta\theta$.

For part (b), you could solve either of the other equations for time. It's easier to use Equation 8.8, because it contains t rather than t^2.

Known: $\omega_0 = 500$ rev/min, $\omega = 200$ rev/min, and $\alpha = -7.08 \times 10^{-3}$ rad/s².

SOLVE (a) Converting the angular velocities to rad/s:

$$\omega_0 = 500 \text{ rev/min} \times \frac{1 \text{ min}}{60 \text{ s}} \times \frac{2\pi \text{ rad}}{\text{rev}} = 52.4 \text{ rad/s}$$

and

$$\omega = 200 \text{ rev/min} \times \frac{1 \text{ min}}{60 \text{ s}} \times \frac{2\pi \text{ rad}}{\text{rev}} = 20.9 \text{ rad/s}$$

Then solving Equation 8.10 for the angular displacement $\Delta\theta$:

$$\Delta\theta = \frac{\omega^2 - \omega_0^2}{2\alpha} = \frac{(20.9 \text{ rad/s})^2 - (52.4 \text{ rad/s})^2}{2(-7.08 \times 10^{-3} \text{ rad/s}^2)}$$

$$= 1.63 \times 10^5 \text{ rad}$$

or about 26,000 revolutions.

(b) Solving Equation 8.8 for time t:

$$t = \frac{\omega - \omega_0}{\alpha} = \frac{20.9 \text{ rad/s} - 52.4 \text{ rad/s}}{-7.08 \times 10^{-3} \text{ rad/s}^2} = 4.45 \times 10^3 \text{ s}$$

REFLECT The answer for part (b) is right on: It's 74 minutes, about the maximum capacity of an audio CD. The answer for part (a) might seem huge—but given that the tracks containing information on a CD are only about 10^{-6} m apart, a lot of turning is required to get the whole disk played.

MAKING THE CONNECTION In part (b), does Equation 8.9 give the same answer for the time t?

ANSWER Yes, but check this yourself! You'll need $\Delta\theta$ from part (a), and you'll have to solve a quadratic equation.

8.3 Rotational and Tangential Motion

Tangential Velocity and Speed

That rotating DVD is a rigid body, so every point on the DVD has the same angular velocity. However, Figure 8.5 shows that *translational* velocities of different points may differ in both direction and magnitude. Each point on the DVD moves in a circle, and as you know from uniform circular motion, the velocity vector is always tangent to the circle. For that reason the translational velocity of a point on a rotating body is called **tangential velocity**, $\vec{v}_t$.

What's the magnitude of this tangential velocity, the **tangential speed** v_t? Since speed is the magnitude of velocity, different points generally have different speeds. Points closer to the rotation move more slowly than points farther out. That's because all take the same time to complete one revolution, regardless of the distance traveled. Suppose the DVD is rotating with constant angular velocity ω. The tangential speed can be computed by taking any arc length and dividing by the time required to traverse it. Taking one complete revolution, the distance is the circumference $2\pi r$, and the time is the period $T = 2\pi/\omega$ (Equation 8.5). Combining these results,

$$\text{Tangential speed } v_t = \frac{\text{distance}}{\text{time}} = \frac{2\pi r}{T} = \frac{2\pi r}{2\pi/\omega} = r\omega$$

Thus,

$$v_t = r\omega \quad \text{(Tangential speed; SI unit: m/s)} \quad (8.11)$$

For example, a point at the 6.0-cm outer radius of the compact disk in Example 8.5 has tangential speed

$$v_t = r\omega = (0.060 \text{ m})(52.4 \text{ rad/s}) = 3.14 \text{ m/s}$$

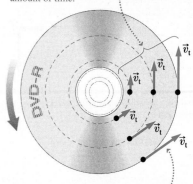

Points farther from the center move faster because they have farther to go in a given amount of time.

The velocity vector of a point on a rotating object is always tangent to the circle and hence is called the *tangential velocity* $\vec{v}_t$.

FIGURE 8.5 Tangential velocity at different points on the rotating DVD.

when the disk is rotating at its maximum 52.4 rad/s (500 rpm). On the other hand, a point just 2.5 cm from the axis—on the innermost information-bearing track—has

$$v_t = r\omega = (0.025 \text{ m})(52.4 \text{ rad/s}) = 1.31 \text{ m/s}$$

Tangential speed is directly proportional to the radius for any rigid rotating body. Although we assumed constant angular speed here, Equation 8.11 holds even if ω isn't constant.

You can now understand why the angular speed of a CD (or DVD) decreases as the laser that "reads" information moves outward. Ideally, the laser pickup should supply information at a constant rate—and that means a constant *tangential* speed. With $v_t = r\omega$ a constant, the angular velocity ω must decrease as r increases. You can easily observe this if your CD player has a clear window; you'll see the CD spinning faster when on the first track (the inner part of the CD) and slower on the last (the outer part of the CD). You can confirm this with the numbers above; you'll find that the 2.5-cm point moves at about 1.3 m/s when the disk spins at 500 rpm and that the 6.0-cm point has the same tangential speed at 200 rpm.

Tangential Acceleration

If the tangential speed changes, then its rate of change is the **tangential acceleration** a_t:

$$a_t = \lim_{\Delta t \to 0} \frac{\Delta v_t}{\Delta t}$$

This expression simplifies, using $v_t = r\omega$, to give

$$a_t = \lim_{\Delta t \to 0} \frac{\Delta(r\omega)}{\Delta t} = r \lim_{\Delta t \to 0} \frac{\Delta \omega}{\Delta t} = r\alpha$$

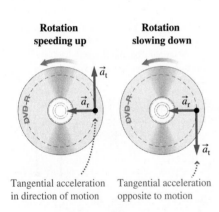

Rotation speeding up   **Rotation slowing down**

Tangential acceleration in direction of motion

Tangential acceleration opposite to motion

for the tangential acceleration of a point a fixed distance r from the rotation axis. In the last step we used the definition of angular acceleration α (Equation 8.7). To summarize,

$$a_t = r\alpha \qquad \text{(Tangential acceleration; SI unit: m/s}^2) \qquad (8.12)$$

On a rigid rotating body, every point has the same angular acceleration α. But the tangential acceleration is proportional to r. That makes sense, because tangential speed is proportional to r (Equation 8.11), and tangential acceleration measures the rate of change of tangential speed. Equation 8.12 allows you to go back and forth between the tangential acceleration a_t and angular acceleration α.

Tangential and Centripetal Acceleration

A point in a rotating object is in circular motion, so, as you learned in Chapter 3, that point has centripetal acceleration $a_r = v_t^2/r$, where r is its distance from the rotation axis. We use the tangential speed v_t here, because that's the translational speed of the rotating point. Using $v_t = r\omega$ (Equation 8.11), the centripetal acceleration becomes

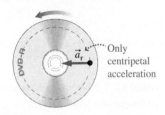

Rotating at constant speed

Only centripetal acceleration

FIGURE 8.6 A rotating object can have both centripetal acceleration and tangential acceleration.

$$a_r = \frac{v_t^2}{r} = \frac{(r\omega)^2}{r} = r\omega^2 \qquad (8.13)$$

In general, a point on a rotating object has an acceleration *vector* with a centripetal component a_r and a tangential component a_t (Figure 8.6). If the rotation rate is constant, the angular acceleration α is zero, and so is the tangential acceleration a_t. However, the centripetal acceleration isn't zero as long as there's rotation.

✓**TIP**

Equations 8.11 through 8.13 are valid only for radian measure.

EXAMPLE 8.6 **Merry-Go-Round**

One child pushes another on a merry-go-round. The rider is on the outer edge, 2.50 m from the rotation axis. What are her centripetal and tangential acceleration at the instant the angular speed and acceleration are 1.35 rad/s and 0.75 rad/s², respectively?

ORGANIZE AND PLAN You're given angular velocity, angular acceleration, and radius (Figure 8.7). The centripetal and tangential acceleration components depend on these quantities: $a_r = r\omega^2$ and $a_t = r\alpha$.

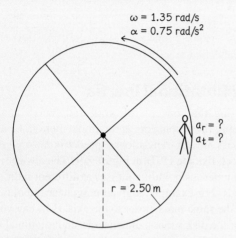

$\omega = 1.35$ rad/s
$\alpha = 0.75$ rad/s²

$a_r = ?$
$a_t = ?$

$r = 2.50$ m

FIGURE 8.7 Finding centripetal and tangential acceleration.

Known: $\omega = 1.35$ rad/s; $\alpha = 0.75$ rad/s²; $r = 2.50$ m.

SOLVE Using the given values, the centripetal acceleration is

$$a_r = r\omega^2 = (2.50 \text{ m})(1.35 \text{ rad/s})^2 = 4.56 \text{ m/s}^2$$

Notice that the radian unit (rad) isn't included in the final answer. It's a dimensionless quantity that doesn't fit in the centripetal acceleration, whose units are m/s².

The tangential acceleration is

$$a_t = r\alpha = (2.50 \text{ m})(0.75 \text{ rad/s}^2) = 1.88 \text{ m/s}^2$$

Again we've dropped the rad.

REFLECT Both answers seem reasonable. The centripetal acceleration is nearly half of g, making it somewhat hard to hold on (for some children, part of the fun!).

MAKING THE CONNECTION If the angular velocity were doubled, by what factor would the centripetal acceleration increase?

ANSWER The relationship between tangential acceleration and angular velocity is $a_r = r\omega^2$. Because the angular velocity is squared, doubling it increases the centripetal acceleration by a factor of *four*. For the numbers given here, that makes $a_r = 18 \text{ m/s}^2$, or nearly twice g. Now could you hold on?

What's the Same?

A potter's wheel rotates with angular velocity ω and angular acceleration α. Figure 8.8 shows blobs of clay at two points, P_1 and P_2, at different distances from the rotation axis. Consider the following quantities: angular displacement, angular velocity, angular acceleration, tangential speed, tangential acceleration, and centripetal acceleration. Which of those quantities are the same and which are different for the two points?

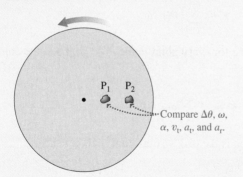

P_1 P_2

Compare $\Delta\theta$, ω, α, v_t, a_t, and a_r.

FIGURE 8.8 Comparing two points on the wheel.

SOLVE In a given time interval, every point in a solid object rotates through the same angular displacement. Therefore, angular velocity ω and angular acceleration α are the same everywhere in the wheel. Tangential speed is $v_t = r\omega$. With angular velocity ω the same everywhere, tangential speed increases in proportion to radial distance r.

Similarly, tangential acceleration is $a_t = r\alpha$, so with angular acceleration α the same everywhere, the tangential acceleration also increases in proportion to r. Finally, centripetal acceleration is $a_r = r\omega^2$, so with angular velocity ω the same everywhere, centripetal acceleration also increases in proportion to r.

REFLECT Note that quantities associated with translational motion—velocity and both acceleration components—depend on radial position. But strictly angular quantities—angular velocity and acceleration—don't. Table 8.3 summarizes rotational quantities and relationships among them.

TABLE 8.3 Some Important Rotational Quantities and Relationships

Quantity	Units	Relationship
Angular displacement $\Delta\theta$	rad	$\Delta\theta = \theta - \theta_0$
Angular velocity ω	rad/s	$\omega = \lim\limits_{\Delta t \to 0} \dfrac{\Delta\theta}{\Delta t}$
Angular acceleration α	rad/s²	$\alpha = \lim\limits_{\Delta t \to 0} \dfrac{\Delta\omega}{\Delta t}$
Tangential speed v_t	m/s	$v_t = r\omega$
Tangential acceleration a_t	m/s²	$a_t = r\alpha$
Centripetal acceleration a_r	m/s²	$a_r = r\omega^2$

GOT IT? Section 8.3 A bicycle wheel rotates clockwise, as shown, and is speeding up. Which diagram correctly represents the tangential and centripetal acceleration at the point indicated on the rim?

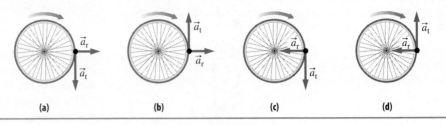

(a)   (b)   (c)   (d)

8.4 Kinetic Energy and Rotational Inertia

In Chapter 5 you learned how using the concepts of work and energy can simplify mechanics problems. Now we'll apply these concepts to rotational motion. Consider first kinetic energy, which for a point mass m with translational speed v is $K = \frac{1}{2}mv^2$. What's the kinetic energy of a rotating object, like the DVD in Figure 8.9? The answer isn't obvious, because the DVD consists of numerous pointlike masses at different radii and therefore with different tangential speeds. For example, a given mass near the outside of the DVD has more kinetic energy than the same mass closer to the axis. How can you find the *total* kinetic energy, as a function of the disk's mass, dimensions, and rotational speed?

Consider a single bit of mass m at radius r from the rotation axis—either of the points shown in Figure 8.9. It moves in a circle of radius r with tangential speed v_t. Therefore, its kinetic energy is

$$K = \tfrac{1}{2}mv_t^2 \qquad \text{(Kinetic energy of point mass } m\text{)}$$

The kinetic energy is the sum of the energies of all the bits of mass that comprise the disk:

$$K = \sum_{i=1}^{n} \tfrac{1}{2}m_i v_{t,i}^2 \qquad \text{(Kinetic energy of a collection of point masses in rotation)}$$

where we consider the disk to comprise n pieces, with individual masses m_i and speeds $v_{t,i}$. From Equation 8.11, you know that $v_{t,i} = r_i\omega$, where r_i is the distance of the ith bit from the axis. Then the kinetic energy becomes

$$K = \sum_{i=1}^{n} \tfrac{1}{2}m_i r_i^2 \omega^2$$

That angular velocity ω is the same for every point in the disk, so it and the common factor $\frac{1}{2}$ can be factored out, leaving

$$K = \tfrac{1}{2}\left(\sum_{i=1}^{n} m_i r_i^2\right)\omega^2$$

The remaining sum in this equation is the disk's **rotational inertia** I. That is,

$$I = \sum_{i=1}^{n} m_i r_i^2 \qquad \text{(Rotational inertia; SI unit: kg} \cdot \text{m}^2\text{)} \qquad (8.14)$$

Rotational inertia has SI units kg $\cdot$ m^2. In terms of rotational inertia, the kinetic energy of a solid rotating object is

$$K = \tfrac{1}{2}I\omega^2 \qquad \text{(Kinetic energy [rotation]; SI unit: J)} \qquad (8.15)$$

So if you know the rotational inertia I and angular velocity ω of a rotating object, you can easily compute its kinetic energy $K = \frac{1}{2}I\omega^2$. The tricky part is the sum in Equation 8.14

Assuming that points m_1 and m_2 have the same mass, m_2 has greater kinetic energy because its greater radius means that it has greater speed v_t:

$$K = \tfrac{1}{2}mv_t^2.$$

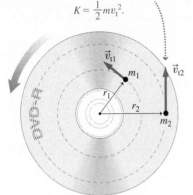

FIGURE 8.9 Velocities of different points on the DVD, used to find kinetic energy.

TABLE 8.4 Rotational Inertia of Some Common Geometrical Shapes

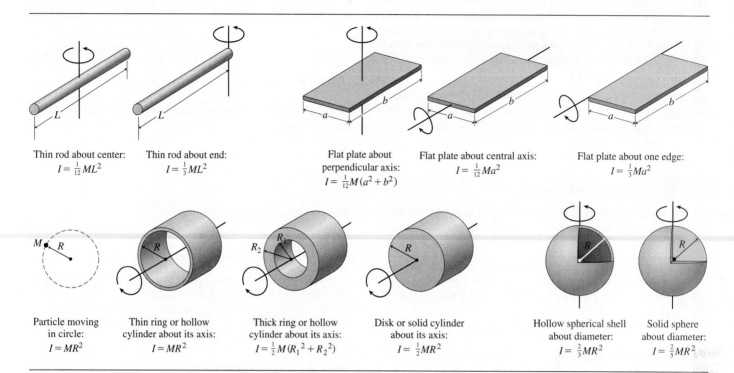

Thin rod about center:
$I = \frac{1}{12}ML^2$

Thin rod about end:
$I = \frac{1}{3}ML^2$

Flat plate about perpendicular axis:
$I = \frac{1}{12}M(a^2 + b^2)$

Flat plate about central axis:
$I = \frac{1}{12}Ma^2$

Flat plate about one edge:
$I = \frac{1}{3}Ma^2$

Particle moving in circle:
$I = MR^2$

Thin ring or hollow cylinder about its axis:
$I = MR^2$

Thick ring or hollow cylinder about its axis:
$I = \frac{1}{2}M(R_1^2 + R_2^2)$

Disk or solid cylinder about its axis:
$I = \frac{1}{2}MR^2$

Hollow spherical shell about diameter:
$I = \frac{2}{3}MR^2$

Solid sphere about diameter:
$I = \frac{2}{5}MR^2$

for rotational inertia. For common symmetric objects, calculus yields I as a function of total mass M and size. For example, a solid cylinder (or disk) with mass M and radius R rotated around its symmetry axis has rotational inertia $I = \frac{1}{2}MR^2$. A solid sphere with mass M and radius R rotated about an axis through its center has $I = \frac{2}{5}MR^2$. These simple formulas result because rotational inertia depends only on the amount of mass present and how that mass is distributed geometrically. Table 8.4 lists the rotational inertia of some common geometric shapes.

Understanding Rotational Inertia

Equation 8.15 for rotational kinetic energy, $K = \frac{1}{2}I\omega^2$, looks a lot like the familiar $K = \frac{1}{2}mv^2$ for translational kinetic energy. In the rotational expression, angular speed ω replaces translational speed v, and rotational inertia I replaces mass m. That's why physicists use the word *inertia*, which, as you saw for mass in Chapter 4, connotes resistance to changes in motion. Just as greater mass makes it more difficult to accelerate objects, greater rotational inertia makes it more difficult to give them angular acceleration. Section 8.6 delves further into the connection between rotational inertia and angular acceleration.

Equation 8.14 and Table 8.4 show that rotational inertia depends not only on an object's mass, but also on how that mass is distributed about the rotation axis. Conceptual Example 8.8 emphasizes this fact. All the bodies shown in Table 8.4 are assumed to have uniform density, and the resulting values of rotational inertia depend on this fact. For example, Earth's core of liquid and solid metal is much denser than the mantle and crust. As a result, our planet's rotational inertia is less than Table 8.4's $I = \frac{2}{5}MR^2$ for a uniform solid sphere.

CONCEPTUAL EXAMPLE 8.8 **Different Rotation Axes**

Consider Table 8.4's two rotational inertias for a thin rod of mass M and length L: $\frac{1}{12}ML^2$ about the center and $\frac{1}{3}ML^2$ about one end. Explain why the second value is greater.

SOLVE Try whirling a meterstick first about its center and then about one end; you'll find the latter is harder. Why? Rotating about the central axis, both halves of the stick contribute the same amount to the

cont'd.

rotational inertia; call this amount I_0. Rotating about the end, the half of the stick closest to the rotation axis contributes the same I_0 to the rotational inertia. But the outer half contributes *more*, because it's farther from the rotation axis and is therefore moving faster (Figure 8.10). This proves that the stick's rotational inertia is higher when it's spun about an end.

REFLECT Why isn't the rotational inertia about the rod's end just twice that about the center? The mass extends twice as far out, but Equation 8.14 shows that its effect scales as the *square* of the distance. That

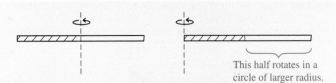

FIGURE 8.10 A rotating thin rod: different axes.

results in a rotational inertia about the end that's four times as great as about the center.

EXAMPLE 8.9   **DVD Rotational Inertia**

An 18-g DVD has inner radius 0.62 cm and outer radius 6.0 cm. Find (a) the disk's rotational inertia about its central axis and (b) its kinetic energy when spinning at 40 rad/s.

ORGANIZE AND PLAN What's the right shape in Table 8.4? It's the "thick ring or hollow cylinder" with inner radius R_1 and outer radius R_2, and rotational inertia $I = \frac{1}{2}M(R_1^2 + R_2^2)$. The kinetic energy is $K = \frac{1}{2}I\omega^2$.

Known: $R_1 = 0.62$ cm; $R_2 = 6.0$ cm; $M = 0.018$ kg; $\omega = 40$ rad/s.

SOLVE Using the numerical values given, the DVD's rotational inertia is

$$I = \tfrac{1}{2}M(R_1^2 + R_2^2) = \tfrac{1}{2}(0.018 \text{ kg})\big((0.0062 \text{ m})^2 + (0.060 \text{ m})^2\big)$$
$$= 3.27 \times 10^{-5} \text{ kg} \cdot \text{m}^2$$

Then the kinetic energy is

$$K = \tfrac{1}{2}I\omega^2 = \tfrac{1}{2}(3.27 \times 10^{-5} \text{ kg} \cdot \text{m}^2)(40 \text{ rad/s})^2$$
$$= 0.026 \text{ kg} \cdot \text{m}^2/\text{s}^2 = 26 \text{ mJ}$$

REFLECT Notice how the units worked in the kinetic energy computation. The radian is dimensionless, so we dropped it. Then $\text{kg} \cdot \text{m}^2/\text{s}^2$ reduces to joule (J), by definition of the joule. The DVD's rotational inertia is small because it's so light, and at a typical rotational speed when playing, its energy is measured in millijoules.

MAKING THE CONNECTION What translational speed would this DVD need for it to have the same kinetic energy?

ANSWER Translational kinetic energy is $K = \frac{1}{2}mv^2$, so with $K = 26$ mJ and $m = 0.018$ kg, solving for speed gives $v = 1.7$ m/s.

8.5 Rolling Bodies

A rolling object, like a wheel, combines translational and rotational motion. We'll assume here that rolling occurs without slipping, which requires friction between the rolling object and the surface on which it rolls. We'll also assume that the rolling object is symmetric about its geometric center. This places the rotation axis at the center of mass, allowing us to break the motion into translation of the center of mass and rotation about the center of mass (Figure 8.11).

The Kinematics of Rolling

Figure 8.12 shows a rolling wheel of radius R. Because the wheel rolls without slipping, it travels a distance $2\pi R$, the wheel's circumference, while the wheel makes one full

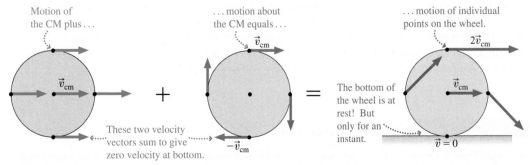

FIGURE 8.11 Motion of a rolling wheel, decomposed into translation of the entire wheel plus rotation about the center of mass.

rotation—an angular displacement $\Delta\theta = 2\pi$. If that rotation takes a time Δt, then the center-of-mass speed is

$$v_{cm} = \frac{\text{distance}}{\text{time}} = \frac{2\pi R}{\Delta t}$$

and the wheel's angular speed is

$$\omega = \frac{\text{angular displacement}}{\text{time}} = \frac{2\pi}{\Delta t}$$

Comparing these two expressions shows that

$$v_{cm} = \omega R$$

For an object that rolls without slipping, then, there's a direct relationship between the translational speed v_{cm} and the rotation rate ω.

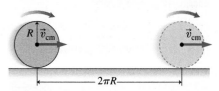

In one revolution, the wheel travels a distance equal to its circumference, $2\pi R$.

FIGURE 8.12 A rolling wheel turning through one full revolution.

Kinetic Energy in Rolling Motion

In Chapter 5 you used energy conservation to provide solutions to problems involving falling bodies. Here we'll do the same for rolling bodies.

Refer again to the rolling object in Figure 8.11. This could be any shape (a solid cylinder, hollow hoop, solid ball, etc.), that's round and rolls. Its motion can be considered a combination of translational motion of the center of mass and rotation about the center of mass—so it has both translational and kinetic energy. Its total energy is then

$$K_{\text{rolling}} = K_{\text{translational}} + K_{\text{rotational}} = \tfrac{1}{2}mv_{cm}^2 + \tfrac{1}{2}I_{cm}\omega^2$$

where we've used the standard equations for the translational and rotational parts of the kinetic energy.

Here the rotational inertia is I_{cm}, because the rotation is about the center of mass. The kinetic energy depends on I_{cm}, so the object's shape—how its mass is distributed—affects its kinetic energy in rolling. We'll illustrate this with some examples.

An experiment (going back to Galileo!) involves rolling a solid ball down an incline. When the ball's center-of-mass speed is v_{cm}, its kinetic energy is $K_{\text{rolling}} = \tfrac{1}{2}mv_{cm}^2 + \tfrac{1}{2}I_{cm}\omega^2$. Table 8.4 lists $I_{cm} = \tfrac{2}{5}mR^2$ for a solid ball of mass m and radius R; since the ball rolls without slipping, $v_{cm} = \omega R$ from above. Thus

$$K_{\text{rolling}} = \tfrac{1}{2}mv_{cm}^2 + \tfrac{1}{2}I_{cm}\omega^2 = \tfrac{1}{2}mv_{cm}^2 + \tfrac{1}{2}\left(\tfrac{2}{5}mR^2\right)\left(\frac{v_{cm}}{R}\right)^2$$

$$= \tfrac{1}{2}mv_{cm}^2 + \tfrac{1}{5}mv_{cm}^2 = \tfrac{7}{10}mv_{cm}^2$$

So the rolling ball's kinetic energy is $\tfrac{7}{10}mv_{cm}^2$, with $\tfrac{1}{2}mv_{cm}^2$ from translational motion and $\tfrac{1}{5}mv_{cm}^2$ from rotation.

The Work-Energy Theorem Revisited

The work-energy theorem, which you learned in Chapter 5, applies to rolling bodies as well as those in pure translation. The following example is an illustration.

EXAMPLE 8.10 **Work-Energy in Rolling**

A solid ball is released from rest at the top of an incline of height h. What's its speed at the bottom?

ORGANIZE AND PLAN This is an ideal problem for the work-energy theorem. Recall that the work done by gravity in this case is mgh as the ball drops a vertical distance h (Figure 8.13), and this net work is equal to the change in the ball's kinetic energy. Using our result for a solid rolling ball, $K_{\text{rolling}} = \tfrac{7}{10}mv_{cm}^2$, will let us find the final speed v_{cm}.

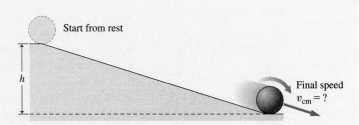

Start from rest

h

Final speed $v_{cm} = ?$

FIGURE 8.13 Ball rolling down an incline.

cont'd.

SOLVE The net work mgh done by gravity equals the change in kinetic energy: $W_{net} = \Delta K$, or

$$mgh = K - K_{initial} = \tfrac{7}{10}mv_{cm}^2 - 0$$
$$mgh = \tfrac{7}{10}mv_{cm}^2$$

The mass cancels, and we solve for speed:

$$v_{cm} = \sqrt{\frac{10gh}{7}}$$

REFLECT This result is independent of mass and the incline angle, as you've come to expect in gravity-driven motion. Interestingly, it doesn't depend on the ball's mass m or radius R. A large and a small ball should keep the same pace rolling down a ramp. You can easily check this!

CONCEPTUAL EXAMPLE 8.11 **The Great Rolling Race**

A solid ball and solid cylinder are released together on an incline. Which wins the race to the bottom?

SOLVE We've already found the ball's final speed: $v_{ball} = \sqrt{\tfrac{10}{7}gh}$. The cylinder's rotational inertia is $I_{cm} = \tfrac{1}{2}mR^2$, so its kinetic energy is

$$K_{cylinder} = \tfrac{1}{2}mv_{cm}^2 + \tfrac{1}{2}I_{cm}\omega^2 = \tfrac{1}{2}mv_{cm}^2 + \tfrac{1}{2}\left(\tfrac{1}{2}mR^2\right)\left(\frac{v_{cm}}{R}\right)^2$$
$$= \tfrac{1}{2}mv_{cm}^2 + \tfrac{1}{4}mv_{cm}^2 = \tfrac{3}{4}mv_{cm}^2$$

Applying the work-energy theorem, $mgh = \tfrac{3}{4}mv_{cm}^2$, so the cylinder's speed is

$$v_{cylinder} = \sqrt{\frac{4gh}{3}}$$

This is slightly less than v_{ball}, so the ball wins by a small margin.

REFLECT Compare the shapes of ball and cylinder; the cylinder has a little more of its mass near its outer radius. This leads to a slightly greater rotational inertia and so more of the work done by gravity goes to the cylinder's rotation, relative to the ball.

GOT IT? Section 8.5 Rank the order of finish (from fastest to slowest) of the following bodies, all rolling without slipping down the same incline. (a) hollow ball; (b) solid cylinder; (c) hollow hoop; (d) solid ball.

8.6 Rotational Dynamics

Chapter 4 introduced dynamics—the study of force and motion. Here we'll expand that subject to include rotational dynamics.

Torque and Rotational Motion

When you push on a door, its response depends on:

how hard you push (magnitude F) ...

... how far from the axis of rotation you push (radius r) ...

... and the angle θ at which you push.

The door's acceleration is proportional to $\sin \theta$.

FIGURE 8.14 Illustrating the concept of torque.

Figure 8.14 shows the simple act of opening a door. You push on the door with force $\vec{F}$, and the door opens. In the language of this chapter, you've put the door into rotational motion by giving it angular acceleration. Your experience should tell you that it's not only the *magnitude* of the force that affects the door's opening, but also the *direction* of that force and *where* on the door it's applied. In accounting for these three factors, we'll develop a Newton's law analog that's the cornerstone of rotational dynamics.

First, the magnitude F of the force matters: push harder, and the door's angular acceleration α increases—showing that α is proportional to F. Second, it matters where you push. For the least effort, you instinctively push far from the door's hinges (Figure 8.14). That's because angular acceleration is proportional to the radial distance $\vec{r}$ from the rotation axis to the point where you apply the force. Finally, the direction of $\vec{F}$ matters: it's most efficient to push perpendicular to the door. Experiment shows that α is proportional to $\sin \theta$, with θ the angle between $\vec{F}$ and a line from the rotation axis to the force application point (Figure 8.14).

We combine these observations to define a quantity called **torque**, τ.

$$\tau = rF \sin \theta \qquad \text{(Torque; SI unit: N} \cdot \text{m)} \qquad (8.16)$$

In SI, torque has units of $N \cdot m$. This is the same combination that defines the joule, but because torque is a different physical quantity from energy and work, we stick with $N \cdot m$ for torque. Equation 8.16 *defines* torque, but what *is* torque? As the opening door illustrates, torque determines angular acceleration. As we'll now show, that's analogous to Newton's second law, in which force determines acceleration.

Consider first a particle of mass m moving in a circle of radius r. A force $\vec{F}$ acts on the particle, with tangential component $F_t = F \sin \theta$ (Figure 8.15). From Newton's second law,

$$F_t = ma$$

Therefore, the torque on the particle is

$$\tau = rF \sin \theta = rF_t = rma_t$$

Equation 8.12 gives $a_t = r\alpha$, so

$$\tau = rm(r\alpha) = mr^2\alpha$$

That's for a single particle. A solid rotator (like the door in Figure 8.14) comprises many particles m_i at different radii r_i, with common angular acceleration α. Thus the torque on the system is

$$\tau = \sum_{i=1}^{n} m_i r_i^2 \alpha = \left(\sum_{i=1}^{n} m_i r_i^2 \right) \alpha$$

The final sum here is just the rotational inertia I (Equation 8.14). Therefore, the torque equation simplifies to

$$\tau = I\alpha \qquad \text{(Torque and angular acceleration; SI unit: N} \cdot \text{m)} \qquad (8.17)$$

Equation 8.17 is the key relationship in rotational dynamics. It relates the applied torque τ to a system's angular acceleration α, in just the way Newton's second law relates force and acceleration: $\vec{F} = m\vec{a}$. Table 8.5 explores this analogy.

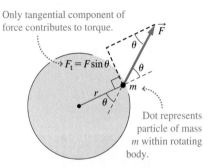

Only tangential component of force contributes to torque.

$F_t = F \sin \theta$

Dot represents particle of mass m within rotating body.

FIGURE 8.15 Accelerating a single particle in rotational motion.

TABLE 8.5 Translational and Rotational Dynamics

Translation	Rotation
Mass m	Rotational inertia I
Acceleration $\vec{a}$	Angular acceleration α
Force $\vec{F}$	Torque τ
Newton's law: $\vec{F} = m\vec{a}$	Newton's law, rotational analog: $\tau = I\alpha$

EXAMPLE 8.12 **Open the Door!**

A large door is a uniform rectangular slab, 2.45 m tall and 1.15 m wide, with mass 36.0 kg. With what minimum force could you push to open the door with angular acceleration 0.30 rad/s²?

ORGANIZE AND PLAN For the greatest effect, the definition of torque suggests pushing perpendicular to the door, as shown in Figure 8.14, so $\theta = 90°$ and $\sin \theta = 1$. Also, pushing on the door's outer edge maximizes the factor r. Doing that makes $r = 1.15$ m.

Torque is related to angular acceleration by $\tau = I\alpha$. Table 8.4 gives the door's rotational inertia as $\frac{1}{3}Ma^2$, where $a = 1.15$ m and $M = 36.0$ kg. So we can find the torque, and from it the force required.

Known: Mass $M = 36.0$ kg; width $a = r = 1.15$ m; angular acceleration $\alpha = 0.30$ rad/s².

SOLVE Using $\tau = I\alpha$, and substituting $\tau = rF \sin \theta$ and $I = \frac{1}{3}Ma^2$,

$$rF \sin \theta = \frac{1}{3}Ma^2\alpha$$

Solving for the force F and inserting numerical values,

$$F = \frac{Ma^2\alpha}{3r \sin \theta} = \frac{(36.0 \text{ kg})(1.15 \text{ m})^2(0.30 \text{ rad/s}^2)}{3(1.15 \text{ m})(1)} = 4.1 \text{ N}$$

REFLECT That's not a large force. However, many doors have a spring-loaded mechanism to keep the door closed, and you'd have to overcome that force and any hinge friction to get the door moving.

MAKING THE CONNECTION What's this door's angular acceleration if you apply the same push (a) halfway out from the hinges and (b) on the door's outer edge but at a 45° angle?

ANSWER (a) Torque is proportional to the distance r from the rotation axis to the force application point. Thus the torque will be half as much, and with torque $\tau = I\alpha$, so will the angular acceleration: 0.15 rad/s². (b) Now the torque (and angular acceleration) are reduced by a factor $\sin 45° \approx 0.707$, so $\alpha = (0.30 \text{ rad/s}^2)(0.707) = 0.21 \text{ rad/s}^2$.

Both children apply a torque that contributes to counterclockwise acceleration.

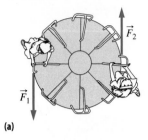

(a)

Children apply competing torques that tend to cancel each other.

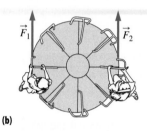

(b)

FIGURE 8.16 The directional nature of torque.

Torque's Direction

So far we've considered only the magnitude of the torque. But torque has direction as well. Consider Figure 8.16a, in which two children push a merry-go-round so as to make it accelerate in the counterclockwise direction. The net torque on the merry-go-round is the sum of the torques from the two children.

However, in Figure 8.16b, one child is trying accelerate the merry-go-round counterclockwise, the other clockwise. The net torque on the merry-go-round is now small, or even zero, as the children's efforts tend to cancel each other.

To compute the net torque in a case like this, choose a direction of positive rotation. (Taking counterclockwise as positive is consistent with our earlier convention on angular displacement.) Assign positive values to torques that by themselves would result in angular acceleration in the positive direction, and negative values to torques that by themselves would result in angular acceleration in the negative direction.

In Figure 8.16b, for example, suppose both children push tangentially to the merry-go-round's rim, 1.8 m from the axis. The child on the left pushes with a 60-N force, the child on the right with 85 N. Then the two torques are

$$\text{Child on left: } \tau = -rF\sin\theta = -(1.8\ \text{m})(60\ \text{N})(1) = -108\ \text{N}\cdot\text{m}$$
$$\text{Child on right: } \tau = +rF\sin\theta = (1.8\ \text{m})(85\ \text{N})(1) = 153\ \text{N}\cdot\text{m}$$

The net torque on the merry-go-round is the sum of these torques:

$$\text{Net torque} = 153\ \text{N}\cdot\text{m} - 108\ \text{N}\cdot\text{m} = 45\ \text{N}\cdot\text{m}$$

The positive sign indicates a net positive torque, tending to accelerate the merry-go-round counterclockwise.

Just as in translational dynamics, where it's the net force that results in acceleration ($\vec{F}_{\text{net}} = m\vec{a}$), it's the net torque that produces angular acceleration. Restating Equation 8.17,

$$\tau_{\text{net}} = I\alpha \quad \text{(Net torque and angular acceleration; SI unit: N·m)} \qquad (8.18)$$

Equation 8.18 is more general, allowing for the fact that there can be multiple torques acting on a rotating body, and it's the net torque that counts.

Applications of Torque

Torque has many applications in physics, engineering, and physiology. Engines produce torque that turns gears and wheels to do mechanical work. A typical automobile engine produces somewhere around 250 N·m of torque. An SUV may have a torque of some 400 N·m for off-road driving and towing, and a sports car produces as much as 500 N·m.

Any change in rotational motion involves torque, even if it's only turning a bolt. Use a wrench, and you're applying torque. A longer wrench gives a greater r in Equation 8.16, producing a larger torque. If a plumber weighing 800 N puts his full weight on the end of a 40-cm wrench, the torque is $(800\ \text{N})(0.40\ \text{m}) = 320\ \text{N}\cdot\text{m}$—about that of an SUV engine!

Your hip, shoulder, and neck joints all allow for rotation. Lift your lower arm and you're rotating it about a rotation axis through your elbow. The following example explores the mechanics of the elbow joint.

EXAMPLE 8.13　**Biceps!**

Hold your forearm horizontal, as shown in Figure 8.17, and assume your biceps muscle exerts an upward force of magnitude F, applied 4.0 cm from the rotation axis at the elbow. If the lower arm has mass 2.25 kg, and the elbow-to-fingertip distance is 50.0 cm, estimate the force F.

ORGANIZE AND PLAN Equation 8.19 gives $\tau_{\text{net}} = I\alpha$. Here the arm is at rest ($\alpha = 0$), so the net torque is zero. Two torques act on the lower arm: One, due to the biceps, pulls up and tends to rotate the arm

counterclockwise in Figure 8.17. The other torque results from the downward pull of the arm's weight, tending to rotate the arm clockwise. For this estimate, assume the lower arm's center of mass is about halfway from the elbow, so $r = 25$ cm in the expression for torque due to the weight. Setting the net torque equal to zero leads to a solution for the biceps force F.

Known: $r_{\text{biceps}} = 4.0$ cm; $r_{\text{arm}} = 25$ cm $= 0.25$ m; $m_{\text{arm}} = 2.25$ kg.

cont'd.

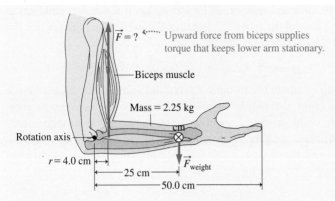

FIGURE 8.17 Torque used to keep the lower arm horizontal.

SOLVE By our sign convention, the biceps torque is positive and the weight torque negative. Then their sum is

$$\tau_{net} = +r_{biceps}F_{biceps}\sin\theta - r_{weight}F_{weight}\sin\theta$$

$$\tau_{net} = +(0.04\ \text{m})(F)(1) - (0.25\ \text{m})(2.25\ \text{kg})(9.80\ \text{m/s}^2)(1) = 0$$

Solving for the unknown F,

$$F = \frac{(0.25\ \text{m})(2.25\ \text{kg})(9.80\ \text{m/s}^2)}{0.04\ \text{m}} = 138\ \text{kg}\cdot\text{m/s}^2 = 138\ \text{N}$$

The upward force is 138 N.

REFLECT To put the answer in perspective, 138 N is the force you'd apply to hold straight overhead a mass of 138 N/g = 14 kg. Keeping your arm extended is more difficult if your hand holds additional weight, which greatly increases the required torque and hence force.

MAKING THE CONNECTION Would more or less force be required if you started with your arm in a horizontal position and then raised it? What if you lowered it?

ANSWER Beginning to raise your arm requires a net positive torque, meaning the biceps has to apply more force. Conversely, beginning to lower your arm requires a net negative torque, meaning less biceps force. If the arm then moves at constant speed, its inclined orientation means less weight torque in both cases, so the biceps force would be lower, too.

Notice in Example 8.13 how we computed the torque due to weight by taking the weight force (magnitude mg) to act at the center of mass. That's a rule that will come in handy in the next section, when we pursue the subject of rotational equilibrium.

Reviewing New Concepts: Torque

- The torque due to an applied force of magnitude F is $\tau = rF\sin\theta$.
- A torque is positive if alone it would lead to a counterclockwise angular acceleration; a torque is negative if alone it would lead to a clockwise angular acceleration.
- Net torque is related to angular acceleration: $\tau_{net} = I\alpha$.

8.7 Mechanical Equilibrium

In Chapter 4 we defined equilibrium as the condition when the net force on an object is zero. In the context of rotation, there's more to equilibrium. Figure 8.18 shows those children from Figure 8.16 pushing on the merry-go-round, with forces of equal magnitude and opposite direction. The net force on the merry-go-round is zero, yet the two forces clearly produce a net torque.

Consider three possible situations:

- **Translational equilibrium** occurs when the net force on an object is zero.
- **Rotational equilibrium** occurs when the net torque on an object is zero.
- **Mechanical equilibrium** occurs when both the net force and net torque are zero.

For an object at rest to be truly in equilibrium, both the net force and net torque on it must be zero. That's mechanical equilibrium—the necessary condition for an object at rest to remain at rest *and* not rotate.

Two children balanced on a seesaw provide an example of mechanical equilibrium (Figure 8.19, next page). The seesaw board rests on a fixed pivot, the *fulcrum*, normally at the board's midpoint. For a uniform board, the center of mass lies on the fulcrum. The fulcrum exerts an upward force that balances the weight of the board and its occupants, so the condition of translational equilibrium is satisfied. But we've still got to worry about rotation.

Pushes are equal in magnitude and opposite in direction, so *net force* is zero . . .

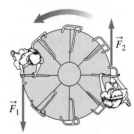

. . . but pushes cause merry-go-round to turn, so the *net torque* they exert is not zero.

FIGURE 8.18 A case of zero net force but nonzero net torque.

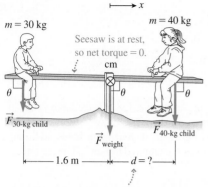

FIGURE 8.19 Example of mechanical equilibrium.

To find distance d, use fact that net torque = 0:

$$\tau_{net} = \tau_{30\text{-kg child}} + \tau_{40\text{-kg child}} + \tau_{weight} = 0$$

Each torque = $rF\sin\theta$. Torque τ_{weight} exerted on board by gravity = 0 because $r = 0$. Factor out the equation and solve for d:

$$\tau_{net} = (1.6\text{ m})(30\text{ kg})(9.8\text{ m/s}^2)(\sin 90°)$$
$$- d(40\text{ kg})(9.8\text{ m/s}^2)(\sin 90°) = 0$$
$$d = 1.2\text{ m}$$

Suppose the children have different masses, 30 kg and 40 kg. If the lighter child sits at one end of the board, where should the other sit? Rotational equilibrium requires that the net torque be zero. Adding the torques about the fulcrum,

$$\tau_{net} = \tau_{30\text{-kg child}} + \tau_{40\text{-kg child}} + \tau_{weight} = 0$$

We'll use $\tau = rF\sin\theta$ to compute the torques. Each child applies a downward force $F = mg$. For the left-hand child, this results in positive torque (counterclockwise rotation), while the right-hand child produces negative torque. The board produces zero torque, because its center of mass is on the rotation axis. Figure 8.19 shows how to solve for the unknown distance—with the obvious result that the heavier child must sit closer to the fulcrum to cancel the torque from the lighter child.

PROBLEM-SOLVING STRATEGY 8.2 **Mechanical Equilibrium**

ORGANIZE AND PLAN

- Draw a sketch of the physical situation showing the object in equilibrium.
- Identify known and unknown forces, to see how translational and rotational equilibrium will be established.
- Note the direction for positive and negative torques.
- Review what quantities you know, and what you're trying to find. Is it necessary to use translational equilibrium, rotational equilibrium, or both?

SOLVE

- Write the conditions for translational and/or rotational equilibrium.
- Solve the resulting equation(s) for the unknown quantity (quantities).
- Insert numerical values and compute the answer(s), using appropriate units.

REFLECT

- Are the dimensions and units of the answer(s) correct?
- If the problem relates to something familiar, think about whether the answer makes sense.

EXAMPLE 8.14 **Gymnastics**

A gymnast in the "cross" position of the rings exercise hangs motionless with his arms extended horizontally, one hand on each ring. The rings are 1.66 m apart, and the gymnast's mass is 62.4 kg. Assume symmetry, with the gymnast's center of mass on a vertical line halfway between the rings. (a) What's the upward force of each ring on the gymnast's hand? (b) Use the answer to part (a) to show that the net torque about each ring is zero, as required for rotational equilibrium.

ORGANIZE AND PLAN There are three forces acting on the gymnast: the downward force of gravity and the upward force from each ring (Figure 8.20). The net force in equilibrium is zero. From symmetry, each ring holds half the gymnast's weight.

With all three forces known, the torque about either ring can be computed using $\tau = rF\sin\theta$ for each force.

Known: $m = 62.4$ kg, ring separation = 1.66 m.

SOLVE The upward force from each ring has magnitude equal to half the gymnast's weight, so

$$F = \frac{mg}{2} = \frac{(62.4\text{ kg})(9.80\text{ m/s}^2)}{2} = 306\text{ N}$$

We'll compute the torque about the left ring. Its force is applied at $r = 0$, so it doesn't contribute to the net torque. The weight is applied

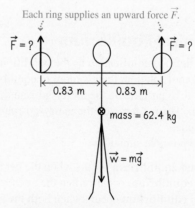

FIGURE 8.20 Analysis of forces on the gymnast.

at $r = (1.66\text{ m})/2 = 0.83$ m, with force $mg = 612$ N; it produces a negative (clockwise) torque. The positive torque from the right-hand ring results from a 306-N upward force applied at $r = 1.66$ m. Therefore, the net torque is

$$\tau_{net} = -(0.83\text{ m})(612\text{ N})(\sin 90°) + (1.66\text{ m})(306\text{ N})(\sin 90°)$$
$$= -508\text{ N}\cdot\text{m} + 508\text{ N}\cdot\text{m} = 0$$

cont'd.

as expected. A similar calculation yields zero torque about the right-hand ring.

REFLECT There is zero net torque about *either* ring. You should convince yourself that the net torque is also zero about the midpoint between the rings. In fact, it's zero about any axis you choose. That's an important fact: For a system that's in translational equilibrium, a net torque of zero about *one arbitrary* axis implies zero net torque about *any* axis. So you only need to check for rotational equilibrium about one axis, and which one is your choice.

EXAMPLE 8.15   **Leaning Ladder**

A ladder leans against a wall, at 16° to the vertical. It has length $L = 3.64$ m and mass $m = 18.2$ kg. There's no friction at the wall, but there is friction at the floor. Find the normal force of the wall on the ladder.

ORGANIZE AND PLAN At the outset, it's not clear whether we should use translational equilibrium, rotational equilibrium, or both. There are four forces acting on the ladder: its weight $\vec{w} = m\vec{g}$, the normal force from the wall $\vec{n}_w$, the normal force from the floor $\vec{n}_f$, and static friction $\vec{f}_s$ at the ground, which keeps the ladder from slipping sideways (Figure 8.21).

Of these, we're given only the ladder's weight. With so many unknown forces, translational equilibrium alone won't solve the prob-

lem. However, you can apply rotational equilibrium, summing the torques about the bottom of the ladder. Then the two forces acting on that point have $r = 0$, so don't contribute to the torque. This leaves a positive torque due to the wall force and a negative torque due to the ladder's weight. They sum to zero in equilibrium.

SOLVE Assuming the ladder is uniform, its weight acts at its center, $L/2$ from the bottom, with the weight vector at 16° to the radius at that point. The wall force acts at distance L from the bottom and makes an angle $90° - 16° = 74°$ with the radius. Therefore, the net torque is

$$\tau_{net} = -(L/2)(mg)(\sin 16°) + (L)(n_w)(\sin 74°) = 0$$

Solving for the wall's normal force gives

$$n_w = \frac{mg \sin 16°}{2 \sin 74°} = \frac{(18.2 \text{ kg})(9.80 \text{ m/s}^2)\sin 16°}{2 \sin 74°} = 25.6 \text{ N}$$

REFLECT Notice how changing the angle affects the answer. Make it smaller, and the normal force decreases—reaching zero for a vertical ladder. But make the angle larger, and the normal force increases—requiring a correspondingly larger frictional force for equilibrium; see "Making the Connection," below.

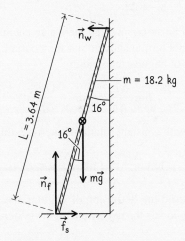

FIGURE 8.21 Forces acting on the ladder.

Equilibrium, Balance, and Center of Mass

From football to ballet to yoga, physical activities depend on equilibrium and balance. You've already seen a worked example from gymnastics. Athletes in many sports assume a common "ready" position, with feet spread wide and the body slightly crouched. The center of mass is on a vertical line between the feet. This leaves the athlete ready to move in any direction, and less likely to fall or be knocked over.

Analyzing rotational equilibrium shows why. As shown in Figure 8.22a (next page), the ground exerts an upward normal force on each foot. There's no problem achieving zero

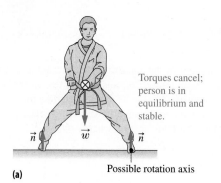

Torques cancel; person is in equilibrium and stable.

$\vec{n}$ $\vec{w}$ $\vec{n}$

(a)   Possible rotation axis

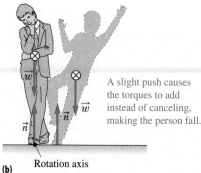

A slight push causes the torques to add instead of canceling, making the person fall.

$\vec{w}$ $\otimes$

$\vec{w}$

$\vec{n}$ $\vec{n}$

(b)   Rotation axis

FIGURE 8.22 Changing feet position can make you lose balance.

torque about each foot, with the weight producing a torque in one direction, the normal force from the ground producing a torque in the other direction. But if the feet are too close together and the body leans a bit (Figure 8.22b), those torques add together instead of canceling. There's a nonzero net torque—and you're going to fall over.

8.8 Angular Momentum

Throughout this chapter you've seen analogies between translational quantities and their rotational counterparts. For example, angular velocity and acceleration are analogous to translational velocity and acceleration in one dimension. Rotational inertia takes the place of mass, and torque takes the place of force. Table 8.6 summarizes these relationships.

The last two rows of Table 8.6 feature new concepts. Chapter 6 introduced momentum $\vec{p} = m\vec{v}$, defined as the product of mass and velocity, and also called *linear momentum* or *translational momentum*. Using the corresponding rotational quantities (rotational inertia and angular velocity), we define the **angular momentum** L of a rotating body analogously:

$$L = I\omega \quad \text{(Angular momentum; SI unit: J·s)} \qquad (8.19)$$

Angular momentum has units $(\text{kg}\cdot\text{m}^2)(\text{rad/s})$, which, since the rad is dimensionless, reduces to $\text{kg}\cdot\text{m}^2/\text{s} = \text{J}\cdot\text{s}$.

Conservation of Angular Momentum

Newton's law equates net force to the rate of change of momentum; stated this way, it reads

$$\vec{F}_{\text{net}} = \lim_{\Delta t \to 0} \frac{\Delta \vec{p}}{\Delta t}$$

It shouldn't be surprising that there's an analogous relationship between net torque and angular momentum. Consider the rate of change of angular momentum:

$$\lim_{\Delta t \to 0} \frac{\Delta L}{\Delta t} = \lim_{\Delta t \to 0} \frac{\Delta(I\omega)}{\Delta t}$$

For an object with constant rotational inertia I, this becomes

$$\lim_{\Delta t \to 0} \frac{\Delta L}{\Delta t} = I \lim_{\Delta t \to 0} \frac{\Delta \omega}{\Delta t} = I\alpha$$

where we've used the definition of angular acceleration. But the product $I\alpha$ equals the net torque on the system, so

$$\tau_{\text{net}} = \lim_{\Delta t \to 0} \frac{\Delta L}{\Delta t} \qquad (8.20)$$

analogous to Newton's second law for translational motion.

In Chapter 6 you used the momentum form of Newton's second law to justify the conservation of momentum in systems with zero net external force. Similarly, if there's zero external torque on a system, then

$$\tau_{\text{net}} = \lim_{\Delta t \to 0} \frac{\Delta L}{\Delta t} = 0$$

TABLE 8.6 Translational Quantities and Their Rotational Counterparts

Translational quantities	Rotational quantities
Position x	Angular position θ
Velocity $v_x = \lim_{\Delta t \to 0} \dfrac{\Delta x}{\Delta t}$	Angular velocity $\omega = \lim_{\Delta t \to 0} \dfrac{\Delta \theta}{\Delta t}$
Acceleration $a_x = \lim_{\Delta t \to 0} \dfrac{\Delta v_x}{\Delta t}$	Angular acceleration $\alpha = \lim_{\Delta t \to 0} \dfrac{\Delta \omega}{\Delta t}$
Force $\vec{F}$	Torque $\tau = rF\sin\theta$
Mass m	Rotational inertia $I = \sum_{i=1}^{n} m_i r_i^2$
Newton's second law $\vec{F}_{\text{net}} = m\vec{a}$	Rotational analog of Newton's second law $\tau_{\text{net}} = I\alpha$
Kinetic energy $K_{\text{trans}} = \frac{1}{2}mv^2$	Kinetic energy $K_{\text{rot}} = \frac{1}{2}I\omega^2$
Momentum $\vec{p} = m\vec{v}$	Angular momentum $L = I\omega$
$\vec{F}_{\text{net}} = \lim_{\Delta t \to 0} \dfrac{\Delta \vec{p}}{\Delta t}$	$\tau_{\text{net}} = \lim_{\Delta t \to 0} \dfrac{\Delta L}{\Delta t}$

showing that **angular momentum is conserved in a system with zero external torque**. This statement is precisely analogous to the conservation of translational momentum, and it can be used analogously in problem solving.

Figure 8.23 shows a spinning skater. The ice is essentially frictionless, so it doesn't produce any torques, and therefore angular momentum is conserved. In the first frame, the skater's arms are extended while she spins. What happens when she brings her arms inward? Her rotational inertia about the rotation axis is reduced, because some of her mass is now rotating in a smaller circle. However, with angular momentum $L = I\omega$ conserved, any decrease in I results in an increase in angular velocity ω, to keep the product $I\omega$ constant. The skater spins faster with her arms tucked in.

Another well-known example is a falling cat, which experiences no external torque, so it can't change its angular momentum. However, it cleverly twists different parts of its body simultaneously, resulting in a feet-first landing.

Arms and leg far from axis: large I, small ω

Mass closer to axis: small I, large ω, same $L = I\omega$

FIGURE 8.23 Because her angular momentum is conserved, the skater spins faster after pulling in her arms.

EXAMPLE 8.16 **Throwing a Pot**

A potter's wheel consists of a uniform stone disk with mass 42 kg and radius 28.0 cm. It's turning freely at 4.10 rad/s, when a 3.2-kg lump of clay drops onto the wheel's outer rim. What happens to the wheel's angular velocity?

ORGANIZE AND PLAN With the wheel turning freely, there's no external torque, so angular momentum is conserved. The initial angular momentum is the product of the wheel's rotational inertia (computable from the information given) and the known angular velocity. Adding the clay changes the rotational inertia, but not the angular momentum, so the product $I\omega$ remains unchanged. If we determine the new rotational inertia, we can find the angular velocity.

Known: Mass $M = 42$ kg; clay mass $m = 3.2$ kg; radius $R = 0.28$ m; $\omega_0 = 4.10$ rad/s.

SOLVE The wheel is a solid cylindrical disk, so Table 8.4 gives $I_0 = \frac{1}{2}MR^2$ for the initial rotational inertia. Dropping the clay adds mR^2 to the rotational inertia (Table 8.4, approximating the clay as a single particle), for a final value $I_f = \frac{1}{2}MR^2 + mR^2 = \left(\frac{1}{2}M + m\right)R^2$.

Conservation of angular momentum means $L_0 = L_f$, so $I_0\omega_0 = I_f\omega_f$. Solving for ω_f gives

$$\omega_f = \frac{I_0\omega_0}{I_f} = \frac{\frac{1}{2}MR^2\omega_0}{\left(\frac{1}{2}M + m\right)R^2}$$

$$\omega_f = \frac{\frac{1}{2}(42\text{ kg})(0.28\text{ m})^2(4.10\text{ rad/s})}{\left(\frac{1}{2}(42\text{ kg}) + 3.2\text{ kg}\right)(0.28\text{ m})^2} = 3.56\text{ rad/s}$$

REFLECT As expected, increasing the rotational inertia reduces the angular velocity in order to conserve angular momentum.

MAKING THE CONNECTION How would your answer change if the clay were dropped closer to the rotation axis, say halfway in?

ANSWER The added mass has a smaller rotation radius, so it adds less to the rotational inertia, and thus the angular velocity doesn't drop as much. Reworking with I_f now $\frac{1}{2}MR^2 + mr^2$, where $r = 0.14$ m, the new angular velocity is 3.95 rad/s.

Torque and Changing Angular Momentum

So far we've considered cases where net torque on an object is zero, so angular momentum is conserved. If there is a nonzero net torque, Equation 8.20 gives $\tau_{net} = \lim_{\Delta t \to 0}(\Delta L/\Delta t)$, implying that angular momentum changes. The changing angular momen-

---

APPLICATION **Sport Strategies and Angular Momentum**

The quarterback throws the football with spin, giving it substantial angular momentum. Torques due to small air currents can't change this angular momentum significantly; the result is a stable trajectory. Similarly, a baseball pitched with spin has a stable flight with a little curve. But the "knuckleball" pitcher deliberately throws with very little spin. With little angular momentum, the ball's flight becomes erratic due to random torques from the air—which makes things difficult for the batter.

tum can be computed and used to understand how the net torque has affected rotation, as illustrated in the following example.

EXAMPLE 8.17   **A Spinning Satellite**

A spacecraft with radius $R = 2.8$ m has a rotational inertia $I = 70$ kg·m^2 about its central axis. It's initially not rotating. Then rockets on the outer edge fire, exerting a 20-N tangential force. After 2.0 s, what are the spacecraft's (a) angular momentum and (b) angular velocity?

ORGANIZE AND PLAN Equation 8.20 relates applied torque to changing angular momentum. Here the torque is $\tau = rF_t$. Once you know the angular momentum, you can get angular velocity from $L = I\omega$.

Known: $R = 2.8$ m; $I = 70$ kg·m^2; $F_t = 20$ N, $\Delta t = 2.0$ s.

SOLVE (a) The torque is $\tau = rF_t$, so by Equation 8.20

$$\tau_{\text{net}} = rF_t = \frac{\Delta L}{\Delta t}$$

With a constant applied force, $\Delta L/\Delta t$ is constant, you don't need the limit. Solving for ΔL,

$$\Delta L = rF_t\Delta t = (2.8 \text{ m})(20 \text{ N})(2.0 \text{ s}) = 112 \text{ J·s}$$

The spacecraft started with $L = 0$, so this is the final angular momentum.

(b) Using $L = I\omega$ gives the angular velocity:

$$\omega = \frac{L}{I} = \frac{112 \text{ J·s}}{70 \text{ kg·m}^2} = 1.6 \text{ rad/s}$$

REFLECT Here we were given the spacecraft's rotational inertia, so we didn't need details of its shape and we didn't consult Table 8.4.

MAKING THE CONNECTION There's another approach to this problem: You could use the information given to compute the torque and then the angular acceleration, because $\tau_{\text{net}} = I\alpha$. Then rotational kinematic equations would give the final angular velocity ω. Check this and see how it works.

ANSWER Following this approach will give $\tau = 56$ N·m and $\alpha = 0.80$ rad/s^2. Then the kinematic equation $\omega = \omega_0 + \alpha t$ gives $\omega = 1.6$ rad/s, as before.

8.9 Rotational Motion with Vector Quantities

Take another look at Table 8.6, showing translational quantities and their rotational counterparts. There's one important difference between the two columns. For translation, we used the *vector* quantities velocity, force, and momentum, but the corresponding rotational quantities are all shown as *scalars*. Why? Shouldn't angular velocity, torque, and angular momentum be vectors, too? In fact, they are, as we'll now show.

So far we've been able to neglect the vector nature of rotational quantities because we've considered only rotation about a fixed axis. Thus, angular velocity can be considered a scalar, whose sign gives the direction of rotation. This is analogous to one-dimensional translational motion along the x-axis, with velocity v_x a scalar whose sign denotes direction. With angular velocity a scalar, there's no need to use vectors for torque or angular momentum, either.

Now we'll consider situations in which the rotation axis might change. This requires us to treat angular velocity, torque, and angular momentum as vectors.

Vector Angular Velocity, Angular Acceleration, and Torque

Figure 8.24 shows a rotating disk. As you know, all points on the disk share a common angular velocity, even though their tangential velocities differ. What points on the disk also have in common is the rotation axis, so it's natural to define the angular velocity vector $\vec{\omega}$ to point along that axis. But that leaves two possible directions. Either would do, but convention is to follow the **right-hand rule** described in Figure 8.24.

With the angular velocity vector defined, the vector angular acceleration follows naturally as a vector version of Equation 8.7, or

$$\vec{\alpha} = \lim_{\Delta t \to 0} \frac{\Delta\vec{\omega}}{\Delta t} \tag{8.21}$$

Consider again the disk in Figure 8.24. If its angular velocity is increasing, then $\Delta\vec{\omega}$ (and hence $\vec{\alpha}$) should be in the same direction as the angular velocity $\vec{\omega}$. But if the angular velocity is decreasing, then $\vec{\alpha}$ is directed opposite to $\vec{\omega}$.

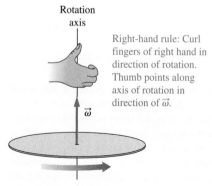

Rotation axis

Right-hand rule: Curl fingers of right hand in direction of rotation. Thumb points along axis of rotation in direction of $\vec{\omega}$.

$\vec{\omega}$

FIGURE 8.24 The right-hand rule gives the direction of the angular velocity vector.

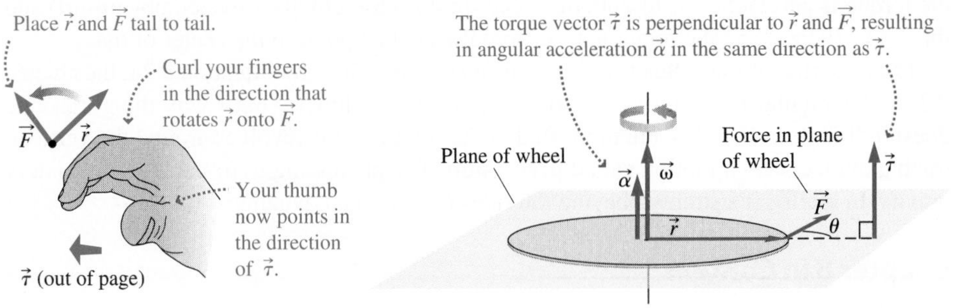

Place $\vec{r}$ and $\vec{F}$ tail to tail.

Curl your fingers in the direction that rotates $\vec{r}$ onto $\vec{F}$.

$\vec{F}$ $\vec{r}$

Your thumb now points in the direction of $\vec{\tau}$.

$\vec{\tau}$ (out of page)

(a) The right-hand rule for the direction of torque

The torque vector $\vec{\tau}$ is perpendicular to $\vec{r}$ and $\vec{F}$, resulting in angular acceleration $\vec{\alpha}$ in the same direction as $\vec{\tau}$.

Plane of wheel

Force in plane of wheel

$\vec{\alpha}$ $\vec{\omega}$

$\vec{\tau}$

$\vec{r}$ $\vec{F}$

θ

(b) The direction of the angular acceleration caused by a torque

FIGURE 8.25 Torque and angular acceleration.

This fits nicely with torque. We'll define the direction of torque using another right-hand rule, illustrated in Figure 8.25a. Remember that torque is due to a force F applied at some radius r, with $\tau = rF \sin \theta$. Then the direction of the torque vector is:

- Perpendicular to both the radius and the force vector
- Given by the right-hand rule of Figure 8.25a

Our newly defined torque and angular acceleration vectors suggest a vector version of the dynamical equation $\tau_{net} = I\alpha$, namely,

$$\vec{\tau}_{net} = I\vec{\alpha} \tag{8.22}$$

(Here rotational inertia I, like its counterpart mass m, remains a scalar.) Imagine starting with the wheel of Figure 8.24 at rest and applying a torque, as shown in Figure 8.25a. The torque vector is in the upward direction. The wheel accelerates counterclockwise, so the directions of its angular velocity and angular acceleration are also upward (Figure 8.25b). This agrees with Equation 8.22, because the vectors on both sides of the equation must point in the same direction. Now with the wheel rotating, apply an opposite torque to slow it. The direction of the torque reverses, and so does the angular acceleration. Equation 8.22 is again satisfied.

Vector Angular Momentum

Angular momentum is defined by the same right-hand rule as angular velocity. This makes sense, because the vector version of $L = I\omega$ is

$$\vec{L} = I\vec{\omega} \tag{8.23}$$

Thus, a rotating object's angular momentum points in the same direction as its angular velocity. The utility of the angular momentum vector comes from rewriting Equation 8.20 with vector quantities:

$$\vec{\tau} = \lim_{\Delta t \to 0} \frac{\Delta \vec{L}}{\Delta t} \tag{8.24}$$

This equation covers the spinning-up and slowing of our wheel, but it also describes cases where the *direction* of the angular momentum changes. This leads to a surprising new phenomenon.

Precession

Place a spinning gyroscope on the floor, with its rotation axis tilted slightly (Figure 8.26). The gyroscope has angular momentum directed along the axis, as shown. Because it's tilted, there's a torque about the gyroscope's contact point on the stand. Remember that

APPLICATION   **Ice Age!**

Earth's rotation causes our planet to bulge at the equator, resulting in a torque due to the Sun's gravity. The planet's tilted rotation axis therefore precesses, with a period around 26,000 years. Earth's tilt is what causes seasons, and because Earth's orbit isn't perfectly circular, precession affects when in the seasonal cycle Earth is closest to the Sun. The upshot is a change in the distribution of solar energy in the polar regions, which, along with other orbital changes, triggers ice ages. This chapter's opening image suggests this effect.

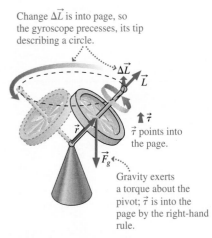

Change $\Delta \vec{L}$ is into page, so the gyroscope precesses, its tip describing a circle.

$\Delta \vec{L}$ $\vec{L}$

$\vec{r}$ $\vec{\tau}$

$\vec{\tau}$ points into the page.

$\vec{F}_g$

Gravity exerts a torque about the pivot; $\vec{\tau}$ is into the page by the right-hand rule.

FIGURE 8.26 Why doesn't the spinning gyroscope fall over?

the torque is perpendicular to both the force (in this case gravity, directed downward) and the radius (here along the rotation axis, from the contact point to the center of mass).

The important thing is that this torque is *horizontal*. Thus, by Equation 8.24, the change $\Delta \vec{L}$ in the angular momentum is horizontal, not up or down. That's why the gyroscope doesn't "fall down" in this situation. Rather, its rotation axis revolves around the vertical, tracing out a cone—a motion called **precession**. The precessing gyroscope seems to defy gravity. In reality, it's simply obeying the rules of rotational dynamics!

Chapter 8 in Context

This chapter introduced *rotational motion*, including the basics of rotational kinematics and dynamics. You've seen how quantities and relationships describing rotation about a fixed axis are analogous to quantities and relationships in one-dimensional translational motion. You've seen how *rolling bodies* combine *translational and rotational motion*, and you've had a brief look at rotational quantities as vectors. Rotational motion will show up again later in several contexts. Here's a brief preview:

Looking Ahead Rotational motion will be important in the next chapter, on gravitation. A primary application is planetary motion, where planets orbit the Sun and simultaneously rotate on their axes. In Chapter 24 you'll see a central role for angular momentum in atomic physics. Rotational motion occurs on the largest and smallest scales!

Rotational Kinematics

(Section 8.1) The **radian** measure of angle is the ratio of arc length to radius. A rotating object is characterized by its **angular position**; **angular displacement** is then the difference between two angular positions, and **angular velocity** is the rate of change of angular position. **Angular acceleration** is the rate of change of angular velocity. In pure rotation, a rigid object rotates around a fixed axis, with all parts of the body having the same angular velocity.

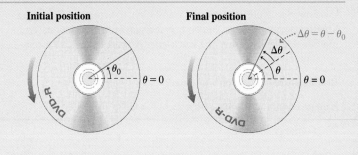

Initial position **Final position**

$\Delta\theta = \theta - \theta_0$

Angular position: θ (in radians) $= \dfrac{\text{arc length}}{\text{radius}} = \dfrac{s}{r}$

Angular velocity: $\omega = \lim\limits_{\Delta t \to 0} \dfrac{\Delta\theta}{\Delta t}$

Angular acceleration: $\alpha = \lim\limits_{\Delta t \to 0} \dfrac{\Delta\omega}{\Delta t}$

Kinematic Equations for Rotational Motion

(Section 8.2) The **kinematic equations for rotational motion** are analogous to the kinematic equations for one-dimensional motion. Rotational variables are analogous to one-dimensional translational variables: angular position θ is analogous to position x; angular velocity ω to velocity v_x, and angular acceleration $\bar{\alpha}$ to acceleration a_x.

Kinematic equations for constant angular acceleration:

$$\omega = \omega_0 + \alpha t \qquad \theta = \theta_0 + \omega_0 t + \tfrac{1}{2}\alpha t^2 \qquad \omega^2 = \omega_0^2 + 2\alpha\Delta\theta$$

Rotational and Tangential Motion

(Section 8.3) Every point in a rotating body has centripetal acceleration and may also have **tangential acceleration**. The translational velocity of a point on a rotating body is its **tangential velocity** $\vec{v}_t$. The tangential acceleration a_t is the rate of change of tangential velocity. At any point in the rotating object, centripetal acceleration points toward the axis of rotation, and tangential acceleration (if nonzero) is tangent to the circular path.

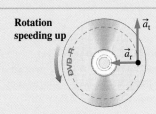

Rotation speeding up $\vec{a}_t$ $\vec{a}_r$

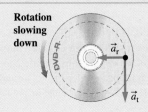

Rotation slowing down $\vec{a}_r$ $\vec{a}_t$

Tangential velocity: $v_t = r\omega$

Tangential acceleration: $a_t = r\alpha$

Kinetic Energy and Rotational Inertia

(Section 8.4) A mass near the outside of a rotating object has a greater kinetic energy than the same mass at a smaller radius. The **rotational kinetic energy** of an object is a function of its mass, dimensions, and rotation speed. **Rotational inertia** depends on mass and how it's distributed. Specific formulas determine rotational inertia for common shapes.

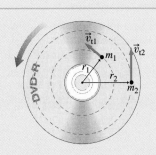

$\vec{v}_{t1}$ m_1 $\vec{v}_{t2}$ r_1 r_2 m_2

Kinetic energy of point mass m: $K = \tfrac{1}{2}mv_t^2$

Kinetic energy of a collection of point masses in rotation:

$$K = \sum_{i=1}^{n} \tfrac{1}{2}m_i v_{t,i}^2 = \tfrac{1}{2}I\omega^2$$

Rotational inertia: $I = \sum_{i=1}^{n} m_i r_i^2$

Rolling Bodies

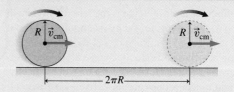

(Section 8.5) **Rolling** combines translation and rotation. The **total kinetic energy** of the rolling body is the sum of its translational and rotational kinetic energy. The **work-energy theorem** holds for rotational motion.

Rolling without slipping: $v_{cm} = \omega R$

Kinetic energy of a rolling object: $K = \frac{1}{2}mv_{cm}^2 + \frac{1}{2}I_{cm}\omega^2$

Work-energy theorem: $W_{net} = \Delta K$

Rotational Dynamics

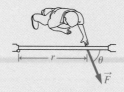

(Section 8.6) **Torque** is the rotational analog of force, and results from a force applied some distance from a rotation axis. A **net torque** results in angular acceleration.

Torque: $\tau = rF \sin\theta$

Net torque: $\tau_{net} = I\alpha$

Mechanical Equilibrium

(Section 8.7) **Translational equilibrium** occurs when the net force on an object is zero; **rotational equilibrium** when the net torque is zero; and **mechanical equilibrium** when both are zero. Rotational equilibrium requires that the sum of the torques about any fixed axis be zero.

Mechanical equilibrium: $\vec{F}_{net} = \vec{F}_1 + \vec{F}_2 + \cdots = \sum \vec{F}_i = 0$
$\tau_{net} = \tau_1 + \tau_2 + \cdots = \sum \tau_i = 0$

Angular Momentum

(Section 8.8) The **angular momentum** of a rotating object is the product of rotational inertia and angular velocity. Angular momentum is conserved if the net torque on the system is zero. Nonzero net torque leads to a change in angular momentum.

Angular momentum: $L = I\omega$

Torque: $\tau_{net} = \lim_{\Delta t \to 0} \frac{\Delta L}{\Delta t}$

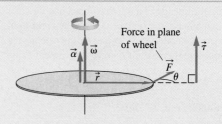

Large I, small ω

Small I, large ω, same $L = I\omega$

Rotational Motion with Vector Quantities

(Section 8.9) If the axis of rotation isn't fixed, angular velocity, torque, and angular momentum must be considered **vectors**. If angular velocity $\vec{\omega}$ is increasing, then angular acceleration $\vec{\alpha}$ is the same direction as $\vec{\omega}$. If angular velocity is decreasing, then $\vec{\alpha}$ is opposite $\vec{\omega}$. Torque and angular acceleration that are not either parallel or antiparallel to $\vec{\omega}$ change the direction of the angular momentum $\vec{L}$; an example is **precession**.

Force in plane of wheel

Angular acceleration vector: $\vec{\alpha} = \lim_{\Delta t \to 0} \frac{\Delta \vec{\omega}}{\Delta t}$

Net torque vector: $\vec{\tau}_{net} = I\vec{\alpha}$

Angular momentum vector: $\vec{L} = I\vec{\omega}$

NOTE: Problem difficulty is labeled as ■ straightforward to ■■■ challenging. Problems labeled BIO are of biological or medical interest.

Conceptual Questions

1. While standing on the rotating Earth, is your centripetal acceleration greater on the equator or at latitude 45° north?
2. A bicycle wheel rotates with increasing angular velocity. Compare the tangential acceleration of a point on the rim with that of a point midway along one spoke.
3. Why does a compact disk turn fastest when information is being read near its inner edge?
4. Explain why the rotational inertia of a hollow ball is greater than that of a uniform solid ball with the same mass and radius.
5. Does a baseball bat have the same rotational inertia when rotated about an axis perpendicular to either end?
6. Is a baseball bat's rotational inertia greater when rotated about an axis perpendicular to one end or when rotated about its axis of symmetry?
7. A wheel rolls without slipping, with center-of-mass speed v_{cm}. What's the instantaneous velocity of the bottom of the wheel (in contact with the ground)? What's the instantaneous velocity of the top of the wheel?
8. Give an example of an object that's in translational equilibrium but not rotational equilibrium.
9. Earth's core is denser than the near-surface layers. Should its rotational inertia be more or less than $\frac{2}{5}MR^2$? Explain.
10. A figure skater spins with arms extended horizontally. Explain what happens, and why, when she brings her arms tight to her body.
11. A spinning top rotates with its point on the floor and its rotation axis slightly tilted. Describe its subsequent motion. Why doesn't it fall over?
12. Why is it easier to balance a basketball on your fingertip if it's spinning?
13. A wheel of mass M and radius R has rotational inertia $I = \frac{9}{10}MR^2$. Is it more like a solid disk, or more like a bicycle wheel with most of the mass at the rim?
14. A common demonstration involves two unopened soup cans with the same dimensions and mass. One soup is thick (e.g., cream of mushroom), the other a watery liquid (e.g., chicken broth). The two are released simultaneously from rest on an incline. Explain why the chicken broth wins the race to the bottom.

Multiple-Choice Problems

15. Earth makes one rotation in 24 hours. What's its angular velocity? (a) 1.16×10^{-5} rad/s; (b) 0.042 rad/s; (c) 1.39×10^{-5} rad/s; (d) 7.27×10^{-5} rad/s.
16. To stop a wheel rotating at 6.0 rev/s in 9.0 s requires average angular acceleration (a) -0.67 rad/s^2; (b) -2.1 rad/s^2; (c) -4.2 rad/s^2; (d) -67 rad/s^2.
17. A potter's wheel starting with angular velocity 2.4 rad/s accelerates in 2.0 s to 3.6 rad/s, with constant angular acceleration. During this time the wheel turns through an angle of (a) 6.0 rad; (b) 12.0 rad; (c) 0.95 rad; (d) 4.8 rad.
18. A compact disk with radius 6.0 cm takes 2.5 s to accelerate from rest to 40 rad/s. What's the tangential acceleration of a point on the CD's edge? (a) 0.49 m/s^2; (b) 0.96 m/s^2; (c) 0.22 m/s^2; (d) 6.0 m/s^2.

19. A wheel rotates with constant angular acceleration. Which of the following is constant? (a) Angular velocity; (b) tangential velocity; (c) tangential acceleration; (d) centripetal acceleration.
20. A disk with radius 1.5 m and rotational inertia 34 kg · m^2 has a 160-N force applied tangentially to its rim. If the disk starts from rest, the angular velocity at the end of 2.0 s is (a) 1.22 rad/s; (b) 6.27 rad/s; (c) 7.06 rad/s; (d) 14.1 rad/s; (e) 25.5 rad/s.
21. A 46-g golf ball has radius 2.13 cm. Assuming uniform density, what's the ball's rotational inertia? (a) 8.3×10^{-6} kg · m^2; (b) 2.1×10^{-5} kg · m^2; (c) 1.3×10^{-5} kg · m^2; (d) 4.3×10^{-6} kg · m^2.
22. What's the rotational kinetic energy of the golf ball in the preceding question when it's rotating at 80 Hz? (a) 0.5 J; (b) 1.0 J; (c) 2.0 J; (d) 4.0 J.
23. A bicycle with wheels 69 cm in diameter is moving at 40 km/h. If the wheels roll without slipping, their angular velocity is (a) 4 rad/s; (b) 9 rad/s; (c) 16 rad/s; (d) 32 rad/s.
24. A solid cylinder has mass 0.55 kg and radius 3.5 cm. It's rolling with center-of-mass speed 0.75 m/s. What's its total kinetic energy? (a) 0.07 J; (b) 0.16 J; (c) 0.23 J; (d) 0.30 J.
25. A heavy machine wheel has a rotational inertia 25 kg · m^2 and radius 0.75 m. It's initially at rest, and a tangential force of 35 N is applied at its edge for 5.0 s. What's the resulting angular velocity? (a) 0.86 rad/s; (b) 1.7 rad/s; (c) 5.3 rad/s; (d) 10.6 rad/s.
26. The Moon (mass 7.35×10^{22} kg) takes 27.3 days to complete an essentially circular orbit of radius 3.84×10^8 m. Estimate the magnitude of the Moon's orbital angular momentum. (a) 7.33×10^{25} kg · m^2/s; (b) 2.81×10^{34} kg · m^2/s; (c) 7.33×10^{41} kg · m^2/s; (d) 7.33×10^{51} kg · m^2/s.
27. In a car's drive train, a flywheel with rotational inertia 26.0 kg · m^2 rotates at 310 rad/s. The clutch engages, pressing the clutch plate—a disk with rotational inertia half that of the flywheel—against the flywheel, so the two rotate as one. Assuming both are otherwise free from torque, what's the rotational speed of the combined system? (a) 155 rad/s; (b) 206 rad/s; (c) 267 rad/s; (d) 310 rad/s.
28. A rotating wheel may have (a) both centripetal and tangential acceleration; (b) neither centripetal nor tangential acceleration; (c) angular but not centripetal acceleration; (d) neither angular nor centripetal acceleration.
29. A wheel has its rotation axis vertical, and turns counterclockwise as viewed from above. The direction of the wheel's angular momentum is (a) straight up; (b) straight down; (c) tangent to the wheel, in the direction of rotation; (d) tangent to the wheel, opposite the direction of rotation.

Problems

Section 8.1 Rotational Kinematics

30. ■ A bicycle wheel with radius 0.79 m is in pure rotation (not rolling) at 4.0 rev/s. What distance does a point on the wheel's rim travel in 1 minute?
31. ■ Jupiter has radius 7.14×10^7 m and makes one rotation every 9 hours, 50 minutes. How far does a point on Jupiter's equator travel each second, due to the planet's rotation?
32. ■ A hydroelectric turbine makes 24,500 revolutions in one day. What's its angular velocity?

33. BIO ■ ■ **Bacterium rotation rate.** A typical lab centrifuge spins at 3000 rpm. How's that compare with the *E. coli* bacterium's flagellum described in Section 8.1?

34. ■ A compact disk spins with angular velocity 43.8 rad/s. The player is turned off, and 2.45 s later the CD has stopped. Find its average angular acceleration.

35. ■ A lawnmower blade accelerates at 98 rad/s². Starting from rest, what's its angular velocity after 2.5 s have elapsed? Answer in rad/s and rpm.

36. ■ ■ A diver jumps from a 10-m-high tower, and hopes to complete $3\frac{1}{2}$ somersaults. What should be his rotation rate?

37. ■ ■ ■ Tides dissipate energy, slowing Earth's rotation. Some 4 billion years ago, Earth's rotation period is estimated to have been 14 hours. Find Earth's average angular acceleration over this 4-billion-year period.

Section 8.2 Kinematic Equations for Rotational Motion

38. ■ A skater spins at 2.50 revolutions per second. She slows with constant angular acceleration, stopping after 1.75 s. (a) What's her angular acceleration? (b) Through how many revolutions did she turn before stopping?

39. ■ ■ A dentist's drill accelerates from rest at 615 rad/s² for 2.10 s and then runs at constant angular velocity for 7.50 s. Through how many total revolutions has the drill turned?

40. ■ Your car's fan belt turns a pulley at 3.40 rev/s. When you step on the gas for 1.30 s, the rate increases steadily to 5.50 rev/s. (a) What's the pulley's angular acceleration? (b) Through what angle did the pulley turn while accelerating?

41. ■ ■ A centrifuge rotating initially at 9000 rpm slows to 5000 rpm with constant angular acceleration over 3.50 s. (a) What's its angular acceleration? (b) Through how many revolutions does it turn while decelerating? (c) Through what distance does a point on the edge of the centrifuge, at radius of 9.40 cm, turn during this time?

42. ■ ■ A machine shop grinding wheel accelerates from rest with a constant angular acceleration of 2.3 rad/s² for 7.5 s and is then brought to rest with a constant angular acceleration of −4.2 rad/s². Find the total time elapsed and the total number of revolutions turned.

43. ■ ■ ■ Earth's rotation rate is slowing due to tidal forces, with the length of the day increasing by about 2.3 ms/century. Find Earth's angular acceleration.

Section 8.3 Rotational and Tangential Motion

44. ■ A 1.75-m-diameter wagon wheel makes one revolution in 3.20 s. What's the tangential speed of a point in its rim?

45. ■ A tornado has wind speed 325 km/h at a rotation radius of 18 m. What is the angular velocity at this point in the tornado?

46. ■ ■ A lab centrifuge with radius 11 cm turns at 75 rev/s. What should be its angular acceleration for a point at the 11-cm radius to have tangential acceleration that is 1% of its centripetal acceleration?

47. ■ ■ A string is wrapped around a pulley of radius 3.50 cm, and a weight hangs from the other end. The weight falls with a constant acceleration 3.40 m/s². (a) What's the angular acceleration of the pulley? (b) If the weight starts from rest 1.30 m above the floor, what's the pulley's angular velocity when the weight hits the floor?

48. ■ ■ A cylindrical space station with diameter 150 m simulates gravity by rotating about its central axis. (a) If an astronaut on the outer edge is to experience a centripetal acceleration $g/2$, what should be the station's angular velocity? (b) What tangential acceleration is required to bring the station to that rate, starting from rest, with a constant acceleration for 60 days?

49. BIO ■ ■ **Human centripetal and angular acceleration.** To simulate the extreme accelerations during launch, astronauts train in a large centrifuge with diameter 10.5 m. (a) If the centrifuge is spinning so the astronaut on the end of one arm is subjected to a centripetal acceleration of 5.5g, what is the astronaut's tangential velocity at that point? (b) Find the angular acceleration needed to reach the velocity in part (a) after 25 s.

50. ■ ■ ■ Early DVD burners operated in the same constant-tangential-speed mode as described for CDs in Example 8.5. For a 6× burn (i.e., six times the normal playback data rate), the rotational speed was highest (8400 rpm) at the innermost data track, 2.6 cm from the rotation axis. (a) What's the corresponding rotation rate near the outer edge, at 5.7 cm from the axis? (b) If the burn takes 9.0 min, what's the DVD's average angular acceleration? (Newer DVD burners use so-called zoned linear velocities, or even constant angular velocity, to avoid very high rotation rates.)

51. ■ ■ ■ Information on a DVD (see preceding problem) is stored on a continuous spiral track in the region from 2.6-cm to 5.7-cm radius. Individual turns of the spiral are 0.74 μm apart. (a) Find the length of the entire track. (b) If one byte of information has average length 2.3 μm, how many bytes are on the DVD?

52. ■ ■ For the situation described in Example 8.6, use the tangential and centripetal components to find the magnitude and direction of the acceleration vector.

53. BIO ■ ■ **Eagle wings.** An eagle with a 2.1-m wingspan flaps its wings back and forth 20 times per minute, each stroke extending from 45° above the horizontal to 45° below. Downward and upward strokes take the same amount of time. On a given downstroke, what's (a) the average angular velocity of the wing and (b) the average tangential velocity of the wingtip?

Section 8.4 Kinetic Energy and Rotational Inertia

54. ■ Using astronomical data from Appendix E, compute Earth's rotational inertia, assuming the planet is a uniform solid ball.

55. ■ ■ Use your answer to the preceding problem to find Earth's rotational kinetic energy.

56. ■ ■ Find the kinetic energy of Earth's orbital motion around the Sun (see Appendix E). Compare your answer with the rotational kinetic energy found in the preceding problem.

57. ■ ■ The circular blade of a power saw has kinetic energy 44 J. If its rotation rate drops to half, what's its new kinetic energy?

58. ■ ■ A revolving door consists of four rectangular glass slabs, with the long end of each attached to a pole that acts as the rotation axis. Each slab is 2.20 m tall by 1.25 m wide and has mass 35.0 kg. (a) Find the rotational inertia of the entire door. (b) If it's rotating at one revolution every 9.0 s, what's the door's kinetic energy?

59. ■ ■ A 145-g baseball has radius 3.7 cm. (a) Assuming uniform density, what's its rotational inertia? (b) The ball is pitched at 22 m/s with spin rate 20 Hz. Find and compare the ball's translational and rotational kinetic energies.

60. ■ ■ A wheel consists of 20 thin spokes, each with mass 0.055 kg, attached to a rim of mass 4.2 kg and radius 0.75 m. Find its rotational inertia.

Section 8.5 Rolling Bodies

61. ■ A bicycle has 69-cm-diameter wheels. If they roll without slipping when the bicycle is traveling at 50 km/h, what's the wheels' angular velocity?

62. ■ ■ A bicycle has 63.5-cm-diameter wheels turning at 11.4 rev/s. How fast is the bicycle moving?

63. ■ ■ The condition for a wheel to roll without slipping is described in the text by $\omega = v_{cm}/R$. Describe what's happening with car tires in the cases (a) $\omega > v_{cm}/r$ and (b) $\omega < v_{cm}/r$.

64. ■ ■ Based on the preceding problem, consider a 32.5-cm-radius automobile tire on a car moving at 10.4 m/s. Describe the motion of the bottom of the tire relative to the road for each of the following angular velocities: (a) $\omega = 25.0$ rad/s; (b) $\omega = 32.0$ rad/s; (c) $\omega = 38.7$ rad/s.

65. ■ A drag racer has 76.2-cm-diameter wheels. Haw fast are they turning when the drag racer is doing 140 km/h?

66. ■ ■ A solid bowling ball with mass 7.2 kg and diameter 22 cm rolls without slipping at 6.5 m/s. Find its translational kinetic energy, rotational kinetic energy, and total kinetic energy.

67. ■ ■ A solid cylinder is released from rest on an incline. When it reaches the bottom, what fraction of its total kinetic energy is translational and what fraction is rotational?

68. ■ ■ Repeat the preceding problem for a solid sphere. Why the difference?

69. ■ ■ A meterstick pivots freely from one end. If it's released from a horizontal position, find its angular velocity when it passes through the vertical. Treat the stick as a uniform thin rod.

70. ■ ■ A 1.00-m-long ramp is inclined at 15° to the horizontal. A solid ball is released from rest at the top of the ramp. Find (a) the ball's speed at the bottom of the ramp and (b) its translational acceleration.

71. ■ ■ ■ Using the work-energy analysis in the text, along with kinematic equations for one-dimensional motion, find the translational acceleration of a solid ball rolling down a ramp inclined at angle θ, expressed in terms of θ and g. Compare with $a = g \sin\theta$, the acceleration of an object sliding without friction down the same ramp.

72. ■ ■ ■ A solid ball is released from rest at the top of a 1.50-m-long ramp inclined at 10°. At the bottom, the ball continues along a flat section that's also 1.50 m long. What's the overall travel time?

73. ■ ■ ■ Consider the "great rolling race" from Example 8.11. Suppose the ball travels 1.00 m along the ramp from top to bottom. (a) How far does the cylinder travel in the same amount of time? (b) Does the answer to part (a) depend on the inclination angle? Explain.

Section 8.6 Rotational Dynamics

74. ■ ■ (a) If you push the outer edge of a 1.05-m-wide door with a 23.0-N tangential force, what torque results? (b) What's the torque if you apply the same magnitude of force, in the same place, but at a 45° angle?

75. ■ An auto mechanic applies a 65-N force near the end of a 35-cm-long wrench. What's the maximum torque?

76. ■ ■ An electric trimmer blade has rotational inertia 1.70×10^{-4} kg·m² (a) What torque is needed to accelerate this blade from rest to 640 rpm in 1.50 s? (b) How much work was done to accelerate the blade?

77. BIO ■ ■ **Bacterial torque.** The cellular motor driving the flagellum in the *E. coli* bacterium exerts a torque of typically 400 pN·nm on the flagellum (see discussion in Section 8.1). If this torque results from a force applied tangentially to the outside of the 12-nm-radius flagellum, what is the magnitude of that force?

78. ■ ■ A meterstick pivots freely from one end. If it's released from a horizontal position and rotates due to gravity, find (a) its angular acceleration just after it's released and (b) the stick's angular acceleration as a function of the angle θ it makes with the vertical. Treat the stick as a uniform thin rod.

79. ■ ■ Two children, with masses 35 kg and 40 kg, sit at opposite ends of a 3.4-m-long seesaw with mass 25 kg, with the fulcrum at its midpoint. With the seesaw horizontal, find (a) the net torque on the seesaw and (b) its angular acceleration.

Section 8.7 Mechanical Equilibrium

80. ■ ■ A meterstick is initially balanced on a fulcrum at its midpoint. You have four identical masses. Three of them are placed atop the meterstick at the following locations: 25 cm, 45 cm, and 95 cm. Where should the fourth mass be placed in order to balance the meterstick?

81. ■ ■ A meterstick of negligible mass has a 0.20-kg mass at its 35 cm mark and a 0.40-kg mass at the 75-cm mark. Where should the fulcrum be so the meterstick is balanced?

82. ■ ■ Repeat the preceding problem if the meterstick has mass 0.15 kg.

83. ■ ■ ■ Consider again the ladder in Example 8.15. Compute the normal force again, assuming a 75-kg man is standing on the ladder (a) at its midpoint and (b) four-fifths of the way up the ladder.

84. ■ ■ ■ A 15-m-long ladder is mounted on a fire truck. The ladder itself has mass 125 kg, and at the top is a 35-kg basket holding a 91-kg firefighter. If the ladder makes a 60° angle with the horizontal, what's the net torque about the ladder's base?

85. ■ ■ In Figure P8.85, the meterstick's mass is 0.160 kg and the string tension is 2.50 N. The system is in equilibrium. Find (a) the unknown mass m and (b) the upward force the fulcrum exerts on the stick.

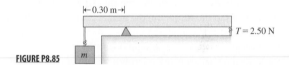

FIGURE P8.85

Section 8.8 Angular Momentum

86. ■ Use data from Appendix E to compute Earth's angular momentum due to its rotation alone.

87. ■ Use data from Appendix E to compute Earth's angular momentum due to its orbital motion, and compare with your answer to the preceding problem.

88. ■ ■ A student sitting on a frictionless rotating stool has rotational inertia 0.95 kg·m² about a vertical axis through her center of mass when her arms are tight to her chest. The stool rotates at 6.80 rad/s and has negligible mass. The student extends her arms until her hands, each holding a 5.0-kg mass, are 0.75 m from the rotation axis. (a) Ignoring her arm mass, what's her new rotational velocity? (b) Repeat if each arm is modeled as a 0.75-m-long uniform rod of mass of 5.0 kg and her total body mass is 65 kg.

89. ■ ■ A turntable with a rotational inertia 0.225 kg·m² is rotating at 3.25 rad/s. Suddenly, a disk with rotational inertia 0.104 kg·m² is dropped onto the turntable with its center on the rotation axis. Assuming no outside forces act, what's the common rotational velocity of the turntable and disk?

90. ■ ■ A grinding wheel has rotational inertia 0.355 kg·m². (a) Find the constant torque needed to bring it from rest to 45.0 rad/s in 3.50 s. (b) Using your answer to part (a), find the wheel's angular momentum change, and show that your answer agrees with the angular momentum computed using $L = I\omega$.

Section 8.9 Rotational Motion with Vector Quantities

91. ■ ■ A merry-go-round with rotational inertia 35 kg·m² rotates clockwise at 1.3 rad/s. Find the magnitude and direction of (a) the merry-go-round's angular momentum and (b) the torque needed to stop the merry-go-round in 10 s.

92. ■ A car drives straight north. What's the direction of its wheels' angular momentum?

93. ■ A wrench handle points straight up, in the $+y$-direction. A mechanic applies a force in the $+x$-direction to the top end of the wrench. What's the direction of the torque on the wrench?

94. ■ ■ In a simple model of the hydrogen atom, an electron orbits a proton at 2.18×10^6 m/s in a circle of radius 5.29×10^{-11} m. Find the magnitude and direction of the electron's angular momentum.

95. ■ ■ Earth's rotation rate is slowing. What's the direction of the torque needed to cause this? Use data from Problem 43 to estimate the torque's magnitude.

General Problems

96. ■ ■ An 83.2-kg propeller blade measures 2.24 m end to end. Model the blade as a thin rod rotating about its center of mass. It's initially turning at 175 rpm. Find (a) the blade's angular momentum, (b) the tangential speed at the blade tip, and (c) the angular acceleration and torque required to stop the blade in 12.0 s.

97. ■ ■ A string is wrapped around the outer rim of a cylindrical disk with mass 500 g and radius 12.5 cm. A student pulls on the string, applying a 23.5-N force tangentially to the cylinder's rim. Find (a) the torque on the cylinder and (b) its angular acceleration, assuming no frictional torque.

98. **BIO** ■ ■ **Torque on arm.** A shot putter holds the 7.26-kg shot still with his arm extended straight, the shot 61.8 cm from his shoulder joint. Find the torque on the athlete's arm due to the shot if the arm (a) is horizontal, (b) makes a 45° angle below the horizontal, and (c) is hanging straight down.

99. ■ ■ Recent advances using microprocessors have caused significant improvements in the safety of power tools such as circular saws. The saw's control unit places a small electric charge on the spinning blade, giving a constant 3-volt signal to the microprocessor. If the blade contacts human skin, the capacitance changes. The microprocessor senses this change and stops the blade in just 5.0 ms. (Capacitance will be discussed in Chapter 16.) Assume constant angular acceleration. (a) If the blade normally spins at 3500 rpm, through what angle does it turn while stopping? (b) What's the change in the blade's kinetic energy, assuming it's essentially a uniform disk 19.0 cm in diameter with mass 0.860 kg? (c) What torque is required to stop the blade?

100. ■ ■ Your car tire has radius 31.0 cm. (a) If it rolls without slipping and has angular velocity 79.3 rad/s, what's your car's speed? Now suppose your car has the speed found in part (a), but different angular velocity. Describe what's happening to the tire where it contacts the road when (b) $\omega = 91.5$ rad/s and (c) $\omega = 52.0$ rad/s.

101. ■ ■ ■ The following objects are released simultaneously from rest at the top of a 1.50-m-long ramp inclined at 3.50° to the horizontal: a solid sphere, a solid cylinder, a hollow cylindrical shell, and a hollow ball. (a) Which wins the race? (b) At the moment the winner reaches the bottom, find the positions of the other three objects.

102. ■ ■ ■ A solid cylinder is released from rest and rolls without slipping down a 4.0° incline. Two photogates are connected to a timer that measures the elapsed time for the ball to roll between them. If the first gate is 1.00 m from the starting point and the second is 0.20 m past the first, what will the timer read?

103. **BIO** ■ ■ **Torque in throwing a ball.** A baseball player extends his arm straight upward to catch a 0.145-kg batted baseball moving horizontally at 42.5 m/s. It's 63.5 cm from the player's shoulder joint to the point where the ball strikes his hand, and his arm remains stiff while it rotates about the shoulder joint during the catch. The player's hand recoils horizontally a distance of 5.00 cm while he stops the ball with constant acceleration. What torque does the player's arm exert on the ball?

104. ■ ■ ■ In Chapter 7 you studied the *simple pendulum*, in which the mass of the string holding the bob was neglected. Now consider the *physical pendulum*, with mass distributed throughout its length. An example is the meterstick in Figure GP8.104. (a) With gravity acting on the stick's center of mass, find the

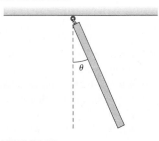

FIGURE GP8.104

torque on the stick as a function of its mass M, length L, the angle θ, and g. (b) Use $\tau = I\alpha$ to find the stick's angular acceleration as a function of the same variables. (c) By analogy with the simple pendulum, show that the period for small oscillations of this physical pendulum is approximately

$$T = 2\pi\sqrt{\frac{2L}{3g}}$$

105. ■ ■ (a) Use the results of the preceding problem to evaluate the period of small oscillations of a 1.00-m long stick. (b) Find the length of a simple pendulum that would have the same period as you found in part (a). Explain why the length is shorter than 1.00 m.

106. ■ ■ Consider your arm to be a uniform rod pivoted about one end. (a) Estimate the period of small oscillations if your 75-cm-long arm hangs freely downward from your shoulder. (b) Should your leg (hanging freely from your hip) have a larger, a smaller, or the same period compared with the period you estimated in part (a)? Explain.

107. ■ ■ ■ Consider a physical pendulum as described in the preceding problems, but not necessarily a uniform stick. Suppose this physical pendulum has one end fixed, with a rotational inertia I about that end and center of mass a distance d from that end. (a) Show that the period for small oscillations is given by

$$T = 2\pi\sqrt{\frac{d}{g}}\sqrt{\frac{I}{Md^2}}$$

(b) Show that this general result reduces to the correct one for the uniform stick and the simple pendulum.

108. ■ ■ ■ If the polar ice caps melt, adding more liquid water to the oceans, Earth's rotational inertia could increase by as much as an estimated 0.3%. Compute the effect such a change would have on the length of 1 day, assuming Earth's angular momentum remains constant.

Answers to Chapter Questions

Answer to Chapter-Opening Question

The rotation axis precesses—changes orientation—over a 26,000-year cycle, due to the torque from the Sun's gravity acting on Earth's equatorial bulge. This alters the relation between sunlight intensity and seasons, triggering ice ages.

Answers to GOT IT? Questions

Section 8.1   (c) ω is negative and α is positive
Section 8.3   (c)
Section 8.5   (d) Solid ball, (b) solid cylinder, (a) hollow ball,
            (c) hollow hoop

9 Gravitation

■ How does this spiral galaxy reveal the existence of unseen "dark matter"?

Gravitation is the fundamental force governing the universe on the large scale, and our understanding of gravity enables space technologies ranging from communications satellites to the Global Positioning System (GPS) to planetary exploration. The study of gravitation brings together key ideas in physics: force, energy, circular motion, and angular momentum.

For nearly all practical applications, gravity is described accurately by **Newton's law of gravitation**, developed in the 17th century. Newton's law is a spectacular achievement, explaining the motions of the planets and their satellites; comets, asteroids, and other astronomical bodies; and effects such as tides. Only in extreme astrophysical situations—such as the neighborhoods of black holes—or when exquisite accuracy is needed—as in the Global Positioning System—is Newton's theory superseded by Einstein's general theory of relativity.

9.1 Newton's Law of Gravitation

Background and History

Our emerging understanding of gravity begins with ancient peoples' interest in the motions of heavenly bodies—phenomena that had both religious and practical significance. Stonehenge is perhaps the best-known ancient structure erected based on astronomical observations, but there were many others throughout the world. Astronomy also enabled reliable calendars that were crucial to agriculture.

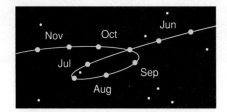

FIGURE 9.1 Retrograde motion of Mars in 2003 as viewed from Earth. In this view looking south, the left of the frame is east and the right is west.

Ancient Greek model for retrograde motion:

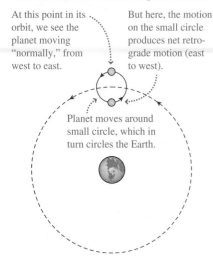

At this point in its orbit, we see the planet moving "normally," from west to east.

But here, the motion on the small circle produces net retrograde motion (east to west).

Planet moves around small circle, which in turn circles the Earth.

FIGURE 9.2 The ancient Greek Earth-centered model.

Ancient astronomers saw the Sun, Moon, planets, and stars circling Earth daily, east to west. The stars remain in fixed patterns, but the others move slightly each day relative to the stars, generally west to east. The Sun moves about 1 degree eastward per day, returning to its starting position in a year; the Moon moves 12 degrees per day, completing its cycle roughly each month. Planets are more complicated: Occasionally, but regularly, they exhibit **retrograde motion**, moving east to west for weeks or months (Figure 9.1).

The ancient Greeks developed an elaborate model for predicting planetary motions. Earth was at the center of their universe, with heavenly bodies describing circles about Earth. This seems sensible: We don't feel we're moving, and heavenly bodies appear to circle above us. Retrograde motion was explained by the planets moving in small circles attached to larger circles that carried them around Earth (Figure 9.2). Second-century CE astronomer Claudius Ptolemy fine-tuned this model so its predictions were quite accurate.

For centuries Ptolemy's work was the standard in planetary astronomy. This was due in part to the theory's mathematical accuracy and in part to medieval European thinkers' reluctance to question the Earth-centered universe that was part of their larger worldview.

Change began in 1543, when Polish astronomer Nicholas Copernicus (1473–1543) published his heliocentric theory, with Earth and the other planets orbiting the fixed Sun. Copernicus's model shared with Ptolemy's the use of nested circles to approximate planetary motions. In fact, Copernicus's system is mathematically equivalent to Ptolemy's, with a change of coordinate system. But it wasn't widely accepted at first, because 16th-century thinkers held to their Earth-centered views.

Early in the 17th century, Galileo became the first to use a telescope in astronomy. His observations of Jupiter's moons, the phases of Venus, and sunspots supported arguments in favor of a Sun-centered universe. Galileo was widely praised in the scientific community but censured by the Catholic Church in his native Italy. He was forced to recant his views publicly and placed under house arrest for his final decade. Not until 1992 was Galileo formally exonerated.

Galileo's contemporary, the German astronomer and mathematician Johannes Kepler (1571–1630), worked with Danish astronomer Tycho Brahe (1546–1601), who had earlier made precise measurements of planetary motion. Kepler used Brahe's data to show that the planets' orbits are better fit by ellipses than the combinations of circles in the Copernican model. Later in this chapter we'll describe Kepler's work in detail.

Thanks to Copernicus, Galileo, Kepler, and others, the view that the planets move in ellipses around a fixed Sun gained wide acceptance by the mid-17th century. But what causes these elliptical orbits? Scientists prior to Newton had no satisfactory answer. Newton succeeded because he had a clear understanding of dynamics (Chapter 4) and because he developed calculus to solve the problem of planetary motion.

Newton's Theory

It's legendary that Newton saw a falling apple and realized that the same force that pulls the apple down also holds the Moon in its orbit. Whether true or not, this legend encapsulates Newton's vital realization that the force of gravity familiar on Earth is also responsible for orbital motions of the Moon around Earth and planets around the Sun.

Newton used his second law, $\vec{F}_{net} = m\vec{a}$, to compare the falling apple's acceleration with that of the orbiting Moon. The kinematics of circular motion (Chapter 3) can be applied to the Moon's nearly circular orbit. Newton knew that comparing these two accelerations could tell him something about the force responsible for both.

| EXAMPLE 9.1 | **Apple and Moon** |

Follow in Newton's footsteps by computing the centripetal acceleration of the orbiting Moon and comparing with the acceleration of the falling apple, $g = 9.80 \text{ m/s}^2$. Assume the Moon's orbit is circular with radius $R = 3.84 \times 10^8$ m (known fairly well in Newton's day) and period $T = 27.3$ days.

ORGANIZE AND PLAN From rotational kinematics, centripetal acceleration is $a_r = v^2/r$. The Moon's orbital speed is distance/time, with distance the circumference $2\pi R$ and time the period T. For SI units, the period should be in seconds.

Known: $g = 9.80 \text{ m/s}^2$; $R = 3.84 \times 10^8$ m; $T = 27.3$ days.

cont'd.

SOLVE Converting the period to seconds,

$$T = 27.3 \text{ days} \times \frac{86,400 \text{ s}}{1 \text{ day}} = 2.36 \times 10^6 \text{ s}$$

Then the Moon's centripetal acceleration is

$$a_r = \frac{v^2}{R} = \frac{(2\pi R/T)^2}{R} = \frac{4\pi^2 R}{T^2}$$

$$= \frac{4\pi^2(3.84 \times 10^8 \text{ m})}{(2.36 \times 10^6 \text{ s})^2} = 2.72 \times 10^{-3} \text{ m/s}^2$$

Comparing with g is straightforward:

$$\frac{g}{a_r} = \frac{9.80 \text{ m/s}^2}{2.72 \times 10^{-3} \text{ m/s}^2} = 3600$$

REFLECT The apple's acceleration is 3600 times greater than the Moon's, and both are directed toward Earth. Clearly, gravity exerts a stronger force on objects closer to Earth.

MAKING THE CONNECTION What role do the masses of the apple and Moon play in the factor 3600? What if the Moon's mass were halved?

ANSWER Masses don't matter. You know that near Earth all objects fall at $g = 9.80 \text{ m/s}^2$, regardless of mass. Similarly, the Moon's mass shouldn't affect its acceleration. Newton's theory has to reflect these facts.

Comparing the accelerations of the apple and Moon told Newton that the force of gravity weakens with increasing distance. Newton reasoned that Earth's center should provide the starting point for distances, because a symmetric body like Earth should attract another body as if Earth's entire mass were concentrated at its center. Thus, the "distance to Earth" for the falling apple is roughly Earth's radius, $R_E = 6.37 \times 10^6$ m. The Moon's distance is just its orbital radius, $R = 3.84 \times 10^8$ m. The distance ratio is

$$\frac{R}{R_E} = \frac{3.84 \times 10^8 \text{ m}}{6.37 \times 10^6 \text{ m}} = 60$$

The apple's acceleration is greater than the Moon's by a factor of $3600 = 60^2$. Newton concluded from this that *the gravitational force varies as the inverse square of the distance*, or

$$F \propto \frac{1}{r^2}$$

From his second law of motion $F_{net} = ma$, Newton realized that the gravitational force should also be proportional to mass. Newton's third law suggests that when two bodies interact gravitationally, they should attract with forces of equal magnitude, and therefore *the gravitational force is proportional to the mass of each of the two objects*:

$$F \propto \frac{m_1 m_2}{r^2}$$

To make this an equation, we need a proportionality constant. That's G, the **universal gravitation constant**, with experimentally determined value $G = 6.67 \times 10^{-11} \text{ N} \cdot \text{m}^2/\text{kg}^2$. Thus Newton's law says that the gravitational force between two masses m_1 and m_2 separated by a distance r is

$$F = \frac{Gm_1 m_2}{r^2} \quad \text{(Newton's law of gravitation; SI unit: N)} \quad (9.1)$$

Although Newton arrived at Equation 9.1 by considering Earth and Moon, he proposed that his law is truly *universal*, implying an attractive force between any two masses in the universe. Strictly speaking, Equation 9.1 applies only to point particles, with r the distance between them. However, Newton used calculus to show that the law is also exact for symmetric spheres, where r is the center-to-center distance (Figure 9.3). This makes Equation 9.1 applicable to many astronomical bodies, including planets and stars.

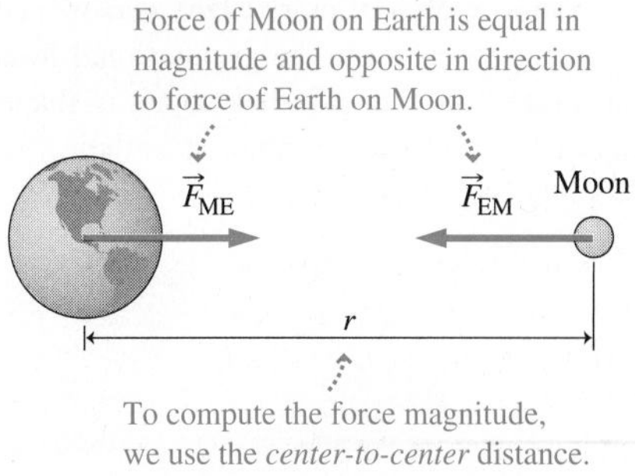

✓ **TIP**

For spherical bodies, use the center-to-center distance to find the gravitational force.

Force of Moon on Earth is equal in magnitude and opposite in direction to force of Earth on Moon.

$\vec{F}_{ME}$ $\vec{F}_{EM}$ Moon

r

To compute the force magnitude, we use the *center-to-center* distance.

FIGURE 9.3 Forces in the Earth-Moon system. Note: drawing is not to scale.

Force is a vector, with direction as well as magnitude. The direction of the gravitational force is always attractive, acting from one body toward the other in a gravitating pair (Figure 9.3). This satisfies Newton's third law, which says that the forces on two interacting objects must have the same magnitude and opposite directions.

EXAMPLE 9.2 **Gravitational Acceleration**

Take Earth to be a sphere with mass $M_E = 5.98 \times 10^{24}$ kg and radius $R_E = 6.37 \times 10^6$ m. Use Newton's law of gravitation to compute the gravitational force on a 0.10-kg apple in free fall near Earth's surface. From this, find the apple's acceleration, assuming that no forces other than gravity are present.

ORGANIZE AND PLAN Newton's law of gravitation, Equation 9.1, gives the gravitational force F. For a near-Earth object, the appropriate distance is Earth's radius R_E (Figure 9.4). Then by Newton's second law of motion, the apple's acceleration is $a = F/m$.

Known: $M_E = 5.98 \times 10^{24}$ kg; $R_E = 6.37 \times 10^6$ m; apple mass $m = 0.10$ kg.

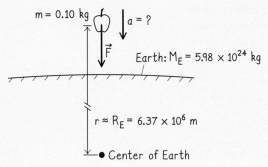

FIGURE 9.4 Apple falling to Earth.

SOLVE Equation 9.1 gives the force:

$$F = \frac{Gm_1 m_2}{R_E^2}$$

$$= \frac{(6.67 \times 10^{-11} \text{ N} \cdot \text{m}^2/\text{kg}^2)(5.98 \times 10^{24} \text{ kg})(0.10 \text{ kg})}{(6.37 \times 10^6 \text{ m})^2}$$

$$= 0.98 \text{ N}$$

Then the apple's acceleration is

$$a = \frac{F}{m} = \frac{0.98 \text{ N}}{0.10 \text{ kg}} = 9.8 \text{ m/s}^2$$

REFLECT As expected, this is the familiar $g = 9.8$ m/s². Also, the gravitational force—the apple's weight—agrees with the familiar $w = mg$, here 0.98 N. That's just about 1 newton. How appropriate!

MAKING THE CONNECTION While it's falling, what gravitational force does the apple exert on Earth, and what's Earth's resulting acceleration?

ANSWER By Newton's third law, the force on Earth has the same magnitude, 0.98 N, directed toward the apple. Earth's acceleration is this force divided by M_E, or about 10^{-25} m/s². That's why you don't see Earth rushing upward to meet the falling apple.

CONCEPTUAL EXAMPLE 9.3 **Nonspherical Bodies**

Newton's law is exact only for point particles or uniform spheres. Why does it work so well for the nonspherical apple, or for a very nonspherical person?

SOLVE It's the relative size that matters. An apple or a person is extremely small compared with Earth's radius, the distance used in the force computation. So to a good approximation these objects behave like point particles.

REFLECT There are cases where the point-particle approximation doesn't work. Consider an astronaut standing on an irregular asteroid 20 m long (Figure 9.5). You might be inclined to measure from the asteroid's center of mass to the person's center of mass and use that distance to compute the gravitational force. That wouldn't be very accurate, however. Neither object is shaped anything like a sphere, and they're too close, relative to their sizes, to be considered point

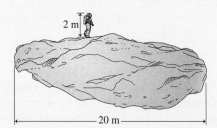

FIGURE 9.5 Equation 9.1 doesn't accurately give the force between astronaut and asteroid, because their sizes aren't small compared with the distance between their centers of mass.

particles. If the astronaut were, say, 1 km from the asteroid, then the point-particle approximation would be good.

Acceleration in Free Fall

The results of Example 9.2 show why near-Earth objects fall with acceleration $g = 9.8$ m/s² regardless of their mass. The gravitational force F of Equation 9.1 is just the weight, which near Earth is mg. Using Newton's law (Equation 9.1), with $m_1 = m$ (the falling body's mass), $m_2 = M_E$ (Earth), and $r = R_E$,

$$F = mg = \frac{GmM_E}{R_E^2}$$

The mass m cancels, showing that free-fall acceleration doesn't depend on mass. That leaves a general expression for g:

$$g = \frac{GM_E}{R_E^2} \quad \text{(Gravitational acceleration } g\text{)} \quad (9.2)$$

The gravitational acceleration depends only on Earth's mass and radius and the universal constant G. As a final check,

$$g = \frac{(6.67 \times 10^{-11} \text{ N} \cdot \text{m}^2/\text{kg}^2)(5.98 \times 10^{24} \text{ kg})}{(6.37 \times 10^6 \text{ m})^2} = 9.8 \text{ m/s}^2$$

as expected. Thus Newton's law of gravitation predicts the free-fall acceleration that Galileo first measured.

The expression for g in Equation 9.2 gives the gravitational acceleration on any other spherically symmetric body, such as the Moon. Using the Moon's mass and radius from Appendix E, the gravitational acceleration at the Moon's surface is

$$g_{\text{Moon}} = \frac{(6.67 \times 10^{-11} \text{ N} \cdot \text{m}^2/\text{kg}^2)(7.35 \times 10^{22} \text{ kg})}{(1.74 \times 10^6 \text{ m})^2} = 1.6 \text{ m/s}^2$$

about one-sixth that of Earth. Apollo astronauts who walked on the Moon verified this value in the late 1960s and early 1970s.

Another prediction of Equation 9.2 is that gravitational acceleration diminishes with distance from Earth's center. Indeed, you've already seen that the Moon, a great distance from Earth, has acceleration only about 1/3600 that at Earth's surface. Example 9.4 shows that variations in g are detectable without going as far as the Moon.

EXAMPLE 9.4   **Higher Elevation, Lower g**

Puget Sound, at sea level, isn't far from Mt. Rainier, with summit elevation $h = 4390$ m. Estimate the difference in gravitational acceleration between those locations.

ORGANIZE AND PLAN Equation 9.2 gives g as a function of r, measured from Earth's center. Let r_1 and r_2 correspond to Puget Sound and Mt. Rainier, respectively (Figure 9.6), so the difference in g is

$$\Delta g = g_1 - g_2 = \frac{GM_E}{r_1^2} - \frac{GM_E}{r_2^2}$$

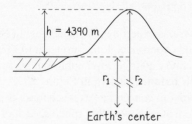

FIGURE 9.6 Relating the mountain's height to the distance to Earth's center.

Known: $h = 4390$ m; $M_E = 5.98 \times 10^{24}$ kg; $R_E = 6.37 \times 10^6$ m.

SOLVE Computing Δg is easier after some algebraic simplification:

$$\Delta g = GM_E\left(\frac{1}{r_1^2} - \frac{1}{r_2^2}\right) = \frac{GM_E}{r_1^2 r_2^2}\left(r_2^2 - r_1^2\right)$$

$$= \frac{GM_E}{r_1^2 r_2^2}(r_2 - r_1)(r_2 + r_1)$$

Now $r_2 - r_1 = h = 4390$ m (Figure 9.6). To an excellent approximation $r_2 + r_1 = 2R_E$, and $r_1^2 r_2^2 = R_E^4$. With these approximations,

$$\Delta g \approx \frac{GM_E}{R_E^4}(h)(2R_E) = \frac{2GM_E h}{R_E^3}$$

$$= \frac{2(6.67 \times 10^{-11} \text{ N} \cdot \text{m}^2/\text{kg}^2)(5.98 \times 10^{24} \text{ kg})(4390 \text{ m})}{(6.37 \times 10^6 \text{ m})^3}$$

$$= 0.0135 \text{ m/s}^2$$

REFLECT You wouldn't notice a difference of 0.01 m/s^2 when dropping a ball, but geologists' *gravimeters* detect much smaller differences. A geologist might use a gravimeter to search for oil, whose low

cont'd.

density makes g above an oil deposit slightly lower than over the surrounding terrain. In this case it's not the variation in the distance from Earth's center that affects g, but rather the local mass difference. Similarly, this example's calculation is only approximate, because we didn't account for Mt. Rainer's mass.

MAKING THE CONNECTION What's the approximate change in g for each kilometer of elevation?

ANSWER Following the same procedure but using $h = 1$ km, the variation in g is 0.0031 m/s²/km —valid for elevation changes much less than Earth's radius.

g and Latitude

Gravitational acceleration varies also with latitude. The sea-level value of g is about 9.78 m/s² at the equator and 9.83 m/s² at the poles. There are several reasons for this variation. First, the rotating Earth bulges slightly at the equator, with its equatorial radius some 21 km greater than its polar radius. As the Mt. Rainer example shows, this reduces gravity at the equator. Second, the bulge puts more mass in the equatorial regions, which tends to increase g—although not as much as the increased radius reduces it. Finally, an object at rest on the equator is in circular motion, so there's a downward centripetal force on it, arising from the vector sum of the upward normal force and downward gravity. That makes the *apparent weight*—as measured by a spring scale—less than the actual weight mg. This effect doesn't actually change g as calculated from Equation 9.2, but it does make the *effective g* lower by the centripetal acceleration v^2/R_E. Because tangential speed v is greatest at the equator, this effect is greatest there. The approximate sizes of these effects are equatorial bulge, -0.034 m/s²; greater equatorial mass, $+0.049$ m/s²; and centripetal effects, -0.066 m/s², for a net effect of -0.051 m/s².

Measuring G

Our computations of gravitational force and acceleration assume knowledge of the universal gravitation constant G and Earth's mass M_E. How do we know both of these quantities? Suppose you measure g near Earth's surface. Then by Equation 9.2,

$$GM_E = gR_E^2$$

This experiment reveals only the product GM_E, not individual values of G and M_E. Similarly, you could use the Moon's orbital radius and period to find GM_E, but not G and M_E separately. Therefore, a different experiment is required to measure G; once you know G and GM_E, you can solve for Earth's mass. Such an experiment would measure the force F between two spheres of known mass m_1 and m_2 a distance r apart. Then Newton's law (Equation 9.1) gives:

$$G = \frac{Fr^2}{m_1 m_2}$$

But this experiment isn't easy to do! Consider the mutual gravitational attraction of two golf balls, each with mass 46 g and radius 2.1 cm. If they're just touching, the center-to-center distance in Newton's law is twice the radius, or 0.042 m. Then the attractive force is

$$F = \frac{Gm_1 m_2}{R_E^2} = \frac{(6.67 \times 10^{-11}\ \text{N} \cdot \text{m}^2/\text{kg}^2)(0.046\ \text{kg})^2}{(0.042\ \text{m})^2} = 8.0 \times 10^{-11}\ \text{N}$$

Measuring such a tiny force presents serious experimental challenges! In Newton's lifetime, G wasn't well known. The first good experimental measurement of G was made by Henry Cavendish in 1798, more than a century after Newton published his work on gravitation. The experiment uses a **Cavendish balance**, shown in Figure 9.7. A rod connects two spheres of mass m, made of lead in order to concentrate as much mass as possible. The rod hangs from a thin fiber. With the system at rest, two other spheres, each of mass M, are placed as shown. The gravitational attraction between pairs of masses m and M produces a torque on the fiber, resulting in a small but measurable twist. The twist is compared with that caused by a known torque, thus revealing the attractive force between the spheres. Knowing that force, the spheres' masses, and their separation lets the experimenter solve for G.

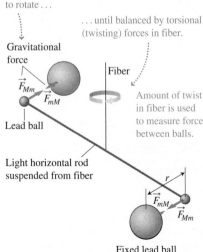

Gravitational attraction between balls causes rod to rotate . . .

. . . until balanced by torsional (twisting) forces in fiber.

Gravitational force

$\vec{F}_{Mm}$

$\vec{F}_{mM}$

Lead ball

Fiber

Amount of twist in fiber is used to measure force between balls.

Light horizontal rod suspended from fiber

r

$\vec{F}_{mM}$

$\vec{F}_{Mm}$

Fixed lead ball

FIGURE 9.7 The Cavendish balance in use.

Jupiter's Mass

How could you determine Jupiter's mass without going there?

SOLVE Jupiter has many natural satellites, or moons. (The four largest moons were discovered by Galileo.) Each moon's orbital period is easily determined by observing it regularly, and the orbital radii follow from the moon's apparent position and the known distance to Jupiter. The rest is dynamics and Newton's law of gravitation. Let Jupiter's mass be M_J and the moon's mass m. For a circular orbit of radius R, gravity provides the centripetal force:

$$\frac{GM_Jm}{R^2} = \frac{mv^2}{R}$$

The moon's mass m cancels, as usual with gravity. The moon's speed is orbital circumference divided by period, or $v = 2\pi r/T$. Therefore,

$$\frac{GM_J}{R} = v^2 = \left(\frac{2\pi R}{T}\right)^2 = \frac{4\pi^2 R^2}{T^2}$$

Solving for M_J,

$$M_J = \frac{4\pi^2 R^3}{GT^2}$$

giving Jupiter's mass in terms of the observed quantities.

REFLECT This method also gives the Sun's mass, using the periods and orbital radii of the planets. How might you determine the mass of a moonless planet like Venus?

Reviewing New Concepts

Some important ideas about gravity:

- The force between two point particles or spherical objects is given by Newton's law of gravitation: $F = \dfrac{Gm_1m_2}{r^2}$.
- The acceleration of gravity near Earth is $g = \dfrac{GM_E}{R_E^2}$, and a similar relationship holds for g near any spherical body.
- G is the universal gravitational constant and can be measured using a Cavendish balance.

Gravity Is Weak!

Today we know the gravitational constant G to five significant digits. That may sound precise, but it's imprecise compared with other fundamental constants. What makes G difficult to measure is that the gravitational force is by far the weakest of nature's fundamental forces. Your experience suggests that gravity is strong, but your experience results from living on a 6×10^{24}-kg planet!

Chapter 4 briefly introduced the four fundamental forces. **Gravitation** is the subject of this chapter. Another is the **electromagnetic force**, subject of much of the second half of this book. Electromagnetic forces are responsible for most everyday interactions; for example, holding materials together. The **nuclear** (or **strong**) **force** binds protons and neutrons within atomic nuclei (Chapter 26). The **weak force** mediates certain nuclear decays (Chapter 26). Although exact comparison of the forces is difficult, Table 9.1 gives an order-of-magnitude comparison for similar particles operating over similar distances.

Action at a Distance and Universal Law

As discussed in Chapter 4, gravity is a noncontact, "action-at-a-distance" force. Equation 9.1 suggests that its range is, in principle, infinite—that is, gravity connects any two objects in the universe through their mutual attraction. If this seems remarkable, it's also puzzled

TABLE 9.1 Relative Strengths of the Fundamental Forces

Force	Relative strength
Nuclear	1
Electromagnetic	10^{-2}
Weak	10^{-10}
Gravitational	10^{-38}

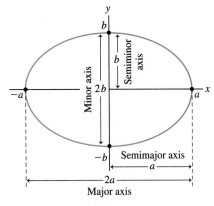

(a) An ellipse centered on an *x-y* coordinate system intersects the axes at ±*a* and ±*b*.

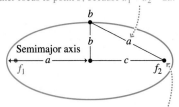

$d_1 + d_2$ (sum of distances to foci) is the same for all points on an ellipse.

(b) The foci f_1 and f_2 of an ellipse

FIGURE 9.8 The geometry of ellipses.

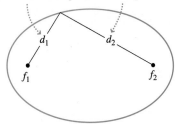

The semimajor axis *a* is also the distance from either focus to point *b*, because $d_1 + d_2 = 2a$.

Semimajor axis

The resulting right triangle locates the focus, because $a^2 = b^2 + c^2$.

(a) Ellipse with larger eccentricity

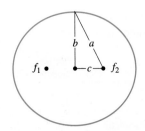

(b) Ellipse with smaller eccentricity

FIGURE 9.9 Two ellipses with different values of eccentricity.

physicists since Newton. Later we'll see a different philosophical approach to action-at-a-distance forces.

The idea of force acting over huge distances led Newton to ponder **universal laws** in physics. Because his gravitation law worked so well in explaining falling apples and orbiting planets, Newton imagined that it applied everywhere in the universe, and this seems to be the case. That means Newton's law describes the motions of galaxies containing billions of stars, and even the motions of whole clusters of galaxies. A physical law that applies both on Earth and in the heavens represented a philosophical breakthrough. Universal laws seem natural today, but in the 17th century religious and philosophical traditions demanded a clear distinction between the earthly realm and the heavens. The idea that through careful observation and reason, humans can understand the fundamental laws governing the universe had a significant impact on scientists and philosophers who followed Newton. Newtonian ideas helped shape the intellectual movement known as the Enlightenment, which dominated 18th-century Europe and contributed to the American Revolution.

GOT IT? Section 9.1 Suppose another planet has twice Earth's mass and twice its radius. If the gravitational acceleration on Earth is *g*, then on the other planet it's (a) 4*g*; (b) 2*g*; (c) *g*; (d) *g*/2; (e) *g*/4.

9.2 Planetary Motion and Kepler's Laws

Until now we've considered only circular orbits for planets and moons. However, their true paths are ellipses.

Ellipse Geometry

Figure 9.8a shows an ellipse with its center at the origin. The quantities *a* and *b* define the **semimajor axis** and **semiminor axis**, respectively. If *a* and *b* are equal, the ellipse is a circle. The more different *a* and *b*, the more elongated the ellipse.

Two special points lie on the major axis: the foci f_1 and f_2. As Figure 9.8b suggests, the ellipse is the set of points such that the sum of distances from the two foci is constant. You might have learned to draw an ellipse this way, with a fixed-length string tacked at its endpoints and tracing the ellipse with a pencil moved around with the taut string. If the foci coincide, the ellipse is a circle; the farther apart the foci, the more elongated the ellipse. Using Figure 9.8b, you can convince yourself that the string's length $d_1 + d_2$ is 2*a*.

As shown in Figure 9.9, the shape of the ellipse depends on the distance *c* from the ellipse's center to each focus. Quantitatively, the ellipse is characterized by its **eccentricity**, defined as

$$e = \frac{c}{a} = \sqrt{1 - \left(\frac{b}{a}\right)^2} \quad \text{(Eccentricity)} \quad (9.3)$$

A circle has $e = 0$, while the maximum possible *e* is 1. An ellipse with *e* near 1 is very long and thin, and, in the limit $e \to 1$, becomes a line.

CONCEPTUAL EXAMPLE 9.6 **Ellipses**

Measure the two ellipses in Figure 9.9 to determine the eccentricity of each.

SOLVE In each case the eccentricity is given by $e = c/a$. Measuring carefully should give an eccentricity of about 0.95 for ellipse (a) and 0.41 for ellipse (b).

REFLECT The second ellipse, with eccentricity around 0.41, appears not much different than a circle. Ellipses with smaller eccentricities—particularly those less than 0.1—are nearly circular.

Kepler's Laws

In using Tycho Brahe's observations to establish that planetary orbits are elliptical, Kepler overcame significant challenges. First, planets' orbital eccentricities are small, making the orbits hard to distinguish from circles. Second, Brahe's observations, while excellent for his time, were limited to naked-eye accuracy, which at best was about 1 minute of arc (1/60 of 1 degree). Third, Kepler's inference was complicated by his position on a planet moving in its own elliptical orbit. Kepler succeeded using some particularly good observations of Mars, which has eccentricity 0.09, higher than most planets. Kepler's description of planetary motion comprises three laws, the first two formulated in 1609:

1. The orbit of each planet is an ellipse, with the Sun at one focus.
2. In a given time, a planet sweeps out the same area regardless of where it is in its orbit.

Figure 9.10a illustrates the first law, showing a planetary orbit with the Sun at one focus. Figure 9.10a also illustrates the second law, showing how a planet moves faster when closer to the Sun, so that it always sweeps out the same area in a given time. Figure 9.10b shows the **perihelion** and **aphelion**—the points closest and farthest from the Sun.

We can thank Kepler's second law for slightly shorter winters in the Northern Hemisphere. Earth reaches perihelion in January and aphelion in July. Since it moves faster at perihelion, winter is several days shorter than summer.

The variation in orbital speeds is consistent with conservation of angular momentum (Chapter 8). With no external torque, a planet's angular momentum relative to the Sun is constant. Recall that angular momentum is $L = I\omega$, where I is the planet's rotational inertia and ω is its angular velocity. When the planet moves closer to the Sun, its rotational inertia decreases. Therefore, its angular velocity increases to keep angular momentum constant.

In 1619 Kepler presented his third law, sometimes called the **harmonic law**:

3. If T is a planet's orbital period and a is the semimajor axis of its orbital ellipse, then

$$\frac{a^3}{T^2} = C$$

where C is a constant for all objects orbiting the Sun. In SI, $C = 3.36 \times 10^{18}\ \mathrm{m^3/s^2}$.

Kepler's third law says that the cube of the semimajor axis is proportional to the square of the period. This holds not only for planets but also for asteroids, comets, and other objects in the solar system—including spacecraft in solar orbits. Later we'll show how C depends on the Sun's mass.

Appendix E lists data on planets and their orbits. You can use any planet's semimajor axis and period to check the value of C in Kepler's third law. Note that most planetary eccentricities are small, implying little deviation from circular orbits. Not shown is the fact that planetary orbits are in essentially the same plane, known as the **ecliptic plane**. However, their major axes aren't aligned.

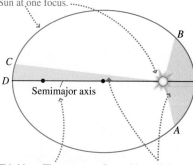

First law: The orbit is elliptical, with the Sun at one focus.

Third law: The square of the orbital period is proportional to the cube of the semimajor axis.

Second law: If the shaded areas are equal, so is the time to go from A to B and from C to D.

(a) Kepler's laws

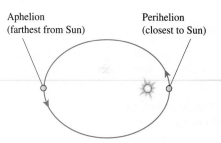

Aphelion (farthest from Sun)

Perihelion (closest to Sun)

(b) Perihelion and aphelion for a planet in an elliptical orbit

FIGURE 9.10 Describing planetary orbits.

EXAMPLE 9.7 **Perihelion and Aphelion**

Use Earth's orbital data to find the perihelion and aphelion distances of Earth from the Sun.

ORGANIZE AND PLAN As Figure 9.11 (next page) shows, the perihelion distance is $d_p = a - c$, and the aphelion distance is $d_a = a + c$. From the definition of eccentricity, $e = c/a$, or $c = ea$. Substituting $c = ea$ into each of the above equations gives the perihelion and aphelion distances in terms of the semimajor axis and eccentricity, which are given in Appendix E.

Known: semimajor axis $a = 1.496 \times 10^{11}$ m; $e = 0.0167$ (Appendix E).

SOLVE Solving for the perihelion distance and inserting numerical values,

$$d_p = a - c = a - ea = (1 - e)a$$
$$= (1 - 0.0167)(1.496 \times 10^{11}\ \mathrm{m}) = 1.47 \times 10^{11}\ \mathrm{m}$$

cont'd.

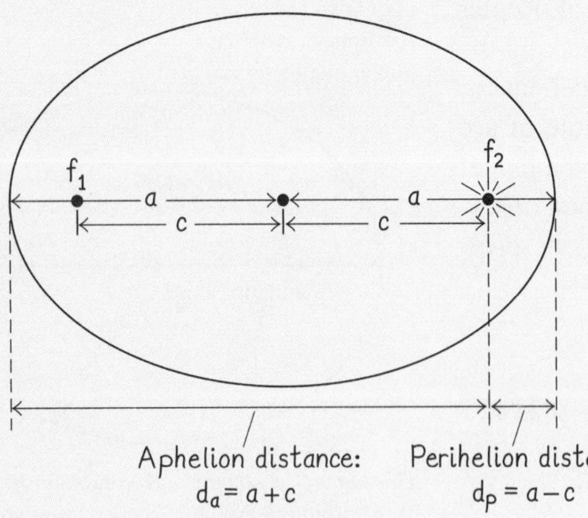

FIGURE 9.11 Finding perihelion and aphelion distances. Note: not to scale.

Similarly for the aphelion,

$$d_a = a + c = a + ea = (1 + e)a$$
$$= (1 + 0.0167)(1.496 \times 10^{11} \text{ m}) = 1.52 \times 10^{11} \text{ m}$$

REFLECT The perihelion and aphelion distances differ by about 5×10^9 m, or 5×10^6 km. That's more than ten times the radius of the Moon's orbit, but it's a difference of only about 3%.

MAKING THE CONNECTION If Earth is that much closer to the Sun in January than in July, why does the Northern Hemisphere have winter in January and summer in July?

ANSWER The seasons are due to the fact that Earth's rotational axis is inclined by 23° away from a line perpendicular to the ecliptic plane. This causes the Northern Hemisphere to get much more sunlight, and more direct sunlight, in summer. Of course, it's the other way around in the Southern Hemisphere.

EXAMPLE 9.8 **Halley's Comet**

Halley's comet orbits the Sun with period $T = 76.1$ years and eccentricity $e = 0.967$. Find the length of the comet's semimajor axis and compare with the planets' orbits.

ORGANIZE AND PLAN Kepler's third law relates the orbital period to the semimajor axis:

$$\frac{a^3}{T^2} = C$$

To use C in SI, we'll have to convert the period to seconds.

Known: $C = 3.36 \times 10^{18}$ m^3/s^2; $T = 76.1$ y; $e = 0.967$.

SOLVE Convert the period:

$$T = 76.1 \text{ y} \times \frac{3.15 \times 10^7 \text{ s}}{1 \text{ y}}$$
$$= 2.40 \times 10^9 \text{ s}$$

Then by Kepler's third law, $a^3 = CT^2$, or

$$a = (CT^2)^{1/3} = \left[(3.36 \times 10^{18} \text{ m}^3/\text{s}^2)(2.40 \times 10^9 \text{ s})^2 \right]^{1/3}$$
$$= 2.68 \times 10^{12} \text{ m}$$

This is between the semimajor axes of Jupiter and Saturn.

REFLECT The orbital eccentricity didn't enter this computation. However, see "Making the Connection" for the shape of Halley's orbit.

MAKING THE CONNECTION Follow Example 9.7 to find the perihelion and aphelion distances for Halley's comet.

ANSWER Using the data in this example, $d_p = (1 - e)a = 8.84 \times 10^{10}$ m and $d_a = (1 + e)a = 5.27 \times 10^{12}$ m. Now you can see the large eccentricity. At perihelion, Halley's comet is inside the orbit of Venus (see Appendix E). But at aphelion, it's beyond Neptune! However, the orbital plane is inclined to the ecliptic, so the comet never passes near the outer planets.

Kepler and Newton

Kepler's laws follow from Newton's law of gravitation, but proof requires calculus—which Newton developed for that reason. In the special case of circular orbits, however, Kepler's third law follows from physics you know.

Consider a planet of mass m in circular orbit of radius R around the Sun, which has mass M. From Chapter 4, the centripetal force on the planet is mv^2/r, and the gravitational attraction between Sun and planet provides this force. Thus:

$$\frac{mv^2}{R} = \frac{GMm}{R^2}$$

The planet's mass cancels, leaving

$$v^2 = \frac{GM}{R}$$

The speed v is the orbital circumference divided by the period: $v = 2\pi r/T$. Thus

$$\left(\frac{2\pi R}{T}\right)^2 = \frac{GM}{R}$$

Rearranging gives something that looks like Kepler's third law:

$$\frac{R^3}{T^2} = \frac{GM}{4\pi^2} \tag{9.4}$$

Recall that a circle is an ellipse with $e = 0$ and $R = a$. Therefore, Equation 9.4 is Kepler's third law for circular orbits. The right-hand side is the constant C, with M the Sun's mass. Check the numerical value to convince yourself.

Newton showed that an inverse-square force from a fixed center (e.g., the Sun) leads to orbits that are *conic sections* (Figure 9.12). While planets follow elliptical paths, some asteroids and comets follow hyperbolic paths. Their orbits aren't closed, so these objects make one pass through the inner solar system and then are gone for good.

Because Newton's law of gravitation is universal, Equation 9.4 governs circular orbits about other bodies. Thus Equation 9.4 describes the Moon's essentially circular orbit, if we take M to be Earth's mass. And it describes spacecraft in circular orbits about Earth, with the same constant C we calculate for the Moon. We'll look further at space flight after exploring gravitational energy.

Reviewing New Concepts: Kepler's Laws

- Kepler's three laws describe the motions of orbiting bodies, including but not limited to planets orbiting the Sun.
- The eccentricity e quantifies the shape of an elliptical orbit.

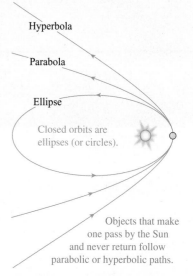

FIGURE 9.12 The orbits of planets and some asteroids and comets are elliptical. However, many asteroids and comets are in hyperbolic orbits.

EXAMPLE 9.9 **Moon Follows Kepler**

The Moon follows Kepler's third law, with the constant $C = GM/4\pi^2$ computed using Earth's mass $M_E = 5.98 \times 10^{24}$ kg. Compute C, then use the Moon's orbital radius $R = 3.84 \times 10^8$ m and period $T = 27.3$ days to verify that the Moon follows Kepler's third law.

ORGANIZE AND PLAN Kepler's third law reads

$$\frac{R^3}{T^2} = \frac{GM}{4\pi^2}$$

The given quantities allow you to compute the two sides separately and thereby see if the Moon follows Kepler's third law.

Known: $M_E = 5.98 \times 10^{24}$ kg; $T = 27.3$ d; $R = 3.84 \times 10^8$ m.

SOLVE The right-hand side of the equation involves Earth's mass:

$$\frac{GM_E}{4\pi^2} = \frac{(6.67 \times 10^{-11}\ \text{N} \cdot \text{m}^2/\text{kg}^2)(5.98 \times 10^{24}\ \text{kg})}{4\pi^2}$$

$$= 1.01 \times 10^{13}\ \text{m}^3/\text{s}^2$$

The left-hand side involves the Moon's orbital radius and period. From Example 9.1, $T = 27.3\ \text{d} \times \dfrac{86{,}400\ \text{s}}{1\ \text{d}} = 2.36 \times 10^6$ s. Then

$$\frac{R^3}{T^2} = \frac{(3.84 \times 10^8\ \text{m})^3}{(2.36 \times 10^6\ \text{s})^2} = 1.02 \times 10^{13}\ \text{m}^3/\text{s}^2$$

Within rounding errors, we've verified Kepler's law.

REFLECT We've assumed a circular orbit. The closeness of our numerical results suggests this is a good approximation. In fact, the Moon's orbital eccentricity is about 0.05.

MAKING THE CONNECTION What would be the period of a satellite in circular Earth orbit with radius half that of the Moon's orbit?

ANSWER Using the numbers available in this example, the new period is $27.3\ \text{d}\ (1/2)^{3/2} = 9.65$ d.

Extrasolar Planets

Imagine a star with a single planet. Absent external forces, this system's center of mass stays fixed—and that means both star and planet must orbit the center of mass. If the star's mass is much larger than the planet's, then the star won't move very much. In our own solar system, Jupiter's mass is about 0.001 times the Sun's. That's enough to make the Sun describe a small orbit, about 0.001 times the size of Jupiter's. An observer viewing the Sun from afar would see it "wobble" slightly with the period of Jupiter's orbit. It's this effect that has led astronomers to discover several hundred planets around other stars, such as 55 Cancri (shown here in artist's conception). Observing the periodic "wobbling" of those stars yields some physical properties of their unseen planets. The prospect of extrasolar planets—particularly those that might support life—is fascinating!

9.3 Gravitational Potential Energy

In Chapter 5 you learned that gravity is a conservative force. The conservative nature of gravity let us develop a gravitational potential energy function $U = mgy$ and then apply conservation of mechanical energy, $E = K + U$.

Now we've expanded our understanding of gravity through Newton's universal gravitation—still a conservative force. So there's a potential energy function U for gravity in the general case. It won't be Chapter 5's $U = mgy$, because we're no longer assuming constant gravitational acceleration g. Deriving this new potential energy function requires calculus, so we'll just give the result: The **gravitational potential energy** U of two point masses m_1 and m_2 separated by a distance r is

$$U = -\frac{Gm_1m_2}{r} \quad \text{(Gravitational potential energy; SI unit: J)} \tag{9.5}$$

Because this expression follows from Newton's law of gravitation, it also holds when one or both objects is spherical. That makes it applicable, for example, to natural and artificial satellites.

✓**TIP**

For spherical bodies, use the center-to-center distance to find the potential energy, just as you do for the gravitational force.

Don't be put off by the fact that the gravitational potential energy in Equation 9.5 is negative. Recall from Chapter 5 that only *changes* in potential energy are meaningful and that you're free to set the zero of potential energy anywhere. It's evident that Equation 9.5 assumes $U = 0$ when two objects are separated by an infinite distance. Otherwise, potential energy is negative and decreases as the objects move closer.

Although gravitational *potential* energy is negative or at most zero, an object's *total* energy can have any value. When the total energy is negative, the object has an elliptical orbit—a closed path that means it's bound gravitationally to another body (Figure 9.12). The Earth-Sun system, for example, has negative total energy, meaning Earth is bound to the Sun. When total energy is positive, the object is unbound, and its orbit is a hyperbola. Zero energy is the intermediate case, and gives a parabolic orbit.

With gravitational potential energy in hand, you can solve problems involving kinetic and potential energy, as you learned in Chapter 5. The new potential energy, Equation 9.5, is applicable universally and isn't limited to situations with g nearly constant, like the near-Earth case leading to $U = mgy$. The strategy for using Equation 9.5 in conservation-of-energy situations is identical to what you learned in Problem-Solving Strategy 5.2; only the form of the potential energy has changed.

A Big Drop

Two satellites with total mass m collide and stick together. They plummet to the ground, starting from rest at an altitude equal to Earth's radius. What's the impact speed of the wreckage, neglecting air resistance?

ORGANIZE AND PLAN This problem is best solved using conservation of energy. Set the initial mechanical energy equal to the total mechanical energy when the wreckage strikes Earth. The initial r, measured from Earth's center, is $r_0 = 2R_E$, and the final value is $r = R_E$ (Figure 9.13).

Known: $M_E = 5.98 \times 10^{24}$ kg, $R_E = 6.37 \times 10^6$ m.

SOLVE From conservation of mechanical energy,

$$K_0 + U_0 = K + U$$

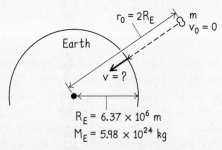

FIGURE 9.13 The wreckage falls.

The wreckage starts from rest, so $K_0 = 0$. Thus

$$K = U_0 - U$$

or

$$\tfrac{1}{2}mv^2 = -\frac{GmM_E}{2R_E} - \left(-\frac{GmM_E}{R_E}\right) = \frac{GmM_E}{2R_E}$$

Mass m cancels, and we solve for the impact speed v:

$$v = \sqrt{\frac{GM_E}{R_E}} = \sqrt{\frac{(6.67 \times 10^{-11}\ \text{N}\cdot\text{m}^2/\text{kg}^2)(5.98 \times 10^{24}\ \text{kg})}{6.37 \times 10^6\ \text{m}}}$$

$$= 7.91\ \text{km/s}$$

REFLECT The gravitational acceleration varies significantly over this drop, so the constant-g approximation wouldn't work. (Check it yourself. You'd get over 11 km/s—a huge error!) However, neglecting air resistance is unrealistic here.

MAKING THE CONNECTION What's the gravitational acceleration g at the satellites' initial altitude?

ANSWER From Section 9.1, $g = GM_E/r^2$. With $r = 2R_E$, g drops by a factor of $2^2 = 4$ from its sea level value, to about 2.5 m/s^2.

Rocket Launch

A rocket is launched straight up from Earth's surface at 2100 m/s. Ignoring air resistance, what's the maximum height it reaches?

ORGANIZE AND PLAN Ignoring air resistance, we again apply conservation of energy. Set the total energy at launch equal to the total energy at the maximum height h. There the rocket is momentarily at rest (Figure 9.14), so its kinetic energy is zero. The potential energy at launch is

$$U_0 = -\frac{GmM_E}{R_E}$$

and at height h (see Figure 9.14), it's

$$U = -\frac{GmM_E}{R_E + h}$$

where M_E and R_E are Earth's mass and radius.

Known: $M_E = 5.98 \times 10^{24}$ kg, $R_E = 6.37 \times 10^6$ m, $v_0 = 2100$ m/s.

SOLVE From conservation of mechanical energy,

$$K_0 + U_0 = K + U$$

At the peak height, $K = 0$. Thus $K_0 = U - U_0$, or

$$\tfrac{1}{2}mv_0^2 = -\frac{GmM_E}{R_E + h} - \left(-\frac{GmM_E}{R_E}\right)$$

Canceling m and solving for h,

$$h = \frac{1}{\dfrac{1}{R_E} - \dfrac{v_0^2}{2GM_E}} - R_E$$

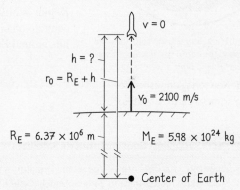

FIGURE 9.14 Determining the rocket's maximum height.

or

$$h = \frac{1}{\dfrac{1}{6.37 \times 10^6\ \text{m}} - \dfrac{(2100\ \text{m/s})^2}{2(6.67 \times 10^{-11}\ \text{N}\cdot\text{m}^2/\text{kg}^2)(5.98 \times 10^{24}\ \text{kg})}}$$
$$- 6.37 \times 10^6\ \text{m}$$

$$= 233\ \text{km}$$

cont'd.

REFLECT That's well above Earth's atmosphere—reasonable, given that the 2100-m/s launch speed is ten times that of a typical commercial aircraft. Note that if we had tried to solve this problem using $U = mgy$ with a constant $g = 9.80 \text{ m/s}^2$, the answer would have been 225 km—a significant error.

> **MAKING THE CONNECTION** When the rocket drops back to Earth, what's its speed at impact?
>
> **ANSWER** Ignoring air resistance, total mechanical energy is conserved. Back on the ground, the rocket has the same potential energy it had at launch, and therefore its kinetic energy must be the same as the initial kinetic energy, so its speed is 2100 m/s. In reality, air resistance would slow the rocket, both going up and coming down, decreasing its maximum height and impact speed.

Newton imagined firing a cannonball horizontally from a mountain with successively greater initial speeds.

At first, the ball falls to Earth farther and farther from the mountain's foot (points D–G).

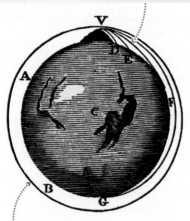

But at a high enough speed, the ball "falls" all the way around the Earth, striking the cannon from the rear! It is in orbit.

FIGURE 9.15 Newton's suggestion for launching artificial Earth satellites.

Shooting a rocket straight up gives only a short time above the atmosphere—although such flights are relatively inexpensive and are widely used in astronomy and atmospheric science. Achieving a continuous orbit, however, provides indefinite time in space.

9.4 Artificial Satellites

After using his law of gravitation to analyze the motion of the Moon and planets, Newton reasoned that it should be possible to launch artificial satellites. Figure 9.15 is Newton's own diagram, showing how he imagined launching a projectile horizontally from a mountaintop. At low launch speeds the projectile crashes to Earth, but with higher speeds—assuming no air resistance—it achieves a continuous orbit. For that orbit to be circular, you saw in Section 9.1, the right speed is one that matches the gravitational force to the centripetal force required. Newton's diagram also shows satellites with elliptical orbits.

Satellite Periods

Kepler's third law (Section 9.2) relates the orbital period and semimajor axis. We'll concentrate on circular orbits, for which the semimajor axis a is the radius R. For circular orbits around Earth, Kepler's third law (Equation 9.4) states

$$\frac{R^3}{T^2} = \frac{GM_E}{4\pi^2}$$

CONCEPTUAL EXAMPLE 9.12 **Geosynchronous Satellites**

A *geosynchronous satellite* orbits above the equator with period 24 hours. Since Earth rotates once in 24 hours, such a satellite remains still with respect to Earth. Why must it orbit above the equator?

SOLVE Any circular orbit around Earth must be centered on Earth's center; you couldn't have a satellite orbiting at, say, a fixed 45° latitude. If the orbit doesn't parallel the equator, the satellite will swing to either side of the equator and won't appear fixed in the sky (Figure 9.16).

A geosynchronous satellite remains fixed in position over Earth's equator. That makes geosynchronous orbit especially useful for communications, because it means transmitting and receiving antennas can be permanently aimed at a satellite. If you've got a TV satellite dish antenna, it's pointing at a satellite 36,000 km above the equator. Television, intercontinental telephone signals, and some Internet traffic routinely go via geosynchronous satellite. Monitoring weather is another use of geosynchronous satellites, and this technology has helped improve weather prediction in recent decades.

REFLECT With geosynchronous satellites in high demand, it's not surprising to learn that it's getting crowded up there in a band 36,000 km

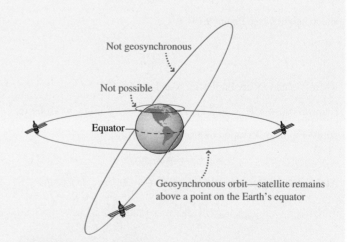

FIGURE 9.16 Why geosynchronous satellites' orbits must be parallel to Earth's equator.

above the equator. International agreements manage this orbital traffic jam, governing where new satellites can be placed.

EXAMPLE 9.13   **Altitude of Geosynchronous Satellites**

What is the altitude of a geosynchronous satellite?

ORGANIZE AND PLAN Kepler's third law states

$$\frac{R^3}{T^2} = \frac{GM_E}{4\pi^2}$$

With known period $T = 24$ h $= 86{,}400$ s, the radius of the orbit can be found. The altitude is then the difference between that radius and Earth's radius, $h = R - R_E$.

Known: $M_E = 5.98 \times 10^{24}$ kg, $R_E = 6.37 \times 10^6$ m, $T = 24$ h $= 86{,}400$ s.

SOLVE Solving Kepler's law for the radius R,

$$R = \left(\frac{GM_E T^2}{4\pi^2}\right)^{1/3}$$

$$= \left(\frac{(6.67 \times 10^{-11}\ \text{N} \cdot \text{m}^2/\text{kg}^2)(5.98 \times 10^{24}\ \text{kg})(86{,}400\ \text{s})^2}{4\pi^2}\right)^{1/3}$$

$$= 4.22 \times 10^7\ \text{m}$$

so the altitude above Earth's surface is

$$h = R - R_E = 4.22 \times 10^7\ \text{m} - 6.37 \times 10^6\ \text{m} = 3.58 \times 10^7\ \text{m}$$

REFLECT That's about 36,000 km, which is more than 5 times Earth's radius. There's no choice here: if you want a geosynchronous satellite, you've got to put it in orbit at that altitude.

MAKING THE CONNECTION   Estimate the smallest possible period of a satellite in a circular orbit around Earth.

ANSWER By Kepler's third law, the smallest period corresponds to the lowest orbit. For Newton's mountain-skimming orbit, $R \approx R_E = 6.37 \times 10^6$ m. With that radius, Kepler's law gives $T = 5060$ s, or about 84 minutes. In reality, the mountaintop orbit suffers from too much atmospheric friction. The shortest practical orbit has a slightly longer period, about 90 minutes. The International Space Station and many other satellites are in so-called low-Earth orbit, a few hundred km up, and these all have orbital periods of about 90 minutes.

Escape Speed

Throw a baseball straight up, and it returns to Earth. The faster you throw it, the higher it goes. But an object launched at **escape speed** or faster will escape Earth's gravity completely.

You can find v_{esc} using energy principles. Say a rocket is launched with initial kinetic energy K_0 (Figure 9.17). Then conservation of energy says

$$K_0 + U_0 = K + U$$

According to Equation 9.5, its potential energy at any point r is

$$U = -\frac{GM_E m}{r}$$

with M_E Earth's mass and m the rocket's mass. For a launch from Earth's surface ($r = R_E$), the initial potential energy is

$$U_0 = -\frac{GM_E m}{R_E}$$

We're interested in the case of a rocket with the minimum energy needed to escape. That is, the rocket must reach a great distance from Earth, giving a final potential energy

$$U = -\frac{GM_E m}{r} \rightarrow 0$$

as r gets arbitrarily large.

What about kinetic energy? That decreases as the rocket gets farther from Earth. At the bare minimum, it will be moving arbitrarily slowly at great distances—so its final kinetic energy approaches zero. Therefore, the rocket's total mechanical energy is

$$K + U = 0 + 0 = 0$$

Because mechanical energy is conserved, the initial energy is also zero. Writing $K_0 = \frac{1}{2}mv_{esc}^2$ and using the initial potential energy given above,

$$K_0 + U_0 = \frac{1}{2}mv_{esc}^2 - \frac{GM_E m}{R_E} = 0$$

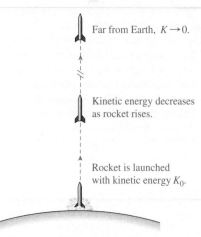

Far from Earth, $K \rightarrow 0$.

Kinetic energy decreases as rocket rises.

Rocket is launched with kinetic energy K_0.

FIGURE 9.17 Determining escape speed from Earth.

Global Positioning System

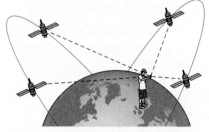

The Global Positioning System (GPS) is rapidly becoming essential to modern life. GPS helps us navigate cars, boats, and aircraft; tracks commercial shipments; saves farmers fuel; lets biologists track wildlife and parents their children; locates cell phones in emergencies; and enables the new sport of "geotracking." The list is long and growing.

GPS uses a "constellation" of some 24 satellites in circular orbits 12,000 miles up; the orbital period is around 12 hours. Orbits are inclined at 55° to the equator, so satellites pass over most inhabited latitudes. Any GPS user "sees" at least four satellites at a given time (as shown here). Exquisite timing of radio signals from the satellites gives the user's position to within meters. In principle, three such signals are sufficient to "triangulate" position; in practice, a fourth is used to correct for errors. By comparing GPS with a known ground location, positions can be measured to within centimeters.

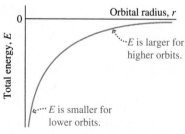

FIGURE 9.18 Total energy E for satellites in orbit.

As usual the mass m cancels, and we solve for v_{esc}:

$$v_{esc} = \sqrt{\frac{2GM_E}{R_E}} \quad \text{(Escape speed)} \tag{9.6}$$

Using Earth's mass and radius,

$$v_{esc} = \sqrt{\frac{2GM_E}{R_E}} = \sqrt{\frac{2(6.67 \times 10^{-11}\,\text{N·m}^2/\text{kg}^2)(5.98 \times 10^{24}\,\text{kg})}{6.37 \times 10^6\,\text{m}}} = 11.2\,\text{km/s}$$

or about 7 miles per second. That's fast, but achievable with rockets carrying smaller payloads. Spacecraft to the outer solar system routinely exceed escape speed, although not right at launch. Apollo Moon missions were kept under escape speed so that if anything went wrong—as with Apollo 13—the spacecraft would remain gravitationally bound to Earth.

Although we imagined launching from Earth in deriving Equation 9.6, the same formula works for other bodies, substituting the appropriate mass and radius. For example, escape speed from our Moon's surface is only 2.4 km/s.

✓**TIP**

The calculation of escape speed ignores air resistance.

Orbital Energy

A satellite with mass m and speed v a distance r from Earth's center has total mechanical energy

$$E = K + U = \tfrac{1}{2}mv^2 - \frac{GM_E m}{r}$$

A simpler expression for the energy in a *circular* orbit results because gravity provides the centripetal force. Therefore,

$$\frac{mv^2}{r} = \frac{GM_E m}{r^2}$$

Multiplying by $r/2$ gives

$$\tfrac{1}{2}mv^2 = \frac{GM_E m}{2r}$$

This is the satellite's kinetic energy; substituting into the equation for total mechanical energy gives

$$E = \tfrac{1}{2}mv^2 - \frac{GM_E m}{r} = \frac{GM_E m}{2r} - \frac{GM_E m}{r} = -\frac{GM_E m}{2r}$$

Thus, the total mechanical energy of a satellite in a circular orbit of radius r about Earth is

$$E = -\frac{GM_E m}{2r} \tag{9.7}$$

Note how the satellite's energy depends on orbital radius (Figure 9.18). As you might expect, a larger orbit requires more energy, rising toward zero (from negative values) as the orbital radius approaches infinity. This is consistent with escape speed, where zero total energy is the minimum needed for escape.

Our analysis leading to Equation 9.7 also gave the satellite's kinetic energy:

$$\tfrac{1}{2}mv^2 = \frac{GM_E m}{2r}$$

This shows that the satellite travels slower in orbits with larger radii. When a satellite moves to a larger orbit, its kinetic energy decreases, but its potential energy increases even more, resulting in an overall energy increase. The same principle governs the speeds of planets orbiting the Sun. Outer planets travel more slowly than inner planets.

GOT IT? **Section 9.4**   The orbital radius for a satellite in a 12-hour circular orbit about Earth is closest to (a) $2R_E$; (b) $3R_E$; (c) $4R_E$; (d) $5R_E$.

9.5 Other Aspects of Gravitation

Here we'll present an assortment of gravity-related topics. Because gravity is universal, it has some pretty far-reaching effects! We'll start on Earth and then look far beyond.

Tides

Newton's gravitation provides a simple explanation for ocean tides. Figure 9.19 shows that the side of Earth facing the Moon is closer to the Moon than an "average" place on Earth, resulting in a stronger gravitational force. At the same time, the side away from the Moon experiences a weaker-than-average force. The difference between these forces creates bulges of water that form high tides on opposite sides of Earth, with low tides halfway between. The Sun also influences tides, but its effect is smaller. This means that tides are extraordinarily high (and low) when Sun, Moon, and Earth are aligned—at new Moon and full Moon. Tides are least extreme at quarter Moons, when the gravitational forces of Sun and Moon are perpendicular.

Binary Stars and Galaxies

Many stars occur in pairs, orbiting their common center of mass (Figure 9.20). You know from Chapter 7 that the center of mass is closer to the more massive star. Therefore, its orbit is smaller. However, both stars orbit with the same period, to keep the center of mass in the same place at all times. Also, as Figure 9.20 shows, there's no torque on the binary system and so its angular momentum is constant.

We live in the Milky Way galaxy, a collection of more than 10^{11} stars. Ours is a spiral galaxy, like the one pictured on this chapter's opening page. Stars in this type of galaxy are attracted gravitationally toward the galactic center and orbit that center, like a planet orbiting a star. The distances are vast and the orbital period is huge—about 200 million years for the Sun—yet it's still Newton's inverse-square law that explains stellar motions in galaxies. Interestingly, the Newtonian explanation requires the presence of mass that we don't see in the galaxies. That's the so-called *dark matter* that makes up much of the universe and about whose properties we know almost nothing.

Apparent Weightlessness

Astronauts float freely about the International Space Station, apparently weightless. A common misconception is that "there's no gravity in space," so of course astronauts and other objects are "weightless." Nonsense! Newton's law shows that Earth's gravitational influence extends to great distances. Furthermore, an astronaut or spacecraft that didn't experience a gravitational force wouldn't be in orbit, but would be moving in a straight line, never to return to Earth's vicinity. In the near-Earth orbits of peopled spacecraft, the gravitational force—weight—is in fact more than 90% of its surface value.

So why "weightlessness"? Given that "weight" means the force of gravity, it's really only **apparent weightlessness**, and it occurs because gravitational acceleration is independent of mass—as we've seen so many times when *m* cancels from an equation. All objects in the vicinity of, say, the space station are in free fall with the same acceleration, and so, absent nongravitational forces, they remain at rest relative to each other and their freely falling reference frame.

The Earth's rotation causes high tides to occur at ≈12-h intervals, with low tides in between.

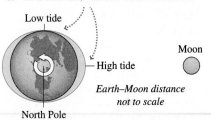

FIGURE 9.19 Tides due to the Moon. In reality, this simple picture is complicated considerably by the effect of continents.

Both stars orbit common center of mass, which is closer to more massive star.

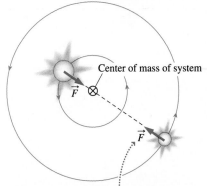

Gravitational forces exert no torque on system because they act along radius to center of mass. Thus, angular momentum is constant.

FIGURE 9.20 Dynamics of a binary star system.

It's not being in space that causes apparent weightlessness—it's being in free fall, with gravity the only force acting. Because of air resistance, that effectively means being in space. Orbits, too, aren't about space but about motion under gravity alone. Absent air resistance, even a baseball would be in orbit—an elliptical orbit with Earth's center at one focus. Orbital motion ends when the ball strikes Earth and nongravitational forces act. But throw a baseball horizontally at the right speed on an airless, smooth planet, and it would follow an unending a circular orbit just above the surface.

Einstein's Gravity

In 1915, Albert Einstein put forth an entirely new theory, which treats gravity not as a force but as the geometry of space and time. Einstein's **general theory of relativity** (GR) makes predictions almost identical to Newton's in the relatively weak gravity of our solar system. In Einstein's day, only a tiny discrepancy in Mercury's orbital motion required GR for its explanation. Even today, Newton's theory works fine except where we need exquisite precision, where gravity is very strong, or where we're considering the entire universe.

For exquisite precision, look to the Global Positioning System (GPS), where timing of signals from multiple satellites determines positions on Earth to within meters or better. So accurate is GPS that it would go off by a kilometer each day were relativity not taken into account.

For strong gravity, look to places where escape speed approaches the speed of light. Neutron stars, collapsed stars with the Sun's mass crammed into a radius of several kilometers, are one example. Rapidly rotating neutron stars—a consequence of angular momentum conservation when their much larger progenitor stars collapsed—are *pulsars* whose highly regular pulses of electromagnetic radiation make them ideal for studying strong gravity. Even more bizarre are black holes, objects so condensed that not even light can escape. Star-mass black holes form when massive stars collapse at the end of their lives. Huge black holes, with millions of stars' mass, lurk at the center of most galaxies, including our Milky Way.

Finally, look to the entire universe, whose large-scale structure and evolution are governed by general relativity. Both GR and Newtonian gravity suggest that the well-known expansion of the universe should be slowing because of the mutual gravitation of its contents. But here we're humbled: A 1998 discovery showed that the expansion is actually accelerating. As a result, understanding the full nature of gravity and the makeup of the universe has become one of 21st-century science's greatest challenges.

Chapter 9 in Context

Newton's law of gravitation is a big idea in physics. As these first nine chapters show, Newtonian gravitation combines with Newton's laws of motion to explain an astonishing array of phenomena from the everyday to the astronomical. In next few chapters we'll turn to Newtonian dynamics to study fluids, waves, sound, and heat.

Looking Ahead In Chapter 15 you'll learn that the electric force, like gravitation, follows an inverse-square law. We'll use this similarity to advantage, for example, in defining electric potential energy and in looking at bound systems such as atoms, which in some ways resemble miniature solar systems.

Newton's Law of Gravitation

(Section 9.1) **Newton's law of gravitation** is a universal description of the attractive force between any two masses in the universe. Newton's law applies exactly for point masses and spherical objects.

The gravitational acceleration varies with altitude; Newton's law of gravitation and his second law of motion give the gravitational acceleration a distance r from the center of a spherical body of mass M. For g at the surface of the body, r becomes the body's radius R.

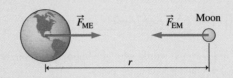

Newton's law of gravitation: $F = \dfrac{Gm_1m_2}{r^2}$ **Gravitational acceleration:** $g = \dfrac{GM}{r^2}$

Planetary Motion and Kepler's Laws

(Section 9.2) **Kepler's laws** describe planetary motion, and can be derived from Newton's law of gravitation and his laws of motion.

Kepler's first law: Planetary orbits are ellipses with the Sun at one focus

Second law: A planet's orbital motion sweeps out equal areas in equal times

Third law: The cube of the semimajor axis is proportional to the period squared

$$\frac{a^3}{T^2} = C$$

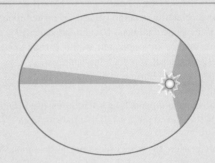

Gravitational Potential Energy

(Section 9.3) The **gravitational potential energy** U of a pair of gravitating objects is negative or at most zero, with $U = 0$ at infinite separation. Gravity is a conservative force, so total mechanical energy is conserved when objects interact only via gravity.

Gravitational potential energy: $U = -\dfrac{Gm_1m_2}{r}$ **Conservation of mechanical energy:** $K_0 + U_0 = K + U$

Artificial Satellites

(Section 9.4) **Escape speed** is the launch speed needed for an object to escape altogether from a gravitating body. Escape speed from Earth's surface is about 11 km/s.

Circular orbits are analyzed quantitatively using dynamics that you already know. A special circular orbit is **geosynchronous orbit**, with 24-hour period, where a spacecraft remains fixed at a point above Earth's equator.

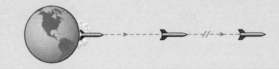

Escape speed, distance r from center of gravitating mass M: $v_{esc} = \sqrt{\dfrac{2GM}{r}}$

Energy in circular orbits:

$$E = K + U = -\frac{GMm}{2r} \qquad K = \tfrac{1}{2}mv^2 = \frac{GMm}{2r} \qquad U = -2K = -\frac{GMm}{r}$$

Other Aspects of Gravitation

(Section 9.5) **Tides** result from the difference in gravitational acceleration from one side of an object to the other. The Moon's gravity is the dominant cause of Earth's ocean tides.

Apparent weightlessness occurs not because there's no gravity in space but because all objects in free fall experience the same acceleration.

Newton's theory of gravity is an approximation to Einstein's **general theory of relativity**, which treats gravity as a geometrical aspect of space and time, and which differs from Newton's theory in regions of strong gravity, where escape speed approaches the speed of light.

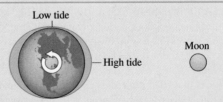

NOTE: Problem difficulty is labeled as ■ straightforward to ■■■challenging. Problems labeled BIO are of biological or medical interest.

Conceptual Questions

1. Explain in your own words why gravitational acceleration g is smaller at the equator than at the poles.
2. A satellite orbits Earth in a circle with radius R. A second satellite orbits Mars in a circle with the same radius. Which (if either) has the longer orbital period?
3. A satellite is in an elliptical orbit around Earth. Does Earth do work on the satellite at any point in its orbit? Repeat for a circular orbit.
4. Earth's core is denser than its outer layers. How does this affect gravitational acceleration g at the surface, compared to an Earth-like planet of the same total mass but uniform density?
5. Taking air resistance into account, would escape speed from Earth increase, decrease, or remain the same?
6. At what time of year is Earth traveling fastest in its orbit? When is it traveling slowest?
7. What's the advantage to launching satellites eastward rather then westward? Is it easier to put a satellite into orbit from Florida or Alaska?
8. How can you measure the mass of a planet with a moon? What if it has no moons?
9. Does escape speed depend on launch angle?
10. A comet can have a total mechanical energy that's positive, negative, or zero. Describe its path in each case.
11. Compare the fuel required for a trip from Earth to Moon with that required for the return trip.
12. Why is it said that the Cavendish balance is used to "weigh the Earth"?
13. Why doesn't a rocket's escape speed depend on its mass?
14. Describe the subsequent motion of a rocket launched with speed twice the escape speed.
15. Does a planet's orbital period depend on the planet's mass?
16. Why does the Moon dominate Earth's tides, even though the Sun dominates Earth's orbital motion?

Multiple-Choice Problems

17. Mars has mass 6.42×10^{23} kg and radius 3.37×10^6 m. What's the gravitational acceleration at the surface of Mars? (a) 3.8 m/s^2; (b) 4.9 m/s^2; (c) 6.2 m/s^2; (d) 9.8 m/s^2.
18. What's the gravitational force between two lead balls with masses 25 kg and 60 kg separated by 0.50 m center to center? (a) 5×10^{-8} N; (b) 1×10^{-7} N; (c) 2×10^{-7} N; (d) 4×10^{-7} N.
19. A 150-kg satellite is in circular orbit with radius 7.1×10^6 m around another planet. If its orbital period is 4 hours, the planet's mass is (a) 10^{27} kg; (b) 10^{26} kg; (c) 10^{25} kg; (d) 10^{24} kg.
20. A satellite is in a 1.5-hour-period circular orbit around Earth. If the satellite moves to a new orbit with radius twice that of the original, the new period is (a) 2.4 hours; (b) 2.8 hours; (c) 3.0 hours; (d) 4.2 hours.
21. A satellite is in a 7000-km-radius circular orbit around Earth. If the satellite moves to a new orbit with period twice that of the original, the new radius is (a) 10,000 km; (b) 11,100 km; (c) 14,000 km; (d) 19,800 km.
22. A rocket launched from Earth at 2400 m/s reaches a maximum height of (a) 300 km; (b) 600 km; (c) 900 km; (d) infinite height (escape).

23. Ignoring air resistance, a ball dropped from rest at altitude 250 km strikes the ground with speed (a) 1500 m/s; (b) 2200 m/s; (c) 2800 m/s; (d) 3400 m/s.
24. The period of a satellite's orbit depends on (a) its mass; (b) the radius of the planet it's orbiting; (c) the planet's mass; (d) all three factors (a), (b), and (c).
25. A planet's closest approach to the Sun is its (a) aphelion; (b) perihelion; (c) minor axis; (d) eccentricity.
26. The total energy of a satellite in elliptical orbit (a) is zero; (b) is positive; (c) is negative; (d) cannot be determined without more information about the orbit.
27. What's the period of a satellite in circular orbit 4000 km above Earth's surface? (a) 1.5 hours; (b) 2.9 hours; (c) 5.2 hours; (d) 7.1 hours.
28. Which of the following is true about ocean tides? (a) They're caused primarily by the Sun. (b) They're most extreme when the Moon is full or new. (c) They're most extreme close to the North Pole. (d) They aren't affected by the Sun.

Problems

Section 9.1 Newton's Law of Gravitation

29. ■ What's the approximate gravitational attraction between two 115-kg football players standing 25.0 m apart?
30. ■ Two identical balls, their centers 25.0 cm apart, experience a mutual gravitational force of 8.8×10^{-6} N. Find each ball's mass.
31. ■ Two stars with masses 2.3×10^{30} kg and 6.8×10^{30} kg form a binary system, with interstellar separation 8.8×10^{11} m. (a) Find the magnitude of the force between these stars. (b) Compare with the force between Earth and Sun.
32. ■■ Compute the net force on Earth due to the Sun and Moon (a) when the Moon is full and (b) when the Moon is new.
33. ■ In a hydrogen atom, proton and electron are separated by 5.29×10^{-11} m. What's the gravitational force between the two particles?
34. ■ Use the data in Appendix E to compute the gravitational acceleration g at the surfaces of Mercury and Venus.
35. ■ Use the data in Appendix E to compute the gravitational acceleration g at the surfaces of Saturn and Jupiter.
36. ■■ A spherical asteroid with radius of 9.50 km has uniform density 3500 kg/m^3. Find the gravitational acceleration at the asteroid's surface.
37. ■■ Deimos is one of Mars's two small moons. It orbits Mars in a nearly circular orbit of radius 23,500 km with a period of 1.26 days. From these data determine the mass of Mars and compare with the value in Appendix E.
38. ■■ Saturn's moon Titan is in a nearly circular orbit of radius 1.22×10^9 m and period 15.9 days. From these data determine the mass of Saturn and compare with the value in Appendix E.
39. ■■ In a typical Cavendish experiment, two small lead balls on a rod are attracted to two large lead balls, as shown in Figure 9.7. The small and large balls are, respectively, 1.0 cm and 3.4 cm in diameter, and the density of lead is 11,350 kg/m^3. (a) Find the attractive force between a small ball and a large one when their *surfaces* are 0.50 cm apart. (b) The rod connecting the small balls is 15 cm long. Find the torque on the apparatus under these conditions.
40. ■■ Earth's orbit is approximately a circle of radius 1.50×10^{11} m (a) Find Earth's centripetal acceleration. (b) Use your answer to

compute the gravitational force exerted on Earth by the Sun. (c) Show that your answer to part (b) agrees with the force computed directly from Newton's law of gravitation.

41. ■■ At what height above Earth's surface is the gravitational acceleration reduced from its sea-level value by (a) 0.1%, (b) 1.0%, and (c) 10%?

42. ■■■ Three solid balls are placed in the *x-y* plane as shown in Figure P9.42. Find the net gravitational force on the ball at the origin.

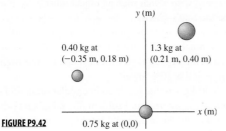

0.40 kg at (−0.35 m, 0.18 m)

1.3 kg at (0.21 m, 0.40 m)

FIGURE P9.42   0.75 kg at (0,0)

y (m)

x (m)

Section 9.2 Planetary Motion and Kepler's Laws

43. ■ Find the eccentricity of an ellipse whose major axis is twice its minor axis.

44. ■ Draw ellipses with the following eccentricities: 0.01, 0.1, 0.5, 0.9.

45. ■■ Use data for Mars in Appendix E to determine the distance of the Sun (at one focus of the ellipse) from the orbit's geometric center.

46. ■ Draw to scale on the same diagram the elliptical orbits of Earth and Mars.

47. ■ Compute the constant $C = a^3/T^2$ for the orbits of Mercury, Venus, and Jupiter.

48. ■■ Find the perihelion and aphelion distances for Pluto. Compare with Neptune's average distance from the Sun.

49. ■■ Satellites A and B are in circular orbits around Earth, with A twice as far as B from Earth's center. How do their orbital periods compare?

50. ■■■ Use conservation of angular momentum to estimate the difference in Earth's orbital speed from perihelion to aphelion.

51. ■■ A satellite is in circular orbit at a height R_E above Earth's surface. (a) Find its orbital period. (b) What height is required for a circular orbit with a period double that found in part (a)?

52. ■■ The Moon orbits Earth in a nearly circular orbit of radius 3.84×10^8 m and period 27.3 days. Compute the constant $C = a^3/T^2$ for this orbit, and use the result to find the period of a satellite in circular orbit 500 km above Earth's surface.

53. ■■■ A *Hohmann ellipse* is the trajectory for a spacecraft going from one planet to another that requires the least energy. This ellipse has its perihelion at one planet's orbit and its aphelion at the other's (Figure P9.53). Find the travel time from Earth to Jupiter along such a trajectory, and compare your answer with the orbital periods of those planets.

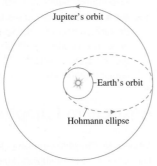

Jupiter's orbit

Earth's orbit

Hohmann ellipse

FIGURE P9.53

Section 9.3 Gravitational Potential Energy

54. ■ Find the potential energy of (a) the Earth-Sun system and (b) the Earth-Moon system.

55. ■ Compare the potential energy of (a) the Sun-Jupiter system and (b) the Sun-Saturn system.

56. ■■ (a) What's the gravitational potential energy of the electron and proton in an atom, 5.29×10^{-11} m apart? (b) How does this compare with the atomic binding energy of about 2×10^{-18} J? (c) Comment on the influence of gravity in the atom.

57. ■■ Ignoring air resistance, find the speed of a 1-kg ball when it strikes the ground after being dropped from rest at heights of (a) 10 km; (b) 1000 km; (c) 10^7 m.

58. ■■ Find the maximum height of a rocket shot straight up from Earth's surface at 2200 m/s.

59. ■■ What's the change in potential energy of the Earth-Sun system from perihelion to aphelion? Use your answer and conservation of energy to find the corresponding difference in Earth's orbital speed.

60. ■■ Halley's comet reaches a maximum speed of 54.6 km/s at perihelion, 8.84×10^{10} m from the Sun. Find its speed at aphelion, 5.27×10^{12} m from the Sun.

61. ■■ What's the total mechanical energy of a 500-kg satellite in circular orbit 1500 km above Earth's surface?

62. ■■■ Two identical asteroids, each with mass *M* and radius *R*, are released from rest a large distance apart. They're attracted by their mutual gravitation and eventually collide. (a) Use conservation of energy to find the speed *v* of each asteroid just before the collision. (b) Evaluate numerically, assuming the asteroids each have mass 2.0×10^{13} kg and radius 1.0 km.

63. ■■ With what speed does a rock strike the Moon when it's dropped from rest from a great distance above the Moon's surface?

64. ■■ One proposal for dealing with radioactive waste is to shoot it into the Sun. Suppose a waste canister were dropped from rest, starting in the vicinity of Earth's orbit. At what speed would it hit the Sun?

Section 9.4 Artificial Satellites

65. ■ What's the orbital radius of a satellite in a 48-hour-period circular Earth orbit?

66. ■ Find the period of a satellite with a circular orbit 150 km above Earth's surface.

67. ■■ For elliptical Earth orbits, the near and far points are called the perigee and apogee. Find the period of a satellite with perigee and apogee 200 km and 1600 km above Earth's surface, respectively.

68. ■ Mars has mass 6.42×10^{23} kg and radius 3.37×10^6 m. Find the escape speed from Mars's surface.

69. ■ Determine the escape speed from (a) Jupiter's moon Callisto, with mass 1.07×10^{23} kg and radius 2.40×10^6 m, and (b) a neutron star with the Sun's mass and a radius of 6.0 km.

70. ■■ A satellite is in geosynchronous orbit. (a) What is the satellite's speed? (b) What additional speed is required for the satellite to escape Earth's gravity completely?

71. ■■ The Moon's rotation period is 27.3 days. What's the height of a "lunosynchronous" satellite that orbits the Moon, appearing fixed in the sky relative to a lunar observer? Compare your answer to the Moon's radius.

72. ■■ (a) Calculate the orbital period of Jupiter's moon Io, which is in a nearly circular orbit of radius 4.22×10^5 km. (b) What's Io's orbital speed?

73. ■■■ A binary system comprises two stars of equal mass *M* separated by distance *d*, orbiting their common center of mass. (a) Find the period of their orbital motion. (b) Compare your result with the period of a small planet ($m \ll M$) orbiting an isolated star of the same mass *M* in a circular orbit with the same radius *d*.

74. ■■■ Suppose a satellite with a 24-hour period over Earth's equator has a noncircular, elliptical orbit. (a) Explain why this satellite doesn't stay above the same point. (b) What is the maximum possible eccentricity of its orbital ellipse?

75. ■■ Find (a) the kinetic energy, (b) the potential energy, and (c) the total mechanical energy of the Moon in its orbit.

76. ■■ Find the energy needed to place a 1.0-kg satellite, initially at rest on Earth's surface, into geosynchronous orbit.

77. ■■■ A satellite in a circular orbit 1000 km above Earth has total mechanical energy -4.0×10^{10} J. Find the satellite's (a) kinetic energy, (b) mass, and (c) speed.

78. ■■ The Moon has no atmosphere, making extremely low orbits possible. Find the period of a circular orbit just above the Moon's surface ($M_{Moon} = 7.35 \times 10^{22}$ kg, $R_{Moon} = 1740$ km).

79. ■■ In 1968, Frank Borman, Jim Lovell, and Bill Anders became the first humans to orbit the Moon. Their orbit was nearly circular, ranging from 59.7 to 60.7 miles above the lunar surface (1 mile = 1.609 km). Find the orbital period, and compare it with the results of the preceding problem.

80. ■■■ (a) To what radius would Earth have to be shrunk, with no loss of mass, in order for escape speed from its surface to double? (b) What would be Earth's average density under those conditions?

Section 9.5 Other Aspects of Gravitation

81. ■ Find the gravitational acceleration 20 km from the center of a 9.9×10^{30}-kg black hole. (This is far enough for Newtonian physics to be applicable.)

82. ■■ Our solar system is 25,000 light years from the center of the Milky Way galaxy. (a) If our solar system orbits the galactic center in a circular orbit with speed 230 km/s, what's its orbital period? (b) What's the approximate mass of the galactic center, assuming it's essentially spherical?

83. ■■ The mean center-to-center Earth-Moon distance is 3.84×10^8 m. Evaluate the Moon's tidal effect by calculating the gravitational acceleration of a water droplet on the side of the Earth (a) closest to and (b) farthest from the Moon. (c) Calculate the difference between these two accelerations—which is a measure of the Moon's tidal effect.

84. ■■ Repeat the preceding problem for the Earth-Sun tidal interaction. You should find a greater solar acceleration in (a) and (b), but a smaller difference in (c), showing why the Sun's tidal effect is smaller than the Moon's despite the Sun's greater gravitational force on Earth.

85. ■■ Tidal forces can break apart objects that orbit too close to a planet, producing rings like those of Saturn. Consider an asteroid 500 km in radius, 75 Mm from Saturn's center. What's the difference in gravitational acceleration between the two sides of the asteroid?

General Problems

86. ■■ A small, spherical asteroid has radius 3.50 km and density 4830 kg/m³. (a) What's the gravitational acceleration at its surface? (b) What's the escape speed from the surface?

87. ■■ Rank in increasing order the gravitational acceleration at the surface of planets that have the given masses and radii, where M_E and R_E are the mass and radius of Earth. Assume spherically symmetric planets. (a) $M = M_E, R = R_E$; (b) $M = 2M_E, R = 2R_E$; (c) $M = 0.5M_E, R = 0.5R_E$; (d) $M = 1.8M_E, R = 1.5R_E$; (e) $M = 0.75M_E, R = 0.90R_E$.

88. ■■ An Earth satellite's elliptical orbit ranges from 230 km to 890 km altitude. At the high point, it's moving at 7.23 km/s. How fast is it moving at the low point?

89. BIO ■■ **High jump.** The winning high jump in the 2004 Summer Olympics was 2.34 m. Assuming the same takeoff conditions, how high could the winner jump on Mars? Repeat for the long jump, for which the winning jump was 8.59 m.

90. ■■ You're tour director for a lunar trip, and want to award your passengers with certificates commemorating their crossing the point where the Moon's gravity becomes stronger than Earth's. How far from Earth should you award the certificate? Express your answer in meters and as a fraction of the center-to-center Earth-Moon distance.

91. ■■ Mercury's orbital speed varies from 38.8 km/s at aphelion to 59.0 km/s at perihelion. (a) If aphelion is 6.99×10^{10} m from the Sun's center, how far is perihelion? (b) Use your answer to compute Mercury's orbital eccentricity, and compare with Appendix E.

92. ■■ Voyager 1 and Voyager 2, launched in 1977, are the first human-made objects to leave our solar system. Find the minimum speed with which such an object must leave Earth's orbit to escape the Sun's gravity.

93. ■■■ A spacecraft is in circular Earth orbit at 5500 km altitude. By how much will its altitude decrease if it moves to a new circular orbit where (a) its orbital speed is 10% higher or (b) its orbital period is 10% lower?

94. ■■■ A 50-kg satellite is placed in circular orbit, 750 m above the surface of the asteroid in Problem 86. What's its orbital period? Why is the answer so different from the periods of Earth-orbiting satellites?

95. ■■■ Tidal effects cause the Moon's orbital period to increase at about 35 ms per century. Assuming a circular orbit, to what rate of change in the Earth-Moon distance does this correspond?

96. ■■■ Two satellites are in geosynchronous orbit but diametrically opposite positions (Figure GP9.96). Into how much lower a circular orbit should one spacecraft descend if it's to catch up with the other after 10 complete orbits? Neglect rocket firing times and the time spent moving from one circular orbit to another.

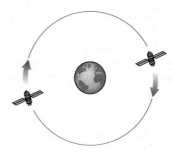

FIGURE GP9.96

97. ■■■ Suppose you can jump 55 cm high from a standing start on Earth. Find the radius of the largest spherical asteroid you could escape by jumping, assuming a uniform asteroid density of 4000 kg/m³.

98. ■■■ (a) Find the change in potential energy of the Earth-Sun system between Earth's aphelion and perihelion. (b) Find the corresponding change in Earth's orbital kinetic energy.

99. ■■■ An object collapses to a black hole when escape speed at its surface becomes the speed of light. At what radius does that occur for an object with (a) the Sun's mass and (b) the mass of a galaxy, approximately 10^{11} solar masses? (You can use Newtonian gravitation for this calculation.)

Answers to Chapter Questions

Answer to Chapter-Opening Question

The rotation rates of stars around the galaxy's center are much faster than those predicted by Newton's law of gravitation, using just the mass of all the visible objects. From this, we can infer a large amount of unseen matter in the galaxy.

Answers to GOT IT? Questions

Section 9.1   (d) $g/2$
Section 9.4   (c) $4R_E$

■ Why does the diver experience increasing pressure at greater depth?

In this chapter we turn to solids and fluids, materials made of so many particles that we can't hope to deal with them individually. Nevertheless, we'll show how ideas from Newtonian dynamics explain their properties. You'll see how solids respond to external forces, and how fluids—liquids and gases—develop pressure that's responsible for such phenomena as buoyancy and fluid flow. We'll relate pressure and fluid flow to the familiar concepts of energy and force, and we'll see how friction affects fluid flow.

10.1 States of Matter

Everyday experience exposes you to the three states of matter: **solids**, **liquids**, and **gases**. They're obviously different on the macroscopic scale, but ultimately, forces at the molecular level distinguish them.

- In **solids**, molecules are locked into place, producing a material that tends to retain its shape, although it's subject to deformation by external forces.
- In **liquids**, intermolecular forces keep molecules close together, but they're free to move. Liquids therefore flow readily to assume the shape of their container, but a liquid's density remains nearly constant.
- In **gases**, molecules are far apart and interact only weakly. Gases flow and change density, and therefore expand to fill their container.

To Learn

By the end of this chapter you should be able to

- Describe the different states of matter.
- Characterize matter by its density.
- Relate stress and strain in materials.
- Describe fluid pressure and its relation to force.
- Explain the pressure-depth relation for a static fluid.
- Explain buoyancy and Archimedes' principle.
- Describe the ideal-fluid approximation.
- Explain the continuity equation for an incompressible fluid.
- Describe Bernoulli's equation and its relation to energy conservation in a fluid.
- Use the continuity equation and Bernoulli's equation to solve problems involving fluid flow.
- Describe surface tension and capillary action in terms of intermolecular forces.
- Describe viscosity and explain quantitatively how it affects fluid flow.

TABLE 10.1 Densities of Common Solids, Liquids, and Gases

Material	Density (kg/m³)	Material	Density (kg/m³)
Solids		**Liquids**	
Ice (near 0°C)	917	Gasoline	680
		Ethanol	790
Concrete (typical)	2000	Benzene Oil (typical)	900
Aluminum	2700	Water (fresh)	1000
Iron or steel	7800	Seawater	1030
Brass	8600	Blood	1060
Copper	8900	Mercury	13,600
Silver	10,500	**Gases** (1 atm, 0°C)	
Lead	11,300	Helium	0.18
Gold	19,300	Air	1.28
Platinum	21,400	Argon	1.78
Uranium	19,100	Water vapor	0.804

The distinction between solid, liquid, and gas is usually clear-cut, with well-defined values of temperature and pressure where transitions occur. (You'll learn more about *phase transitions* in Chapter 13.) But there are exceptions. At high pressure, liquid and gas phases merge into *superfluid*. Over long time periods, solid glass actually deforms by flowing. And gases of electrically charged particles behave so unusually that they're often considered a fourth state of matter, called *plasma*. And in an unusual quantum state called a *Bose-Einstein condensate,* all the individual particles act essentially in unison.

An important property of materials is **density**, ρ, defined in Chapter 1 as mass per volume: $\rho = m/V$. The densities of some common materials are shown in Table 10.1. The SI density unit is kg/m³, although chemists often use g/cm³. The intermolecular forces that determine phase also affect density, with the strongly bound solid phase usually the densest, liquid a bit less dense, and gas much less dense. An important exception is water, for which solid (ice) is less dense than liquid. As you'll see, that's why ice floats.

Because of their similar densities, solids and liquids are sometimes grouped as **condensed matter**. Their ability to flow groups liquids and gases as **fluids**. We'll consider the elastic properties of solids in the next section and then devote the remainder of the chapter to fluids.

GOT IT? Section 10.1 A solid cube of metal labeled O is split into two pieces, A and B, as shown, with A twice the size of B. Which of the following correctly ranks in order the density of the original cube and the two pieces? (a) O > A > B; (b) O > A = B; (c) O = A = B.

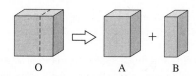

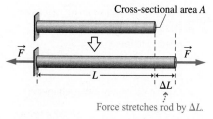

(a) Tensile stress

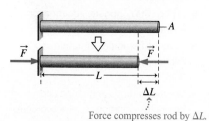

(b) Compressional stress

FIGURE 10.1 The two kinds of stress on a rod. Each results in strain $\Delta L/L$.

10.2 Solids and Elasticity

A block of wood seems solid enough. But stand at the end of a wooden diving board, and it bends under your weight. Apply an increasing force, and it bends further—and then breaks. The karate expert knows this and can break a stack of wood with one blow (see Conceptual Example 10.3). Some solids are stronger and bend or break less easily than wood. Here we'll characterize and quantify the elastic properties of solids.

Stress and Strain

We'll first consider elasticity in one dimension. Figure 10.1a shows a solid rod of length L and cross-sectional area A. A force of magnitude F is applied to both ends as shown, stretching the rod an amount L. The force is applied across the rod's entire cross section. The force per unit area determines the stretch; that quantity is the **stress**, F/A.

TABLE 10.2 Young's Modulus and Bulk Modulus of Selected Materials

Material	Young's modulus (N/m^2)	Bulk modulus (N/m^2)
Aluminum	7×10^{10}	7×10^{10}
Concrete	3×10^{10}	
Copper	11×10^{10}	14×10^{10}
Mercury		3×10^{10}
Steel	20×10^{10}	16×10^{10}
Cortical bone (tension)	1×10^{10}	
Cortical bone (compression)	2×10^{10}	
Trabecular bone (tension)	0.3×10^{10}	
Trabecular bone (compression)	0.1×10^{10}	
Water		0.2×10^{10}
Iron	15×10^{10}	12×10^{10}

The rod responds with a fractional change in length, $\Delta L/L$, called **strain**. For the outward pull shown in Figure 10.1a, the material is under **tension** and exhibits **tensile stress**. We could also push inward (Figure 10.1b), putting it under **compression**, resulting in **compressional stress**.

For small stress, strain is proportional to stress. That's because molecular bonds in the solid act as miniature Hooke's law springs (Figure 10.2, top). But for stresses greater than the so-called **elastic limit**, strain is no longer proportional to stress, and for large enough stress, the solid breaks apart (Figure 10.2, bottom).

Within the elastic limit, stress and strain follow a linear equation: stress = $Y \times$ strain, where Y is **Young's modulus**, a constant for a particular solid (Table 10.2). A larger Y implies a stronger material. Using the definitions of stress and strain, our stress-strain equation becomes

$$\frac{F}{A} = Y\frac{\Delta L}{L} \quad \text{(Young's modulus; SI unit: N/m}^2\text{)} \quad (10.1)$$

For example, applying a 1-kN force to a 10-cm-long aluminum rod with 1-cm^2 cross-sectional area causes the rod to stretch by

$$\Delta L = \frac{FL}{YA} = \frac{(1000 \text{ N})(0.1 \text{ m})}{(7 \times 10^{10} \text{ N/m}^2)(10^{-4} \text{ m}^2)} = 1.4 \times 10^{-5} \text{ m} = 0.014 \text{ mm}$$

That's a small stretch for such a large force, showing that aluminum is a very stiff material. Equation 10.1 works for both tension and compression, because both F and ΔL are magnitudes, and hence both are positive for both cases.

Usually, Young's modulus is about the same under tension and compression. An exception is given in Table 10.2: Young's modulus for bone under tension differs from that under compression. Trabecular bone is the softer bone tissue in the interior of many bones, while denser cortical bone forms the harder exterior. Trabecular bone has a larger Y under tension, while the harder cortical bone has a larger Y under compression. The measured Young's modulus of a real bone, comprising both types, lies between the listed values.

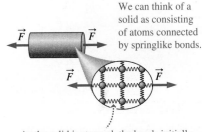

We can think of a solid as consisting of atoms connected by springlike bonds.

As the solid is stressed, the bonds initially respond like Hooke's law springs: Stress is proportional to strain.

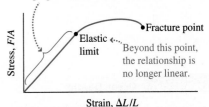

FIGURE 10.2 A solid's response to applied force is linear, but only up to the elastic limit. Beyond that the response is nonlinear, and fracture will occur.

EXAMPLE 10.1 **Weight Lifter**

A weight lifter performing a bench press holds a 340-kg bar overhead. Compute the stress, strain, and Young's modulus in each humerus (upper arm) bone, which compresses by 0.15 mm in this position. The humerus is 25 cm long, with average diameter 3.0 cm.

ORGANIZE AND PLAN Here the force on each arm is half the weight, or $F = mg/2$. The bone's cross-sectional area is $A = \pi r^2$, so you can find the stress F/A. Computing the strain $\Delta L/L$ then allows you to compute Young's modulus as the ratio of stress to strain.

Known: $\Delta L = 0.15$ mm; $L = 25$ cm; bone diameter $= 3.0$ cm.

SOLVE Using the given numerical values, with radius half the diameter, or $r = 0.015$ m,

$$\text{stress} = \frac{F}{A} = \frac{mg/2}{\pi r^2} = \frac{(340 \text{ kg})(9.8 \text{ m/s}^2)}{2\pi(0.015 \text{ m})^2} = 2.4 \times 10^6 \text{ N/m}^2$$

The strain is

$$\text{strain} = \frac{\Delta L}{L} = \frac{1.5 \times 10^{-4} \text{ m}}{0.25 \text{ m}} = 6.0 \times 10^{-4}$$

Then Young's modulus is

$$Y = \frac{\text{stress}}{\text{strain}} = \frac{2.4 \times 10^6 \text{ N/m}^2}{6.0 \times 10^{-4}} = 4.0 \times 10^9 \text{ N/m}^2$$

REFLECT Note that the bone is under compression. The value found for Young's modulus is reasonable, being between the compression values for the two bone types.

MAKING THE CONNECTION Is stress increased, decreased, or unchanged while the weight is being accelerated upward?

ANSWER The net upward force on the bar must be larger than mg, so stress increases.

Volume Compression and Bulk Modulus

Stress and strain also occur in three dimensions. Figure 10.3 shows a solid block pushed inward in all directions. We'll still call the stress F/A, the force per unit area on each surface. But here we consider **volume strain** $\Delta V/V$. As in one dimension, this strain is dimensionless.

Molecular bonds respond linearly to the applied force, up to an elastic limit, giving a stress-strain relation analogous to Equation 10.1:

$$\frac{F}{A} = -B\frac{\Delta V}{V} \quad \text{(Bulk modulus; SI unit: N/m}^2) \qquad (10.2)$$

where B is the **bulk modulus**. The negative sign in Equation 10.2 reflects the negative volume change resulting from an inward force. The three-dimensional bulk modulus plays the same role as Young's modulus in one dimension. Table 10.2 includes bulk moduli.

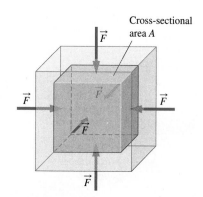

FIGURE 10.3 A cube undergoing volume change ΔV in response to compressional stress.

Cross-sectional area A

✓**TIP**

When under compression, the volume of a material shrinks, so ΔV is negative.

EXAMPLE 10.2 **Ocean Depths**

An iron block near the ocean bottom experiences a 3.0×10^7-N/m² stress. What's its fractional change in volume?

ORGANIZE AND PLAN The fractional change is the volume strain $\Delta V/V$, given by Equation 10.2, using the known stress and B from Table 10.2.

Known: $F/A = 3.0 \times 10^7$ N/m²; $B = 12 \times 10^{10}$ N/m².

SOLVE Equation 10.2 gives the volume strain:

$$\frac{\Delta V}{V} = -\frac{1}{B}\frac{F}{A} = -\frac{1}{12 \times 10^{10} \text{ N/m}^2}(3.0 \times 10^7 \text{ N/m}^2)$$
$$= -2.5 \times 10^{-4}$$

REFLECT The force from the water may seem large, but water's substantial density means it exerts large forces, increasing with depth, as you'll learn in the next section. Despite the large stress, the volume strain is much less than 1%.

MAKING THE CONNECTION For an iron cube 1 m on each side, how much does the length of each side change under these conditions?

ANSWER The original volume was exactly 1 m^3, so the new volume is $V = 1 \text{ m}^3 - 2.5 \times 10^{-4} \text{ m}^3 = 0.99975 \text{ m}^3$. The cube's side length is the cube root of its volume, or 0.999917 m. Each side has shrunk by $1 \text{ m} - 0.999917 \text{ m} = 8.3 \times 10^{-5} \text{ m}$.

CONCEPTUAL EXAMPLE 10.3 **Fracturing Boards**

How does a karate expert break a stack of boards with a single blow?

SOLVE Several concepts apply here. The boards are supported at the ends, so when the hand strikes them they're under tension. Wood breaks at lower stress in tension than in compression. The expert's hand bones, with higher stress limits, don't fracture. As a board bends, the maximum tension occurs at the bottom, so that's where it breaks; the hand itself doesn't break the board directly. Each board pushes down into the next one as it breaks, transferring the force from the hand. Therefore, it's *not* twice as hard to break two boards, three times as hard to break three, and so on. As long as the force transmits between boards, the impressive string of multiple breaking continues.

REFLECT Concrete is harder than wood, with a higher elastic limit and bulk modulus. The novice is well advised to start with wood.

GOT IT? Section 10.2   Rods of copper, aluminum, and steel are the same size and shape. If the same outward force is applied to the ends of each, rank in increasing order each rod's stretch.

Reviewing New Concepts

Important ideas about elasticity:

- In one dimension, stress F/A and strain $\Delta L/L$ are related by Young's modulus Y:

$$\frac{F}{A} = Y\frac{\Delta L}{L}$$

- In three dimensions, stress F/A and strain $\Delta V/V$ are related by the bulk modulus B:

$$\frac{F}{A} = -B\frac{\Delta V}{V}$$

10.3 Fluid Pressure

Everyday experience makes you aware of pressure. You inflate tires, rubber rafts, and air mattresses. You dive or fly, feeling pressure on your eardrums. You hear meteorologists reporting atmospheric pressure and its implications for weather. You use pressurized spray cans, and perhaps you've cooked with a pressure cooker.

Poke a hole in a container of liquid or gas, and the fluid escapes. This implies that the fluid exerts a force on the container walls. The fluid force per unit area is what we call **pressure** P:

$$P = \frac{F}{A} \tag{10.3}$$

The SI pressure unit is N/m^2, which defines the **pascal** (Pa). Standard atmospheric air pressure at sea level is 1.013×10^5 Pa. Hence, the atmosphere (atm) is a common non-SI pressure unit, with 1 atm $= 1.013 \times 10^5$ Pa.

Pressure is a scalar quantity, even though force has direction. A fluid exerts a pressure-associated force on any material it contacts, such as its container walls, an object immersed in the fluid, or even adjacent fluid (Figure 10.4). For a **static fluid**—one that's not flowing—the force of the fluid must be perpendicular to the surface. By Newton's third law, the surface exerts a force on the fluid that's equal in magnitude but opposite in direction to the force the fluid exerts on the surface. If these forces weren't perpendicular to the surface, then the tangential component of the force on the fluid would cause it to flow, and it wouldn't be static.

✓**TIP**

The pressure units Pa and N/m^2 are equivalent and therefore interchangeable.

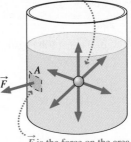

The fluid exerts pressure internally as well as on the container. The internal pressure is the same in all directions.

$\vec{F}$ is the force on the area A, so the pressure is $P = F/A$.

FIGURE 10.4 Pressure, the force per unit area, is exerted equally in all directions.

CONCEPTUAL EXAMPLE 10.4 **Internal Fluid Pressure**

A fluid exerts pressure on materials it contacts—including adjacent fluid. What direction is that force?

SOLVE The fluid exerts a force perpendicular to any surface it contacts, such as the arbitrary surface shown on the container wall in Figure 10.4. Because the fluid force is perpendicular to any surface it contacts, we conclude that fluid pressure exerts forces in all directions.

REFLECT Why doesn't the force of fluid on adjacent fluid cause the fluid to accelerate? In a static fluid, any volume of fluid has zero net force on it. That force comes from the surrounding fluid, container walls, or other materials the fluid volume contacts.

Pressure and Depth

Fluid pressure increases with depth. Your ears sense that increase when you dive underwater or drive down a mountain road. Figure 10.5 shows how the fluid's weight is responsible for the increasing pressure. Liquid with density ρ fills a tank to depth h. We've highlighted a fluid column with cross-sectional area A. The column's weight is mg, creating pressure $P = F/A = mg/A$ at the bottom. The fluid mass is $m = \rho V$, and the column has volume $V = Ah$. Combining these results,

$$P = \frac{mg}{A} = \frac{\rho(Ah)g}{A} = \rho g h$$

There's one correction to make here. There may already be pressure at the top of the column, perhaps atmospheric pressure from the air above. What we've calculated is actually the pressure *difference* between bottom and top of the column. If the pressure at the top is P_0, then the pressure at the bottom becomes

$$P = P_0 + \rho g h \qquad \text{(Liquid pressure at depth } h\text{; SI unit: Pa)} \qquad (10.4)$$

Our calculation assumes h is depth measured positive going downward. As h increases, so does pressure P. Move upward through the liquid and h decreases; so, too, does pressure.

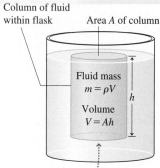

Column of fluid within flask   Area A of column

Fluid mass $m = \rho V$

Volume $V = Ah$

h

Pressure at bottom of column is due to weight of overlying fluid:

$$P = \frac{F}{A} = \frac{mg}{A}$$

FIGURE 10.5 Finding fluid pressure as a function of depth.

EXAMPLE 10.5 **Deep-Sea Diver**

Scuba divers typically don't go deeper than about 40 m. What's the pressure in freshwater at that depth?

ORGANIZE AND PLAN Equation 10.4 gives pressure as a function of depth. Here the density is that of freshwater, 1000 kg/m^3. The surface pressure is atmospheric pressure, 1.01×10^5 Pa.

Known: $P_0 = 1.01 \times 10^5$ Pa; $\rho = 1000$ kg/m^3; $h = 40$ m.

SOLVE Equation 10.8 gives

$$P = P_0 + \rho g h$$
$$= 1.01 \times 10^5 \text{ Pa} + (1000 \text{ kg/m}^3)(9.80 \text{ m/s}^2)(40 \text{ m})$$
$$= 4.93 \times 10^5 \text{ Pa}$$

REFLECT We used SI throughout, so the units of the product $\rho g h$ must be Pa, but verify this for yourself. Our answer is almost 5 atm. The application box describing decompression sickness shows why greater depths—and therefore pressures—are risky.

MAKING THE CONNECTION How much does air pressure increase when you descend 40 m?

ANSWER This problem is the same as in the example, except that the density of air is only 1.28 kg/m^3 (Table 10.2). Therefore, pressure change in air is only $\rho g h = 500$ Pa, or just 0.005 atm. Pressure changes depend strongly on fluid density.

Decompression Sickness

Divers who venture to significant depths experience great pressure—enough that nitrogen gas from the air dissolves into body fluids, including blood. If the diver surfaces too quickly, dissolved nitrogen emerges suddenly as gas, just like the bubbles you see when opening a bottle of soda. The result can be painful or even fatal. To prevent this, the diver can surface slowly, or use a decompression chamber, in which the pressure is reduced slowly to 1 atm. Either way, dissolved nitrogen is released slowly and painlessly.

Pascal's Principle

Any external force—not only gravity—can increase fluid pressure. Figure 10.6 shows a fluid-filled cylinder with a movable piston. Push on the piston with force F, and the pressure throughout the fluid increases by F/A, where A is the cylinder's cross-sectional area. This is an illustration of **Pascal's principle**:

Pascal's principle: Any external pressure applied to a confined fluid is transmitted throughout the entire fluid.

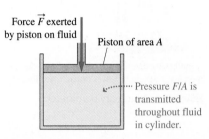

FIGURE 10.6 The applied force increases the fluid pressure.

Pascal's principle is about increased pressure applied from outside the fluid. The resulting pressure increase is in addition to any other pressure variation that may exist, such as a $\rho g h$ increase in pressure with depth.

A common application of Pascal's principle is in the hydraulic lift, shown in Figure 10.7. The key to this device is the difference in cross-sectional areas of the two pistons. Since force $F = PA$, a relatively small force applied to the left piston can lift a heavy car on the right piston. This application of Pascal's principle requires a liquid, which is essentially incompressible; a gas would change volume and therefore would not give much lift.

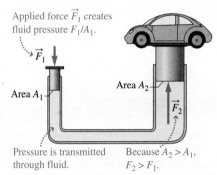

FIGURE 10.7 A hydraulic lift.

Garage Lift

In Figure 10.7, take the area of the piston holding the car to be 20 times that of the other piston. With the two pistons at the same height, what force must be applied to the left-hand piston to lift a 1200-kg car?

ORGANIZE AND PLAN With the two pistons at the same height, there's no $\rho g h$ pressure difference, so the pressures on the two sides are equal. That pressure is atmospheric pressure P_0 plus the pressure F/A due to the piston.

Known: area ratio = 20; car mass = 1200 kg.

SOLVE Equating pressures at left (1) and right (2), $P_0 + F_1/A_1 = P_0 + F_2/A_2$. Atmospheric pressure P_0 cancels, leaving $F_1/A_1 = F_2/A_2$, which reduces to $F_1 = (A_1/A_2)F_2$. F_2 is the car's weight, so with $A_1/A_2 = 1/20$,

$$F_1 = (A_1/A_2)F_2 = \tfrac{1}{20}(mg) = \tfrac{1}{20}(1200 \text{ kg})(9.80 \text{ m/s}^2) = 590 \text{ N}$$

REFLECT The required force is just 1/20 of the car's weight. You can raise the car by sitting on the other piston if your mass is at least 60 kg!

MAKING THE CONNECTION Aren't you getting something for nothing here? Is there some violation of the principle of energy conservation?

ANSWER No. You have to move the smaller piston a lot farther than the car rises. The work you do—force times the distance you move the piston—is equal to the larger force on the larger piston times the smaller distance it moves. That work ends up increasing the car's potential energy. Once the car starts rising, you'll need extra force to raise the fluid above the level of the left-hand piston.

Hydraulic systems are widely used. One important application is your car's brakes. When you step on the brake, you apply pressure to hydraulic fluid in tubing running to the wheels. There the pressure pushes brake pads against spinning disks. The resulting friction slows the wheels' rotation.

Pressure Gauges

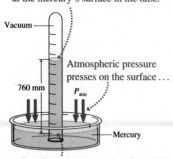

A vacuum has zero pressure, so $P_0 = 0$ at the mercury's surface in the tube.

Vacuum

760 mm

P_{atm}

Atmospheric pressure presses on the surface . . .

Mercury

. . . and pushes mercury up the tube until the mercury's weight balances the pressure force.

FIGURE 10.8 A mercury barometer.

The pressure-versus-depth relationship (Equation 10.4) is the principle behind some pressure gauges, including the **mercury barometer** shown in Figure 10.8. The pressure P_0 on the top of the mercury in the column is zero, while atmospheric pressure pushes on the open pool. By Equation 10.4, the height h of the mercury column is related to atmospheric pressure: $P_{atm} = P_0 + \rho g h = 0 + \rho g h$, or $P_{atm} = \rho g h$. For standard atmospheric pressure $P_{atm} = 1.013 \times 10^5$ Pa, and mercury's density 13,600 kg/m^3,

$$h = \frac{P_{atm}}{\rho g} = \frac{1.013 \times 10^5 \text{ Pa}}{(13,600 \text{ kg/m}^3)(9.80 \text{ m/s}^2)} = 0.760 \text{ m}$$

The mercury barometer gives us the unit millimeters of mercury (mm Hg, also called the *torr*) for pressure, with standard atmospheric pressure being 760 mm.

You often hear atmospheric pressure given in a weather report. That's because higher pressure is generally associated with fair weather, lower pressure with storms. Air pressure also varies with altitude. But we can't use the pressure-depth relation of Equation 10.4—derived for incompressible liquids—because air's density varies with altitude. As an example, normal air pressure in Denver, at 1.6-km altitude, is 17% lower than at sea level.

Other pressure gauges rely on mechanical springs or electronic sensors; you may have used either type to measure tire pressure. Because we live under fairly constant atmospheric pressure, the **gauge pressure** read by a tire gauge is the difference between the **absolute pressure** of the air inside the tire and atmospheric pressure. If your tires specify a 206-kPa inflation pressure, for example, that's the gauge pressure. With atmospheric pressure of 101 kPa, the absolute tire pressure is 206 kPa + 101 kPa = 307 kPa.

CONCEPTUAL EXAMPLE 10.7   **Bicycle Tire**

The recommended gauge pressure for bicycle tires is normally much higher than that for automobile tires. While most automobile tires specify gauge pressures around 200 kPa, bicycle tires can go as high as 600 kPa. Why should bicycle tire pressures be so much higher?

SOLVE The key is that pressure is force per unit area ($P = F/A$). Car tires support much more weight than bicycle tires, but bicycle tires are much smaller (Figure 10.9). For a 1000-kg car (weight 10,000 N), each tire supports 2500 N. Therefore the tire surface area contacting the road must be $A = F/P \approx 0.013$ m^2. For a bicycle, with cycle and rider totaling, say, 100 kg (weight 1000 N), each of the two tires supports 500 N. The surface contacting the road has area $A = F/P \approx 0.00083$ m^2. It's the smaller surface area that correlates to higher pressure, even with the much lower mass.

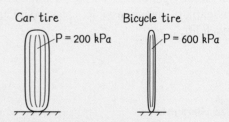

Car tire $P = 200$ kPa

Bicycle tire $P = 600$ kPa

FIGURE 10.9 Car and bicycle tires.

REFLECT Manufacturers recommend pressures for the tires on their vehicles. If you look at your tire, the "maximum recommended pressure" is normally somewhat higher, to allow for changing load and temperature.

Reviewing New Concepts

Important ideas about pressure:

- Pressure is force per unit area: $P = F/A$.
- In a fluid with uniform density ρ, pressure increases linearly with depth: $P = P_0 + \rho g h$.
- Pascal's principle says that any pressure applied to a confined fluid is transmitted throughout the entire fluid.

Blood Pressure

Your doctor might tell you your blood pressure is, for example, "120 over 70." Those numbers are gauge pressures in millimeters of mercury. Figure 10.10 is a graph of blood pressure versus time and shows why there are two numbers. Most of the time the pressure is near the lower **diastolic pressure**. When the heart ventricles contract to force blood through the arteries, the pressure rises to a peak called **systolic pressure**. Higher than normal blood pressure means larger forces on blood vessel walls, which damages blood vessels over time and contributes to heart disease and stroke.

Blood pressure measurement uses an inflatable cuff on the upper arm, pinching off blood flow. The cuff is gradually deflated, and a stethoscope or electronic sensor at the wrist detects the returning pulse as the cuff pressure drops below the systolic blood pressure. The cuff pressure continues to drop, and when it's below diastolic the blood flow becomes smooth and the pulse diminishes. Thus the device determines both systolic and diastolic pressures. Today's blood pressure monitoring equipment is often fully automatic, with the cuff inflating periodically and pressures detected electronically.

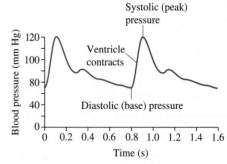

FIGURE 10.10 Systolic and diastolic blood pressure.

CONCEPTUAL EXAMPLE 10.8 **Where to Measure Blood Pressure**

When measuring blood pressure, health professionals place the cuff on your arm at a vertical position near the heart. Why?

SOLVE Although your blood is flowing, its average pressure is still given approximately by Equation 10.4: $P = P_0 + \rho g h$. Table 10.1 gives blood's density as 1060 kg/m^3, so from head to toe a 1.8-m-tall person has a blood pressure difference of about

$$\Delta P = \rho g h = (1060 \text{ kg/m}^3)(9.80 \text{ m/s}^2)(1.8 \text{ m}) = 19 \text{ kPa}$$

That's about 140 mm of mercury, a huge difference! For accuracy, it's important to place the blood pressure cuff within a few centimeters of heart level.

REFLECT Gravity is one reason your measured blood pressure can vary between measurements if you aren't in the same position each time. There's also some uncertainty due to indecision about when an audible pulse starts and stops. Electronic blood pressure sensors eliminate this guesswork.

GOT IT? Section 10.3 Rank in order, from lowest to highest, the following pressures: 1 atm, 1 mm of mercury, 1 Pa, 1 torr, 1 kPa.

10.4 Buoyancy and Archimedes' Principle

The force that supports a floating object is the **buoyant force**. The buoyant force also acts on submerged objects, but it's not always enough to overcome gravity. There's even a buoyant force acting upward on you from the surrounding air. Here we'll explore the origin of the buoyant force.

Archimedes' Principle

The buoyant force acts as an upward force on any object immersed fully or partially in a fluid. Figure 10.11 shows the origin of the buoyant force on a submerged cylinder. Fluid pressure exerts forces on all sides of the cylinder, but since the pressure is greater at the bottom of the cube, there's a net upward force. Fluid forces on the sides cancel in pairs, leaving a net (upward) buoyant force $F_B = F_{bottom} - F_{top}$. Since force is pressure times area, this becomes $F_B = P_{bottom}A - P_{top}A = (P_{bottom} - P_{top})A$. In a liquid, the pressure difference $P_{bottom} - P_{top}$ follows from Equation 10.4: $P_{bottom} - P_{top} = \rho g h$, where h is the cube height. Although Equation 10.4 itself doesn't apply to a gas like the atmosphere, it does hold to a very good approximation for pressure *differences* across small heights. So in either case we have $F_B = \rho_{fluid} g h A$. Now hA is just the volume of the immersed cube. Put another way, it's the volume V of fluid that the cube has displaced. With

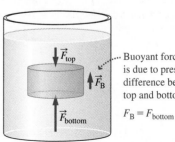

FIGURE 10.11 The buoyant force arises because fluid pressure increases with depth.

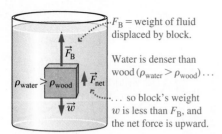

F_B = weight of fluid displaced by block.

Water is denser than wood ($\rho_{water} > \rho_{wood}$) . . .

. . . so block's weight w is less than F_B, and the net force is upward.

(a) Wood block in water

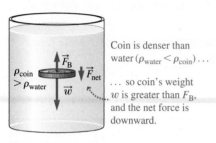

Coin is denser than water ($\rho_{water} < \rho_{coin}$) . . .

. . . so coin's weight w is greater than F_B, and the net force is downward.

(b) Coin in water

FIGURE 10.12 Forces on two submerged objects.

mass = density $\times$ volume, you can see that $\rho_{fluid}hA = \rho_{fluid}V = m_{displaced\ fluid}$. Finally, since weight = mg,

$$F_B = w_{displaced\ fluid} = \rho_{fluid}gV \quad \text{(Archimedes' principle; SI unit: N)} \quad (10.5)$$

Equation 10.5 is **Archimedes' principle**, after the 3rd-century BCE Greek mathematician Archimedes, who first proposed it. Stated in words,

> Archimedes' principle is the buoyant force on an object submerged in a fluid equals the weight of the fluid displaced by that object.

According to Archimedes' principle, every object experiences a buoyant force. Whether it floats or sinks depends on the object's density relative to the fluid. Push a wood block underwater and you'll feel the upward buoyant force (Figure 10.12a). By Archimedes' principle, the buoyant force equals the weight of water displaced; since water is denser than wood, the buoyant force is greater than the wood's weight, and that's why wood floats.

Now submerge a coin (Figure 10.12b). It's denser than water, so the coin's weight is greater than the weight of the displaced water. Hence the coin's weight is greater than the upward buoyant force, and it sinks.

EXAMPLE 10.9 | **Helium Balloon**

(a) What's the buoyant force on a 30.0-cm-diameter helium balloon? (b) If the mass of the rubber balloon skin is 7.0 g, what's the net upward force on it?

ORGANIZE AND PLAN Our sketch is shown in Figure 10.13. By Archimedes' principle, the buoyant force on the balloon equals the weight of air displaced; the net force is the difference between that and the downward weight of the balloon, including helium. The given diameter of the balloon lets you find its volume, $V = \frac{4}{3}\pi r^3$ for a sphere. Mass = density $\times$ volume, and weight $w = mg$. From Table 10.1, the density of air is $\rho_{air} = 1.28 \text{ kg/m}^3$, and the density of helium is $\rho_{He} = 0.18 \text{ kg/m}^3$.

Known: $\rho_{air} = 1.28 \text{ kg/m}^3$; $\rho_{He} = 0.18 \text{ kg/m}^3$; balloon diameter = 30.0 cm (radius 15 cm), rubber balloon mass = 7.0 g.

SOLVE (a) The balloon's volume is

$$V = \frac{4}{3}\pi r^3 = \frac{4}{3}\pi(0.150 \text{ m})^3 = 0.0141 \text{ m}^3$$

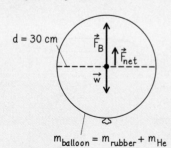

d = 30 cm

$m_{balloon} = m_{rubber} + m_{He}$

FIGURE 10.13 Forces on the balloon.

From Archimedes' principle, the buoyant force is F_B = weight of air displaced = $m_{air}g = \rho_{air}Vg$, or

$$F_B = \rho_{air}Vg = (1.28 \text{ kg/m}^3)(0.0141 \text{ m}^3)(9.80 \text{ m/s}^2) = 0.177 \text{ N}$$

(b) The weight of helium plus rubber balloon is

$$w = \rho_{He}Vg + m_{rubber}g$$
$$= (0.18 \text{ kg/m}^3)(0.0141 \text{ m}^3)(9.8 \text{ m/s}^2)$$
$$+ (7.0 \times 10^{-3} \text{ kg})(9.8 \text{ m/s}^2) = 0.093 \text{ N}$$

Then the net upward force on the balloon is the difference: $F_{net} = 0.177 \text{ N} - 0.093 \text{ N} = 0.084 \text{ N}$

REFLECT Note that the answer to part (a) didn't depend on the balloon's mass. The upward buoyant force on *any* 30-cm-diameter sphere—even a lead one—would be the same. But the *net* force would be very different—and downward—for a sphere that's denser than air.

MAKING THE CONNECTION Use the data in this problem to estimate the balloon's acceleration at the moment you release it. What's wrong with this calculation?

ANSWER The helium's mass $m = \rho_{He}V = 0.0025 \text{ kg}$, and adding the mass (0.0070 kg) of the balloon's rubber skin gives a total mass $m = 0.0095 \text{ kg}$. With $F_{net} = ma = 0.084 \text{ N}$, the acceleration is 8.8 m/s^2. This answer for acceleration is reasonable for an initial value, but drag forces on the light balloon reduce its acceleration, and soon the balloon approaches some terminal velocity (in the upward direction).

Measuring Density

One way to measure an object's density is to weigh it underwater—a procedure that determines its *apparent weight*, the difference between its actual weight mg and the magnitude of the buoyant force F_B (Figure 10.14). Here the scale supports a denser-than-water block in equilibrium, with an upward force equal to the block's apparent weight w_a, giving zero net force on the block. Summing the vertical force components, $F_{net} = F_B + w_a - mg = 0$. As usual, the buoyant force is $F_B = \rho_{water}Vg$, so the net force becomes $F_{net} = \rho_{water}Vg + w_a - mg = 0$. Solving this equation for the block's volume V gives

$$V = \frac{mg - w_a}{\rho_{water}g} \quad \text{(Submerged volume)} \quad (10.6)$$

Knowing the actual weight mg measured in air then gives the mass, so you can find the density $\rho = m/V$.

Try lifting a heavy brick underwater, and you'll find it a lot easier than doing so in air. But even in air, an object's apparent weight is actually slightly less than mg—and it's a lot less for objects of very low density.

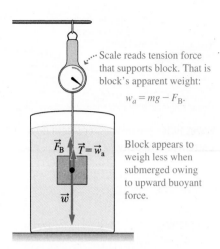

Scale reads tension force that supports block. That is block's apparent weight:

$w_a = mg - F_B.$

Block appears to weigh less when submerged owing to upward buoyant force.

FIGURE 10.14 Measuring density with an underwater scale.

EXAMPLE 10.10 Body Fat

Body fat's density is about 900 kg/m³, and nonfat "lean mass" tissue averages about 1100 kg/m³, so body density provides an indication of a person's body fat composition. Suppose a 65.4-kg woman has an apparent weight $w_a = 36.0$ N when underwater. Find (a) her volume and (b) her density.

ORGANIZE AND PLAN Figure 10.15 shows our sketch for the situation. Equation 10.6 gives volume in terms of known and measured quantities. The density of water is $\rho_{water} = 1000$ kg/m³. Once the woman's volume V is known, her density follows from $\rho = m/V$.

Known: fat density = 900 kg/m³; lean density = 1100 kg/m³; $m = 65.4$ kg; $w_a = 36.0$ N.

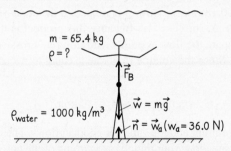

$m = 65.4$ kg
$\rho = ?$

$\vec{F_B}$

$\rho_{water} = 1000$ kg/m³

$\vec{w} = m\vec{g}$

$\vec{n} = \vec{w_a}\,(w_a = 36.0$ N)

FIGURE 10.15 Determining body density using an underwater scale.

SOLVE With the values given, the woman's volume is

$$V = \frac{mg - w_a}{\rho_{water}g} = \frac{(65.4 \text{ kg})(9.80 \text{ m/s}^2) - 36.0 \text{ N}}{(1000 \text{ kg/m}^3)(9.80 \text{ m/s}^2)} = 0.0617 \text{ m}^3$$

Then her density is

$$\rho = \frac{m}{V} = \frac{65.4 \text{ kg}}{0.0617 \text{ m}^3} = 1060 \text{ kg/m}^3$$

REFLECT This value is much closer to the lean mass density, indicating that she's fairly lean.

MAKING THE CONNECTION It's important for the person being weighed underwater to expel as much air as possible from her lungs. Why?

ANSWER There are two reasons. First, most people's bodies normally float, because of low-density air in the lungs. Underwater weighing clearly requires a submerged object. Expelling as much air as possible from the lungs leaves only a small residual air volume that's not enough to make you float. Second, air's density is so low that any air in the lungs makes the body's average density appear lower than it is. Thus, expelling more air makes the measurement more accurate.

10.5 Fluid Motion

So far we've considered static fluids, but you often experience fluids that move. Winds carry air across Earth's surface, while rivers and ocean currents move vast volumes of water. Your heart drives blood flow through your circulatory system, and hot air or water circulates to keep your house warm in winter. The remainder of this chapter explores fluid motion.

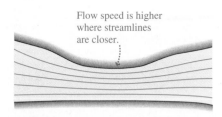

FIGURE 10.16 Streamlines represent flow pattern and velocity.

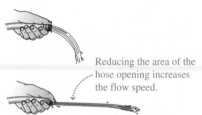

FIGURE 10.17 How opening size affects flow speed.

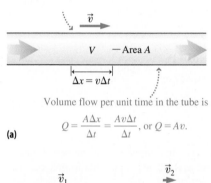

(a)

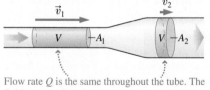

(b)

FIGURE 10.18 Fluid flow in (a) a hose of fixed diameter and (b) a hose of varying diameter.

Ideal Fluids

We began our study of kinematics and dynamics by assuming frictionless systems. Similarly, we need some simplifying assumptions to understand fluid flow. We define an **ideal fluid** as having the following characteristics:

1. The fluid is *incompressible*. This is generally a good assumption for liquids, with their large bulk modulus (Section 10.2). Although gases are more easily compressed, they can be treated as essentially incompressible as long as flow speeds remain well below the speed of sound in the gas.
2. The flow is *steady*, meaning the velocity at each point in the fluid doesn't change with time. A lazily flowing river exhibits nearly steady flow, but the constantly changing flow of a whitewater rapid isn't steady. The whitewater flow is *turbulent*, meaning it not only changes with time but also does so in an erratic way.
3. The fluid is *irrotational*. This excludes the "whirlpool" motion you see as water goes down your sink drain. The precise definition of irrotational is beyond the scope of this book, but you can imagine the following test: Drop a small leaf into the fluid. The flow is irrotational if the leaf doesn't rotate as it's carried along.
4. The fluid is *inviscid*. That means **viscosity**, or fluid friction, doesn't affect the flow. Viscosity depends on the fluid and the size of the flow region. A milk shake flowing through a narrow straw is an example of viscous flow; water flowing through the same straw is essentially inviscid. We'll consider viscosity in Section 10.6.

Steady flow doesn't mean that every point in the flow has the same velocity. It does, however, mean that the flow follows continuous lines, called **streamlines**, as shown in Figure 10.16. The fluid velocity at any point is tangent to a streamline, so streamlines help you visualize the flow pattern.

The Continuity Equation

Put your thumb partially across the end of a garden hose, and the water comes out faster and squirts farther (Figure 10.17). Why?

An incompressible fluid like water doesn't change volume, so in steady flow the volume of water passing each point in the hose per unit time must be the same. That quantity is the **volume flow rate**, Q. Figure 10.18a shows how Q is related to the cross-sectional area A and flow speed v: $Q = Av\Delta t/\Delta t$, or $Q = Av$. Because this product is the same everywhere in the hose, you can compare the flow at points with different cross-sectional areas, as illustrated in Figure 10.18b. Since the volume flow rate is equal to Av and is the same throughout the flow,

$$Q = Av = \text{constant} \quad \text{(Continuity equation; SI unit: m}^3\text{/s)} \quad (10.7)$$

Equation 10.7 is the **continuity equation**, and it's valid for all ideal flows. The continuity equation explains the garden hose situation: Your thumb over the end of the hose gives it a smaller cross-sectional area A there, so the water's speed v is proportionally higher to keep the product Av constant. Another example is the water faucet in Figure 10.19. As the water falls, its speed v increases, so the area A must decrease—and thus the stream becomes narrower.

EXAMPLE 10.11 **Fire Hose**

A fire hose with inside diameter 12.7 cm delivers water at 340 L/min. What's the speed of the water (a) in the hose and (b) as it exits a 1.91-cm-diameter nozzle?

ORGANIZE AND PLAN The continuity equation $Q = Av$ relates the volume flow rate to the cross-sectional area A and the flow speed v. For both parts, we can find the area from the diameter and then use the continuity equation to find flow speeds. We'll need the flow rate in SI.

Known: hose diameter $d_1 = 12.7$ cm; $Q = 340$ L/min; nozzle diameter $d_2 = 1.91$ cm.

cont'd.

SOLVE (a) Converting the flow rate to SI,

$$\frac{340 \text{ L}}{\text{min}} \times \frac{1 \text{ min}}{60 \text{ s}} \times \frac{10^{-3} \text{ m}^3}{\text{L}} = 5.67 \times 10^{-3} \text{ m}^3/\text{s}$$

The hose's radius is $d_1/2$, or 6.35 cm. Then with $Q = Av$, the flow speed is

$$v_1 = \frac{Q}{A_1} = \frac{5.67 \times 10^{-3} \text{ m}^3/\text{s}}{\pi(0.0635 \text{ m})^2} = 0.448 \text{ m/s}$$

(b) The relation $Q = Av$ holds everywhere, so at the nozzle we have

$$v_2 = \frac{Q}{A_2} = \frac{5.67 \times 10^{-3} \text{ m}^3/\text{s}}{\pi(0.00955 \text{ m})^2} = 19.8 \text{ m/s}$$

REFLECT This answer seems reasonable. Water flowing at about 20 m/s could reach across a street or several stories up a building.

MAKING THE CONNECTION What's the maximum height the water from this nozzle could reach?

ANSWER Using basic kinematics, a stream shot straight up with a speed of 19.8 m/s reaches a maximum height of $v^2/2g = 20.0$ m. That's five or six stories on a typical building.

Bernoulli's Equation

Suppose a fire hose is attached to a ground-level hydrant, while firefighters carry the other end to a building's second floor to attack a fire. Based on your study of static fluids, you should expect the elevation change to reduce water pressure in the hose. And considering conservation of mechanical energy, you might expect a lower flow speed on the second floor than at ground level. Our goal now is to relate these quantities—pressure, speed, and height—for an ideal fluid in motion.

Figure 10.20 shows a schematic of a narrow tube of varying diameter carrying a flow of ideal fluid. A fluid volume is shown entering and then leaving the tube. An ideal fluid is incompressible, so the fluid volume V stays the same. Thus $V = A_1x_1 = A_2x_2$.

We'll now apply work and energy arguments to this fluid flow. Work is done from outside, by the forces from adjacent fluid on the left ($\vec{F}_1$) and right ($\vec{F}_2$), as shown in Figure 10.20. The net work W_{ext} done by these external forces equals the change in the fluid's total mechanical energy:

$$W_{ext} = \Delta E = \Delta K + \Delta U \qquad (10.8)$$

The external work W_{ext} is the sum of the work W_1 done by force $\vec{F}_1$ and work W_2 done by force $\vec{F}_2$. Because pressure is force per area, these are $W_1 = F_1\Delta x_1 = P_1A_1\Delta x_1 = P_1V$ and $W_2 = -P_2V$, with W_2 negative because here force and fluid displacement are in opposite directions. Therefore, the external work done on the fluid element is $W_{ext} = W_1 + W_2 = P_1V - P_2V$. That's one piece of Equation 10.8. Next, the difference in the fluid's kinetic energy between points 1 and 2 is

$$\Delta K = \tfrac{1}{2}mv_2^2 - \tfrac{1}{2}mv_1^2$$

Since mass = density × volume

$$\Delta K = \tfrac{1}{2}\rho V v_2^2 - \tfrac{1}{2}\rho V v_1^2$$

Finally, the potential energy difference is

$$\Delta U = mgy_2 - mgy_1 = \rho V g y_2 - \rho V g y_1$$

Putting these results into Equation 10.8,

$$P_1V - P_2V = \tfrac{1}{2}\rho V v_2^2 - \tfrac{1}{2}\rho V v_1^2 + \rho V g y_2 - \rho V g y_1$$

The volume V cancels, and rearranging gives the following:

$$P_1 + \tfrac{1}{2}\rho v_1^2 + \rho g y_1 = P_2 + \tfrac{1}{2}\rho v_2^2 + \rho g y_2 \qquad \text{(Bernoulli's equation; SI unit: Pa)} \qquad (10.9)$$

Equation 10.9 is the relationship we're after. It involves fluid pressure, speed, and height at two points, along with the density ρ, which is constant for an incompressible fluid.

FIGURE 10.19 By the continuity equation, the water stream's diameter decreases as its speed increases.

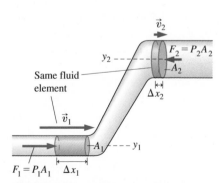

FIGURE 10.20 A flow tube showing the same fluid element entering and leaving. The work done by external forces equals the change in mechanical energy of the fluid element.

APPLICATION **Streamlined Cars**

Engineers designing new cars use smoke to trace streamlines (see photo) as they study how the vehicles interact with air. Lower air pressure on vehicle surfaces means smaller drag forces, and thus better fuel efficiency.

This is **Bernoulli's equation**, after the Swiss mathematician Daniel Bernoulli, who proposed it in 1738.

Bernoulli's equation may at first appear imposing, but it reduces to familiar results in two special cases. First, for static fluid, $v_1 = v_2 = 0$, and thus $P_1 + \rho g y_1 = P_2 + \rho g y_2$, which is equivalent to Equation 10.4 for static pressure. Second, if pressures P_1 and P_2 are equal, then no external work is done on the fluid, so mechanical energy is conserved. With $P_1 = P_2$, and using $\rho = mV$, Bernoulli's equation then becomes $\frac{1}{2}mv_1^2 + mgy_1 = \frac{1}{2}mv_2^2 + mgy_2$, which is our familiar statement of mechanical energy conservation (Chapter 5).

EXAMPLE 10.12 **Fire on the Third Floor!**

The hose of Example 10.11 is attached to a street-level hydrant, where the (gauge) water pressure is 75 psi (515 kPa). The firefighters have taken the nozzle to the third floor, 7.80 m above street level. The hydrant still delivers water at 340 L/min. Find (a) the flow speed through the nozzle and (b) the pressure at the nozzle.

ORGANIZE AND PLAN The continuity equation still holds, so the flow speeds are as found in Example 10.11. With the speeds known, Bernoulli's equation relates the third-floor pressure (Figure 10.21) to the other parameters.

Known: Street-level pressure = 515 kPa; $\Delta y = 7.80$ m; $v_1 = 0.448$ m/s; $v_2 = 19.8$ m/s.

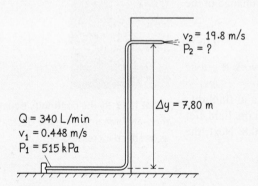

$v_2 = 19.8$ m/s
$P_2 = ?$

$\Delta y = 7.80$ m

$Q = 340$ L/min
$v_1 = 0.448$ m/s
$P_1 = 515$ kPa

FIGURE 10.21 Finding the pressure at the hose nozzle.

SOLVE To find the water pressure P_2 at the third floor, use Bernoulli's equation. The hydrant pressure is $P_1 = 515$ kPa, and the flow speeds are $v_1 = 0.448$ m/s and $v_2 = 19.8$ m/s. Solving Bernoulli's equation for P_2,

$$P_2 = P_1 + \frac{1}{2}\rho v_1^2 + \rho g y_1 - \frac{1}{2}\rho v_2^2 - \rho g y_2$$
$$= P_1 - \frac{1}{2}\rho(v_2^2 - v_1^2) - \rho g(y_2 - y_1)$$

Inserting numerical values, with water's density 1000 kg/m³,

$$P_2 = 5.15 \times 10^5 \text{ Pa}$$
$$- \frac{1}{2}(1000 \text{ kg/m}^3)\left[(19.8 \text{ m/s})^2 - (0.448 \text{ m/s})^2\right]$$
$$- (1000 \text{ kg/m}^3)(9.80 \text{ m/s}^2)(7.80 \text{ m})$$
$$= 2.43 \times 10^5 \text{ Pa}$$

REFLECT This is less than half the street-level pressure. Both the increased flow speed and higher elevation contribute to the reduced pressure at the third-floor level. Since the street-level pressure was a gauge pressure, so is our answer—that is, it's the excess over atmospheric pressure.

MAKING THE CONNECTION What's the maximum height at which this hose could deliver water, assuming the same street-level pressure?

ANSWER Maximum height occurs when the flow ceases, so the nozzle's gauge pressure and flow speeds are zero. Then Bernoulli's equation becomes $P_1 + \rho g y_1 = \rho g y_2$. Solving for the height $h = y_2 - y_1$ gives $h = 52.6$ m.

As presented in Equation 10.9, Bernoulli's equation relates pressure, speed, and height at two points in a fluid. But there's another way to think of it—namely, as a conservation law. Bernoulli's equation says that the quantity $P + \frac{1}{2}\rho v^2 + \rho gh$ is the same *anywhere* in an ideal flow. That is,

$$P + \tfrac{1}{2}\rho v^2 + \rho gh = \text{constant}$$

In this form, Bernoulli's equation is a statement about conservation of energy in an ideal fluid. You can convince yourself that all terms in the equation have the units of energy density—that is, J/m³. The equation shows that if one of the quantities P, v, or h increases, then one or both of the others must decrease.

Bernoulli's Principle

Earlier we considered the two special cases of zero flow speed and equal pressures at two points. A third special case occurs when two points in the fluid are at the same height. Then $\rho g h$ cancels, leaving

$$P_1 + \tfrac{1}{2}\rho v_1^2 = P_2 + \tfrac{1}{2}\rho v_2^2$$

This form embodies **Bernoulli's principle**, which shows a tradeoff between pressure and flow speed: Increase the flow speed v and the pressure P drops, and vice versa. Bernoulli's principle explains many fluid phenomena, some of them counterintuitive (Figure 10.22), and is widely used in instrumentation to measure fluid flow. The principle also helps explain how airplanes fly and curve balls curve. However, as Figure 10.23 shows, a complete explanation necessarily invokes Newton's third law.

Fluid flow is found throughout living systems, such as the human body's circulatory system. The aorta is the body's main blood vessel, a garden hose-sized artery running vertically through the torso. An aneurysm is a weakening of the aortic wall, and under the influence of blood pressure the result is a bulge that widens the artery. By the continuity equation, the blood flow rate at the aneurysm is reduced—and by Bernoulli's principle that means increased blood pressure. That, in turn, results in further pressure-induced bulging, exacerbating the aneurysm. Unfortunately, most people with aortic aneurysms show no symptoms until the aneurysm bursts—a usually fatal event.

FIGURE 10.22 A ping-pong ball supported by downward-flowing air. High-velocity flow is inside the narrow part of the funnel.

GOT IT? Section 10.5 A buried water pipe has a diameter of 3 cm. It's connected to an above-ground pipe with a 2-cm diameter. Which of the following correctly compares the pressures of the water below ground (P_1) and above ground (P_2)? (a) $P_1 > P_2$; (b) $P_1 < P_2$; (c) $P_1 = P_2$; (d) this cannot be determined from the information given.

Reviewing New Concepts

Some important ideas about buoyancy and fluid flow:

- Archimedes' principle says that the buoyant force on an object in a fluid equals the weight of the fluid displaced by that object: $F_B = w_{\text{displaced fluid}} = \rho_{\text{fluid}}gV$.
- For an incompressible fluid, the continuity equation relates flow speed and cross-sectional area: $A_1v_1 = A_2v_2$.
- The pressure, speed, and height of an incompressible fluid are related by Bernoulli's equation: $P_1 + \tfrac{1}{2}\rho v_1^2 + \rho gy_1 = P_2 + \tfrac{1}{2}\rho v_2^2 + \rho gy_2$.

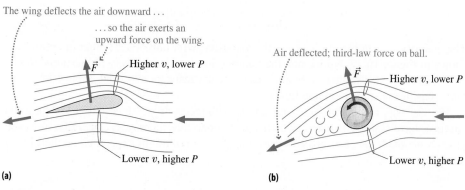

(a) (b)

FIGURE 10.23 Bernoulli's principle and Newton's third law help explain both an airplane's flight and the path of a curve ball. (a) Side view of a wing, showing the force $\vec{F}$ resulting from higher pressure below, lower above, associated with different flow speeds. Downward air deflection confirms that there must be an upward force on the wing. (b) Top view of a spinning baseball showing a similar effect, which deflects the ball sideways. Higher flow speed on one side results from the ball's spin.

10.6 Surface Tension and Viscosity

So far we've considered only ideal fluids. A thorough study of nonideal flows—for example, compressible fluids—requires advanced mathematics beyond the scope of this book. In this section we'll introduce two important nonideal phenomena that can be described fairly simply.

Surface Tension

You've probably seen insects walking across a calm body of water. Why don't they fall in? Look closely, and you'll see the insect's legs depressing the water surface, as though it were an elastic membrane (Figure 10.24). When a fluid behaves this way, we say it exhibits **surface tension**.

Figure 10.25 shows the physics behind surface tension. Water molecules attract weakly through electrical interactions called *van der Waals forces*. A molecule within the water is in equilibrium, attracted equally by neighbors in all directions. A molecule at the surface is also in equilibrium, but with no water above it, forces parallel to the surface dominate. Those surface forces act like springs that stretch under tension, making the surface elastic and able to support small objects. Surface tension is also the reason water and other liquids tend to form spherical droplets, as the surface forces tug on any bulges or distortions from spherical shape.

Measure liquid volume with a graduated cylinder, and you'll notice the liquid curving upward at the edges. This *meniscus* is due to **capillary action**, a phenomenon related to surface tension. Attractive forces between liquid and glass are stronger than surface tension. Thus the liquid is pulled up around the edges. Capillary action can also cause a liquid to creep through a narrow tube (hence the name capillary), as the liquid's leading edge is attracted to the tube's inner surface.

Viscosity

The ideal fluid approximation neglects friction. However, real liquids experience **viscosity**, a fluid friction that impedes their flow. Figure 10.26 shows viscous flow through a tube. Friction between the fluid and the tube's inner walls reduces the flow speed at the walls. Fluid farther from the walls is less affected, but it's still slowed by its interaction with adjacent fluid. Flow speed is therefore highest at the tube's center.

Viscosity saps the fluid's energy, causing pressure to drop as the flow progresses. This viscous pressure drop is why the heart's systolic pressure is needed to force blood through the circulatory system, or why an oil pipeline needs to be pressurized to drive the oil flow. In many cases, the volume flow rate Q through a tube is proportional to the pressure difference between its ends. **Poiseuille's law** describes this relation:

$$Q = \frac{\pi R^4 (P_1 - P_2)}{8\eta L} \quad \text{(Poiseuille's law)} \quad (10.10)$$

The quantity η in Poiseuille's law is the **viscosity**, with units Pa·s; other quantities are shown in Figure 10.27. Notice that the volume flow rate is inversely proportional to viscosity, showing that higher viscosity means greater resistance to flow. Table 10.3 shows the viscosities of some common fluids.

The viscosity of liquids decreases rapidly with increasing temperature. Most gases, in contrast, have slightly higher viscosities as temperature increases. Generally, however, gases have much lower viscosities than liquids—as you might expect, given their lower densities.

FIGURE 10.24 Surface tension supports the water strider. Note the depressions where the insect's legs contact the water surface.

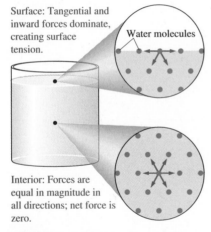

Surface: Tangential and inward forces dominate, creating surface tension.

Water molecules

Interior: Forces are equal in magnitude in all directions; net force is zero.

FIGURE 10.25 Origin of surface tension.

Fluid flows slowest near the tube's walls, because of friction, . . .

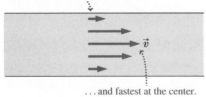

$\vec{v}$

. . . and fastest at the center.

FIGURE 10.26 Viscous flow.

EXAMPLE 10.13   Arterial Blood Flow

One section of a patient's femoral artery is 0.32 m long with inner radius 3.6 mm. If the pressure difference between the two ends of the artery is 1.2 mm Hg (160 Pa), what's the blood flow rate through the artery?

ORGANIZE AND PLAN Poiseuille's law (Equation 10.10) gives the volume flow rate as a function of pressure difference, viscosity, and the tube's length and radius—all known quantities, with the viscosity given in Table 10.3.

Known: $\Delta P = 1.2$ mm Hg (160 Pa); $L = 0.32$ m; $R = 3.6$ mm; $\eta = 1.7 \times 10^{-3}$ Pa·s.

SOLVE Using the given values,

$$Q = \frac{\pi R^4 (P_1 - P_2)}{8\eta L} = \frac{\pi (3.6 \times 10^{-3} \text{ m})^4 (160 \text{ Pa})}{8(1.7 \times 10^{-3} \text{ Pa·s})(0.32 \text{ m})}$$

$$= 1.9 \times 10^{-5} \text{ m}^3/\text{s}$$

REFLECT Is this answer reasonable? It's about 1 L/min—and a major artery can lose a liter of blood in a minute, so the answer is indeed reasonable.

MAKING THE CONNECTION   If plaque on the arterial walls decreases the artery's effective radius by 10%, by what factor is blood flow reduced?

ANSWER Volume flow rate Q depends on the radius to the *fourth* power. The reduction in blood flow is by a factor of $(1.1)^4 = 1.46$, almost a 50% reduction. The body can restore the flow rate with an increase in pressure, but that's not healthy for the artery walls.

Chapter 10 in Context

In this chapter you've used your understanding of forces to learn how solids expand and compress with the application of outside forces. You then explored characteristics of fluids, focusing on how gravity and other forces affect fluid pressure and flow. You've seen many real-world applications of these forces, from ocean depths to your own body.

Looking Ahead You've learned here that pressure is a fundamental quantity for static and moving fluids. So far we've concentrated on incompressible liquids. In Chapters 12–14, you'll see pressure's role in gases, and how pressure relates to volume and temperature.

Owing to viscosity, pressure drops as fluid moves through tube: $P_1 > P_2$.

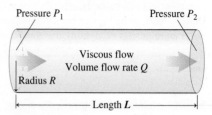

FIGURE 10.27 Pressure difference in viscous flow.

TABLE 10.3 Viscosities of Some Common Fluids

Fluid	Viscosity (Pa · s)
Glycerin (20°C)	1.5
Motor oil, SAE 20 (20°C)	0.13
Water (20°C)	1.0×10^{-3}
Water (100°C)	2.8×10^{-4}
Ethanol (20°C)	1.2×10^{-3}
Blood (37°C)	1.7×10^{-3}
Mercury (20°C)	1.6×10^{-3}
Air (20°C)	1.8×10^{-5}
Air (100°C)	2.2×10^{-5}

States of Matter

(Section 10.1) **Density** describes a material's mass per unit volume. Density tends to be highest for solids, slightly lower for liquids (with water an important exception), and much lower for gases.

Density: $\rho = \dfrac{m}{V}$

Solids and Elasticity

(Section 10.2) Solids expand and compress under the influence of external forces, as described by their **Young's modulus** (one-dimensional changes) or **bulk modulus** (three-dimensional changes).

Young's modulus: $\dfrac{F}{A} = Y\dfrac{\Delta L}{L}$ **Bulk modulus:** $\dfrac{F}{A} = -B\dfrac{\Delta V}{V}$

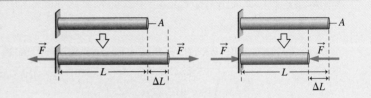

Fluid Pressure

(Section 10.3) **Pressure** describes the force per unit area acting in a fluid. Normally, pressure is the same in all directions.

Pressure of an incompressible fluid increases with depth:
$P = P_0 + \rho g h$

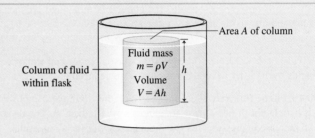

Area A of column

Fluid mass
$m = \rho V$

Column of fluid within flask

Volume
$V = Ah$

Buoyancy and Archimedes' Principle

(Section 10.4) **Archimedes' principle** states that the upward buoyant force on an object immersed in a fluid is equal to the weight of displaced fluid. The principle shows that an object floats or sinks depending on whether its density is less or greater, respectively, than that of the fluid.

Archimedes' principle: $F_B = w_{\text{displaced fluid}} = \rho_{\text{fluid}} g V$

Fluid Motion

(Section 10.5) The **continuity equation** states that the volume flow rate Q of an incompressible fluid is the same throughout a flow tube. The flow speed v therefore changes with changing tube area A.

Bernoulli's equation describes energy conservation in an ideal fluid, relating fluid pressure, flow speed, and height.

Continuity equation: $Q = Av$

Bernoulli's equation: $P_1 + \frac{1}{2}\rho v_1^2 + \rho g y_1 = P_2 + \frac{1}{2}\rho v_2^2 + \rho g y_2$

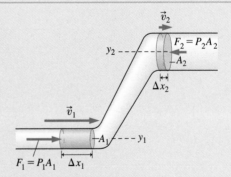

Surface Tension and Viscosity

(Section 10.6) **Viscosity**, or fluid friction, impedes fluid flow and means that a pressure difference is required to drive a steady flow through a tube.

Poiseuille's law: $Q = \dfrac{\pi R^4 (P_1 - P_2)}{8\eta L}$

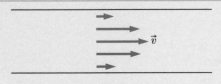

NOTE: Problem difficulty is labeled as ■ straightforward to ■■■ challenging. Problems labeled BIO are of biological or medical interest.

Conceptual Questions

1. Explain why you would expect a gas to have a lower density than the same substance in liquid form.
2. You have a 30-cm-long solid copper rod. If you saw it into two pieces, 10 cm and 20 cm long, how do the densities of those pieces compare with the density of the original rod?
3. The maximum height water can be raised with the aid of a vacuum pump is about 10 m. Why? How could you pump water to a height of 15 m?
4. You have a glass of water with several floating ice cubes. The glass is filled to the rim, and the floating ice cubes stick above that level. When the ice melts, does the cup overflow?
5. Why does a steel ship float, given that steel is much denser than water?
6. Is it easier for you to float on a lake or on the ocean? Assume calm water.
7. If a submarine is completely submerged, does the buoyant force depend on its depth?
8. You're swimming underwater and exhale an air bubble, which floats to the surface. As the bubble rises, does its size increase, decrease, or remain the same?
9. When you are sitting in a car that accelerates forward suddenly, you feel yourself thrown back in your seat. What happens at the same time to a helium balloon that's floating in the car?
10. Describe the behavior of the balloon in the preceding problem when you're (a) driving along a straight road at a constant speed and (b) rounding a curve at constant speed.
11. Give three examples of nonideal fluids, and in each case identify the nonideal behavior.
12. During a storm, the wind blows horizontally across the surface of a flat roof. What's the direction of the resulting net force on the roof? *Hint:* Consider Bernoulli's principle.
13. Hold a piece of paper by one end and let the other end flop down. Then blow across the top of the paper, and you'll see it rise. Why?
14. Melting of arctic sea ice and mountain glaciers is among the most dramatic evidence for global climate change. Do both contribute equally to sea-level rise? Explain.

Multiple-Choice Problems

15. Aluminum's density is 2700 kg/m^3. What's the radius of a solid aluminum sphere with mass 10 kg? (a) 0.9 mm; (b) 2.3 cm; (c) 5.4 cm; (d) 9.6 cm.
16. A cylindrical metal rod 1.0 cm in diameter and 25 cm long is compressed by 0.10 mm when a force of 4.7 kN is applied. What's Young's modulus for this material? (a) 2.5×10^{11} N/m^2; (b) 1.5×10^{11} N/m^2; (c) 1.0×10^{11} N/m^2; (d) 0.5×10^{11} N/m^2.
17. A cylindrical metal rod 1.0 cm in diameter and 25 cm long is pulled outward with a 650-N force. What's the stress on the rod? (a) 2.1×10^6 N/m^2; (b) 8.3×10^6 N/m^2; (c) 3.3×10^7 N/m^2; (d) 6.6×10^7 N/m^2.
18. A steel cube with bulk modulus 1.6×10^{11} N/m^2 is submerged 150 m deep in pure water. By what fraction is the cube's volume reduced compared with its volume just above the water? (a) 9×10^{-6}; (b) 6×10^{-6}; (c) 4×10^{-6}; (d) 1×10^{-6}.
19. Seawater's density is 1030 kg/m^3. The pressure 1.5 km deep in the ocean is (a) 1.5×10^5 Pa; (b) 1.5×10^6 Pa; (c) 1.5×10^7 Pa; (d) 1.5×10^8 Pa.
20. Copper's density is 8900 kg/m^3, and water's is 1000 kg/m^3. What is the buoyant force on a solid copper cube, 3.9 cm on a side, submerged in water? (a) 0.06 N; (b) 0.63 N; (c) 4.5 N; (d) 6.1 N.
21. A submarine's exterior can withstand a pressure of 4.2 MPa before collapsing. What's the maximum possible depth for this submarine? (a) 4200 m; (b) 2100 m; (c) 840 m; (d) 430 m.
22. A spherical helium balloon is 30 cm in diameter. Helium's density is 0.179 kg/m^3, and air's is 1.28 kg/m^3. The net upward force on the balloon is (a) 0.15 N; (b) 0.20 N; (c) 0.25 N; (d) 0.30 N.
23. Water flows through a 1.5-cm-diameter hose at 5.2 L/min. What's the flow speed? (a) 0.24 m/s; (b) 0.37 m/s; (c) 0.49 m/s; (d) 0.67 m/s;.
24. Water flows through a 1.5-cm-diameter hose into a 12-L bucket, which it fills in 90 s. What's the speed of the water in the hose? (a) 0.19 m/s; (b) 0.36 m/s; (c) 0.75 m/s; (d) 0.85 m/s.

Problems

Section 10.1 States of Matter

25. ■ If you boil away 1 L of water, what volume of water vapor is produced (at the temperature and pressure given in Table 10.1)?
26. ■ If you fill your tank with 56 L of gasoline, by how much does the car's mass increase?
27. ■ Uranium, depleted of its fissile isotope, is used in armor-penetrating bullets because it's hard and dense. (It's also mildly radioactive, making this a controversial use.) How many times heavier is a uranium bullet than a lead bullet of the same size and shape?
28. ■■ What volume of water has the same mass as 1 L of gasoline?
29. BIO ■■■ **Air and density.** A 65-kg person's density is 990 kg/m^3 with 2.4 L of air in the lungs. What volume of air would the person have to expel to bring the density to that of water, 1000 kg/m^3? (You can neglect the mass of the air in this calculation.)

Section 10.2 Solids and Elasticity

30. ■ How much does a 0.35-m-long copper rod with a 7.0-mm diameter stretch when one end is fixed and the other is pulled with a 1.2-kN force?
31. ■ A 95-kg man climbs on top of a cube-shaped concrete block 28 cm on each side. By how much does the block compress vertically?
32. ■ A copper rod hangs vertically from a fixed support. The rod is 1.05 m long and 1.50 mm in diameter. If a 10.0-kg mass is hung from the bottom of the rod, how far does the rod stretch?
33. ■ A steel rod hangs vertically from a fixed support. The rod is 1.5 m long and 1.2 mm in diameter. What mass hung from the rod will stretch it by 0.50 mm?
34. BIO ■■ **Bone compression.** A woman's femurs (upper-leg bones) are 29 cm long, with average diameter 3.8 cm. Her upper-body mass (i.e., the mass supported by her two legs) is 48 kg. Assuming the femur has a Young's modulus of 5.0×10^9 N/m^2 under compression, by how much is each femur compressed when the woman is standing?
35. ■ A steel guitar string is 73 cm long and 0.15 mm in diameter. If it's under tension of 1.5 kN, how much is it stretched?
36. ■ Find the distance by which a 45-cm-diameter, 8.9-m-tall concrete column is compressed when it supports a 12,500-kg load.

Section 10.3 Fluid Pressure

37. ■■ If you use water instead of mercury in a barometer, what's the height of the water column when the pressure is 1 atm? Is a water barometer practical?

38. ■ The deepest ocean trench is about 10.9 km deep. (a) Find the pressure at this depth. (b) How many times larger than atmospheric pressure is this?

39. ■■ After a shipwreck, a solid steel spoon lies at the bottom of the ocean, 5.75 km below the surface. (a) What's the water pressure at that depth? (b) Find the fractional volume change in the spoon due to compression forces.

40. ■■ Assuming constant 1.28-kg/m^3 air density, by how much does air pressure decrease for an altitude increase of 100 m?

41. ■ What is the ocean depth at which a diver experiences a pressure of 3 atm?

42. ■ A small town stores its drinking water in a tower. What should be the tower's height if the gauge pressure of the water in the town below is to be 4 atm?

43. ■■■ At what ocean depth is the volume of a steel ball reduced by 0.15%?

44. ■■ A 3.0-m-deep swimming pool measures 25 m by 15 m. It's filled with fresh water. What force does the water exert on the bottom of the pool?

45. BIO ■■ **Blood transfusion.** During a transfusion, it's best for the pressure of the incoming blood to be equal to the body's diastolic pressure. If that's 70 mm Hg, how high above the insertion point should the blood supply be placed? See Table 10.1 for blood's density.

46. ■■ Repeat Example 10.6 but with the car 2.0 m above the other piston and area $A_1 = 0.10$ m^2. Assume the system is filled with hydraulic fluid of density 850 kg/m^3.

47. ■■ A hydraulic lift has pistons with areas 0.50 m^2 and 5.60 m^2, and they're at the same height. With a 2.0-kN force on the smaller piston, how much mass can the larger piston support?

48. ■■ A 65-cm-tall graduated cylinder is filled with 30 cm of glycerin (density 1260 kg/m^3) and 35 cm of water. Find the pressure difference between the top and bottom of the cylinder.

Section 10.4 Buoyancy and Archimedes' Principle

49. ■ Find the buoyant force on a submarine with volume 185 m^3 when it's completely submerged in the ocean.

50. ■ What is the buoyant force on a helium balloon in air if the balloon is spherical with diameter 17.5 cm?

51. ■■ A 70-kg parachutist whose density is 1050 kg/m^3 is in free fall. Find the buoyant force due to the air, and compare with the parachutist's weight.

52. ■■ A 2.50-g penny with density 7140 kg/m^3 is released just below the surface of a pool of water. What's its initial downward acceleration?

53. ■■■ A 6500-kg iceberg with density 931 kg/m^3 is afloat in seawater with density 1030 kg/m^3. Find (a) the buoyant force on the iceberg, (b) the volume of water displaced by the iceberg, and (c) the fraction of the iceberg's volume that is below the water line.

54. ■■■ Ice ($\rho = 931$ kg/m^3) is floating in pure water. What fraction of the ice's volume is above the water's surface?

55. ■■ A solid wood ball floats in pure water with exactly half its volume above the water line. What's the wood's density?

56. ■■ An 89.2-kg person with density 1025 kg/m^3 stands on a scale while completely submerged in water. What does the scale read?

57. ■■ A 69.5-kg person completely submerged in water sits on a scale, which reads 22.0 N. What's that person's density?

58. ■■ An air mattress is 1.90 m long, 0.75 m wide, and 0.11 m deep, and its mass (not including air) is 0.39 kg. What's the maximum mass this mattress can support with the top of the mattress at water level?

59. ■■ Aluminum's density is 2700 kg/m^3. An aluminum cube 5.0 cm on a side is placed on a scale. What does the scale read when the cube is entirely (a) in air; (b) under water?

60. ■■ How big would a spherical helium balloon have to be to lift a 60-kg person? The mass of the balloon's skin is 2.0 kg.

61. ■■ Estimate the percentage body fat of the person in Example 10.10.

62. ■■ Expanded polystyrene—the familiar white "Styrofoam"—has density 16 kg/m^3. By what fraction is a piece of Styrofoam's apparent weight less than its actual weight due to the buoyant force of air?

63. ■■■ A submarine remains submerged by holding excess seawater in its bilge tank. Suppose a submarine with volume 135 m^3 is submerged at rest when it expels 1.5 m^3 of seawater from its tank. What is its subsequent upward acceleration?

64. BIO ■■■ **Buoyant fish.** Fish control their buoyancy with a gas-filled organ called a *swim bladder*. The average density of a particular fish's tissues, not including gas in the bladder, is 1050 kg/m^3. If the fish's mass is 9.5 kg, what volume of gas in its swim bladder will keep it in neutral buoyancy—neither sinking nor rising—at a depth where the density of the surrounding seawater is 1028 kg/m^3? Neglect the mass of the bladder gas.

Section 10.5 Fluid Motion

65. ■ Water flows through a 2.75-cm-diameter hose at 0.450 m/s. What's the volume flow rate?

66. ■ Water flows through a 4.00-cm-diameter hose at 1.20×10^{-4} m^3/s. What's the flow speed?

67. ■■ Water flows through a 2.25-cm-diameter hose at 0.320 m/s. Find (a) the volume flow rate and (b) the speed of the water emerging from a 0.30-cm-diameter nozzle.

68. ■■ Blood flows at 2.65 cm/s through an artery with inside diameter 1.45 mm. What's the flow speed in a section where the artery narrows to 1.36 mm in diameter?

69. ■■ Water flows at 1.20×10^{-4} m^3/s through a 2.0-cm-diameter pipe, which then branches into two 1.0-cm-diameter pipes. What's the flow rate in the smaller pipes?

70. ■■ Water flows through a 1.2-cm-diameter pipe into a 250-L bathtub, which it fills in 6.0 min. What is the speed of the water in the pipe?

71. ■■ Oil flowing through a pipeline passes point A at 1.55 m/s with gauge pressure 180 kPa. At point B, the pipe is 7.50 m higher in elevation and the flow speed is 1.75 m/s. Find the gauge pressure at B.

72. ■■ Water flows at 0.850 m/s from a hot water heater, through a 450-kPa pressure regulator. The pressure in the pipe supplying an upstairs bathtub 3.70 m above the heater is 414 kPa. What's the flow speed in this pipe?

73. ■■■ A large cylindrical container is full of water. At a point 0.75 m below the water level, a small hole is punctured in the side of the container. At what speed does water emerge from the hole?

74. ■■■ A large container is full of liquid. A distance h below the fluid surface, a small hole is punctured in the side of the container. Show that flow speed through this hole is $v = \sqrt{2gh}$. This result is known as Torricelli's law.

75. ■■ On a stormy day a wind of 90 km/h blows parallel to the surface of a picture window with area 4.5 m^2. What are the magnitude and direction of the force on the window?

76. ■■■ The Boeing 777 aircraft has a takeoff mass of 230,000 kg and a takeoff speed of 75 m/s. Assume that's the speed of air across the wing's bottom. The total surface area of both wings is 427 m^2.

What airflow speed across the top of the wing is necessary for the plane to fly?

Section 10.6 Surface Tension and Viscosity

77. **BIO** ■ ■ **Arterial blood flow.** By what fraction would the inside diameter of an arterial wall have to decrease in order for blood flow to be reduced by 10%?

78. **BIO** ■ ■ **Artery size and blood pressure.** Over time, plaque decreases the inner diameter of a person's artery by 5%. If the person's initial systolic blood pressure was 120 mm Hg, what systolic pressure is required to maintain the blood flow when the artery size is reduced?

79. ■ ■ Water at 20°C flows from a pumping station to a home 2.50 km away, traversing a 10-cm-diameter pipe at 12 L/min. Find the pressure difference between the ends of the pipe.

80. ■ ■ To make it flow more easily through a pipeline, crude oil is warmed to 50°C, at which its viscosity is only 0.016 Pa · s. What pressure difference will drive a 0.50-m³/s flow through a 20-km pipeline with diameter 0.76 m?

General Problems

81. ■ ■ When two people with total mass 130 kg lie on a waterbed, the pressure increases by 4.7 kPa. What's the total surface area of the two bodies in contact with the bed?

82. ■ ■ An SUV loaded with passengers has total mass 3800 kg. If the tires are inflated to a gauge pressure of 240 kPa, what's the surface area of each tire in contact with the road?

83. ■ ■ Oil comes in standard 42-gallon barrels that are 32 inches high. Find (a) the mass of the oil in such a barrel, (b) the inside diameter of the barrel, and (c) the pressure difference between the top and bottom of the barrel.

84. ■ ■ A vertical tube 1.0 cm in diameter and open at the top contains 25 g of oil on top of 25 g of water. (a) Find the height of each column (oil and water) in the tube. (b) Find the gauge pressure at the bottom of the oil and at the bottom of the water.

85. ■ ■ An airplane's emergency escape window is a rectangle measuring 90 cm by 50 cm. If the cabin pressure is 0.75 atm and the external pressure is 0.25 atm, what force would be required to pull the window inward? Is it likely that a single passenger could do this?

86. **BIO** ■ ■ **Intravenous drip.** A patient is given an intravenous drip from a bottle of fluid through a needle in the patient's arm. At what height above the arm should the bottle be placed so that the gauge pressure of the fluid entering the vein matches the patient's diastolic blood pressure of 80 mm Hg? Assume that the fluid has the same density as water.

87. **BIO** ■ ■ ■ **Hypodermic syringe.** A hypodermic syringe has a plunger with a circular cross section and diameter of 1.2 cm. (a) With how much force must a health provider push on the plunger in order to create a fluid pressure that matches the patient's 130-mm Hg blood pressure? (b) The injection is administered at a rate of 1.5 mL of fluid per second. What is the speed at which the plunger is moving? (c) The opening in the syringe needle is circular with diameter 220 μm. At what speed does fluid emerge from the needle?

88. ■ ■ ■ One section of the Columbia River is 1.3 km wide and 4.5 m deep, with mass flow rate 1.5×10^7 kg/s. (a) What's the volume flow rate? (b) What's the flow speed? (c) If 5% of the river's kinetic energy could be harnessed as electricity, how much power would be produced?

89. ■ ■ (a) How much helium is required to lift a balloon if the total payload (basket, people, supplies, and mass of the balloon's skin) is 340 kg? (b) What's the diameter of such a balloon, assuming it is spherical?

90. **BIO** ■ ■ **Blood flow.** The aorta is the main artery from the heart. A typical aorta has an inside diameter of 1.8 cm and carries blood at speeds of up to 35 cm/s. What is the speed of the blood if plaque on the artery's walls has reduced its diameter by 50%?

91. **BIO** ■ ■ **Blood pressure.** The gauge pressure of the healthy aorta described in the preceding problem is 120 mm Hg. What's the gauge pressure at the site of the plaque buildup?

92. **BIO** ■ ■ ■ **Capillaries.** Consider a typical aorta with an inside diameter of 1.8 cm. All the blood flowing through the aorta must eventually pass through capillaries, which have an average diameter of 10 μm. Blood flows through the aorta at about 1.0 m/s and through the capillaries at 1.0 cm/s. (a) How many capillaries does your body have? (b) If your body contains 5.5 L of blood, how much time does it take for the blood to circulate completely though the body?

93. ■ ■ ■ Archimedes is believed to have suggested weighing objects underwater to determine whether they were pure metals as claimed, such as gold or silver. Suppose a 25.0-N crown is weighed on a scale under water. What does the scale read if the crown is (a) pure gold; (b) 90% gold and 10% silver by weight?

94. ■ ■ ■ A pencil is weighted so it floats vertically with length L submerged. Show that if it's pushed slightly downward it experiences an upward force proportional to the downward displacement, and use this result to show that when released it undergoes simple harmonic motion with period $T = 2\pi\sqrt{L/g}$.

95. ■ ■ ■ The flowmeter shown in Figure GP10.95 measures the flow rate of water in a solar collector system. The flowmeter is inserted in a pipe with inside diameter 1.9 cm; at the constriction the diameter is reduced to 0.64 cm. The thin tube contains oil with density 0.82 times that of water. If the difference in oil levels on the two sides of the tube is 1.4 cm, what is the volume flow rate?

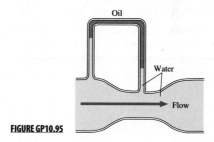

FIGURE GP10.95

Answers to Chapter Questions

Answer to Chapter-Opening Question
As depth increases, there's more fluid pushing down on the diver from above, causing the pressure to increase linearly with depth.

Answers to GOT IT? Questions
Section 10.1 (c) O = A = B
Section 10.2 ΔL (steel) < ΔL (copper) < ΔL (aluminum)
Section 10.3 1 Pa < 1 torr = 1 mm Hg < 1 kPa < 1 atm
Section 10.5 (a) $P_1 > P_2$

■ What's this, and how does it reduce our dependence on fossil fuels?

This chapter is about waves, including familiar water waves and sound waves. We'll begin with common wave properties, including period, frequency, wavelength, and amplitude. Next you'll see what happens when two or more waves interfere as they overlap at the same place.

Much of the chapter is devoted to sound. We'll show two measures of loudness. Then you'll see how the properties of sound are exploited in the design of musical instruments. Finally, we'll discuss the Doppler effect—the observed change in frequency when a wave source moves relative to you. The Doppler effect has many applications, for both sound and other waves, including light and radio.

11.1 Wave Properties

Water waves are probably the wave phenomenon most familiar to you. Toss a rock into a pond and you see waves—ripples—moving in ever-widening circles. Visit the ocean and observe the waves that wash endlessly to shore.

Sound and light are also waves. Sound is obviously important in our lives, and we'll devote much of this chapter to it. Light is crucial, too, but understanding light waves requires knowledge of electromagnetism. You'll get that in Chapters 15–20. Chapters 21–23 are devoted to light and optics.

Fundamentally, **a wave is a traveling disturbance that transports energy but not matter**. A buoy bobs up and down as a water wave passes, but the buoy doesn't move shoreward with the wave. As you speak, you cause changes in air pressure that propagate as waves to a listener's ears. But air itself doesn't move from you to the listener. Although

If you give a spring a transverse twitch …

… you create a transverse wave that moves along the spring with speed *v*.

Each loop of the spring is displaced *transversely* as the wave passes.

(a) Transverse wave

If you give a spring a back-and-forth twitch …

… you create a compression wave that travels along the spring with speed *v*.

Each loop of the spring is displaced *longitudinally* as the wave passes.

(b) Longitudinal wave

FIGURE 11.1 Transverse and longitudinal waves.

they don't carry matter, clearly these waves carry energy. As this chapter's opening photo shows, we can extract energy from ocean waves to generate electricity, and your ears absorb sound energy that's ultimately sensed and processed by the brain.

Transverse and Longitudinal Waves

Figure 11.1 shows two fundamentally different wave geometries. In a **transverse wave** (Figure 11.1a), the disturbance is perpendicular to the direction of wave travel. In a **longitudinal wave** (Figure 11.1b), the disturbance is in the direction of travel. You can generate both kinds of wave on a coiled spring like a slinky.

We find both wave types in nature. Sound is longitudinal; light is transverse. Water waves are a combination, as Figure 11.2 suggests. As water waves approach shore, contact with the bottom causes them to "break," meaning they become more longitudinal. You may notice pieces of seaweed near shore sliding mainly back and forth horizontally—the longitudinal direction.

Periodic Waves

Many waves are **periodic**, comprising long trains of identical disturbances. Figure 11.3 shows a mechanical oscillator generating periodic transverse waves on a taut string. We'll consider transverse waves for now, because it's easier to visualize their wave properties. The wave has a series of **crests** (tops) and **troughs** (bottoms). The wave height, relative to the undisturbed string, is the wave's **amplitude** *A*. Amplitude here has essentially the same meaning it did for the simple harmonic oscillator. In fact, the amplitude of the mechanical oscillator is equal to the amplitude of the periodic wave it produces (Figure 11.3a).

A periodic wave has a regular **frequency**. You've encountered frequency before, in uniform circular motion (Chapter 3) and simple harmonic motion (Chapter 7). Frequency is the number of complete cycles per unit time, measured in SI units of hertz (Hz), with 1 Hz = 1 cycle per second. For the wave in Figure 11.3, the wave frequency is the same as the oscillator frequency. For a periodic wave, you can think of frequency as the number of complete wave cycles that pass a fixed location per unit time (Figure 11.3b). You could time the passage of successive wave crests to measure the wave's frequency. As you've seen before, frequency f is the reciprocal of the period T. That is, $f = 1/T$.

Another important property is **wavelength** λ (Greek lowercase lambda), the distance between successive wave crests (Figure 11.3b). You'll see that our perception of both sound and light depends significantly on wavelength.

A final property, related to frequency and wavelength, is the wave **speed** *v*. In a uniform medium, like the stretched string of Figure 11.3, wave speed is constant. (Later we'll discuss how wave speed depends on the string's tension and density.) Because speed in this case is simply distance/time, the speed is the length, λ, of one complete wave cycle, divided by the time T for the complete cycle to pass: $v = \lambda/T$.

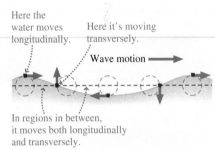

Here the water moves longitudinally. Here it's moving transversely.

In regions in between, it moves both longitudinally and transversely.

FIGURE 11.2 A water wave has both longitudinal and transverse components.

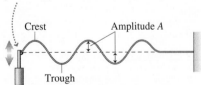

Oscillator vibrates up and down in simple harmonic motion with constant frequency, generating periodic waves on the string.

Crest   Amplitude *A*

Trough

(a) Generating periodic waves on a string

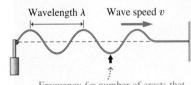

Wavelength λ Wave speed *v*

Frequency f = number of crests that pass a fixed position per unit time.

(b) Wavelength, wave speed, and frequency

FIGURE 11.3 Periodic waves.

Because $f = 1/T$, this means that the wave speed is

$$v = \lambda f \quad \text{(Speed of a periodic wave; SI unit: m/s)} \qquad (11.1)$$

Equation 11.1 is a fundamental relationship for all periodic waves, linking wave speed, wavelength, and frequency.

EXAMPLE 11.1   **Wavelengths of Sound**

Human hearing covers frequencies from about 20 Hz to 20 kHz. Sound travels at 343 m/s in air under standard atmospheric conditions. What are the wavelengths associated with those minimum and maximum frequencies?

ORGANIZE AND PLAN Equation 11.1 gives the relationship between speed, wavelength, and frequency: $v = \lambda f$.

Known: $v = 343$ m/s; $f_{min} = 20$ Hz; $f_{max} = 20$ kHz.

SOLVE Solving Equation 11.1 for wavelength, $\lambda = v/f$. Then with $f = f_{min} = 20$ Hz,

$$\lambda = \frac{v}{f} = \frac{343 \text{ m/s}}{20 \text{ Hz}} = 17 \text{ m}$$

Note that the seconds cancelled because 1 Hz is, formally, one inverse second.

For $f = f_{max} = 20$ kHz,

$$\lambda = \frac{v}{f} = \frac{343 \text{ m/s}}{2.0 \times 10^4 \text{ Hz}} = 0.017 \text{ m} = 17 \text{ mm}$$

REFLECT In this calculation, the time units cancel ($\text{Hz} = \text{s}^{-1}$), leaving meters for the wavelength unit. Note that high frequencies go along with short wavelengths and low frequencies with long wavelengths.

MAKING THE CONNECTION   Compare these wavelengths, at the ends of the human hearing scale, to the dimensions of the ear, which gathers and processes the sound waves.

ANSWER The short wave corresponding to the high frequency has a wavelength of just 1.7 cm, comparable to the ear's dimensions. The longer, low-frequency wave has a wavelength much larger than the person doing the hearing. The human ear is remarkable in its sensitivity to this huge wavelength range. We'll discuss sound and hearing in more detail shortly.

GOT IT? Section 11.1   Two waves, A and B, move through the same medium with the same speed. Wave A has wavelength λ_A and frequency f_A. Wave B has wavelength $\lambda_B = 3\lambda_A$. What's wave B's frequency? (a) $f_A/9$; (b) $f_A/3$; (c) f_A; (d) $3f_A$; (e) $9f_A$.

11.2 Interference and Standing Waves

We introduced waves by considering a rock dropped into a pond. Now drop two rocks simultaneously, a small distance apart. Watch the two sets of waves; they seem to pass through one another. But while they're actually meeting (Figure 11.4), they **interfere**, producing quite a different wave pattern. How can we explain this?

Wave Interference and the Principle of Superposition

Interference is fundamental behavior of all waves. The two-dimensional water waves in Figure 11.4 create a complex **interference pattern**. It's easier to visualize and understand interference with one-dimensional waves, like those on the string in Figure 11.5.

We start with two identical wave pulses that meet and interfere. Figure 11.5a shows the resulting **constructive interference**—a momentarily larger pulse that combines the two. But when the wave disturbances are oppositely directed, then **destructive interference** results (Figure 11.5b) as the pulses momentarily cancel. In both cases the pulses continue unchanged after their interaction.

Our example of interfering wave pulses illustrates a general principle that holds for many waves:

Principle of superposition: When two or more waves interfere, the resulting wave disturbance is equal to the sum of the disturbances of the individual waves.

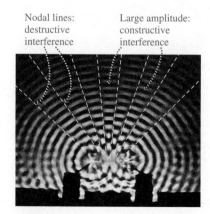

Nodal lines: destructive interference

Large amplitude: constructive interference

FIGURE 11.4 Wave interference in a water tank.

You can see superposition in Figure 11.5, where the wave disturbances are displacements of the stretched string. In Figure 11.5a, both displacements are positive, so they add to make a larger displacement—that's constructive interference. In Figure 11.5b, one displacement is positive and the other negative; adding them results in cancellation, or destructive interference.

The superposition principle also holds for the two-dimensional waves in Figure 11.4. Where the waves overlap, the net displacement of the water is the sum of displacements of the two waves. At different points, and at different times, the interference could be constructive, destructive, or in between. Note that the meeting of two troughs also results in constructive interference, because the combined wave there is the sum of two negative displacements—a net large negative displacement.

All waves interfere, and in most cases they obey the superposition principle. (An exception occurs in materials that respond nonlinearly to wave disturbances.) In Chapter 23 you'll learn about the interference of light, while the following example explores interference of sound waves.

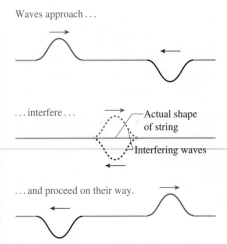

(a) Constructive interference

(b) Destructive interference

FIGURE 11.5 Wave superposition results in constructive interference and destructive interference.

CONCEPTUAL EXAMPLE 11.2   **Dead Spots**

You're in a room with two loudspeakers that produce sound waves with the same constant frequency. Explain why there are "dead spots" in the room, where the sound you hear is particularly faint. Why is it important that the waves have the same frequency?

SOLVE The dead spots result from destructive interference. If a wave crest from speaker A arrives at your ear at the same time as a trough arrives from speaker B (Figure 11.6), the resulting destructive interference leads to a drop in amplitude (volume).

With equal frequencies, the wave crest from A always reaches your ear at the same instant as the trough from B, continuing the pattern of destructive interference. If the frequencies were different, then crests and troughs would gradually get out of sync, and the interference would eventually become constructive. If the frequencies were unequal but very close, you'd notice the amplitude growing louder and softer periodically. This phenomenon, known as *beats*, will be discussed shortly.

Sound waves are longitudinal disturbances in air pressure, so we can't illustrate them as easily as we do the transverse waves of Figures 11.4 and 11.5. Nevertheless, the superposition principle applies just as well, with the crests and troughs being regions of higher and lower pressure in the sound wave.

REFLECT How large is the dead spot? That depends on wavelength. Example 11.1 shows that sound's wavelength varies from centimeters to meters. The size of a dead spot is roughly comparable to the wavelength. Incidentally, this example also shows why it's a good idea to rotate food cooking in a microwave oven. The wavelength of the microwaves is about 12 cm, and interference of reflected waves sets up a pattern of "hot spots" and "cold spots" in the oven; rotating the food ensures it's evenly heated.

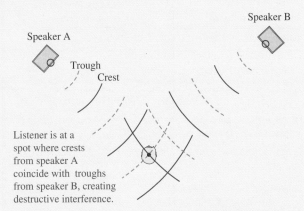

FIGURE 11.6 Destructive interference leads to dead spots.

In the room described in Conceptual Example 11.2, it's difficult to find a spot where you can't hear the sound coming from the speakers at all. One thing that helps prevent complete cancellation of the waves is reflection from walls or other objects. You hear not only waves coming directly from the speakers, but also these reflections. Reflected waves from multiple points are unlikely to cancel exactly. Reflections and the subsequent interference can lead to some interesting effects, as you'll see when we describe interference on a vibrating string.

Beats

A fixed interference pattern requires that the interfering waves have *exactly* the same frequency. If they don't, then their wave crests will gradually shift relative to one another, and interference at a given point will gradually swing between constructive and destructive. Figure 11.7 shows that when frequencies are close, the result is a slow variation in the overall amplitude of the combined wave. You can show in Problem 95 that the frequency of this variation is half the difference in the individual waves' frequencies. If these are sound waves, you hear the slow amplitude variation—a phenomenon called **beats**. Note that there are *two* amplitude peaks for each cycle of the slow oscillation, so the beat frequency heard is simply the difference of the two individual frequencies.

Beats provide a way of detecting and correcting small frequency differences. You've probably heard beats on a two-engine aircraft, as the engines operate at nearly but not quite the same frequency. Reducing the beat frequency lets pilots bring the engines arbitrarily closer to the same rotation rate. Musicians use beats to tune instruments, and beats are the basis of some extremely precise measurement techniques using electromagnetic waves.

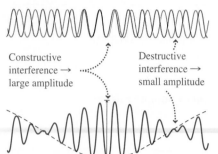

Constructive interference → large amplitude

Destructive interference → small amplitude

FIGURE 11.7 Beats result from the super-position of two waves with slightly different frequencies.

Standing Waves on a String

Reflection is a basic wave behavior. You use reflection when you look in a mirror, and indeed any time you view an illuminated object. We'll discuss reflection of light waves specifically in Chapter 22. For now, consider the reflection of a single wave pulse on a string. Figure 11.8a shows what happens when the wave pulse is reflected from a fixed end, and Figure 11.8b shows reflection from an end that's free to move.

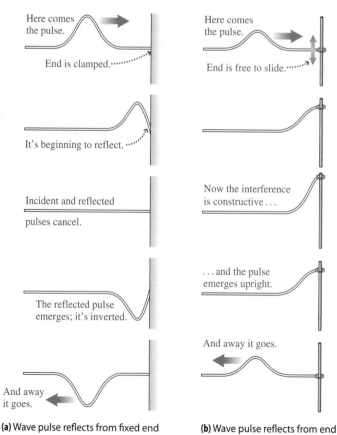

Here comes the pulse.

End is clamped.

It's beginning to reflect.

Incident and reflected pulses cancel.

The reflected pulse emerges; it's inverted.

And away it goes.

Here comes the pulse.

End is free to slide.

Now the interference is constructive . . .

. . . and the pulse emerges upright.

And away it goes.

(a) Wave pulse reflects from fixed end

(b) Wave pulse reflects from end that is free to move tranversely

FIGURE 11.8 Reflected wave pulses.

Next, consider reflection of a periodic wave from a fixed end. Each individual crest or trough is reflected and inverted, just like the single pulse in Figure 11.8a. As a result, the wave incident from the left interferes with the reflected wave, which is traveling in the opposite direction. Because these waves have the same speed, frequency, and wavelength, their superposition is also periodic. So you see an alternating sequence of constructive and destructive interference.

When a string vibrates between two fixed ends, a special and visually impressive situation arises: a **standing wave**, shown in Figure 11.9. A standing wave is characterized by the presence of **nodes**, points where there's no string movement whatsoever, and **antinodes**, points halfway between each pair of nodes where the amplitude is a maximum.

As Figure 11.9 suggests, standing waves are possible only for particular wavelengths, related to the string's length L. Physically, that's because the two ends of the string are fixed and therefore must be nodes. The distance between nodes is then some fraction L/n of the string's length, where n is an integer, giving a standing wave pattern with n antinodes.

Figure 11.9 also shows that the distance between adjacent nodes is just half a wavelength, or $\lambda/2$. But we've just seen that the node spacing is L/n, so $L/n = \lambda/2$. Therefore, the allowed wavelengths for standing waves are

$$\lambda = \frac{2L}{n} \qquad \text{(Wavelength of standing waves on a string, with } n = 1, 2, 3, \ldots; \text{ SI unit: m)} \qquad (11.2)$$

The longest possible wavelength is the **fundamental wavelength** λ_f. This corresponds to the $n = 1$ standing wave, so $\lambda_f = 2L$. All possible wavelengths (called **harmonics**) are equal to the fundamental wavelength divided by an integer. The first harmonic ($n = 1$) has the longest wavelength, the second harmonic ($n = 2$) has the next longest, and so on. Vibrations with shorter wavelengths than the fundamental are **overtones**. Stringed instruments such as guitars and violins vibrate with a combination of fundamental and overtones—more in Section 11.4.

Figure 11.10 shows standing wave patterns when one end of the string is free. Here the fixed end is a node and the free end an antinode. This makes the string length L an odd multiple of a quarter wavelength ($\lambda/4$). You won't see this pattern on a stringed musical instrument, where both ends are clamped. But you get an analogous pattern in an organ pipe that has one closed and one open end.

CONCEPTUAL EXAMPLE 11.3 **Fundamental Frequencies**

What fundamental frequency corresponds to the fundamental wavelength for a string fixed at both ends? How are the frequencies of the harmonics related to the fundamental frequency?

SOLVE Equation 11.1 relates wavelength and frequency: $v = \lambda f$. The fundamental wavelength is given by $\lambda_f = 2L$, so the corresponding fundamental frequency is

$$f_f = \frac{v}{\lambda_f} = \frac{v}{2L}$$

Thus, the fundamental frequency depends on *both* the string length and the wave speed on the string. The harmonics have wavelengths $\lambda = 2L/n$, so their frequencies are

$$f = \frac{v}{\lambda} = \frac{v}{2L/n} = n\frac{v}{2L} = nf_f$$

That is, the frequency of each harmonic is an integer multiple of the fundamental frequency.

REFLECT Note that the fundamental frequency depends inversely on the string's length. The guitarist knows this: By pushing down on part of the string, she effectively shortens it, raising the fundamental frequency.

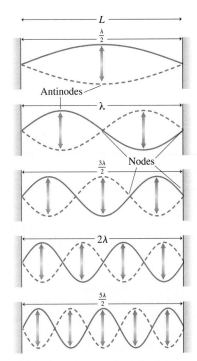

FIGURE 11.9 Standing waves on a string clamped at both ends; shown are the fundamental and four overtones. Note that the distance between antinodes is always $\lambda/2$.

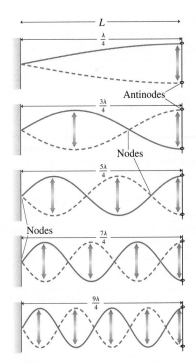

FIGURE 11.10 When one end of the string is fixed and the other is free, the string can accommodate only an odd number of quarter-wavelengths.

- A standing wave pattern alternates nodes and antinodes.
- At a node, there's no oscillation; at an antinode, oscillation has a maximum amplitude.
- The fundamental wavelength is the longest possible wavelength for a standing wave; the corresponding fundamental frequency is the lowest possible frequency.

Wave Speed, Tension, and Density

You've seen how the fundamental frequency of a vibrating string depends on the wave speed. But what determines the speed of waves on a particular string? Both experimentally and through calculus, we find that two factors affect wave speed: the string's tension, T, and its linear mass density (mass per unit length), μ. The resulting wave speed is

$$v = \sqrt{\frac{T}{\mu}} \tag{11.3}$$

In SI units, tension is in N and linear mass density in kg/m. You can verify that this combination gives a speed in m/s.

APPLICATION **Seismic Waves**

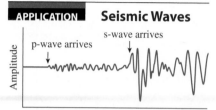

s-wave arrives

p-wave arrives

Amplitude

Travel time

An earthquake occurs when tectonic plates shift suddenly. Energy radiates outward in waves from the underground epicenter where the shift occurs. Primary waves (p-waves) are longitudinal, and secondary waves (s-waves) are transverse. These waves travel at different speeds through Earth's crust, p-waves normally at about 6 km/s and s-waves at 4 km/s. Accordingly, there's a delay between the arrival of the p- and s-waves, as shown here in a seismograph trace. By comparing readings from various seismic stations, geologists can determine the location (including depth) of the original event.

CONCEPTUAL EXAMPLE 11.4 **Violin Strings**

Compare the fundamental frequencies on a violin's four strings, which have approximately the same length and tension but different thicknesses. Does the thinnest string or thickest string have the lowest fundamental frequency?

SOLVE Thicker strings have higher mass densities. Because $v = \sqrt{T/\mu}$, a thicker string has a *slower* wave speed. The strings have the same length and therefore the same fundamental wavelength. Since $f = v/\lambda$, the string with the lowest speed—the thickest one—has the lowest fundamental frequency.

REFLECT The other main string instruments in the orchestra—viola, cello, and bass—all have longer strings. With longer fundamental wavelengths, their corresponding frequencies are lower.

EXAMPLE 11.5 **String Frequency and Tension**

A 0.750-m-long steel string has uniform mass density of 2.51×10^{-4} kg/m. (a) Find the wave speed on this string. (b) If you wish to make middle C ($f = 256$ Hz) its fundamental frequency, what should be the string tension?

ORGANIZE AND PLAN Our sketch of the vibrating string is shown in Figure 11.11. As shown in Conceptual Example 11.3, the fundamental frequency is $f_f = v/\lambda_f = v/2L$. Therefore, the wave speed is $v = 2Lf_f$. Also, the wave speed depends on tension and mass density: $v = \sqrt{T/\mu}$. So in terms of the speed, which we'll find first in part (a), the tension is $T = \mu v^2$.

Known: $L = 0.750$ m; $\mu = 2.51 \times 10^{-4}$ kg/m; $f = 256$ Hz.

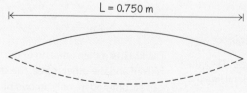

L = 0.750 m

FIGURE 11.11 Our sketch for Example 11.5.

SOLVE (a) Inserting the given values, the wave speed is

$$v = 2Lf_f = 2(0.750 \text{ m})(256 \text{ s}^{-1}) = 384 \text{ m/s}$$

(b) Then the required tension is

$$T = \mu v^2 = (2.51 \times 10^{-4} \text{ kg/m})(384 \text{ m/s})^2$$
$$= 37.0 \text{ kg} \cdot \text{m/s}^2 = 37.0 \text{ N}$$

REFLECT The SI units do indeed reduce to units of force (N). The answer is reasonable: This tension could be produced by pulling with a tensioning screw or by hanging a 37-N (3.8-kg) mass from one end of the string.

MAKING THE CONNECTION What's the string's diameter?

ANSWER Mass per unit length (which is given) is mass density multiplied by cross-sectional area, πR^2. Table 10.1 gives the mass density for steel: 7800 kg/m³. This leads to a diameter of 0.20 mm, which is thin wire, #32 gauge.

So far we've considered standing waves on strings and hinted at musical implications. We'll return later in this chapter to musical instruments and harmony. But first we need to cover the basics of sound, giving a context for understanding both music and hearing.

GOT IT? Section 11.2   A stretched string with length L, tension T, and linear mass density μ is vibrating at its fundamental frequency. Which of the following changes will increase the fundamental frequency? (More than one is correct.) (a) increase T; (b) decrease T; (c) increase L; (d) decrease L; (e) increase μ; (f) decrease μ.

11.3 Sound Waves

Sound fills our lives, from the gentle rustling of leaves to the pounding beat of a rock concert. Here we'll introduce important properties of sound waves and provide several applications.

Sound Speed

We generally experience sound as it travels through air. However, sound also travels through other gases, as well as liquids and solids. Whatever the medium, sound waves are always longitudinal, as illustrated in Figure 11.12. The waves originate in whatever is causing the sound and consist of compressions and rarefactions of the medium. Periodic sound waves have definite wavelength (the distance from one rarefaction to another or from one compression to another), frequency, and speed. Table 11.1 gives the speeds of sound in selected media. The relationship between speed, wavelength, and frequency in sound waves is the same as that for any periodic waves, as you learned in Equation 11.1: $v = \lambda f$.

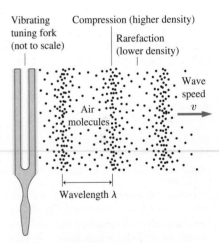

FIGURE 11.12 Sound waves produced by a vibrating tuning fork.

✓**TIP**

As with any periodic wave, use the basic relation $v = \lambda f$ for sound.

Table 11.1 shows that the speed of sound in air increases with increasing temperature. For everyday temperatures, the dependence is approximately linear:

$$v(T) = 331 \text{ m/s} + 0.60T \tag{11.4}$$

with T in Celsius.

✓**TIP**

Sound speed depends on air temperature: The lower the temperature, the slower the speed.

TABLE 11.1 Speeds of Sound in Selected Media

Medium	Speed (m/s)
Air (0°C)	331
Air (20°C)	343
Air (100°C)	387
Helium (0°C)	970
Oxygen (0°C)	316
Ethanol	1170
Water	1480
Copper	3500
Glass	5200
Granite	6000
Aluminum	6420

Note: Speeds are given at a temperature of 20°C unless otherwise noted.

Sound Frequency

The way we perceive sound's frequency is called **pitch**. We say that high-frequency sounds, like a soprano's voice or referee's whistle, are high pitched, while low-frequency sounds, like a baritone's voice or a tuba, are low pitched. As mentioned earlier, most people can hear sound with frequencies from about 20 Hz to 20 kHz. Typically, however, the ear's response near those extremes isn't as good as it is throughout the rest of the frequency range. Most people are good at distinguishing fairly small differences in frequency, meaning our ears are good frequency sensors.

Frequencies above 20 kHz are **ultrasonic**. Some animals use ultrasonic waves for both communication and range finding (**echolocation**). The bat emits ultrasonic waves from its mouth or nose. By detecting the reflected waves, it can find small objects (like insects) and fly between small cracks in a cave, even in the dark. Marine mammals including the dolphin and whale also use echolocation. They emit a train of clicking sounds from the melon, a fat-filled organ in the head, and detect reflected clicks using another fat-filled detector in their jaws.

APPLICATION **Ultrasound Images**

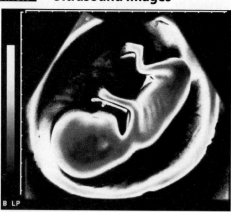

B LP

Ultrasound images provide a noninvasive medical diagnostic. The high-frequency waves pass readily through most body tissue, but some frequencies are reflected more than others by different kinds of tissue, fluid, or bone. The region being examined is scanned with a range of frequencies, and reflected waves are processed by computer to generate images like the one shown.

Humans use ultrasonic waves in technological applications, including ultrasound imaging of the body and range finding. Sound waves below 20 Hz are **subsonic** or **infrasonic**. At high amplitudes, these can cause dangerous resonances within the body. (Recall Chapter 7's discussion of resonance.)

EXAMPLE 11.6 **Sounds of the Game**

You're watching baseball from the upper deck beyond right field, 138 m from home plate. Assuming a 20°C day, how much time elapses from when you see the ball hit the bat to when you hear it?

ORGANIZE AND PLAN The speed of light is 3.00×10^8 m/s, nearly a million times faster than sound. For the purposes of this example, it's therefore safe to assume that light reaches your eye instantaneously. Sound, on the other hand, travels at 343 m/s. For this straight-line travel, speed = distance/time or $v = d/t$.

Known: $d = 138$ m; $v = 343$ m/s.

SOLVE The travel time for the sound is

$$t = \frac{d}{v} = \frac{138 \text{ m}}{343 \text{ m/s}} = 0.402 \text{ s}$$

By comparison, the light takes essentially zero time to reach you, so the time difference between seeing and hearing the ball hit the bat is 402 ms.

REFLECT You can easily detect this degree of time difference in a big arena, whether you're watching a sporting event or a concert.

MAKING THE CONNECTION If the time between seeing lightning and hearing the thunder is 5.0 s, how far away is the lightning?

ANSWER Distance is speed × time, so the lightning is about 1.7 km or 1 mile away—close enough that you should take cover. A good rule of thumb is that every 5-s difference between lightning and thunder means 1 mile between you and the lightning.

At a distance $r = R$, intensity $I = \dfrac{P}{A} = \dfrac{P}{4\pi r^2}$. As distance r increases, area A increases as its square: $A = 4\pi r^2$. Therefore, $I \propto \dfrac{1}{r^2}$.

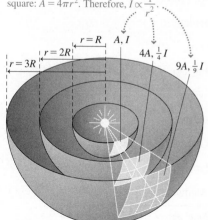

FIGURE 11.13 Sound intensity decreases in proportion to $1/r^2$.

Sound Intensity

Another obvious property of sound is its **loudness**. Waves carry energy, and loudness is related to the rate of energy flow—specifically, the *power per unit area*, also called **intensity**, I (Figure 11.13). Thus,

$$I = \frac{P}{A} \quad \text{(Sound intensity; SI unit: W/m}^2\text{)} \quad (11.5)$$

where power P is spread over area A. Recall that power is energy per time, measured in watts (from Chapter 5, 1 W = 1 J/s)—making the units of intensity W/m². Sound intensity is usually a more useful measurement than the total power, because it's power per unit area that dictates what our ears—with their fixed area—can hear. As Figure 11.13 shows, intensity for a point source falls as the square of the distance from the source. If you move twice as far from the source, then the sound intensity drops by a factor of 4.

✓**TIP**

Sound intensity from a point source decreases as the inverse square of the distance.

Sound Intensity Level and Decibels

Humans with good hearing can generally hear sounds with intensities as low as 10^{-12} W/m². (The exact threshold depends on frequency.) We hear sounds over a wide intensity range, with the maximum comfortable intensity before pain ensues being about 1 W/m². However, our sensation of loudness doesn't correspond directly to intensity. Rather, we sense loudness increasing by the same amount every time the intensity increases by a given factor. For this reason, sound intensity is often measured on a logarithmic scale called **sound intensity level**. This intensity level β is defined as $\beta = \log{(I/I_0)}$, where $I_0 = 10^{-12}$ W/m², the approximate threshold of human hearing. Although β is dimensionless, it's given a unit—the bel (B), named for Alexander Graham Bell. This "unit" is used in the same way that the radian (rad) is used to denote a dimensionless angular measure. In addition to inventing the telephone, Bell conducted research in human hearing and worked to improve the lives of hearing-impaired people.

More familiar is the unit 0.1 B = 1 decibel (dB). With 10 dB for every B, the definition of intensity level becomes

$$\beta \text{ (in dB)} = 10 \log{\left(\frac{I}{I_0}\right)} \qquad (11.6)$$

For example, intensity $I = 1.0 \times 10^{-6}$ W/m² corresponds to intensity level

$$\beta = 10 \log{\left(\frac{I}{I_0}\right)} = 10 \log{\left(\frac{10^{-6} \text{ W/m}^2}{10^{-12} \text{ W/m}^2}\right)} = 10 \log{(10^6)} = 10 \cdot 6 = 60 \text{ dB}$$

With the logarithmic decibel scale, the intensity level increases by 10 dB for every 10-fold increase in actual intensity. For example, if the intensity given above increases to 1.0×10^{-5} W/m², the intensity level rises to 70 dB. Table 11.2 lists typical intensity levels, while Figure 11.14 shows the response of the human ear.

TABLE 11.2 Typical Sound Intensity Levels

Sound intensity level (dB)	Description of sound
0	Barely audible sound
20	Whisper
40	Soft conversation heard at a distance
60	Television at normal level in closed room
80	Busy city street
100	Rock band at 4 m
120	Jet aircraft at takeoff (listener standing beside runway)
160	Eardrum ruptures

TACTIC 11.1	**Intensity Level and Decibels**

Equation 11.6 connects the two measures of loudness, intensity I and intensity level β:

$$\beta \text{ (in dB)} = 10 \log{\left(\frac{I}{I_0}\right)}$$

with $I_0 = 10^{-12}$ W/m². This definition lets you work back and forth between intensity and intensity level.

1. Converting from intensity to intensity level is straightforward: Just plug the intensity I into the equation.
2. To convert from intensity level to intensity, remember how to work with base 10 logarithms. By definition, $10^{\log x} = x$. Therefore, to find intensity I in terms of intensity level β, first rearrange Equation 11.6 as $\beta/10 = \log{(I/I_0)}$. Since the two sides of the equation are equal, so are the quantities resulting from raising 10 to the power of each side: $10^{\beta/10} = 10^{\log(I/I_0)} = I/I_0$. Therefore, $I = I_0 10^{\beta/10}$.

For example, if the intensity level is $\beta = 90$ dB, then
$I = I_0 10^{\beta/10} = (1.0 \times 10^{-12} \text{ W/m}^2)10^{90/10} = (1.0 \times 10^{-12} \text{ W/m}^2)10^9 = 1.0 \times 10^{-3} \text{ W/m}^2$.

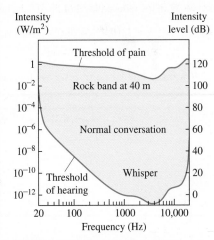

FIGURE 11.14 Human hearing at different frequencies.

Loudspeaker

At 4.0 m from a loudspeaker, the sound intensity is 6.2×10^{-5} W/m^2. (a) What's the intensity level at that point? (b) What's the total power emitted by the speaker?

ORGANIZE AND PLAN The intensity level is given by Equation 11.6: $\beta = 10 \log (I/I_0)$. The intensity is power/area, or $I = P/A$. Assuming a point source, the power radiates outward through a spherical surface of area $A = 4\pi r^2$, with r the distance from the point source.

Known: $I = 6.2 \times 10^{-5}$ W/m^2; $r = 4.0$ m.

SOLVE (a) The intensity level is

$$\beta = 10 \log \left(\frac{I}{I_0} \right) = 10 \log \left(\frac{6.2 \times 10^{-5} \text{ W/m}^2}{1.0 \times 10^{-12} \text{ W/m}^2} \right)$$

$$= 10(7.79) = 77.9 \text{ dB}$$

(b) Using the given values, the speaker's power output is

$$P = IA = I(4\pi r^2) = (6.2 \times 10^{-5} \text{ W/m}^2) 4\pi (4.0 \text{ m})^2$$

$$= 0.0125 \text{ W}$$

REFLECT The intensity level is sensible; 78 dB is loud, but not painful (see Table 11.2). You may be surprised that the power is so low, given that your stereo amplifier might be rated at 100 W or more. However, that's the electrical power output; typical speakers are only about 1% efficient at converting electrical energy to sound. Furthermore, that huge power capacity is needed only for the loudest musical passages.

MAKING THE CONNECTION Find the intensity and intensity level at half the distance (2 m) from this source.

ANSWER Intensity is proportional to the inverse square of the distance ($I = P/A = P/4\pi r^2$). Therefore, at half the distance, intensity I rises by a factor of 4, to about 2.5×10^{-4} W/m^2. But the intensity level β increases by only 6 dB, to 84 dB.

Give Me Some Quiet!

In the preceding example, how far should you move from the speaker to reduce the intensity level to a modest 58 dB?

ORGANIZE AND PLAN Tactic 11.1 shows that the intensity at the new distance is $I = I_0 10^{\beta/10}$. With power P from the preceding example, the distance can be found using $P = IA = 4\pi r^2 I$.

Known: $\beta = 58$ dB; $P = 0.0125$ W.

SOLVE With the reduced level of 58 dB, the intensity is

$$I = I_0 10^{\beta/10} = (1.0 \times 10^{-12} \text{ W/m}^2) 10^{58/10}$$

$$= 6.31 \times 10^{-7} \text{ W/m}^2$$

Then using $P = IA = 4\pi r^2 I$, we can find the unknown distance r:

$$r = \sqrt{\frac{P}{4\pi I}} = \sqrt{\frac{0.0125 \text{ W}}{4\pi (6.31 \times 10^{-7} \text{ W/m}^2)}} = 40 \text{ m}$$

REFLECT That's a 10-fold increase in distance for a 20-dB drop. This makes sense: A 20-dB reduction corresponds to intensity dropping by a factor of $10^2 = 100$. With intensity proportional to the inverse square of the distance, this requires that the distance increase 10-fold.

MAKING THE CONNECTION How close to the speaker would you have to move for the intensity level to increase 10 dB from its original value?

ANSWER By the same reasoning, the distance has to be reduced by a factor of $\sqrt{10}$, to 4 m/$\sqrt{10}$ = 1.26 m.

GOT IT? Section 11.3 Rank the loudness of the following intensities and intensity levels, from loudest to softest: (a) 54 dB; (b) 61 dB; (c) 5.6×10^{-7} W/m^2; (d) 1.0×10^{-6} W/m^2; (e) 60 dB.

11.4 Musical Instruments and Harmony

In Section 11.2 you learned how standing waves form on strings. Here we'll look more closely at standing waves, on both strings and in air columns. These serve as the basis for many musical instruments, so we'll introduce some concepts of musical sound.

Stringed Instruments

Stringed musical instruments are configured as shown in Figure 11.9, with strings fixed at both ends. The possible standing waves are the fundamental, shown in the upper frame of Figure 11.9, and the higher harmonics, or overtones.

When you pluck a string, you typically hear some combination of the fundamental frequency and harmonics. Each musical instrument has a unique sound, resulting from a distinctive pattern of harmonics. If two different instruments play the same fundamental note, your ear can easily distinguish them by their harmonic patterns. The surrounding structure—usually a wooden box—enhances an instrument's harmonic pattern. In instruments of the violin and guitar families, strings pass over openings in the box. In a piano, the strings are within the box, which may be closed or partially open. Different harmonics also result from how the strings are set oscillating. In a piano they're hammered, but the violin strings are usually bowed. The fibers of the bow produce a particular set of harmonics, which is evident from the unique sound you hear when a violin string is plucked.

Musical Harmony

Stringed instruments, such as the lyre and lute, have ancient origins. Members of the Pythagorean school of philosophy in Greece were fascinated with the harmonies of vibrating strings as long ago as 2500 years. The Pythagoreans noticed that when they simultaneously plucked two strings that were identically prepared except for length, pleasing "harmonious" sounds resulted when the string lengths had certain whole number ratios, such as 2 : 1. We now define "identically prepared" as having the same tension and density, giving equal wave speeds on the two strings. Thus, if the strings' lengths have a 2 : 1 ratio, their fundamental frequencies have the same 2 : 1 ratio.

A 2 : 1 frequency interval is an **octave**. On the standard musical scale, the set of seven major notes (A through G) repeats with each octave. For example, the concert A note has frequency 440 Hz. One octave above is another A with $f = 880$ Hz. One octave below is A with $f = 220$ Hz.

The Pythagoreans discovered other harmonious ratios. The ratio 3 : 2 is the musical "fifth," separating C from the next higher G. The "major third" (the smaller step from C to E) has frequency ratio 5 : 4. If you play C, E, and G together (the "tonic chord"), you enjoy several ratios simultaneously!

Musicians have adjusted the standard note frequencies to fit a total of 12 steps (including flats and sharps) into each octave. The whole scale, with common sharps and flats, is C, C-sharp, D, E-flat, E, F, F-sharp, G, A-flat, A, B-flat, B, and C. To make the frequency ratio of any two successive notes equal, that ratio has to be the 12th root of $2(\approx 1.05946)$. Therefore, the step from C to G is $2^{7/12} \approx 1.4983$—not exactly the perfect fifth, but very close.

TIP

Musical harmonies result when fundamental sound frequencies in certain whole-number ratios are played together.

Wind Instruments

The resonant vibrations of air in tubes are responsible for the music of brass and woodwind sections, some percussion instruments, and the stately organ. Figure 11.15 shows simple models of three types of tubes. These support longitudinal standing sound waves, rather than the transverse standing waves on a string. However, the rules governing allowed wavelengths are still simple. At any closed end, the air's displacement is zero—making a node. At an open end, there's an antinode, with maximum air displacement. As with standing waves on strings, each case has a fundamental frequency, which depends on the tube length, and then higher-frequency harmonics.

Real musical instruments aren't as simple as the uniform columns in Figure 11.15, but one that comes close is the flute. It's approximately 70 cm long, with one open end. Several centimeters from the other end is a hole into which the musician blows to create standing waves. By selectively closing some of the many holes along the length of the flute, the

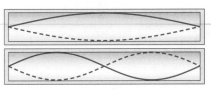

(a) Tube with two closed ends has nodes at both ends

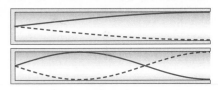

(b) Tube with one closed end has a node at the closed end and an antinode at the open end

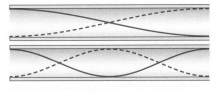

(c) Tube with two open ends has antinodes at both ends

FIGURE 11.15 Illustration of standing wave patterns in musical instruments. At each closed end is a node, and at each open end is an antinode. Note that these are graphs of sound wave patterns, which are longitudinal, not transverse waves.

musician can play a wide range of notes. With all holes closed, the flute approximates a uniform column open at each end. The standing wave pattern in Figure 11.15c suggests that the fundamental wavelength is twice the flute's length L: $\lambda = 2L$. The note played in this configuration is B below middle C, with frequency $f = 247$ Hz. With $v = f$ and sound speed $v = 343$ m/s at room temperature, this implies an effective air column length of

$$L = \frac{\lambda}{2} = \frac{v}{2f} = \frac{343 \text{ m/s}}{2(247 \text{ s}^{-1})} = 0.694 \text{ m}$$

which is in fact the flute's approximate length. Note that opening more holes creates standing waves with shorter wavelengths, and hence higher frequencies. The flute's highest note is more than two octaves above this low B!

As you might expect, other woodwinds have lower registers. Lengthening the air column leads to longer wavelengths, hence lower pitches. At the extreme is the bassoon, over 2 m long and with much lower pitch. Brass instruments also rely on standing waves, but their convoluted twists and turns mean a brass instrument can't be modeled as a uniform air column.

The physics of waves also affects musical sound after it leaves the instrument. Designing concert halls challenges acoustic engineers, who need to ensure there are no dead spots (see Conceptual Example 11.2). That requires sound reflection off multiple objects to reduce destructive interference. But too many reflections produces annoying echoes. Sound is most easily reflected by objects comparable in size to its wavelength. Thus, a high-pitched piccolo note interacts with a small light fixture, while a bass note might scatter from the facing on the balcony.

EXAMPLE 11.9 **Measuring Sound Speed**

One technique for measuring sound speed uses a vertical column partially filled with water, as shown in Figure 11.16. Changing the water level alters the length of the air column. A tuning fork is struck over the top of the column, and a noticeable increase in volume results when the height of the air column permits a standing wave. In a measurement using a 510-Hz tuning fork, a maximum volume is heard each time the water level drops by 33.5 cm. What's the sound speed?

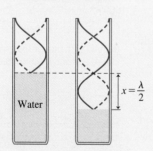

FIGURE 11.16 Lowering the water level changes the standing wave pattern.

ORGANIZE AND PLAN As the water level drops, each subsequent maximum in volume means the standing wave pattern contains one more node and antinode (Figure 11.16). Therefore, the additional length x of the air column corresponds to one-half of a wavelength: $x = \lambda/2$. Speed, wavelength, and frequency are related by the familiar $v = \lambda f$.

Known: $x = 33.5$ cm; $f = 510$ Hz.

SOLVE With $\lambda = 2x$, the speed becomes $v = \lambda f = 2xf$.
Inserting the given values,

$$v = 2xf = 2(0.335 \text{ m})(510 \text{ s}^{-1}) = 342 \text{ m/s}$$

REFLECT This is quite close to the value for 20°C, 343 m/s. Given the temperature dependence of the sound speed, the speed may well have been 342 m/s when this measurement was done.

MAKING THE CONNECTION If the sound speed is indeed 342 m/s, is the temperature higher or lower than 20°C? What temperature results in this sound speed?

ANSWER Equation 11.4 shows that sound speed drops with decreasing temperature. With $v = 342$ m/s, solving Equation 11.4 for temperature gives $T = 18$°C.

CONCEPTUAL EXAMPLE 11.10   **Half-Open Column**

An air column has one open end and one closed end. For waves of given frequency f and speed v, what column lengths will permit standing waves?

SOLVE Figure 11.15b shows that L is the distance from one node to the next antinode, or one-fourth of the fundamental wavelength. Thus $L = \lambda_1/4$. With $\lambda = v/f$, this gives $L = v/4f_1$. For the next harmonic, the tube length must be three times as great as the fundamental, because an entire node-to-node distance has been added. Then $L = 3v/4f$. The third harmonic adds another node-to-node distance, so $L = 5v/4f$. There's a pattern here; the column length must be $L = nv/4f$, where n is an *odd integer*.

REFLECT The change in distance ΔL between any two successive harmonics is $v/2f$, consistent with the results of Example 11.9.

GOT IT? Section 11.4   In an air column closed at both ends, the ratio of the second harmonic frequency to the fundamental is (a) 1.5; (b) 2; (c) 3; (d) 4.

11.5 The Doppler Effect

Stand by the roadside as a fire truck goes by, and you'll hear the pitch of its siren drop from higher to lower as it passes. This shift in frequency is the **Doppler effect**, and it results from the motion of a sound source relative to the observer. Here we'll show how the Doppler effect arises.

Figure 11.17 shows two fixed observers, Joe and Linda, and between them a source emitting sound waves with frequency f. When the source is stationary, both listeners hear that same frequency (Figure 11.17a). If the source moves toward Linda, Figure 11.17b shows that for her the distance between successive wave crests—the wavelength λ—is shortened. Because $f = v/\lambda$, and the waves travel through air with constant speed v, shorter wavelength means higher frequency. Conversely, for Joe the wavelength is longer and the frequency lower.

The Doppler Effect—Quantitative

The Doppler frequency shift follows from the different wavelengths measured by Linda and Joe. The source emits waves with frequency f, hence period $T = 1/f$—the time interval between emission of successive wave crests. Let the source move toward Linda with speed v_s. Then the distance between wave crests, which would normally be vT, becomes $vT - v_sT = (v - v_s)T$. Call that distance λ', the wavelength for Linda. She then hears a frequency $f' = v/\lambda'$ or

$$f' = \frac{v}{\lambda'} = \frac{v}{(v - v_s)T} = \frac{v}{(v - v_s)}f$$

Dividing the numerator and denominator by v leads to

$$f' = \frac{f}{1 - v_s/v} \quad \text{(Doppler effect, source approaching; SI unit: Hz)} \quad (11.7)$$

When $v_s < v$, Equation 11.7 shows that the perceived frequency f' is greater than the source frequency f. Joe's situation is analogous, now with the wavelength *increased* by v_sT, giving a perceived frequency

$$f' = \frac{f}{1 + v_s/v} \quad \text{(Doppler effect, source receding; SI unit: Hz)} \quad (11.8)$$

Here $f' < f$, so Joe hears a lower frequency. All this agrees with your experience with the passing fire truck.

When source is stationary, wavelength and frequency are the same for both listeners.

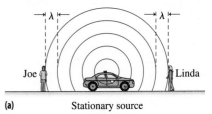

(a)   Stationary source

When source is moving to right, Joe hears a lower frequency (longer wavelength) than does Linda.

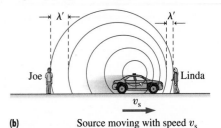

(b)   Source moving with speed v_s

FIGURE 11.17 The Doppler effect causes Joe and Linda to hear the sound at different frequencies.

Ambulance!

An ambulance zooms down the street at 80.0 mph (35.8 m/s), its siren blaring at 1.20 kHz. The ambulance is approaching Linda and receding from Joe. What frequency does each hear? The air temperature is 20°C.

ORGANIZE AND PLAN Linda hears a higher frequency, according to Equation 11.7: $f' = f/(1 - v_s/v)$, while the frequency Joe hears is lower, from Equation 11.8: $f' = f/(1 + v_s/v)$.

Known: $v = 343$ m/s (air at 20°C); $f = 1.20$ kHz.

SOLVE Using the given values, Linda and Joe perceive the following frequencies, respectively:

$$f' = \frac{f}{1 - v_s/v} = \frac{1.20 \text{ kHz}}{1 - (35.8 \text{ m/s})/(343 \text{ m/s})} = 1.34 \text{ kHz}$$

$$f' = \frac{f}{1 + v_s/v} = \frac{1.20 \text{ kHz}}{1 + (35.8 \text{ m/s})/(343 \text{ m/s})} = 1.09 \text{ kHz}$$

REFLECT This frequency shift would be obvious to your ear. Notice that the shift depends only on motion, not on distance from source to observer.

MAKING THE CONNECTION Are the frequency shifts greater or less on a hot day, with $T > 20°C$?

ANSWER Equation 11.4 shows that sound speed increases with rising temperature. This makes the fraction v_s/v in the Doppler equations smaller than at 20°C, resulting in a smaller frequency shift.

Moving Observers

What happens to the frequency of sound for an observer moving relative to a source that's at rest with respect to the air?

SOLVE Figure 11.18 shows that an observer approaching the stationary source encounters wave crests more frequently; hence she perceives a higher frequency. A receding observer, in contrast, perceives a lower frequency.

Quantitatively, wave crests pass the approaching observer at a higher speed $v + v_o$, where v is the sound speed and v_o the observer's speed. But because the source is at rest relative to the air, the wavelength is unchanged. Using $f = v/\lambda$, the observer's perceived frequency is therefore $f' = (v + v_o)/\lambda$. But λ is related to the nonshifted frequency f by $\lambda = v/f$, so this becomes $f' = (v + v_o)/(v/f) = f(1 + v_o/v)$. For a receding observer, you can similarly show that $f' = f(1 - v_o/v)$.

REFLECT These formulas are slightly different from those for a moving source. They become essentially identical when the speed of the moving source or observer is small compared with the wave speed. Note that we had to specify that the source is stationary *with respect to the air*. That's because air is the medium in which sound waves travel. Light and other electromagnetic waves have no medium, so there's no way to distinguish whether it's source or observer that's moving. The Doppler formulas for light are different, although they reduce to the formulas we've derived for speeds that are small compared with the speed of light. For light, it's truly only the relative motion that counts—a key idea in Einstein's relativity.

Observer approaching source encounters wave crests more frequently than if stationary, and thus hears a higher frequency: $f' > f$.

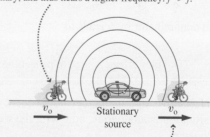

Observer moving away from source encounters wave crests less frequently than if stationary, and thus hears a lower frequency: $f' < f$.

FIGURE 11.18 The Doppler effect for a stationary source and moving observer.

| EXAMPLE 11.13 | **Tornado Alarm!** |

A stationary horn blows at 470 Hz to warn of a tornado in the area. While driving toward the horn, you perceive the frequency as 510 Hz. How fast are you driving? Assume air at 20°C.

ORGANIZE AND PLAN Conceptual Example 11.12 shows that the shifted frequency in this case is given by $f' = f(1 + v_o/v)$. This equation can be solved for your car's speed v_o.

Known: $v = 343$ m/s (air at 20°C); $f = 470$ Hz; $f' = 510$ Hz.

SOLVE Rearranging the Doppler equation gives $f' - f = f(v_o/v)$, or

$$v_o = \left(\frac{f'}{f} - 1\right)v = \left(\frac{510 \text{ Hz}}{470 \text{ Hz}} - 1\right)(343 \text{ m/s}) = 29.2 \text{ m/s}$$

REFLECT That's a reasonable highway speed of 65 miles per hour.

MAKING THE CONNECTION Is there any limit to the maximum perceived frequency in this case? What if you're moving away from the source?

ANSWER From the results of Example 11.12 you can see that there's no limit for an approaching observer, because f' increases continuously as v_o increases. However, the result for a receding observer suggests a problem when $v_o > v$, because you get a nonsensical negative frequency. You're actually "outrunning" the sound in this case!

The Doppler effect has many applications. Physicians use it to measure blood flow. A device emits ultrasonic waves with frequencies ranging from about 1 MHz to 10 MHz, and the frequencies of the waves that reflect from particles in the flowing blood are Doppler shifted. Measuring that shift allows computation of the blood flow speed, and can help diagnose vascular disorders such as arteriosclerosis. Doppler blood flow analysis is also useful in monitoring recovery from vascular or orthopedic surgery, as well as organ transplants.

Some other applications—for example, radar guns used by police to detect speeding cars and the Doppler radar systems used to forecast weather—involve the Doppler effect for light, which is similar to that for sound but mathematically a bit different. However, the Doppler formulas we've developed for sound also apply to light in the approximation that the source and observer speeds are much less than the speed of light. We'll discuss light waves in Chapters 20–22, along with more Doppler applications.

Shock Waves and Sonic Booms

What happens when a sound source moves *faster* than the sound speed? With $v_s > v$, Equation 11.7 predicts a negative frequency, which is physically meaningless. A closer look at Equation 11.7 shows that the frequency f' grows without bound as source speed approaches the sound speed, becoming infinite right at the sound speed and then turning negative.

Figure 11.19 shows what happens when a source—say, an airplane—approaches and then exceeds the sound speed. As the plane's speed v_s approaches the sound speed v, sound waves pile up in front of the aircraft, creating a **sound barrier** with increased drag that's

(a) Flying at normal speed

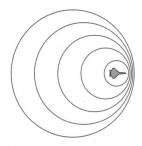

(b) Flying at just below speed of sound

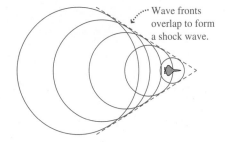

Wave fronts overlap to form a shock wave.

(c) Flying at supersonic speed

FIGURE 11.19 Formation of shock waves.

difficult for the plane to penetrate. Once it does, the waves superpose to form a very intense, cone-shaped **shock wave**. When the shock wave passes an observer on the ground, the sudden **sonic boom** is quite distinctive; it carries a lot of energy and can even do damage.

Shocklike phenomena occur whenever an object moves through a medium at a speed greater than the wave speed in that medium. The bow wave trailing diagonally from a boat is one example. Although nothing can travel faster than light *in a vacuum*, it's possible to accelerate subatomic particles to speeds greater than that of light in particular media such as water. The resulting shock waves are one way to generate intense light beams.

Chapter 11 in Context

This chapter introduced important wave properties and phenomena, including *interference* and *superposition*. You've learned to distinguish *transverse* and *longitudinal* waves. You've seen how the familiar concepts of period and frequency also apply to waves—and lead to a simple, universal relationship between speed, wavelength, and frequency of a periodic wave. We've also shown how the basic physics of sound applies to the music you hear and to the design and use of musical instruments.

Looking Ahead Much of this chapter considered sound waves—longitudinal waves in air and other media. You'll encounter waves again, most importantly in your study of light and related electromagnetic waves in Chapters 20–22. Light is of fundamental importance not only for its application to optics and vision, but also in helping us understand atomic and quantum phenomena in Chapters 23–26.

Wave Properties

(Section 11.1) A **wave** is a traveling disturbance that carries energy but not matter. In a **transverse wave**, the disturbance is perpendicular to the direction of wave travel. In a **longitudinal wave**, the disturbance consists of a density variation in the direction of travel.

Periodic waves consist of repeated, identical disturbances. Wave **speed**, **wavelength**, and **frequency** are fundamentally related.

Speed of a periodic wave: Speed = wavelength × frequency, or $v = \lambda f$

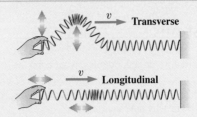

Interference and Standing Waves

(Section 11.2) **Wave interference** occurs when two or more waves meet. Interfering waves follow the **superposition principle**: the total disturbance is the sum of the individual disturbances. **Constructive interference** increases overall wave amplitude; **destructive interference** reduces it. **Standing waves** result from interference of waves reflecting on confined structures.

Wavelength of standing waves on a string fixed at both ends:

$$\lambda = \frac{2L}{n} \ (n = \text{integer})$$

Wave speed on a string: $v = \sqrt{\dfrac{T}{\mu}}$

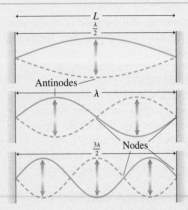

Sound Waves

(Section 11.3) Sound waves are longitudinal waves in air and other media. Most people hear sound with frequencies between 20 Hz and 20 kHz. Frequencies above 20 kHz are called **ultrasonic**; those below 20 Hz are called **subsonic** or **infrasonic**. The frequency of a sound wave is often referred to as its **pitch**.

Sound intensity is power per unit area: $I = \dfrac{P}{A}$

Sound intensity level $\beta = 10 \log\left(\dfrac{I}{I_0}\right)$ (measured in decibels, dB)

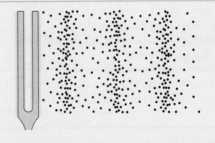

Musical Instruments and Harmony

(Section 11.4) Whole-number ratios of sound frequencies lead to musical harmonies. The 2 : 1 frequency interval is an **octave**, containing the seven major notes. The size and shape of a musical instrument affect its pitch by determining how and where standing waves form, and the mix of harmonics determines the instrument's unique sound.

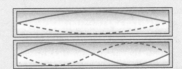

The Doppler Effect

(Section 11.5) The **Doppler effect** is a shift in the frequency of a wave resulting from the relative motion of source and observer. When the two approach, frequency increases. When they recede, frequency decreases.

Source approaching: $f' = \dfrac{f}{1 - v_s/v}$

Source receding: $f' = \dfrac{f}{1 + v_s/v}$

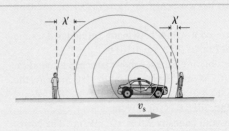

NOTE: Problem difficulty is labeled as ▪ straightforward to ▪▪▪ challenging. Problems labeled BIO are of biological or medical interest.

Conceptual Questions

1. If the frequency of waves in a water tank doubles, what happens to their wavelength?
2. Based on your experience, do you think the speed of sound in air depends on frequency? Why or why not?
3. If the amplitude of water waves suddenly increases by 50% due to a passing boat, how are the frequency and wavelength affected?
4. You shout to your friend who's swimming under water. How do you expect the sound waves to change in frequency and wavelength once they enter the water?
5. At the moment two waves overlap with complete destructive interference (Figure 11.5b), the wave amplitude is zero. Where did the energy of the two waves go?
6. If a standing wave on a string has n nodes (counting the two fixed ends), how many antinodes are there?
7. A climbing rope hangs vertically from the ceiling. If you shake the bottom of the rope to send a single wave pulse upward, how does the speed of the pulse change as it rises? What happens to the speed of the reflected pulse on the way down?
8. Brass musicians "warm up" their instruments by playing for a while before a concert. For what physical reason might this be a good idea?
9. You have two steel strings of equal length under equal tension. If they're to vibrate with fundamental frequencies exactly one octave apart, what should be the ratio of the strings' diameters?
10. Based on Table 11.1, do you expect that the sound speed in nitrogen is faster or slower than in air?
11. If you double your distance from a sound source, how does the sound intensity change? What about the intensity level?
12. The flute's lowest note plays when all the holes are covered. Why does uncovering any hole raise the pitch?
13. How does sliding the trombone's slide change the pitch?
14. One car follows behind another on the freeway at the same speed. If one emits a sound, do passengers in the other perceive it as Doppler shifted?

Multiple-Choice Problems

15. A water wave has frequency 0.40 Hz and wavelength 2.0 m. Its speed is (a) 5.0 m/s; (b) 1.3 m/s; (c) 1.0 m/s; (d) 0.80 m/s.
16. A sound wave with speed 360 m/s has wavelength 1.5 m. Its frequency is (a) 540 Hz; (b) 360 Hz; (c) 240 Hz; (d) 140 Hz.
17. A water wave has frequency 0.50 Hz, amplitude 0.35 m, and wavelength 2.6 m. If the amplitude doubles, the wavelength becomes (a) 5.2 m; (b) 2.6 m; (c) 1.3 m; (d) 1.0 m.
18. An FM radio signal with frequency 98.1 MHz and speed $c = 3.00 \times 10^8$ m/s has wavelength (a) 33 m; (b) 3.1 m; (c) 1.0 m; (d) 0.33 m.
19. For a standing wave on a string, how many antinodes does the third harmonic have? (a) 2; (b) 3; (c) 4; (d) 6.
20. You're 5.0 m from a blaring trumpet and experience a sound intensity of 4.0×10^{-5} W/m^2. If you retreat to 10.0 m away, the intensity becomes (a) 4.0×10^{-5} W/m^2; (b) 2.0×10^{-5} W/m^2; (c) 1.0×10^{-5} W/m^2; (d) 5.0×10^{-6} W/m^2.
21. You hear a clarinet with intensity 2.0×10^{-7} W/m^2. The corresponding intensity level is (a) 50 dB; (b) 53 dB; (c) 56 dB; (d) 59 dB.
22. The intensity level 25 m from a band is 72 dB. At 50 m, the level will be (a) 66 dB; (b) 62 dB; (c) 36 dB; (d) 18 dB.
23. A train approaches at 20 m/s, its whistle sounding a 1.13-kHz note. You perceive the whistle's frequency as (a) 1.15 kHz; (b) 1.20 kHz; (c) 1.25 kHz; (d) 1.30 kHz.
24. An airplane flies directly away from you. If you perceive its engine sound at 80% of the emitted frequency, the plane's speed is (a) 66 m/s; (b) 72 m/s; (c) 78 m/s; (d) 86 m/s.
25. An organ pipe with one end closed has fundamental frequency 220 Hz. The frequency of its first overtone is (a) 110 Hz; (b) 330 Hz; (c) 440 Hz; (d) 660 Hz.
26. An organ pipe with both ends open has fundamental frequency 440 Hz. The frequency of its first overtone is (a) 660 Hz; (b) 880 Hz; (c) 1320 Hz; (d) 1760 Hz.

Problems

Section 11.1 Wave Properties

27. ▪ A wave has a fixed wavelength 1.55 m. (a) Find its speed if its frequency is 0.365 Hz. (b) Repeat for a doubled frequency.
28. ▪ A sound wave with speed 343 m/s has wavelength 1.10 m. (a) What's its frequency? (b) Repeat for a halved wavelength.
29. ▪▪ Seismic p-waves travel at about 6 km/s and s-waves at 4 km/s. If a seismometer records the arrival of these two waves 24 s apart, how far away was the earthquake that produced them?
30. ▪▪ Waves in shallow water with depth d have speed given approximately by $v = \sqrt{gd}$. You observe waves in a ripple tank with wavelength 1.10 cm and frequency 23.5 Hz. (a) Find the water depth. (b) What happens to the wavelength as the waves pass into a region with twice that depth?
31. ▪▪▪ A transverse sinusoidal wave traveling in the $+x$-direction has the form $y(x, t) = A \cos(kx - \omega t)$, where A, k, and ω are constants. (a) Graph the wave displacement y at time $t = 0$. (b) Show that the constant k (called the wave number) is related to the wavelength by $k = 2\pi/\lambda$. (c) Show that the constant ω is given by $\omega = 2\pi/T = 2\pi f$. (d) Express the wave speed in terms of k and ω.
32. ▪▪ For the wave in the preceding problem, graph the wave displacement at the following times: $T/4$, $T/2$, $3T/4$, and T.
33. ▪▪▪ Using the results of the preceding two problems, write an equation that describes a transverse sinusoidal wave traveling in the $-x$-direction. Let the wave have wavelength λ and period T, with amplitude A in the y-direction.
34. BIO ▪▪ **Medical ultrasound.** Ultrasound with $f = 4.8$ MHz is used in a medical imager. Find the wavelength (a) in air, with sound speed 343 m/s, and (b) in muscle tissue, with $v = 1580$ m/s. (c) The ultrasound image can't show detail less than about one wavelength in size. Comment on the quality of the image you might expect to see in muscle.

Section 11.2 Interference and Standing Waves

35. ▪ Verify that the SI units for $\sqrt{T/\mu}$ in Equation 11.3 are m/s.
36. ▪ The main cables supporting New York's George Washington Bridge have linear mass density 4100 kg/m and are under tension of 250 MN. At what speed does a transverse wave travel on these cables?
37. ▪▪ You're midway between two loudspeakers producing sound with the same 86.0-cm wavelength, and you experience constructive interference. How far must you move toward one speaker in order to find a dead spot?

38. ■■ A 75-cm-long metal string under 36.5-N tension has a linear mass density 2.10×10^{-4} kg/m. (a) Find the wave speed on this string. (b) Find the frequencies of the first three harmonics.

39. ■■ A violin G string is 60.0 cm long and vibrates with fundamental frequency 196 Hz. (a) What's the wave speed on this string? (b) If the string tension is 49 N, what's its linear mass density?

40. ■■ Fundamental frequencies and tensions for the four violin strings are tabulated at right. Each string is 60 cm long. Find the linear mass density of each.

String	Fundamental frequency (Hz)	Tension (N)
G	196	49
D	294	53
A	440	60
E	659	83

41. ■■ Using the data from the preceding problem, find each string's first two overtone frequencies.

42. ■■ A standing wave on a 76.0-cm-long string has three antinodes. (a) What's its wavelength? (b) If the string has linear mass density 3.86×10^{-5} kg/m and tension 10.2 N, what are the wave's speed and frequency?

43. ■■ The maximum stress before a copper wire breaks is 331 MPa. Copper's density is 8890 kg/m^3. (a) Find the speed of transverse waves on the wire just at the breaking point. (b) Why doesn't your answer depend on the wire's diameter?

44. BIO ■■■ **Vocal ligament.** The vocal ligament is a structure in the human "voice box" that helps produce high-pitched sounds. The ligament is sometimes modeled as a string under tension, clamped at both ends—although current research suggests that this is not a particularly accurate model. A 1.42-cm-long vocal ligament has density 1040 kg/m^3 and is under 17.5-kPa stress. (a) Show that Equation 11.3 can equally well be written as $v = \sqrt{\sigma/\rho}$, where σ is the stress (force per area; see Chapter 10) and ρ is the volume mass density. (b) What fundamental frequency does the string model predict for this ligament?

45. ■■■ Two violins are supposed to be playing concert A, at 440 Hz. One of them is right on, but the other's frequency is 439.6 Hz. What beat frequency do the violinists hear?

Section 11.3 Sound Waves

In this section assume the temperature is 20°C unless directed otherwise.

46. BIO ■ **Dog hearing.** Find the wavelength of the highest-frequency sound that a dog can hear, about 50 kHz.

47. ■■ A submarine 65 m deep emits a sonar (sound) wave. The wave reflects from the bottom and returns to the submarine 0.86 s after it was emitted. How deep is the ocean at that point?

48. ■■ You shout toward a cliff 175 m distant. How much later do you hear the echo?

49. ■■ A sound wave takes 35.0 s to travel between two points in 20°C air. (a) What's the distance between the points? (b) What would be the travel time at 0°C?

50. ■■ At 25 m from a sound source, the intensity level is 55 dB. What's the level at distances of (a) 50 m? (b) 250 m?

51. BIO ■■ **Eardrum.** The human eardrum is roughly circular with a diameter of approximately 1.0 cm. Find the total power impinging on the eardrum when it's subject to 85-dB sound.

52. ■■ Find the wavelength of 550-Hz sound waves (a) in air and (b) in glass.

53. ■■ By approximately what common fraction does sound intensity decrease when the intensity level drops by 3 dB?

54. ■■ What's the intensity of sound with intensity level 95 dB?

55. ■■■ The band at an outdoor frat party pumps out sound energy at the rate of 6.5 W, and neighbors 25 m away complain about the din. (a) What's the intensity level at the neighbor's place? (b) To what percent of its original value must the band reduce its power in order to drop the neighbor's sound level by 15 dB? Assume the sound spreads in all directions.

56. ■■ At a rock concert you're 15 m from the stage, where the intensity level is 105 dB. How far away should you move in order to reduce the intensity level to a slightly more bearable 92 dB?

57. ■■ How much does the intensity level change if you increase your distance from the source by a factor of (a) 2; (b) 10; (c) 100?

58. ■■■ A listener perceives one clarinet at an intensity level of 60 dB. How many clarinets playing at the same volume and same distance from the listener would it take to produce a 70-dB level?

59. ■■■ A loudspeaker outputs sound energy at a rate of 100 mW. (a) How far away could a person with excellent hearing barely hear this sound if it's a pure 1000-Hz tone? (b) Repeat if the sound frequency is 100 Hz.

60. BIO ■■ **Sensitivity of hearing.** Most people can detect an intensity level difference of 1.0 dB. What's the ratio of the two sound intensities that differ by 1.0 dB in intensity level?

61. BIO ■■ **Hearing loss.** Person A can barely hear a sound at a particular frequency with an intensity level of 2.4 dB. Person B, who has hearing loss, can barely hear a 9.4-dB tone with the same frequency. Find the ratio of sound intensities at these two hearing thresholds.

Section 11.4 Musical Instruments and Harmony

62. ■ For the flute described in the text, playing the low B, how would the frequency be affected if the flute were made (a) 1.0 cm longer; (b) 1.0 cm shorter?

63. ■ For the flute described in the text, playing the low B, how would the frequency be affected if the flute were being played outside on a 0°C-day?

64. ■ A 4.30-m-long organ pipe is open on both ends. Find its fundamental and first three overtone frequencies.

65. ■ Repeat the preceding problem if the same pipe has one end open and the other closed.

66. ■■ An organ pipe, open on both ends, has fundamental wavelength 2.16 m. Find (a) the fundamental frequency and (b) the distance between standing-wave nodes for the second harmonic.

67. ■■ Find the lengths of organ pipes with one closed end needed to play the following fundamental frequencies: (a) 56 Hz; (b) 262 Hz (middle C); (c) 523 Hz (C above middle C); (d) 1200 Hz.

68. ■■ The concert A note has frequency exactly 440 Hz. Find the frequencies of (a) the note one step higher (B-flat); (b) the C note, three steps above A; and (c) the A note one octave above concert A.

69. ■■ An organ pipe closed on one end has fundamental frequency 512 Hz. If the closed end is then opened, what's the frequency of the first overtone?

Section 11.5 The Doppler Effect

In this section assume the sound speed in air is 343 m/s.

70. ■ A jet aircraft's engines emit an 850-Hz roar. What frequency do you hear if the aircraft (a) approaches you at 520 mph (232 m/s) or (b) recedes at the same speed?

71. ■ A runner approaching the start/finish line hears the bell signaling one lap to go. The bell emits a 352-Hz tone, and the runner hears it at 359 Hz. How fast is she going?

72. ■ A classic example of the Doppler effect involves a train whistle. Suppose the train whistle sounds at 1.20 kHz. Find the frequency you hear when the train is approaching and when it's receding for (a) a 27.0-m/s freight train and (b) the 250-km/h Japanese Shinkansen bullet train.

73. ■■ A goose flying toward you at 13.0 m/s emits a squawk, which you hear at 257 Hz. (a) What frequency did the goose emit?

(b) What frequency will you hear if the goose emits the same squawk while flying away at the same speed?

74. ■ ■ For the goose in the preceding problem, find the wavelength of the sound you hear in both cases.

75. ■ ■ A parachutist leaps from a hovering helicopter, and after 4.0 s of free fall shouts back toward the helicopter. If the shout is emitted at 425 Hz, what frequency is heard on the helicopter?

76. ■ ■ ■ A fire truck approaches and passes you at constant speed. If the siren frequency you hear drops from an initial 686 Hz (approaching) to a final 628 Hz (receding), what's the truck's speed?

77. ■ ■ A jet is flying at 99% the speed of sound, its engines emitting a 1200-Hz tone. Find the frequency and wavelength of the tone you hear when the jet is flying (a) toward you and (b) away from you.

78. ■ ■ How fast would a source have to move toward you for you to hear sound at twice the emitted frequency?

79. ■ ■ How fast would you have to move toward a stationary sound source for you to hear sound at twice the emitted frequency?

80. ■ ■ ■ Consider what happens when the source and observer of sound are *both* in motion toward one another. Let the source's speed with respect to the air be v_s, and the observer speed with respect to the ground v_o. Show that if the source emits sound with frequency f, the received frequency is

$$f' = \left(\frac{1 + v_o/v}{1 - v_s/v}\right)f$$

81. ■ ■ ■ Repeat the preceding problem if the source and receiver are moving apart.

General Problems

82. ■ ■ Show that *any* doubling of sound intensity corresponds to approximately a 3-dB increase in intensity level.

83. ■ ■ Suppose you double your original distance from a sound source. (a) By what factor does the sound intensity decrease? (b) By how many decibels does the intensity level decrease?

84. ■ ■ You're directly between two loudspeakers emitting 720-Hz tones. How fast would you have to walk or run on a direct line between the speakers to hear a difference of 2.0 Hz between the two tones?

85. ■ ■ ■ Having grown tired of hearing loud music from passing cars at 3 AM and the boisterous yelling of bar patrons at closing, a professor buys a new home in a quiet neighborhood. She is, however, concerned about noise from freeway traffic. Using a sound intensity level meter, she stands at an overpass 10 m above the passing cars and measures a maximum intensity level of 80 dB. If her new home is 2.6 km from the freeway, what's the intensity level there? Assume no decrease due to absorption by air, trees, or other objects. Compare your result with a quiet whisper, 20 dB.

86. ■ ■ A 1.12-m-long organ pipe has one end open. Among its possible standing-wave frequencies is 225 Hz. The next higher frequency is 375 Hz. Do *not* assume that the speed of sound is 343 m/s. Find (a) the fundamental frequency and (b) the sound speed.

87. **BIO** ■ ■ ■ **Bat echolocation.** A bat emits a 52.0-kHz ultrasound burst as it flies toward a cave wall at 7.50 m/s. At what frequency does the bat receive the reflected pulse? *Hint:* Consider the Doppler-shifted frequency of the emitted waves striking the wall and then a second Doppler shift of the reflected pulse received by the bat. Assume air at 20°C.

88. ■ ■ A piano's A string ($f = 440$ Hz) is 38.9 cm long and clamped at both ends. If the string is under 667 N tension, what's its mass?

89. **BIO** ■ ■ **Vocal tract.** A crude model of the human vocal tract treats it as a pipe closed at one end. Find the effective length of the vocal tract in a person whose fundamental tone is 620 Hz. Assume air at body temperature. *Note:* Variations in the shape of the mouth and position of the tongue significantly alter this simple model.

90. ■ ■ An organ pipe is designed to play at 22 Hz, near the frequency threshold for human hearing. Find the pipe's length if it's open (a) at both ends and (b) at one end.

91. ■ ■ A loudspeaker at a rock concert produces sound intensity of 90 dB in the front row of the audience, 8.0 m from the speaker. What's the speaker's total power output?

92. ■ ■ ■ A musical scale has A fixed at 440 Hz. (a) Find the frequency of middle C, nine notes below A. (b) How long should you make an organ pipe, closed at one end, so that it will produce a C note two octaves above middle C?

93. **BIO** ■ ■ ■ **Fetal heartbeat.** Obstetricians use ultrasound to monitor fetal heartbeat. If 5.0-MHz ultrasound reflects off the moving heart wall with a 100-Hz frequency shift, what's the speed of the heart wall?

94. ■ ■ ■ Find the frequency shift of a 60-GHz police radar signal when it reflects off a speeding car traveling at 130 km/h. Radar travels at the speed of light, and the speeding car is traveling directly toward the stationary police car.

95. ■ ■ ■ Consider the superposition of two sinusoidal waves with the same amplitude A but different frequencies f_1 and f_2. At a given point, an observer sees sinusoidal oscillations in time, with the displacement in the combined wave given by $y(t) = A\cos 2\pi f_1 t + A\cos 2\pi f_2 t$. Use the trig identity $\cos\alpha + \cos\beta = 2\cos\left[\frac{1}{2}(\alpha - \beta)\right]\cos\left[\frac{1}{2}(\alpha + \beta)\right]$ to show that $y(t)$ can be written as a product of two cosines, one oscillating at half the difference between the two frequencies, the other at half the sum. The first of these factors describes beats.

Answers to Chapter Questions

Answer to Chapter-Opening Question

It's a device that extracts energy from ocean waves and converts that energy to electricity. This unit produces energy at a rate of 750 kW. Multiple-unit "wave farms" operating off the coast of Europe now generate several megawatts each.

Answers to GOT IT? Questions

Section 11.1 (b) $f_A/3$
Section 11.2 (a) Increase T, (d) decrease L, (f) decrease μ
Section 11.3 (b) > (d) = (e) > (c) > (a)
Section 11.4 (c) 3

12 Temperature, Thermal Expansion, and Ideal Gases

■ Rising sea level is one danger of global warming, shown here as it might impact a coastal area. What's the dominant cause of sea-level rise?

This is the first of three chapters on thermal physics. We'll begin with the fundamental concepts of temperature and thermal energy—the energy of random molecular motions. You'll learn the common temperature scales and how they relate to physical properties, such as the melting and boiling points of water. Next, we'll explore thermal expansion of solids and liquids, noting the usual behavior of water near its melting point. Then you'll learn the behavior of ideal gases, including the relation between gas pressure, volume, and temperature as described by the ideal-gas law. Finally, we'll relate the motion of individual molecules in a gas to the behavior of the gas as a whole.

12.1 Temperature and Thermometers

Everyday experience makes you familiar with **temperature**. You know hot from cold; you've experienced fevers; you freeze food in your freezer, boil it on the stove, and bake it in the oven—all at different temperatures. Globally, you're aware of concerns about Earth's rising temperature. Although you understand temperature intuitively, if asked to define it you would probably struggle. Physically, temperature is a subtle concept.

To Learn

By the end of this chapter you should be able to

■ Understand the meaning of temperature and its relation to thermal energy.

■ Describe temperature measurement and the Celsius, Kelvin, and Fahrenheit scales.

■ Describe how solids and liquids expand with increasing temperature.

■ Explain water's unusual thermal expansion.

■ Explain the ideal-gas approximation.

■ Describe the ideal-gas law.

■ Discuss how kinetic theory explains the ideal-gas law.

■ Describe the distribution of molecular speeds in an ideal gas.

■ Describe the process of diffusion.

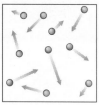

Thermal energy of a simple gas is due to the random motion of molecules.

(a)

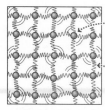

Thermal energy of a solid is due to random motion of atoms . . .

. . . and to the potential energy of molecular bonds.

(b)

FIGURE 12.1 Thermal energy in gases and solids.

What Is Temperature?

To understand temperature, we first introduce **thermal energy**. This is the kinetic and potential energy associated with individual molecules or atoms, as Figure 12.1 shows for simple gases and solids. Kinetic energy results from random atomic and molecular motions, potential energy from the stretching of molecular bonds. Although all or part of the thermal energy is due to molecular motion, thermal energy is distinct from an object's kinetic energy. A moving ball has kinetic energy $K = \frac{1}{2}mv^2$, as you learned in Chapter 5, where v is the ball's speed. The ball also has thermal energy, from the random motion of its molecules—motion that's in addition to the overall motion of the ball in flight. Thermal energy is present even when the ball is at rest, with zero kinetic energy.

With the concept of thermal energy in hand, here are several ways to think of temperature:

- In general, temperature is closely related to the average thermal energy per atom or molecule. **The higher the average thermal energy**, **the higher the temperature**. In Section 12.4 you'll see a direct relationship between temperature and thermal energy in ideal gases. For other materials, the connection is harder to quantify but still exists.
- Materials that can exchange thermal energy are said to be in **thermal contact**. When materials with different temperatures are put into thermal contact, energy flows from the warmer material to the cooler one.
- After two materials have been in thermal contact for a long time, they reach the same temperature and are then in **thermal equilibrium**.

✓**TIP**

Temperature is related to average thermal energy—the higher the average thermal energy, the higher the temperature.

It's important to distinguish *temperature* from *heat*. You'll see in Chapter 13 that heat is thermal energy that flows between bodies at different temperatures. Temperature and heat are physically different quantities, with different units. You'll explore their relationship in Chapter 13.

Temperature Scales

The **Fahrenheit temperature scale** (°F) is in everyday use in the United States. The German physicist Daniel Fahrenheit devised this scale in 1724, with 0°F set as the freezing point of a saturated solution of salt water and 32°F as the melting point of freshwater. Later the scale was adjusted, with 32°F kept as one reference point but water's boiling point, 212°F, used as the other. On this scale, normal body temperature is 98.6°F. The actual resting body temperature of a healthy human typically varies between 98.0°F and 98.6°F.

Scientists use the **Celsius temperature scale** (°C), with water's freezing and boiling points (under normal atmospheric pressure) at 0°C and 100°C, respectively. The conversion from Fahrenheit to Celsius is

$$T \text{ in } °C = \tfrac{5}{9}(T \text{ in } °F - 32°)$$

For example, on a warm 77°F day the Celsius temperature is

$$T \text{ in } °C = \tfrac{5}{9}(77° - 32°) = \tfrac{5}{9}(45°) = 25°C$$

Another important scale is the **Kelvin scale** (K). (That's the SI unit kelvin, *not* degrees kelvin or °K.) The kelvin and the Celsius degree are the same size, but the zero points differ by 273.15 K. Thus, the conversion between degrees Celsius and kelvins is

$$(T \text{ in } K) = (T \text{ in } °C) + 273.15$$

The kelvin scale honors Scottish physicist William Thomson (1824–1907). Thomson proposed an absolute temperature scale in 1848 and was rewarded for his scientific work with the title Baron Kelvin.

APPLICATION **Ear Thermometer**

The ear thermometer detects infrared radiation (IR) from the eardrum, which emits IR at a rate dependent on temperature (as you'll see in the next chapter). The eardrum is connected thermally to the hypothalamus, the brain region that regulates body temperature. The eardrum method is very quick and is less invasive than oral or rectal measurements.

The kelvin scale is defined so that the lowest possible temperature is 0 K, known as **absolute zero**. At absolute zero, a material can have no more thermal energy removed. Quantum mechanics shows that some thermal energy remains at 0 K and can't be removed, but for most situations this energy is negligible. By the early 21st century, physicists had achieved temperatures as low as 10^{-9} K (or 1 nK).

Figure 12.2 compares the three temperature scales. Although the kelvin is the SI temperature unit, scientists often work in °C so they can use smaller numbers for everyday temperatures—for example, 20°C rather than 293 K for room temperature. As with other units used in science, it's good for you to develop some "feel" for temperatures in °C and K.

Kelvins and Celsius degrees are larger than Fahrenheit degrees by 9/5:

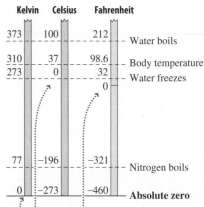

The three scales also have different zero points.

FIGURE 12.2 Relationships among three temperature scales.

EXAMPLE 12.1   **Body Temperature**

You've known since childhood that a healthy body temperature is about 98.6°F. What's this in °C and kelvins?

ORGANIZE AND PLAN The conversion from Fahrenheit to Celsius is:

$$T \text{ in } °C = \tfrac{5}{9}(T \text{ in } °F - 32°)$$

The conversion from Celsius to kelvin is T in K = T in °C + 273.15.

Known: $T = 98.6°F$.

SOLVE The first conversion gives

$$T \text{ in } °C = \tfrac{5}{9}(98.6° - 32°) = \tfrac{5}{9}(66.6°) = 37.0\,°C$$

Then T in K = $(37.0 + 273.15)$ K = 310 K.

REFLECT If you fall ill in another country, don't be surprised to find that your body temperature is in the high 30s!

MAKING THE CONNECTION Find the Celsius equivalents of the lowest and highest temperatures recorded on Earth in modern times, $-128.6°F$ (Antarctica) and 134.0°F (Death Valley, CA).

ANSWER Using the conversion formula, these extreme temperatures are found to be $-89.2°C$ and 56.7°C.

Thermometers

A **thermometer** is a device that exploits some physical property that changes with temperature. Examples include thermal expansion of materials, gas pressure, or electrical properties. Most modern thermometers use electrical resistance and have digital readouts.

Any thermometer requires *calibration* for accuracy. The freezing and boiling points of water at 0°C and 100°C provide good standards. If a thermometer uses a property that varies linearly with temperature, then calibration at two points is enough to ensure accuracy over a reasonable range.

GOT IT? Section 12.1   Rank in order the following temperatures, from highest to lowest: (a) 270 K; (b) $-2.0°C$; (c) 25°F; (d) water's freezing point.

12.2 Thermal Expansion

Think "thermometer" and you probably picture the device illustrated in Figure 12.3, which has a glass bulb connected to a narrow tube marked to indicate temperature. The **liquid-bulb thermometer** was invented by Fahrenheit in 1709. He first used alcohol but soon replaced it with mercury, which became the standard until recently. Today, alcohol is again more common because of mercury's toxicity. The concept behind the liquid-bulb thermometer is simple: As temperature increases, **thermal expansion** increases the volume

Bulb is placed in thermal contact with object whose temperature you want to measure. Liquid expands as T increases.

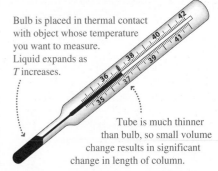

Tube is much thinner than bulb, so small volume change results in significant change in length of column.

FIGURE 12.3 A liquid-bulb thermometer.

of liquid in the bulb, pushing it farther into the narrow tube. Thermal expansion results in a nearly linear increase in volume with temperature, and that translates into evenly spaced temperature markings. We'll now take a more general look at thermal expansion.

Linear Thermal Expansion

Stroll down the sidewalk, and you'll notice cracks etched across the concrete every meter or so. Why? The concrete undergoes thermal expansion with increasing temperature, and that expansion needs a place to go. Check out the ends of a bridge, and you'll see finger-like steel joints that allow the bridge to expand and contract.

Most materials expand when heated. Over a wide temperature range, expansion is approximately proportional to temperature change, which is expressed as the following:

$$\frac{\Delta L}{L} = \alpha \Delta T \quad \text{(Linear thermal expansion)} \tag{12.1}$$

Here L is the object's original length and ΔL the change resulting from temperature change T. The parameter α is the **coefficient of linear expansion**, which is characteristic of a given material. With temperature in °C, α has units °C^{-1}. Materials expand at different rates, so values differ; see Table 12.1. Numerical values of α are small, on the order of 10^{-5}°C^{-1} for most solids. However, that's large enough to require careful engineering, as in the sidewalk and bridge discussed above.

TABLE 12.1 Thermal Expansion Coefficients*

Material	Coefficient of linear expansion (°C^{-1})	Coefficient of volume expansion (°C^{-1})
Solids		
Aluminum	2.4×10^{-5}	7.2×10^{-5}
Brass	2.0×10^{-5}	6.0×10^{-5}
Copper	1.7×10^{-5}	5.1×10^{-5}
Concrete	1.2×10^{-5}	3.6×10^{-5}
Glass (common)	4.0×10^{-6} to 9.0×10^{-6}	1.2×10^{-5} to 2.7×10^{-5}
Glass (Pyrex)	3.3×10^{-6}	9.9×10^{-6}
Lead	2.9×10^{-5}	8.7×10^{-5}
Quartz	4.0×10^{-7}	1.2×10^{-6}
Silver	1.9×10^{-5}	5.7×10^{-5}
Steel (typical)	1.2×10^{-5}	3.6×10^{-5}
Liquids		
Ethanol		7.5×10^{-4}
Glycerin		4.9×10^{-4}
Mercury		1.8×10^{-4}
Methanol		1.2×10^{-3}
Water (1°C)		-4.8×10^{-5}
Water (20°C)		2.1×10^{-4}
Water (50°C)		5.0×10^{-4}

*at 20°C unless otherwise noted.

EXAMPLE 12.2 **Working on the Railroad**

A standard 39-foot (11.9-m) length of steel rail expands and contracts as temperature varies. How much expansion occurs as the track temperature changes from −20°C in winter to 45°C in hot summer sunlight?

ORGANIZE AND PLAN Figure 12.4 is our sketch of the situation. Equation 12.1 gives the linear thermal expansion: $\Delta L/L = \alpha \Delta T$. Here the temperature change is $\Delta T = 65°C$, and the thermal expansion coefficient for steel is given in Table 12.1 as $1.2 \times 10^{-5}°C^{-1}$.

Known: $L = 11.9$ m, $\Delta T = 65°C$, $\alpha = 1.2 \times 10^{-5}°C^{-1}$ (for steel, Table 12.1).

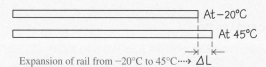

Expansion of rail from −20°C to 45°C ⋯⋯▸ ΔL

FIGURE 12.4 Thermal expansion of a railroad rail.

SOLVE Solving Equation 12.1 for ΔL gives

$$\Delta L = \alpha L \Delta T = (1.2 \times 10^{-5}°C^{-1})(11.9 \text{ m})(65°C)$$
$$= 0.0093 \text{ m} = 9.3 \text{ mm}$$

REFLECT Older railroads accommodate this expansion with gaps between rail sections. These gaps result in the characteristic "clickety-clack" you hear in the train. With modern welded rails, clamps hold the rail rigidly in place, allowing stress to build without significant expansion.

MAKING THE CONNECTION What gap is needed between track sections when the temperature is −20°C, to accommodate a maximum temperature of 45°C?

ANSWER The rail expands equally in each direction, by half the amount calculated in the example. The adjacent rail expands the same amount. With a 9.3-mm gap size equal to ΔL computed in the example, the rails will just meet when the temperature reaches 45°C.

Note in Table 12.1 that the coefficient for Pyrex is much smaller than that for common glass. You shouldn't pour boiling water into an ordinary glass container, because the sudden uneven thermal expansion will crack it. But Pyrex doesn't expand nearly as much, so it's safer to use in the kitchen or chemistry lab.

Volume Thermal Expansion

With increasing temperature, solids expand in every direction, and liquids expand as permitted by their containers. The volume change is proportional to the temperature change and is given by the equation

$$\frac{\Delta V}{V} = \beta \Delta T \quad \text{(Volume thermal expansion)} \tag{12.2}$$

where V is the volume before expansion, ΔT the temperature change, and β the **coefficient of volume expansion**. Equation 12.2 for volume expansion is similar to Equation 12.1 for linear expansion. With temperature measured in °C, β has units of $°C^{-1}$. Table 12.1 includes values of β. Conceptual Example 12.4 shows that $\beta \approx 3\alpha$ for small expansions.

EXAMPLE 12.3 **Mercury Thermometer**

A mercury thermometer contains exactly 1 cm³ of mercury at 20°C. How much does the mercury's volume change when the temperature drops to −30°C?

ORGANIZE AND PLAN Mercury is a liquid, with its thermal expansion coefficient given in Table 12.1. Equation 12.2 gives the volume change: $\Delta V/V = \beta \Delta T$. Here the temperature drops, so ΔT is negative: $\Delta T = -50°C$. Accordingly, the volume change ΔV is also negative (Figure 12.5).

Known: $\Delta T = -50°C$, $\beta = 1.8 \times 10^{-4}°C^{-1}$ (Table 12.1), $V = 1.00$ cm³ $= 1.00 \times 10^{-6}$ m³.

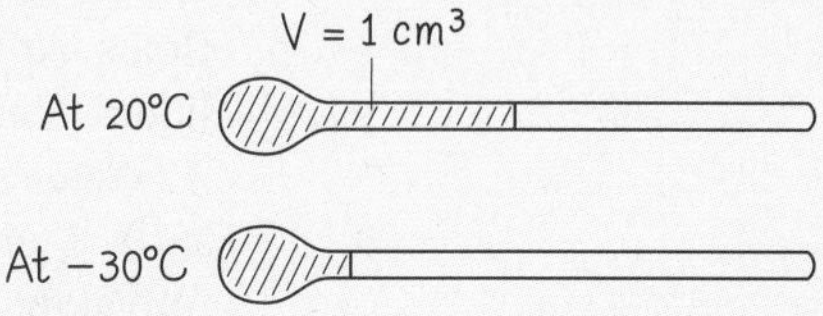

$V = 1$ cm³

At 20°C

At −30°C

FIGURE 12.5 Contraction of liquid mercury in a thermometer.

cont'd.

SOLVE From Equation 12.2, the volume change is

$$\Delta V = \beta V \Delta T = (1.8 \times 10^{-4} \text{°C}^{-1})(1.00 \times 10^{-6} \text{ m}^3)(-50\text{°C})$$
$$= -9.00 \times 10^{-9} \text{ m}^3$$

That is, the volume decreases by $9.00 \times 10^{-9} \text{ m}^3$.

REFLECT This result seems tiny because it's in m^3. Instead, think of it as $9.00 \times 10^{-3} \text{ cm}^3$, or a decrease of just under 1% of the original volume. Now it doesn't seem so insignificant.

MAKING THE CONNECTION Suppose the thermometer in this example has degree markings every 2 mm along the tube. If the tube's interior has a circular cross section, what's the inside diameter?

ANSWER For cross-sectional area A, the change in volume is $\Delta V = A\Delta L$, where ΔL is the distance the mercury expands along the tube. With each 1-degree change corresponding to 2 mm, our 50-degree change corresponds to $\Delta L = 100 \text{ mm} = 0.10 \text{ m}$. The cross-sectional area is $A = \Delta V/\Delta L = 9.00 \times 10^{-8} \text{ m}^2$, which requires an inside diameter of 0.34 mm. You might wonder about the glass; doesn't it expand, too? It does, but Table 12.1 shows that glass's expansion coefficient is much less than mercury's, so this effect is negligible.

Expansion Joints in Buildings and Bridges

Bridges and skyscrapers are particularly vulnerable to thermal expansion. Because of their extreme size, a modest fractional change in length translates into a large absolute change. The bridge or building must respond to temperature changes by expanding and contracting with flexible joints like those shown, in order to avoid cracking that could damage the structure.

CONCEPTUAL EXAMPLE 12.4 **Volume and Length**

How are the linear and volume expansion coefficients α and β related for a solid?

SOLVE We can consider a solid cube of side L, and therefore volume L^3, made from material with linear expansion coefficient α. Under a temperature change ΔT, Equation 12.1 shows that each side expands by $\Delta L = L\alpha \Delta T$. The new volume is then $(L + \Delta L)^3 = (L + L\alpha \Delta T)^3$. Factoring out L and expanding gives $L^3(1 + \alpha \Delta T)^3 = L^3(1 + 3\alpha \Delta T + 3(\alpha \Delta T)^2 + (\alpha \Delta T)^3)$. Now, the length change ΔL is small compared with the original length, and that means $\alpha \Delta T$ is much less than 1. Therefore, the terms $(\alpha \Delta T)^2$ and $(\alpha \Delta T)^3$ are even smaller (say $\alpha \Delta T = 0.01$; then its square is 10^{-4} and its cube is 10^{-6}). So we'll neglect these tiny terms and approximate the new volume as $L^3(1 + 3\alpha \Delta T)$. Since the original volume was $V = L^3$, the volume increase is $\Delta V = 3L^3\alpha \Delta T = 3V\alpha \Delta T$. So we can write

$$\frac{\Delta V}{V} = 3\alpha \Delta T$$

This is identical to Equation 12.2, with $\beta = 3\alpha$. Thus the volume expansion coefficient for a solid is just three times the linear expansion coefficient.

REFLECT Why doesn't this work for liquids? Because a liquid takes the shape of its container and isn't free to expand in all directions. In a liquid-bulb thermometer, for example, the volume expansion is constrained to take place in one dimension—along the thin tube. That's why Table 12.1 doesn't list α for liquids.

Thermal Expansion of Water

Water is unusual: It doesn't always expand with increasing temperature. You can see this from the negative expansion coefficient for water at 1°C in Table 12.1, and in more detail in Figure 12.6. From 4°C to 100°C, water behaves like other substances, increasing in volume with rising temperature (Figure 12.6b). However, from 0°C to 4°C, this behavior is reversed. At 4°C, a sample of water is at its minimum volume and therefore its maximum density (Figure 12.6c).

This unorthodox behavior is related to the structure of ice, in which hydrogen bonds create an open configuration whose density is lower than liquid water's; that's why ice floats. At temperatures just above freezing, a residue of these bonds results in the lower liquid density, increasing with temperature up to 4°C as the bonds break.

Water's unusual thermal behavior has important implications for freshwater life. As surface temperature drops toward 0°C in the fall, the coldest water is less dense and therefore at the surface. Ice forms, and since it's less dense than liquid water, it floats. The ice insulates the water below, slowing the growth of the ice layer and allowing aquatic life to survive. In the summer the hottest water is on top, and the densest water at the bottom is near 4°C. Twice a year, in spring and fall, temperatures are nearly equal throughout and the lake "overturns," mixing surface and bottom water. This brings up nutrients from the depths, helping sustain aquatic life.

Temperature (°C)	Volume of 1 g of water (cm³)	Density (g/cm³)
0	1.0002	0.9998
4	1.0000	1.0000
10	1.0003	0.9997
20	1.0018	0.9982
50	1.0121	0.9881
75	1.0258	0.9749
100	1.0434	0.9584

(a) Volume of 1 g of water as a function of temperature

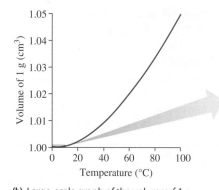

(b) Large-scale graph of the volume of 1 g of water as a function of temperature

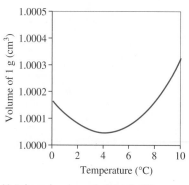

(c) Enlarged and rescaled detail of the lower-left end of the curve in (a)

FIGURE 12.6 Water contracts slightly as it warms from the freezing point (0°C) to 4°C, then expands with increasing temperature.

Reviewing New Concepts

- Most materials expand with increasing temperature.
- Both linear and volume expansion are proportional to temperature change.
- Water behaves unusually, with maximum density at 4°C.

GOT IT? Section 12.2 Increasing the volume of mercury in a thermometer by 1% requires a temperature increase of about (a) 24°C; (b) 36°C; (c) 44°C; (d) 56°C.

12.3 Ideal Gases

A gas differs from a liquid or solid in that it always expands to fill its container. We'll describe gases using **state variables** including volume V, pressure P, and temperature T; these characterize the gas's macroscopic state rather than its individual molecules, which would be difficult to observe. Typically, pressure and temperature are consistent throughout a gas, for containers that aren't too large. That's because individual molecules transfer thermal energy throughout the container, creating an equilibrium condition with uniform pressure and temperature.

Here we'll consider the **ideal gas**—a gas with a particle density low enough that forces between molecules are negligible (Figure 12.7). The significant interactions are between molecules and the container walls. In fact, pressure on the walls results from those collisions. In equilibrium, the collisions are elastic, so the gas doesn't gain or lose energy. However, if containers with gases at different temperatures are placed in thermal contact, then thermal energy is exchanged through the walls, and the temperatures change. We'll describe how that works in Section 12.4. For now, we'll just consider ideal gases in equilibrium.

Most common gases exhibit nearly ideal behavior at everyday temperatures and pressures. Gases become less ideal when cooled or compressed, which increases their density. A gas just above its boiling point may not behave ideally. The most common gases in air—nitrogen, oxygen, and argon—all have boiling points below 100 K, so they're very nearly ideal at room temperature and atmospheric pressure.

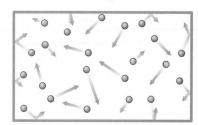

FIGURE 12.7 Gas molecules confined to a rectangular box. The molecules don't interact with one another, but they do collide with the container walls.

The Amount of Gas

In addition to pressure, volume, and temperature, another state variable is the amount of gas. There are several ways to describe this. You might consider the total mass, but it's more convenient to describe gases using the number of molecules N, or the number of moles n. Recall that one **mole** (mol) is Avogadro's number N_A of anything, where $N_A = 6.022 \times 10^{23}$. Therefore, N and n are related by $N = N_A n$. For example, the number of molecules in a 2.85-mol sample is $N = N_A n = (6.022 \times 10^{23})(2.85) = 1.72 \times 10^{24}$ molecules.

The mass of an ideal gas depends on the number of molecules or moles and the **molar mass**. For example, one mole of oxygen (O_2) has $m_{molar} = 32.0$ g. You get this from the periodic table, which lists atomic oxygen (O) at 16.0 g per mole, so O_2 has twice that, or 32.0 g/mol. Two moles of oxygen gas have a mass of 64.0 g, and so on. In general, $m = nm_{molar}$.

✓**TIP**

Use grams rather than kilograms for molar masses, so you can use the atomic masses in the periodic table.

Reviewing New Concepts

- The amount of gas can be described by the number of moles n, number of molecules N, or mass m.
- Moles and molecules are related by $N = N_A n$.
- Mass m is related to the molar mass m_{molar} and the number of moles by $m = nm_{molar}$.

EXAMPLE 12.5   **Carbon Dioxide**

How many molecules are in a 77.0-g sample of carbon dioxide (CO_2)?

ORGANIZE AND PLAN The molar mass comes from the periodic table and the formula CO_2: It's the sum of one mole of carbon atoms and two moles of oxygen atoms. Then the number of moles and number of molecules follow from the relations $m = nm_{molar}$ and $N = N_A n$.

Known: mass $m = 77.0$ g.

SOLVE From the periodic table (or Appendix D), the molar masses for carbon and oxygen are 12.0 g and 16.0 g, respectively. Therefore, CO_2's molar mass is

$$m_{molar} = 12.0 \text{ g} + 2(16.0 \text{ g}) = 44.0 \text{ g}$$

With mass $m = 77.0$ g, our sample contains

$$n = \frac{m}{m_{molar}} = \frac{77.0 \text{ g}}{44.0 \text{ g/mol}} = 1.75 \text{ mol}$$

Then the number of molecules is $N = N_A n = (6.022 \times 10^{23} \text{ molecules/mol})(1.75 \text{ mol}) = 1.05 \times 10^{24}$ molecules.

REFLECT This large number is one reason scientists use moles to measure the amount of a substance; the smaller numbers of moles are much more manageable.

MAKING THE CONNECTION   Air is 78% nitrogen (N_2), 21% oxygen (O_2), and 1% argon (Ar). What's the average molar mass of air?

ANSWER The three molar masses are, respectively, 28.0 g, 32.0 g, and 39.9 g. Taking their weighted average gives 29.0 g.

The Ideal-Gas Law

Pressure, volume, temperature, and amount of gas are all related. If you change one state variable, one or more of the others will likely change as well. Relationships between state variables are **equations of state**.

Historically, equations of state were developed experimentally by systematically changing one variable while allowing only one of the others to change. Figure 12.8 shows a piston-cylinder system used in such experiments. The movable piston traps gas within the cylinder, so the amount of gas remains constant. Then, for example, you might hold the piston in place to keep the volume constant. If you raise the temperature, you'll discover how pressure varies with temperature.

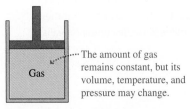

The amount of gas remains constant, but its volume, temperature, and pressure may change.

FIGURE 12.8 A piston-cylinder system, used to perform experiments that reveal the properties of gases.

We now summarize the results of experiments including the one just described.

Boyle's law: If the amount of gas and temperature are fixed, pressure and volume vary inversely:

$$PV = \text{constant} \quad \text{(Boyle's law)}$$

Charles's law: If the amount of gas and pressure are fixed, volume is proportional to temperature:

$$V \propto T \quad \text{(Charles's law)}$$

Gay-Lussac's law: If the amount of gas and volume are fixed, pressure is proportional to temperature:

$$P \propto T \quad \text{(Gay-Lussac's law)}$$

Finally, at constant volume and temperature, pressure is proportional to the amount of gas: $P \propto n$.

Combining these results gives the **ideal-gas law**, a single equation relating all four variables:

$$PV = nRT \quad \text{(Ideal-gas law)} \tag{12.3}$$

The quantity R in Equation 12.3 is the **molar gas constant**. Its value, determined experimentally, is $R = 8.315\,\text{J}\cdot\text{mol}^{-1}\cdot\text{K}^{-1}$.

With R expressed in $\text{J}\cdot\text{mol}^{-1}\cdot\text{K}^{-1}$, the right side of the equation (nRT) works out to joules (J). This implies SI units for pressure (Pa) and volume (m^3). It's also important to express temperatures in kelvin (K) when using the ideal-gas law. That's because the direct proportionalities in Charles's law and Gay-Lussac's law hold only if temperature is measured from absolute zero. Further, the ideal-gas law wouldn't make sense for negative temperatures, because none of the other quantities is ever negative.

EXAMPLE 12.6   Molar Volume

Find the volume of one mole of ideal gas under standard conditions ($T = 0°C$ and $P = 1$ atm).

ORGANIZE AND PLAN The ideal-gas law, $PV = nRT$, gives volume V as a function of the other parameters.

Known: $n = 1$ mol, $T = 20°C = 293$ K, and $P = 1$ atm $= 1.013 \times 10^5$ Pa.

SOLVE Solving the ideal-gas law for the volume V and inserting the SI values,

$$V = \frac{nRT}{P} = \frac{(1\,\text{mol})(8.315\,\text{J}\cdot\text{mol}^{-1}\cdot\text{K}^{-1})(273\,\text{K})}{1.013 \times 10^5\,\text{Pa}} = 0.0224\,\text{m}^3$$

Here we were careful to convert temperature to kelvins and pressure to pascals; the volume is then in SI: m^3.

REFLECT Note that we never mentioned the type of gas. The volume of one mole of ideal gas under standard conditions is the same, regardless of the type of gas.

MAKING THE CONNECTION   Express the molar volume of ideal gas in liters, and find the diameter of a spherical balloon containing this much gas.

ANSWER With 1 $\text{m}^3 = 1000$ L, the volume is 22.4 L—a number you may recognize from chemistry. In terms of spherical radius r, $V = 4\pi r^3/3$, giving $r = 0.175$ m. The balloon's diameter is then 35.0 cm.

Constant-Volume Gas Thermometer

Gay-Lussac's law is explored using a *constant-volume gas thermometer*, in which gas pressure is measured as a function of temperature. Figure 12.9a shows some representative data from this instrument. Do the data follow the ideal-gas law? How can this instrument be used to determine absolute zero? Why is this instrument called a *thermometer*?

SOLVE According to the ideal-gas law, pressure is proportional to temperature for a sample with constant amount and volume of gas. That is, $P = (nR/V)T$. With the quantity in parentheses constant, a graph of P versus T should be a straight line with slope nR/V and intercept at $T = 0$ K. Therefore, extrapolating the graph back to zero pressure (Figure 12.9b) determines $T = 0$ K, absolute zero. The graph shows this point at $T = -273°C$, consistent with our earlier definition of absolute zero. The straight-line nature of the graph shows that the data do indeed follow the ideal-gas law.

This instrument is a thermometer because measuring the pressure gives a value of the temperature, using the equation above.

REFLECT This may not seem like a very practical thermometer, compared with a liquid-bulb or electronic thermometer. However, the constant-volume gas thermometer works over an extremely broad temperature range, and because all ideal gases behave the same, it provides a readily reproducible temperature standard.

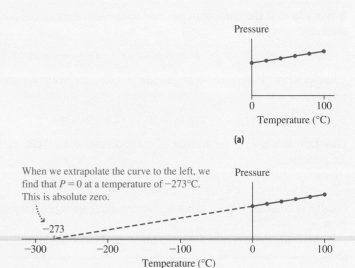

FIGURE 12.9 (a) Pressure versus temperature for the gas at constant volume. (b) Extrapolation to absolute zero.

EXAMPLE 12.8   **Density of Helium**

Use the ideal-gas law to compute the density of helium gas at $T = 25°C$ and $P = 1$ atm.

ORGANIZE AND PLAN Density = mass/volume, or $\rho = m/V$. The ideal-gas law doesn't directly involve density, but it does let you find the number of moles per unit volume. Moles can then be converted to mass using $m = nm_{molar}$. For helium (a monatomic gas), Appendix D gives $m_{molar} = 4.00$ g.

Known: $T = 25°C = 298$ K and $P = 1$ atm $= 1.013 \times 10^5$ Pa.

SOLVE Using the ideal-gas law $PV = nRT$, the number of moles per unit volume is

$$\frac{n}{V} = \frac{P}{RT} = \frac{1.013 \times 10^5 \text{ Pa}}{(8.315 \text{ J} \cdot \text{mol}^{-1} \cdot \text{K}^{-1})(298 \text{ K})} = 40.9 \text{ mol/m}^3$$

With $m = nm_{molar}$, the density becomes

$$\rho = \frac{m}{V} = \frac{nm_{molar}}{V} = \frac{n}{V}m_{molar}$$

Before inserting numerical values, note that the quantity n/V is expressed in terms of mol/m^3. Therefore, we'd better express the molar mass in kilograms, to avoid mixing SI and cgs units. Then $m_{molar} = 4.00 \times 10^{-3}$ kg, and the density is

$$\rho = \frac{n}{V}m_{molar} = (40.9 \text{ mol/m}^3)(4.00 \times 10^{-3} \text{ kg/mol})$$

$$= 0.164 \text{ kg/m}^3$$

That's very low, as expected, because helium is a light gas.

REFLECT The density of air at this temperature is much larger, about 1.2 kg/m^3. That's why helium-filled balloons are buoyant.

MAKING THE CONNECTION Table 10.1 gives the density of helium, to two significant figures, as 0.18 kg/m^3. How do you explain the difference between that and the result of this example?

ANSWER Look carefully at Table 10.1, and you'll see that tabulated densities are at 0°C = 273 K. Reworking this example at that lower temperature gives $\rho = 0.179$ kg/m^3, in agreement with the tabulated value to two significant figures.

The Ideal-Gas Law: Molecular Version

Sometimes it's more convenient to express the ideal-gas law in terms of the number of gas molecules N instead of moles n. Because $n = N/N_A$, where N_A is Avogadro's number, the ideal-gas law becomes $PV = nRT = NRT/N_A$.

The quantity R/N_A defines **Boltzmann's constant**, $k_B = R/N_A$, so the ideal-gas law becomes

$$PV = Nk_BT \quad \text{(Ideal-gas law, molecular version)} \qquad (12.4)$$

Numerically, Boltzmann's constant is $k_B = 1.38 \times 10^{-23}$ J/K. You can think of it as the **molecular gas constant**, which plays the same role in this version of the ideal-gas law as the molar gas constant R plays in the version $PV = nRT$.

✓**TIP**

Think of Equation 12.4 as the molecular version of the ideal-gas law, and Equation 12.3 as the molar version.

Physicists seldom use moles, so they usually express the ideal gas as Equation 12.4. Example 12.9 shows one use for this form: calculating the **number density**, or number of molecules per cubic meter.

A hot-air balloon is a good example of the ideal-gas law in action. What makes a hot-air balloon rise? Hot air is less dense than cool air, so the balloon is buoyant. But the ideal-gas law requires only that the product PV increase with temperature. So why is it volume V that increases, thus decreasing density? It's because the balloon is very nearly in pressure equilibrium with the surrounding air, so in this case P hardly changes with increasing T—making V proportional to T. With lower-density air inside, there's an upward buoyancy force that lifts the balloon and its basket.

EXAMPLE 12.9   **Number Density of Air**

Calculate the number density for air at room temperature (20°C) and 1 atm pressure. Use your result to find the average volume occupied by each molecule.

ORGANIZE AND PLAN The air temperature is high enough that the ideal-gas law applies. The number density—molecules per unit volume—is the quantity N/V. Both N and V appear in the ideal-gas law of Equation 12.4, $PV = Nk_BT$, and we can solve for N/V. The volume associated with each molecule is V/N, the reciprocal of the number density.

Known: $T = 20°C = 293$ K and $P = 1$ atm $= 1.013 \times 10^5$ Pa.

SOLVE Solving Equation 12.4 for the number density N/V gives

$$\frac{N}{V} = \frac{P}{k_BT} = \frac{1.013 \times 10^5 \text{ Pa}}{(1.38 \times 10^{-23} \text{ J/K})(298 \text{ K})}$$

$$= 2.46 \times 10^{25} \text{ molecules/m}^3$$

The average volume occupied by each molecule is then

$$\frac{V}{N} = \frac{1}{N/V} = \frac{1}{2.46 \times 10^{25} \text{ molecules/m}^3}$$

$$= 4.06 \times 10^{-26} \text{ m}^3/\text{molecule}$$

REFLECT The number of molecules in a cubic meter is extremely large, as you might expect. It's interesting to compare air at sea level with intergalactic space, where the number density is only one hydrogen atom per cubic meter! Because "molecule" and "atom" are dimensionless, we often express number densities simply in m^{-3}.

MAKING THE CONNECTION Use the results of this example to estimate the average distance between air molecules, then compare with the size of a molecule, on the order of 10^{-10} m.

ANSWER Imagine each molecule occupying a cube with the volume we just found. Then the cube's side length is a typical distance between molecules. That length is $(4.06 \times 10^{-26} \text{ m}^3)^{1/3}$, or about 3×10^{-9} m. That's much greater than the size of a molecule, which is why air behaves as ideal gas with negligible interaction between molecules.

GOT IT? Section 12.3 A closed container of ideal gas is compressed so its volume is halved while its temperature doubles. During this process the gas pressure (a) decreases by a factor of 4; (b) decreases by a factor of 2; (c) is unchanged; (d) increases by a factor of 2; (e) increases by a factor of 4.

12.4 Kinetic Theory of Gases

State variables and the ideal-gas law describe the macroscopic behavior of a gas, but they don't say anything about individual gas molecules. Yet it's the molecules that constitute the gas and that ultimately determine its properties and behavior. Here, as elsewhere in physics, we'd like to connect the macroscopic properties revealed in the state variables to the microscopic behavior of individual molecules. We'll use this connection to understand why ideal gases behave as they do. We'll exploit the macro/micro connection again when we study heat capacity in Chapter 13.

The **kinetic theory of gases** provides the connection we're after. Assumptions of kinetic theory include

- The ideal-gas assumptions, namely, dilute gas with negligible intermolecular interactions and frequent, energy-conserving collisions with container walls.
- Pressure on the container walls resulting from collisions between gas molecules and the walls.
- A sample consisting of many molecules, moving in random directions with a range of speeds.

Here you'll see how gas pressure is related to molecular masses and speeds, and how those speeds relate to temperature. You'll also see how the range of molecular speeds varies with temperature.

✓ TIP

Kinetic theory assumes gases are ideal. However, it's an excellent approximation for most gases under typical conditions.

Pressure, Kinetic Energy, and Temperature

A gas molecule collides elastically with the walls of a rigid container, rebounding with no change in kinetic energy. But its momentum *does* change, because it's moving in a different direction after colliding (Figure 12.10). By Newton's second law, the wall must have exerted a force to change the molecule's momentum, and by Newton's third law, the molecule must have exerted a force on the wall. It's that force, averaged over the vast number of molecules, which gives rise to gas pressure. The pressure depends on the number of molecules, their speeds, their masses, and how widely spaced they are—equivalently, the volume they occupy. We'll leave the calculation to the end-of-chapter problems and state the result:

$$P = \frac{Nm\overline{v^2}}{3V} \qquad \text{(Gas pressure; SI unit: Pa)} \qquad (12.5)$$

Here V is the container volume, N the number of gas molecules, and m the molecular mass. The quantity $\overline{v^2}$ is called the **mean-square speed**. As usual, the bar designates an average, and here it refers to the average of the squares of the molecular speeds—hence, *mean square.* The square root of $\overline{v^2}$ is called **root-mean-square speed** (or **rms speed**): $v_{\text{rms}} = \sqrt{\overline{v^2}}$. The root-mean-square speed is a typical speed for a gas molecule.

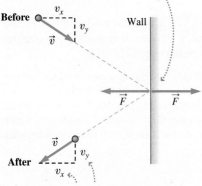

When a molecule collides elastically with a container wall, the molecule and wall exert forces on each other.

Before v_x v_y $\vec{v}$ Wall

$\vec{F}$ $\vec{F}$

$\vec{v}$ v_y **After** v_x

The force exerted on the molecule reverses the sign of the x-component of the velocity but does not change the y-component.

FIGURE 12.10 A molecule undergoes an elastic collision with one container wall.

EXAMPLE 12.10 **RMS Speed**

Find the rms speed of oxygen molecules at 20°C and 1 atm pressure.

ORGANIZE AND PLAN The square of the rms speed appears in Equation 12.5, $P = Nm\overline{v^2}/3V$, so we could solve for it given the other quantities. We aren't given N or V. However, they enter the equation as N/V—the number density, which in Example 12.9 we found to be $N/V = 2.46 \times 10^{25}\,m^{-3}$ under the given conditions.

Known: $P = 1\,atm$, $T = 20°C$, giving $N/V = 2.46 \times 10^{25}\,m^{-3}$.

SOLVE We rearrange to find $\overline{v^2}$, then take the square root to get v_{rms}:
$\overline{v^2} = 3PV/Nm$, so $v_{rms} = \sqrt{\overline{v^2}} = \sqrt{3PV/Nm}$.

Given O_2's 32-g molar mass, the molecular mass is $m = 32\,g/6.022 \times 10^{23} = 5.31 \times 10^{-23}\,g$, or $5.31 \times 10^{-26}\,kg$. Since we know N/V, we rewrite our equation with that as a single quantity, to get

$$v_{rms} = \sqrt{\frac{3P}{m(N/V)}} = \sqrt{\frac{3(1.013 \times 10^5\,Pa)}{(5.31 \times 10^{-26}\,kg)(2.46 \times 10^{25}\,m^{-3})}}$$
$$= 482\,m/s$$

REFLECT That's high, but it's a typical speed for gases at room temperature.

MAKING THE CONNECTION Would a helium molecule (He) have a higher or lower rms speed than oxygen (O_2) under the same conditions?

ANSWER The molecular mass is in the denominator of the formula for v_{rms}. Therefore, a lighter molecule like helium travels faster, on average. For helium, you'll find that $v_{rms} = 1360\,m/s$.

Note that temperature never appeared in Example 12.10, even though it would seem important. But the other three quantities in the ideal-gas law—pressure, volume, and number of molecules—do appear, so temperature is implicitly there. We can make it explicit by comparing the molecular form of the ideal-gas law with Equation 12.5. The latter can be rearranged to give $PV = Nm\overline{v^2}/3$, while the ideal-gas law (Equation 12.4) reads $PV = Nk_BT$. Equating these two expressions for PV and rearranging gives $m\overline{v^2} = 3k_BT$.

The quantity on the left side looks a lot like kinetic energy $\frac{1}{2}mv^2$. In fact, it's just twice the **average molecular kinetic energy** $\overline{K}$, because $\overline{K} = \frac{1}{2}m\overline{v^2}$. Therefore, the average molecular kinetic energy is

$$\overline{K} = \tfrac{3}{2}k_BT \quad \text{(Average molecular kinetic energy; SI unit: J)} \quad (12.6)$$

This is a striking result. It says that the average kinetic energy of ideal gas molecules depends directly and only on temperature. That is, for an ideal gas, temperature is essentially a measure of molecular energy. Note that there's no reference to molecular mass in Equation 12.6. Thus the average molecular kinetic energy for any ideal gas is the same at a given temperature. This is true even for molecules of different gases in the same container.

TIP

When you use Equation 12.6 for average kinetic energy, you must use absolute temperature (in kelvins). The average kinetic energy approaches zero as temperature approaches zero.

EXAMPLE 12.11 **Molecular Kinetic Energy**

Find the average kinetic energy of an ideal gas molecule at 20°C. Compare with the ionization energy of hydrogen, $2.18 \times 10^{-18}\,J$. (The ionization energy is the energy needed to remove the electron from the atom.)

ORGANIZE AND PLAN The average kinetic energy of any gas molecule is given by Equation 12.6: $\overline{K} = \tfrac{3}{2}k_BT$.

Known: $T = 20°C = 293\,K$.

SOLVE With Boltzmann's constant $k_B = 1.38 \times 10^{-23}\,J/K$, the average kinetic energy is

$$\overline{K} = \tfrac{3}{2}k_BT = \tfrac{3}{2}(1.38 \times 10^{-23}\,J/K)(293\,K) = 6.07 \times 10^{-21}\,J$$

This is much lower than the 2.18×10^{-18}-J ionization energy.

REFLECT The fact that the average thermal energy is much lower than the ionization energy means that thermal energy alone won't ionize many atoms at room temperature.

MAKING THE CONNECTION How high a temperature is needed for the average kinetic energy to be equal to hydrogen's ionization energy?

ANSWER Solving Equation 12.6 for the needed temperature gives $T = 1.1 \times 10^5\,K$. Such temperatures occur inside stars, where most atomic electrons are completely separated from their nuclei. Such an ionized gas is called *plasma*.

Thermal Energy

The kinetic theory we've developed so far establishes a significant macro/micro connection. If you know the average molecular kinetic energy of a gas (Equation 12.6), then the total thermal energy E_{th} of a sample containing N molecules is N times the average kinetic energy:

$$E_{th} = N\overline{K} = \tfrac{3}{2}Nk_BT \quad \text{(Total thermal energy; SI unit: J)} \quad (12.7)$$

One caution: Equation 12.7 is valid only for monatomic ideal gases like He or Ar. For diatomic gases, including O_2 and N_2, there can be thermal energy in the molecules' rotations and oscillations. We'll address this issue in Chapter 13. But that complication doesn't change our most important conclusion here: For any ideal gas, total thermal energy is directly proportional to temperature. What does change with different types of molecules is the factor 3/2 in Equation 12.7.

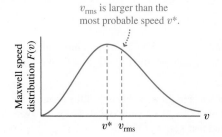

FIGURE 12.11 Maxwell speed distribution for an ideal gas.

Distribution of Molecular Speeds

The root-mean-square speed implicit in Equation 12.5 and the kinetic energy in Equation 12.6 are *averages*. The actual speeds and energies of individual molecules vary substantially—one of the assumptions of kinetic theory given at the beginning of this section. The distribution of molecular speeds was derived in about 1870 by the Scottish physicist James Clerk Maxwell and is called the **Maxwell distribution**.

The Maxwell distribution is a function $F(v)$ that gives the relative probability of a molecule having a particular speed v:

$$F(v) = 4\pi\left(\frac{m}{2\pi k_B T}\right)^{3/2} v^2 e^{-\frac{mv^2}{2k_B T}} \quad (12.8)$$

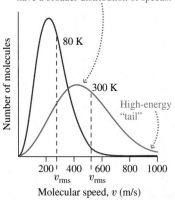

FIGURE 12.12 Maxwell distributions for nitrogen (N_2) gas at temperatures 80 K and 300 K.

where m is the mass of a molecule in a gas at temperature T. The Maxwell distribution is mathematically complicated, so it's more instructive to graph it, as in Figure 12.11. The **most probable speed**, v^*, occurs at the peak of the distribution; its value is $v^* = \sqrt{2k_B T/m}$. But the distribution isn't symmetric about its peak, so, as Figure 12.11 indicates, the most probable speed is lower than the rms speed.

It's instructive to compare the Maxwell distributions for the same gas at different temperatures. As temperature rises, the overall distribution shifts to the right, toward higher speeds (Figure 12.12). That makes sense, because higher temperature implies greater molecular kinetic energy, meaning the gas molecules are generally moving faster.

EXAMPLE 12.12   **Speedy Oxygen**

Find the most probable speed for an oxygen molecule at room temperature ($T = 20°C = 293$ K). Compare with the root-mean-square speed computed in Example 12.10.

ORGANIZE AND PLAN The most probable speed is given by $v^* = \sqrt{2k_B T/m}$. The mass m of a single molecule is its molar mass—here 32 g—divided by Avogadro's number of molecules: $m = m_{molar}/N_A$. Once again, it's important to use SI: temperature in kelvin and mass in kg.

Known: $T = 20°C = 293$ K.

SOLVE The molecular mass is

$$m = \frac{m_{molar}}{N_A} = \frac{0.032 \text{ kg}}{6.022 \times 10^{23}} = 5.31 \times 10^{-26} \text{ kg}$$

giving a most probable speed of

$$v^* = \sqrt{\frac{2k_B T}{m}} = \sqrt{\frac{2(1.38 \times 10^{-23} \text{ J/K})(293 \text{ K})}{5.31 \times 10^{-26} \text{ kg}}} = 390 \text{ m/s}$$

This is somewhat less than the 482-m/s rms speed from Example 12.10.

REFLECT You should verify that the units combine to give speed in m/s. Although the most probable speed is significantly lower than the rms speed, it's still fast. Most gas molecules under normal conditions travel hundreds of meters per second.

MAKING THE CONNECTION At what temperature would oxygen's most probable speed double to 780 m/s?

ANSWER Because of the square root, the temperature must increase by a factor of 4, to 1172 K.

Most Probable Speed and RMS Speed

Explain why the rms speed is always greater than the most probable speed in a gas subject to the Maxwell distribution. What's the ratio of these speeds?

SOLVE The explanation lies in the shape of the Maxwell distribution curve. It's not symmetric, but rather has a longer "tail" extending to high speeds. Thus the averaging that gives the rms speed is skewed toward higher speeds than the peak, and that's why the rms speed is greater than the most probable speed.

Finding the ratio of the rms to the most probable speed is straightforward, given that we know how both depend on temperature. As shown in the text, $v^* = \sqrt{2k_BT/m}$, and by Equation 12.6, $v_{rms} = \sqrt{3k_BT/m}$. Therefore, the ratio is

$$\frac{v_{rms}}{v^*} = \frac{\sqrt{\dfrac{3k_BT}{m}}}{\sqrt{\dfrac{2k_BT}{m}}} = \sqrt{\frac{3}{2}} \approx 1.225$$

REFLECT It's interesting that this ratio is independent of temperature; rather, it depends solely on the shape of the Maxwell distribution.

 TIP

The root-mean-square speed is greater than the most probable molecular speed, regardless of temperature.

The molecules move randomly, but the density difference means that more molecules move from the high- to the low-density region than vice versa, until the density difference is erased.

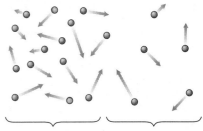

Higher density   Lower density

FIGURE 12.13 Why diffusion results in uniform concentration.

Diffusion

Pop the cork on a perfume bottle, or open the oven while you're baking bread, and soon your friends across the room catch the scent. What's happening is that molecules responsible for the characteristic smells move through the air by **diffusion**.

Molecules in a gas are in constant motion throughout their container. In our ideal-gas model, we neglected interactions between molecules. Actually, intermolecular collisions do occur—although point-particle collisions don't alter our ideal-gas conclusions. But you can see that those interactions play a big role in diffusion. When you open that perfume bottle, it takes several seconds or more for the scent to travel across a room—despite the fact that at room temperature gas molecules typically move at hundreds of meters per second. If the perfume vapor traveled without obstruction, it would cross the room in milliseconds. But vapor molecules do suffer collisions with air molecules, which slows the vapor's progress.

Diffusion tends to even out the concentration of a substance. Although molecular motions are random, there are more molecules moving *from* a region of higher concentration than there are moving *to* that region (Figure 12.13). The net result is a movement from regions of higher concentration to those of lower concentration, which eventually results in a uniform concentration.

Diffusion also occurs in liquids, a process more easily pictured than diffusion in gases. Figure 12.14 shows an ink drop introduced into water. The ink molecules slowly diffuse, mixing with the water as they go. Eventually, the ink reaches a uniform concentration. Diffusion in a gas is similar. If that perfume bottle is opened briefly and then closed, the scent fills the room fairly uniformly within minutes. After a few hours, it's diffused away through open doors or windows.

Diffusion can be controlled using a porous barrier, as shown in Figure 12.15. If the pores are small, then smaller or faster molecules will diffuse more rapidly, leading to a higher concentration of that species on the other side. A historically important application is the separation of uranium isotopes. Natural uranium is about 99.3% U-238 and 0.7% U-235, but in most nuclear reactor and weapons designs only the U-235 undergoes fission. The uranium fission weapon that destroyed Hiroshima at the end of World War II was

FIGURE 12.14 Diffusion of ink in water.

Faster molecules encounter the barrier more often than slower ones and therefore pass through it more frequently.

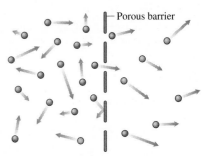

Porous barrier

FIGURE 12.15 Diffusion through a porous barrier.

made by first converting uranium to the gas uranium hexafluoride (UF_6), which was then forced through porous barriers. As you know from kinetic theory, UF_6 molecules formed from the lighter uranium isotope move slightly faster, and thus their concentration increases as they cross the barrier. Repeating this process hundreds of times resulted in a weapons-grade mix of predominantly U-235. Today, centrifuges have superseded gaseous diffusion for uranium isotope separation, but in the United States the production of nuclear reactor fuel still relies on gaseous diffusion plants.

Delivery of medication often involves diffusion. Orally administered drugs diffuse into the blood through the walls of the digestive system. The drugs then diffuse throughout the bloodstream and circulate to target areas. A hypodermic injection delivers medication faster by putting the full dose directly into the bloodstream, eliminating the first of the several diffusion processes. A transdermal patch delivers medications gradually into the body. It works by controlled diffusion through a membrane and into the skin, where diffusion continues as the medication is absorbed into the bloodstream. Its chief advantage is that the medication can be administered continuously, keeping the dose fairly constant. Medications delivered transdermally include contraceptives, nitroglycerin (for heart pain), nicotine (to ease withdrawal symptoms), and pain medication.

GOT IT? **Section 12.4** Rank in order from lowest to highest the rms speeds of the following molecules at the same temperature: (a) nitrogen (N_2); (b) oxygen (O_2); (c) helium (He); (d) argon (Ar).

Chapter 12 in Context

Chapter 12 is the first of three chapters on thermal physics. These chapters expand on the important concept of energy, first introduced in Chapter 5. You've seen that *temperature* relates to the average molecular energy, and you've learned the common *temperature scales*. You then explored *thermal expansion* of liquids and solids and noted the unusual thermal behavior of water near its melting point. Finally, you learned how the *ideal-gas law* describes the behavior of common gases and how that behavior follows from *kinetic theory* applied to gas molecules.

Looking Ahead The concepts of temperature and thermal energy introduced here will be fundamental in our continuing study of thermal physics. Chapter 13 focuses on the flow of thermal energy called *heat*. There you'll learn mechanisms of heat flow and how that flow relates to temperature differences. Chapter 14 presents the first and second laws of thermodynamics. The first law generalizes our statement of energy conservation, while the second places significant limitations on the efficiency of machines such as engines, electric power plants, refrigerators, and air conditioners. In a world concerned with energy consumption and its environmental consequences, understanding the second law of thermodynamics is fast becoming a civic responsibility.

Temperature and Thermometers

(Section 12.1) **Thermal energy** is the energy of random molecular motion. **Temperature** is a measure of thermal energy. SI temperature measurements use the **kelvin** scale, with its zero at absolute zero. The **Fahrenheit temperature scale** (°F) is common in the United States, while scientists regularly use the **Celsius scale** (°C). Thermometers use thermal expansion or changes in electrical properties to measure temperature.

Fahrenheit to Celsius conversion: T in °C $= \frac{5}{9}(T$ in °F $- 32°)$

Celsius to kelvin conversion: T in K $= T$ in °C $+ 273.15$

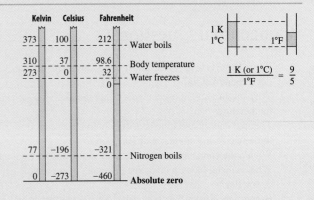
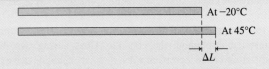

Thermal Expansion

(Section 12.2) Most solids and liquids undergo **thermal expansion** that's proportional to their temperature change. Lengthwise expansion is **linear thermal expansion**, while **volume thermal expansion** occurs when solids expand in every direction and liquids expand as permitted by their containers. Each material has its own **coefficient of thermal expansion**.

Water exhibits unusual behavior, decreasing in volume from its 0°C freezing point until it reaches maximum density at 4°C.

Linear thermal expansion: $\frac{\Delta L}{L} = \alpha \Delta T$, where α is the coefficient of linear expansion.

Volume thermal expansion: $\frac{\Delta V}{V} = \beta \Delta T$, where β is the coefficient of volume expansion.

At −20°C
At 45°C
ΔL

Ideal Gases

(Section 12.3) **Ideal gases** are so dilute that intermolecular forces are negligible. The pressure of an ideal gas results from collisions between gas molecules and container walls.

The **ideal-gas law** relates the **state variables** pressure, volume, temperature, and amount of a gas. The latter is described by the number of molecules N, or the number of **moles** n.

Moles and molecules:
$N = N_A n$, where 1 mol = Avogadro's number $N_A = 6.022 \times 10^{23}$.

Ideal-gas law: $PV = nRT$ or $PV = Nk_B T$

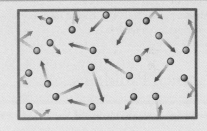

Kinetic Theory of Gases

(Section 12.4) The **kinetic theory of gases** connects macroscopic and microscopic gas properties. The **root-mean-square speed** is the square root of the mean of the squares of molecular speeds.

The **average kinetic energy** of an ideal gas molecule is proportional to temperature, and the gas's **total thermal energy** also increases linearly with temperature. The **Maxwell-Boltzmann distribution** describes the molecular speeds in an ideal gas. Gas molecules spread in a process known as **diffusion**.

Pressure of an ideal gas: $P = \dfrac{N m \overline{v^2}}{3V}$

Average kinetic energy for gas molecules: $\overline{K} = \frac{3}{2} k_B T$

Total thermal energy: $E_{\text{th}} = N\overline{K} = \frac{3}{2} N k_B T$

Maxwell speed distribution: $F(v) = 4\pi \left(\dfrac{m}{2\pi k_B T} \right)^{3/2} v^2 e^{-\frac{mv^2}{2k_B T}}$

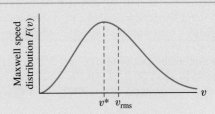

NOTE: Problem difficulty is labeled as ■straightforward to ■■■challenging. Problems labeled BIO are of biological or medical interest.

Conceptual Questions

1. Rank from largest to smallest the Fahrenheit degree, Celsius degree, and kelvin.
2. Why might you choose Celsius rather than kelvins for everyday temperatures?
3. Thermal energy and kinetic energy both involve motion. How do these two forms of energy differ?
4. What's the connection between temperature and thermal energy?
5. When a metal block with a hole is heated, does the hole get larger or smaller?
6. How might life on Earth be different if water expanded consistently with increasing temperature? What if ice were denser than water?
7. To strengthen concrete structures, steel rods called *rebar* are inserted throughout the concrete. Does having two different materials create a problem when thermal expansion occurs? *Hint:* See Table 12.1.
8. One way to loosen a jar lid is to run it under water. Should you use cold or hot water?
9. Most common gases are essentially ideal at room temperature and atmospheric pressure. Why would you expect a gas to cease behaving ideally as it's cooled toward its boiling point?
10. Describe an experiment to illustrate each of the following: Boyle's law, Charles's law, and Gay-Lussac's law.
11. A scuba diver exhales an air bubble. How does the bubble's volume change as it rises?
12. Rank the rms speeds of air's major components: nitrogen, oxygen, argon, and water vapor.
13. If typical speeds of ideal gas molecules are on the order of hundreds of meters per second, why does it take several seconds for an odor to permeate a room?

Multiple-Choice Problems

14. A 95°F temperature is equivalent to (a) 30°C; (b) 35°C; (c) 40°C; (d) 45°C.
15. Nitrogen boils at 77 K. This is closest to (a) −162°C; (b) −179°C; (c) −187°C; (d) −196°C.
16. On a winter day, Seattle is 36°F warmer than Chicago. What's this difference in °C? (a) 32°C; (b) 20°C; (c) 18°C; (d) 2°C.
17. Starting at room temperature, increasing a steel rod's length by 1% requires a temperature of about (a) 650°C; (b) 850°C; (c) 1050°C; (d) 1250°C.
18. Heating an aluminum block from 20°C to 440°C causes its density to decrease by about (a) 0.5%; (b) 1%; (c) 2%; (d) 3%.
19. Water's density is greatest at (a) 0°C; (b) 4°C; (c) 8°C; (d) 100°C.
20. An ideal gas at 20°C and 1.0×10^5 Pa pressure occupies a constant-volume container. If its temperature increases to 80°C, the pressure becomes (a) 1.0×10^5 Pa; (b) 1.2×10^5 Pa; (c) 1.5×10^5 Pa; (d) 4.0×10^5 Pa.
21. A sealed balloon occupies 120 cm³ at 1.00 atm pressure. If it's squeezed to a volume of 110 cm³ without its temperature changing, the pressure in the balloon becomes (a) 0.92 atm; (b) 1.00 atm; (c) 1.09 atm; (d) 1.19 atm.
22. You fill your car tires with air on a cold morning (0°C) to a gauge pressure of 210 kPa. As you drive, friction and the warming day increase the air temperature inside the tire to 45°C, while the volume remains constant. What's the new gauge pressure? (a) 210 kPa; (b) 233 kPa; (c) 244 kPa; (d) 261 kPa.
23. What's the rms speed of nitrogen (N_2) molecules at 273 K? (a) 465 m/s; (b) 492 m/s; (c) 510 m/s; (d) 560 m/s.
24. What's the thermal energy of one mole of ideal gas at 0°C? (a) 0 J; (b) 1700 J; (c) 2200 J; (d) 3400 J.

Problems

Section 12.1 Temperature and Thermometers

25. ■ On a cold day it's 5°F. What's the Celsius temperature?
26. ■ Two rooms in your house differ in temperature by 4.5°F. What's the temperature difference in (a) Celsius and (b) kelvin?
27. ■ Calculate the Fahrenheit and Kelvin equivalents of (a) nitrogen's boiling point, −196°C and (b) lead's melting point, 327°C.
28. ■ A dog's body temperature is 1.5°C higher than a human's. Find the dog's temperature in °C and °F.
29. ■ Climatologists project a 21st-century global temperature rise of around 3°C, largely resulting from human greenhouse gas emissions. What's that rise in °F?
30. ■ In 2005 the Huygens probe landed on Saturn's moon Titan, where the average temperature is −292°F. (a) What's this temperature in °C? (b) How far above absolute zero is this (in K)?
31. ■ The *natural greenhouse effect*, resulting from atmospheric water vapor and carbon dioxide, keeps Earth's surface some 33°C warmer than it otherwise would be (the temperature rise of Problem 32 is in addition to this natural effect). (a) Express the natural greenhouse effect in °F. (b) If Earth's average temperature is 15°C, what would it be without the natural greenhouse effect? Answer in both °C and °F.
32. ■ What's absolute zero in Celsius and in Fahrenheit?
33. BIO ■ **Fever!** You're traveling in Europe when you fall ill with a fever of 38.2°C. What's that in °F?
34. ■ Engineers in the United States sometimes express temperatures in *degrees Rankine*, where a Rankine degree is the same size as a Fahrenheit degree, but with the zero of the Rankine scale at absolute zero. What's room temperature (68°F) in Rankine?
35. ■■ (a) At what point are Fahrenheit and Celsius temperatures the same? (b) What's that temperature in kelvins?
36. ■■ (a) At what point are Fahrenheit and Kelvin temperatures the same? (b) What's that temperature in Celsius?

Section 12.2 Thermal Expansion

37. ■ A 50.00-m-long steel measuring tape is calibrated for use at 20.0°C. How long is this tape under the following conditions: (a) a hot day with $T = 32$°C and (b) a cold day with $T = -10$°C?
38. ■ A 246-m-tall skyscraper has a steel frame. By how much does the building's height on a cold day (−20°C) differ from its height on a hot day (40°C)?
39. BIO ■■ **Bone.** Bone is an *anisotropic* material, because its expansion coefficients are different in different directions. One experimental measurement gives 8.9×10^{-5}°C^{-1} for the linear expansion coefficient along bone's long dimension and 5.4×10^{-5}°C^{-1} for that along the short dimension. An individual's femur is normally

43.2 cm long and 2.75 cm in diameter. Find the change in each dimension when the individual suffers a high 104.5°F fever.

40. ■■ Gasoline's thermal expansion coefficient is $9.5 \times 10^{-4}\,°C^{-1}$. A truck's 100-gallon (378.5 L) gas tank is full on a 12°C summer morning. The truck drives into the desert, where the afternoon temperature reaches 39°C. How much gasoline spills from the tank due to thermal expansion? (Ignore possible expansion of the tank itself. Modern vehicles have expansion tanks to prevent such spillage.)

41. BIO ■■■ **Hypothermia.** A biological cell is mostly water. It's 5.0 μm in diameter at normal body temperature of 37.0°C. If a hypothermia victim's temperature drops to 32°C, what's the change in the cell diameter? Interpolate an approximate expansion coefficient from Table 12.1.

42. ■■ Using the graph in Figure 12.6c, plot the density of water from 0°C to 10°C.

43. ■■■ A pendulum clock has an aluminum pendulum exactly 1 m long when the clock is calibrated perfectly. If the temperature increases by 5°C, does the clock run fast or slow? By how much is it off at the end of one day? Assume a simple pendulum in which the aluminum rod's mass is negligible compared to the pendulum bob.

44. ■■ A beaker with a capacity of exactly 100 mL contains 99.8 mL of ethanol at 25°C. If its temperature rises, at what temperature will it overflow? (Ignore expansion of the beaker itself.)

The next four problems deal with thermal expansion in *two dimensions*.

45. ■■ Imagine a circular hole in a piece of metal, such as a cylinder in an engine block. When the temperature increases, does the metal expand into the hole, making it smaller, or does the hole expand outward, making it larger? Explain your reasoning.

46. ■■■ Suppose a hole of area A is cut into a piece of metal with linear thermal expansion coefficient α. Show that the expansion of the hole with temperature increase ΔT is given approximately by

$$\frac{\Delta A}{A} = (2\alpha)\Delta T$$

47. ■■ A flat copper sheet has a hole with area 0.250 m² at room temperature (25°C). If it's heated to 400°C, what's the hole's new area?

48. ■■ A machine has a 2.00-cm-diameter copper cylinder fitted into a hole in a steel block. At 25°C there's a uniform 0.0525-mm gap between the cylinder and the steel block. (a) At what temperature will the copper and steel just make contact? (b) How would the situation change if the copper cylinder were replaced with a steel one?

Section 12.3 Ideal Gases

49. ■ Find the mass of (a) 1 mol of argon (Ar); (b) 0.25 mol of carbon dioxide (CO_2); (c) 2.6 mol of neon (Ne); (d) 1.5 mol of UF_6.

50. ■■ Suppose you have a spherical balloon filled with air at room temperature and 1.0 atm pressure; its radius is 12 cm. You take the balloon in an airplane, where the pressure is 0.85 atm. If the temperature is unchanged, what's the balloon's new radius?

51. ■ How many air molecules are in a classroom measuring 8.0 m by 7.0 m by 2.8 m, assuming 1 atm pressure and a temperature of 22°C?

52. BIO ■■ **Taking a breath.** Taking a deep breath, a person inhales 5.5 L of air at atmospheric pressure and $T = 15°C$. By volume, air is about 78% nitrogen (N_2), 21% oxygen (O_2), and 0.93% argon (Ar). Find the number of molecules and mass of each of those substances in that deep breath.

53. ■■ A spherical balloon with radius 10.0 cm contains a gas at 1.05 atm pressure. The balloon is put into a hyperbaric (high-pressure) chamber at 1.75 atm. Assume that the balloon's temperature remains constant. (a) Does the balloon's size increase or decrease? (b) Compute its new radius.

54. ■■ A closed flask with fixed volume contains a gas at 25°C and 1 atm pressure. After heating over a Bunsen burner, the pressure is 1.65 atm. What's the new temperature?

55. ■■ (a) Compute the densities of each of the noble gases (elements in the last column of the periodic table, starting with helium) at $T = 25°C$ and $P = 1$ atm. (b) Which are lighter than air?

56. ■■ A good laboratory vacuum has pressure 10^{-8} torr. (a) What's the number density of air molecules at this pressure, assuming room temperature 20°C? (b) Compare your result with the number density under standard conditions, computed in Example 12.9.

57. ■■ You fill your tires with air on a cold morning ($-5°C$) to 220-kPa gauge pressure, then drive into a 32°C desert. (a) Assuming the volume of air in the tires remains constant, what's the new gauge pressure? (b) What would be the gauge pressure if the volume of air had expanded by 3%?

58. ■■■ A blimp typically contains about 5000 m³ of helium. Suppose the helium's pressure and temperature are 1.1×10^5 Pa and 15°C, respectively. (a) What's the mass of helium in the blimp? (b) What's the buoyant force on the blimp? (c) What's the maximum possible mass for the rest of the blimp (skin and payload) if it's neutrally buoyant (neither rising nor sinking)?

59. ■■ Your bicycle tire, with volume 3.1×10^{-4} m³, calls for a 600-kPa gauge pressure. But you measure the pressure at only 250 kPa. (a) What mass of air do you need to add to reach the specified pressure? Assume the temperature doesn't change during inflation. (b) If you've ever inflated a tire, you know that it warms in the process. Suppose in this case the air temperature rises from 15°C to 22°C. Now how much additional air is required to reach the specified pressure?

60. ■■ A compressed-air cylinder stands 100 cm tall and has internal diameter 20.0 cm. At room temperature, its pressure is 180 atm. (a) How many moles of air are in the cylinder? (b) What volume would this air occupy at room temperature and 1 atm pressure?

61. ■■■ A scuba diver is 12.5 m below the ocean surface, and seawater's density is 1030 kg/m³. The diver exhales a 25.0-cm³ bubble. What's the bubble's volume as it reaches the surface? Assume uniform water temperature.

62. ■■■ A scuba diver is 14.0 m below the surface of the lake, where the water temperature is 8.60°C. The density of fresh water is 1000 kg/m³. The diver exhales a 22.3-cm³ bubble. What's the bubble's volume as it reaches the surface, where the water temperature is 13.6°C?

63. ■■■ The *Hindenburg*, a famous German airship that exploded spectacularly in 1937 as it moored at a New Jersey air station, carried 2.12×10^5 m³ of hydrogen (H_2) for buoyancy. (a) How did the mass of the *Hindenburg*'s hydrogen compare with the mass of an equal volume of less flammable helium (He) under identical conditions? (b) If the gas pressure was 1.05×10^5 Pa and temperature was 10°C, what was the total mass of the *Hindenburg*'s hydrogen?

64. ■■■ One problem facing fuel cell cars is storing enough hydrogen for a reasonable driving distance. Hydrogen's energy density is 142 MJ/kg, higher than gasoline's 44 MJ/kg. But gasoline is a liquid (density approximately 720 kg/m³), and hydrogen is a gas. Suppose you want to store the energy equivalent of a full tank of gasoline in a tank with the same volume, but containing hydrogen

gas (H_2). What pressure would you need, assuming a temperature of 20°C? Is this practical?

Section 12.4 Kinetic Theory of Gases

65. ■ (a) Find the rms speed in hydrogen (H_2) at 0°C (273 K). (b) How much does the rms speed change when the temperature doubles to 546 K?

66. ■■ Compute the ratio of the rms speeds of air's major components, N_2 and O_2, at 273 K.

67. ■■ An ideal gas has rms speed v_{rms} at a temperature of 293 K. At what temperature is the rms speed doubled?

68. ■ What's the average kinetic energy per molecule in (a) helium and (b) oxygen at $T = 273$ K?

69. ■■ If the temperature of an ideal gas increases from 20°C to 80°C, by what factor is the rms speed increased?

70. ■ Venus's atmosphere is mostly CO_2. If the rms speed of a carbon dioxide molecule at Venus's surface is 652 m/s, what's the temperature there?

71. ■■ The Sun's surface temperature is about 5800 K. At this temperature, hydrogen is in its atomic state (H), rather than its molecular state (H_2). (a) What's the average thermal energy of hydrogen atoms at the solar surface? (b) Compare with hydrogen's ionization energy, 2.18×10^{-18} J.

72. ■■ (a) Find the most probable speed for a hydrogen molecule (H_2) at 293 K. (b) Graph the Maxwell distribution for H_2 at this temperature. (c) Use your graph to compare relative numbers of molecules at the following speeds: the most probable speed, 200 m/s, and 600 m/s.

73. ■■■ (a) Compute the most probable and rms speeds for helium (He) at room temperature (293 K). (b) There's essentially no helium in our atmosphere, and any helium released to the atmosphere eventually escapes to space. However, the speeds you found in part (a) are significantly lower than Earth's 11-km/s escape speed (Chapter 9). Why then does helium escape Earth's atmosphere? *Hint:* Think about the shape of the Maxwell distribution at higher speeds.

74. ■■■ Derive Equation 12.5 by following these steps. (a) Consider a single molecule of mass m traveling in the $+x$-direction with velocity v_x. Show that when this molecule collides and rebounds elastically from the container wall in a time Δt, it exerts a force $F = 2m v_x/\Delta t$ on the wall. (b) Assume that the container is a cube of side L, so the average time between collisions on a particular wall is $\Delta t = 2L/v_x$. Show therefore that the average force in part (a) can be written $F = mv_x^2/L$. (c) Let the area of each wall of the container be A. Use the fact that pressure $P = F/A$, along with the fact that the container's volume is $V = AL$, to show that the average pressure is $P = Nmv_x^2/V$. (d) Use the fact that $v^2 = v_x^2 + v_y^2 + v_z^2$ to argue that the average pressure is $P = Nm\overline{v^2}/3V$. *Hint:* You may assume from symmetry that, on average, $\overline{v_x^2} = \overline{v_y^2} = \overline{v_z^2}$.

General Problems

75. ■■ An ideal gas is maintained at $P = 1$ atm. By what percentage does the density of the gas change from a cold day ($-10°C$) to a hot day ($32°C$)?

76. ■■ Venus's average temperature is 730 K, and its pressure is 100 times that of Earth. Find the volume of 1 mole of Venus's atmosphere.

77. ■■ Steel rails 20.0 m long are laid end to end with gaps between them to account for thermal expansion. (a) If the track is laid when it's 15°C, how large should the gaps be to allow temperatures up to 38°C? (b) If the track is laid using the gaps you found in part (a), how large will the gaps be on a $-20°C$ winter morning?

78. ■■ The Sun's outer atmosphere, or corona, is a hot, diffuse gas with approximate temperature $T = 2 \times 10^6$ K and pressure $P = 0.03$ Pa. What's the number density of particles in the corona?

79. ■■ A steel measuring tape is exactly correct at 22°C. (a) On a cold day ($-5°C$), the tape measures the length of an aluminum beam to be 19.357 m. What's the beam's actual length? (b) On a hot day (33°C), what will be the actual length of the beam, and what will the tape measure?

80. ■■ A copper cylinder has diameter 1.000 cm and height 7.000 cm at 18°C. If the cylinder is immersed in ice water at 0°C, what are its dimensions?

81. ■■ An aerosol can of whipped cream is at gauge pressure 440 kPa when refrigerated at 3°C. The can warns against temperatures exceeding 50°C What's the maximum safe pressure for this can?

82. ■■ An aluminum block measures 1.000 cm by 2.000 cm by 3.000 cm. Find its volume after a 100°C temperature increase in two ways: (a) using aluminum's linear expansion coefficient on each side and then determining the new volume and (b) using the volume expansion coefficient. Your results should verify the relation $\beta = 3\alpha$.

83. BIO ■■ **Medical oxygen.** The M6 medical oxygen cylinder supplies 165 L of oxygen gas at 20°C and 1 atm pressure. Internally, it measures 28 cm high by 6.8 cm in diameter. What's the pressure in a full M6 cylinder at 20°C?

84. ■■■ A steel ball bearing exactly 1 cm in diameter fits tightly into a Pyrex cube at 330 K. At what temperature will there be a 1.0-μm clearance on all sides?

85. ■■■ As described in the text, one method for separating the uranium isotopes U-235 and U-238 involves diffusion of the gas UF_6; this works because the lighter U-235 moves faster and so diffuses more readily. Treating UF_6 as ideal, find the ratio of the rms speeds of UF_6 molecules containing the different isotopes at 25°C.

86. ■■■ A 3000-mL flask is initially open to air at 20°C and 1 atm pressure. It's then closed and immersed in boiling water. When it has reached equilibrium, the flask is opened and air is allowed to escape. Then it's closed and cooled back to 20°C. (a) What's the maximum pressure reached in the flask? (b) How many moles escape when the air is released? (c) What's the final pressure?

87. ■■■ One danger of global warming is a rise in sea level that could inundate coastal areas. In addition to the melting polar ice caps, a primary cause of this rise is thermal expansion of water. Estimate the sea-level rise resulting from each 1°C rise in average ocean temperature. Assume a uniform ocean depth of 3.8 km and water temperature 20°C. Your answer is an underestimate; among other things, it doesn't include the effects of salinity changes or processes in the cold depths where, as Table 12.1 shows, water's expansion coefficient is much different than that at 20°C.

Answers to Chapter Questions

Answer to Chapter-Opening Question

It's not melting ice, but thermal expansion of ocean water that's responsible for most of the recent sea-level rise.

Answers to GOT IT? Questions

Section 12.1 (d) Water's freezing point > (b) $-2.0°C$ > (a) 270 K > (c) 25°F

Section 12.2 (d) 56°C

Section 12.3 (e) Increases by a factor of 4

Section 12.4 (d) Argon (Ar) < (b) oxygen (O_2) < (a) nitrogen (N_2) < (c) helium (He)

■ Hurricane Katrina approaches the Gulf Coast in this 2005 satellite image. What's the energy source that powers the hurricane?

This chapter is about heat, the transfer of thermal energy resulting from a temperature difference. You'll learn about heat capacity and specific heat, quantities that determine how heat flow results in temperature changes, and you'll see how calorimetry is used to measure these quantities. Next we'll explore transitions among solid, liquid, and gas phases—including conditions under which they occur and energies involved. Finally, we'll explore three important heat-transfer mechanisms: conduction, convection, and radiation. Understanding heat transfer helps you solve practical problems such as keeping your house warm, scientific questions such as the temperatures of stars, and the pressing societal issue of climate change.

13.1 Heat and Thermal Energy

"Heat" is a word used in everyday speech, as in "There's a lot of heat coming off that stove." As with many other scientific words in common use—including velocity, force, and energy—we need a precise definition of heat:

Heat is energy being transferred from one object to another due to a temperature difference between the two.

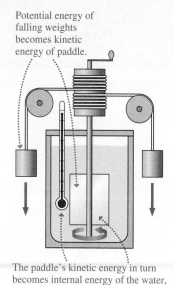

Potential energy of falling weights becomes kinetic energy of paddle.

The paddle's kinetic energy in turn becomes internal energy of the water, indicated by rising temperature.

FIGURE 13.1 Joule's apparatus for measuring what he called "the mechanical equivalent of heat."

We use the symbol Q for heat, to distinguish it from other types of energy. Because heat is energy *in transit*, it makes no sense to say that an object "contains" some amount of heat. What you can talk about is the *thermal energy* an object contains; as you saw in Chapter 12, that's the energy of random molecular motion. Many people use the terms *heat* and *thermal energy* synonymously, but they're not the same. An object can acquire thermal energy as a result of a heat flow, but, as you'll see, there are other ways to acquire thermal energy that don't involve heat.

The Mechanical Equivalent of Heat

The English physicist James Joule (1818–1899) first explored the relation between mechanical and thermal energy—for which he's honored with the SI energy unit. Joule developed the device shown in Figure 13.1. Here falling weights turn a paddle wheel in a bucket of water. Agitation from the wheel raises the water temperature, ultimately transforming gravitational potential energy of the falling weights into thermal energy of the water.

Joule found that the water's temperature increase was proportional to the potential energy change of the falling weights, thus establishing a mechanical equivalent of thermal energy. According to Joule's analysis, 817 pounds falling through 1 foot (i.e., 817 foot-pounds of energy) raises the temperature of 1 pound of water by 1°F. In SI, we would say that 4186 joules of mechanical energy raises the temperature of 1 kg of water by 1°C.

Note that Joule's experiment doesn't involve heat at all. That's because the energy that warms the water comes from mechanical agitation, not a temperature-driven energy flow (the paddle wheel isn't warmer than the water). But Joule also knew that the water could be warmed by contact with something hotter, thus showing that the transfer of mechanical energy has the same effect as a heat flow. This establishes a **mechanical equivalent of heat**, making credible the idea that heat is a transfer of energy.

Units for Thermal Energy and Heat

It would make sense to use the SI energy unit, the joule, for thermal energy, mechanical energy, and heat. However, the non-SI **calorie** (cal) is often used for thermal energy and heat. The calorie was originally defined as the energy needed to warm 1 gram of water by 1°C. Today we often use the equivalent 1 cal = 4.186 J. (Technically this is called the "15° calorie," because it's the energy needed to raise 1 gram of water from 14.5°C to 15.5°C.)

To complicate matters further, the energy content of food is sometimes given in "food calories" (abbreviated Cal) that are actually kilocalories (kcal). That is, 1 food calorie = 1 Cal = 1 kcal = 1000 cal = 4186 J. We'll generally use joules for all forms of energy. But because you may be familiar with calories from chemistry or biology, we'll sometimes give equivalent values in calories. In many countries you'll see food energy measured in joules.

✓**TIP**

Food "calories" aren't calories, but rather kilocalories, or 1000 cal.

EXAMPLE 13.1 **Food Calories**

A typical human consumes 2000 food calories per day. Find the average rate of food energy consumption, in watts.

ORGANIZE AND PLAN Remember that 1 W = 1 J/s. Therefore, we need to convert those 2000 food calories into joules, and the time (1 day) into seconds. The average energy intake rate is then the total energy ingested in 1 day divided by the number of seconds per day.

Known: Daily intake = 2000 Cal, 1 Cal = 4186 J.

SOLVE Converting 2000 food calories (Cal, kcal) into joules,

$$2000 \text{ Cal} \times \frac{4186 \text{ J}}{\text{Cal}} = 8.372 \times 10^6 \text{ J}$$

Converting 1 day into seconds,

$$1 \text{ d} \times \frac{24 \text{ h}}{\text{d}} \times \frac{60 \text{ min}}{\text{h}} \times \frac{60 \text{ s}}{\text{min}} = 8.64 \times 10^4 \text{ s}$$

cont'd.

The average rate of energy intake is then

$$\frac{8.372 \times 10^6 \text{ J}}{8.64 \times 10^4 \text{ s}} = 96.9 \text{ W}$$

That's about the same as a 100-W lightbulb!

REFLECT The energy you take in is converted to many forms: kinetic energy of your motion, blood flow, and the thermal energy associated with your body temperature. Your temperature is usually higher than your surroundings, and we'll see in Section 13.4 how that results in a substantial energy loss that's replaced as you metabolize food. It's important to note that our result is an average. About one-third of the time you're asleep and require much less energy. Sitting passively

requires a bit more energy, and vigorous exercise much more. Researchers in physiology and exercise science study these energy uses extensively.

MAKING THE CONNECTION Compare the answer to this example with the energy required for a 70-kg person to climb stairs at a rate of one floor (4 vertical meters) every 10 seconds.

ANSWER The change in gravitational energy (mgy) occurs at a rate of 2740 J every 10 s, or 274 J/s = 274 W. Not surprisingly, your energy consumption rate while climbing stairs is much larger than your average rate.

13.2 Heat Capacity and Specific Heat

Your coffee's too hot and you can't wait for it to cool, so you dump in cold milk. Soon the mixture reaches a uniform temperature—it's then in **thermal equilibrium**. Microscopically, what's happened is that faster-moving coffee molecules shared their energy through collisions with molecules in the milk. Macroscopically, there's been heat transfer from the hotter coffee to the cooler milk. In Chapter 14 we'll have a lot more to say about why heat always flows from warmer to cooler. For now, we want to know what determines that equilibrium temperature.

 TIP

Heat always flows from a warmer to a cooler body.

Heat Capacity

When an object absorbs heat, its temperature generally increases (Figure 13.2). (We say "generally" because it might instead undergo a phase change, such as melting; more on that later.) The temperature change ΔT is proportional to the heat Q absorbed:

$$Q = C\Delta T \quad \text{(Definition of heat capacity } C; \text{ SI unit: J/K)} \quad (13.1)$$

where the constant C is the object's **heat capacity**. Equation 13.1 shows that the SI units of heat capacity are J/K. Because heat capacity involves a temperature *difference*, and because the kelvin and Celsius degree are the same size, that's equally well expressed as J/°C.

 TIP

Be careful with symbols. Don't confuse the C for heat capacity with the temperature unit °C.

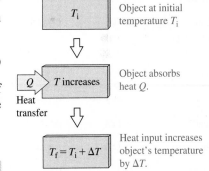

FIGURE 13.2 Absorbing heat raises an object's temperature.

EXAMPLE 13.2 **Heat Capacity**

A piece of metal absorbs 1.86 kJ of heat, raising its temperature by 12°C. Find (a) its heat capacity and (b) the heat required to raise its temperature by 60°C.

ORGANIZE AND PLAN Equation 13.1 relates temperature change to heat absorbed. We can solve the equation for the heat capacity C. Once that's known, we can find the heat required for any temperature change.

Known: $Q = 1860$ J, $\Delta T = 12°C$.

SOLVE (a) Solving Equation 13.1 for C gives

$$C = \frac{Q}{\Delta T} = \frac{1860 \text{ J}}{12°C} = 155 \text{ J/°C}$$

(b) Using $C = 155$ J/°C, the heat required for a 60°C temperature increase is $Q = C\Delta T = (155 \text{ J/°C})(60°C) = 9300$ J. Naturally, a larger ΔT requires more heat.

cont'd.

REFLECT The heat required is proportional to the temperature change, as Equation 13.1 shows. That 60°C temperature change in part (b) is five times the 12°C change in part (a). Thus the heat Q required is five times greater: $5 \times 1860 \text{ J} = 9300 \text{ J}$.

MAKING THE CONNECTION Would the heat capacity of a larger sample of the same material be the same (155 J/°C), or larger?

ANSWER Heat capacity measures the energy required per degree temperature increase. A larger sample would need more energy for the same temperature increase, so its heat capacity is larger.

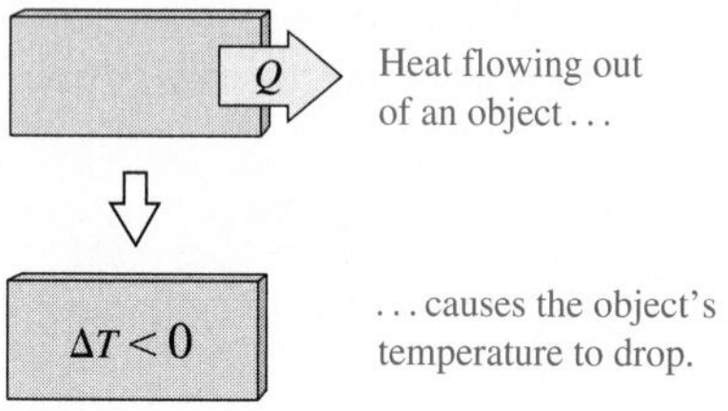

FIGURE 13.3 Heat flowing out results in a temperature decrease.

Reversing the Heat Flow

Suppose the metal piece in Example 13.2 is now immersed in cold water, dropping its temperature by 10°C. What's the heat flow involved in this case? Equation 13.1 still works just fine, and we've already found $C = 155 \text{ J/°C}$. Here the temperature *drops*, so $\Delta T = -10°C$. Then $Q = C\Delta T = (155 \text{ J/°C})(-10°C) = -1550 \text{ J}$. This *negative* value means that heat flows *from* the object to its surroundings (Figure 13.3).

Specific Heat

Last time you ordered a medium coffee, but today your coffee is an extra large, and it's again too hot. Clearly, you'll need more milk to cool it. Why? Microscopically, there are more fast-moving molecules to be slowed. Macroscopically, that makes heat capacity proportional to mass, so Equation 13.2 can be written

$$Q = mc\,\Delta T \qquad \text{(Definition of specific heat } c\text{; SI unit: J/(kg·°C))} \qquad (13.2)$$

where m is the mass and c the **specific heat**. Formally, specific heat measures the energy required to raise a material's temperature on a per-mass basis. Whereas heat capacity C applies to a particular sample of material, specific heat is a property of the material itself; all samples of that material have the same specific heat. Table 13.1 lists specific heats of common materials. Note the lowercase c for specific heat and uppercase C for heat capacity. Our definitions (Equations 13.1 and 13.2) show that they're related by $C = mc$.

SI units of specific heat are J/(kg·K); equivalently, J/(kg·°C). In Table 13.1 we list these units and a common alternative, cal/(g·°C).

TABLE 13.1 Specific Heat of Selected Materials (at $T = 20°C$ unless indicated)

Material	Specific heat c, J/(kg·°C)	Specific heat c, cal/(g·°C)
Aluminum	900	0.215
Beryllium	1970	0.471
Copper	385	0.092
Ethanol	2430	0.581
Human body (average, $T = 37°C$)	3500	0.840
Ice (0°C)	2090	0.499
Iron	449	0.107
Lead	128	0.031
Mercury	140	0.033
Silver	235	0.056
Water	4186	1.000
Wood (typical)	1400	0.33
Steel (typical)	500	0.12

✓ **TIP**

Think of specific heat as heat capacity per unit mass.

EXAMPLE 13.3 **Identifying the Unknown Material**

Weighing the sample in Example 13.2 gives $m = 403 \text{ g}$. Which material in Table 13.1 is it likely to be?

ORGANIZE AND PLAN Example 13.2 gave $C = 155 \text{ J/°C}$. Knowing the mass lets you find the specific heat of this material, because $C = mc$. The result can be checked against Table 13.1 to identify the material.

Known: $C = 155 \text{ J/°C}$, $m = 403 \text{ g} = 0.403 \text{ kg}$.

SOLVE With $C = mc$, the specific heat is

$$c = \frac{C}{m} = \frac{155 \text{ J/°C}}{0.403 \text{ kg}} = 385 \text{ J/(kg·°C)}$$

This matches copper's entry in Table 13.1.

REFLECT The specific heat is only one clue to the nature of the material. Copper has a characteristic reddish-brown color, so color would be another good key in this case.

MAKING THE CONNECTION What mass of aluminum has the same heat capacity as the copper sample in this example?

ANSWER The specific heat of aluminum is $c = 900 \text{ J/(kg·°C)}$. With $C = mc$ and $C = 155 \text{ J/°C}$, the mass is $m = C/c = 0.172 \text{ kg}$—less than half that of the copper.

EXAMPLE 13.4 **Mixing Water**

You draw a 35-L bath, but at 47°C it's too hot. So you start adding 9.0°C water from the cold tap. How much cold water do you need to add to bring the bath to a comfortable 39°C?

ORGANIZE AND PLAN Heat flows from hot water to cold, until the water is at a uniform temperature. Assuming no heat lost to the environment, the sum of the heat lost by the hot water and the heat gained by the cold water is zero. For both hot and cold water, $Q = mc\Delta T$ (Equation 13.2), with Q_{cold} positive and Q_{hot} negative. This lets us solve for the unknown mass.

Water's density is 1000 kg/m^3, or 1 kg/L. Therefore, the mass of the initial hot water is 35 kg.

Known: $c = 4186$ J/(kg·°C) for water, from Table 13.1; $m_{hot} = 35$ kg, $T_{hot} = 47$°C, $T_{final} = 39$°C.

SOLVE For both the hot and cold water, $Q = mc\Delta T$. The sum of the heat lost by the hot water and the heat received by the cold water is zero: $Q_{hot} + Q_{cold} = 0$, or $m_{hot} c\Delta T_{hot} + m_{cold} c\Delta T_{cold} = 0$. Solving for the unknown m_{cold},

$$m_{cold} = -\frac{m_{hot} c\Delta T_{hot}}{c\Delta T_{cold}} = -\frac{m_{hot}\Delta T_{hot}}{\Delta T_{cold}}$$

Now $\Delta T_{hot} = 39$°C − 47°C = −8°C, and $\Delta T_{cold} = 39$°C − 9°C = 30°C. Therefore,

$$m_{cold} = -\frac{m_{hot}\Delta T_{hot}}{\Delta T_{cold}} = -\frac{(35\,\text{kg})(-8°C)}{30°C} = 9.3\,\text{kg}$$

With a density of 1 kg/L, that means 9.3 L of cold water.

REFLECT Because the added water is so cold relative to the final temperature of the mixture, you might have guessed that the amount of cold water required is less than the initial amount of hot water.

MAKING THE CONNECTION What happens to the water temperature when a 62-kg person enters the tub? Assume the person starts with normal body temperature, 37°C.

ANSWER The water has mass of 44.3 kg and initial temperature 39°C. The average specific heat of a person (from Table 13.1) is 3.5 kJ/(kg·°C). Following the procedure of this example, the final temperature of the person + water mixture is about 37.9°C. This neglects energy released in metabolism, which tends to keep body temperature constant.

Calorimetry

Experiments with heat flow from one substance to another can reveal properties of those substances. Immerse an unknown solid at one temperature in a known quantity of water at a different temperature. Measuring the final temperature lets you determine the specific heat of the unknown. If you suspect that the unknown sample is of some pure substance, you could then compare your calculated specific heat with known values.

The method used in such experiments is **calorimetry**, named for the calorie. For calorimetry to be successful, it's important to account for all the heat. Figure 13.4 shows schematically how a **calorimeter** works. If the calorimeter is well insulated, you can assume that the only heat exchange is between the two substances, in this case the metal and the water. The following example illustrates this.

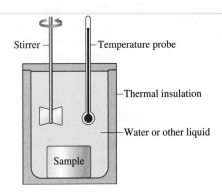

FIGURE 13.4 A calorimeter in use.

EXAMPLE 13.5 **Calorimetry in Action**

You have a 115-g metal cylinder that you suspect is aluminum. You heat it to 80°C and then immerse it in a calorimeter containing 250 g of water initially at 20°C. The final temperature is 25.4°C. Is the material aluminum?

ORGANIZE AND PLAN As in the preceding example, the sum of the heat Q_M lost by the metal and the heat Q_W gained by the water is zero. That is, $Q_M + Q_W = 0$, or, using Equation 13.2, $m_M c_M \Delta T_M + m_W c_W \Delta T_W = 0$.

Known: $c_W = 4186$ J/(kg·°C), $m_M = 0.115$ kg, $m_W = 0.250$ kg, initial temperatures $T_M = 80$°C and $T_W = 20$°C, final temperature 25.4°C.

SOLVE Writing the two expressions for ΔT in terms of the initial temperatures and the single final temperature T_f: $\Delta T_M = T_f - T_M$ and $\Delta T_W = T_f - T_W$. Using these in our equation, $m_M c_M \Delta T_M + m_W c_W \Delta T_W = 0$, and solving for c_M gives

$$c_M = -\frac{m_W c_W \Delta T_W}{m_M \Delta T_M}$$

$$= -\frac{(0.250\,\text{kg})(4186\,\text{J/(kg·°C)})(25.4°C - 20°C)}{(0.115\,\text{kg})(25.4°C - 80°C)}$$

$$= 900\,\text{J/(kg·°C)}$$

Table 13.1 confirms that it's aluminum.

cont'd.

REFLECT In practice you should watch the water temperature as it rises. It eventually levels off at a maximum, which you take as T_f. If you wait too long, heat loss from the calorimeter to its surroundings will spoil your measurement.

If the cylinder had been silver, with the same mass, would the final temperature have been higher or lower than 25.4°C?

ANSWER Table 13.1 gives $c = 235$ J/(kg·°C) for silver. That's a lot lower than for aluminum, meaning the silver has less thermal energy to give up in a specified temperature drop. Therefore, the water can't warm as much. Work the numbers and you'll find a final temperature of 21.5°C. Conceptual Example 13.6 makes this point in a slightly different way.

CONCEPTUAL EXAMPLE 13.6 **Available Thermal Energy**

You've got equal-mass samples of two different materials with specific heats c_1 and c_2, where $c_1 < c_2$. You begin with the samples at different temperatures, and then place them in thermal contact. Which sample's temperature changes more—the one with the larger or the one with the smaller specific heat? Does it matter which is hotter initially?

SOLVE Suppose $T_1 < T_2$ initially, so heat flows from substance 2 to substance 1 (Figure 13.5). Then the net heat flow in the system is, as in the preceding examples, $m_1 c_1 \Delta T_1 + m_2 c_2 \Delta T_2 = 0$, with $\Delta T_1 > 0$ and $\Delta T_2 < 0$.

The equal masses cancel, giving $c_1 \Delta T_1 = -c_2 \Delta T_2$. For this equality to hold with $c_1 < c_2$, the magnitude of ΔT_1 must be *larger*. If the initial temperatures were reversed, so $T_2 < T_1$ initially, our analysis wouldn't change. The material with the smaller specific heat still undergoes a temperature change of greater magnitude.

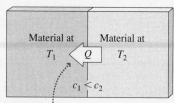

$T_1 < T_2$; heat flows from hotter to cooler object

FIGURE 13.5 Heat flowing between two materials with different specific heats.

REFLECT Substance 1's heat capacity $C = mc$ is lower because it has the same mass but smaller specific heat. Heat capacity measures the heat absorbed per unit temperature change. A smaller heat capacity means that the same amount of heat gives a larger temperature change.

Specific Heat of Gases

Gases have heat capacity and specific heat, but they're expressed differently than for solids and liquids. That's because a gas can change pressure and volume when heated, while solids and liquids undergo much less of those changes.

When a gas is heated, its temperature change depends on how much the pressure and volume change. For this reason, there are two measures of a gas's specific heat: at constant volume and at constant pressure. Another difference is that gas specific heats are usually given on a molar basis, rather than per unit mass. Thus, the equations relating heat flow Q to temperature change contain the number of moles n of the gas. For a constant-volume process,

$$Q = nc_V \Delta T \qquad \text{(Specific heat of a gas at constant volume; SI unit: J/(mol·°C))} \qquad (13.3)$$

where c_V is the **molar specific heat at constant volume**. Similarly, for a constant-pressure process,

$$Q = nc_P \Delta T \qquad \text{(Specific heat of a gas at constant pressure; SI unit: J/(mol·°C))} \qquad (13.4)$$

where c_P is the **molar specific heat at constant pressure**. Table 13.2 shows some values of c_V and c_P for selected gases. The units for both are J/(mol·°C). Therefore, with n in mol and ΔT in °C, the heat Q is in J.

TABLE 13.2 Molar Specific Heats of Selected Gases

Gas	c_V in J/ (mol·°C)	c_P in J/ (mol·°C)
Monatomic gases		
He	12.5	20.8
Ne	12.5	20.8
Ar	12.5	20.8
Diatomic gases		
H_2	20.4	28.7
N_2	20.8	29.1
O_2	20.9	29.2
Air (a predominantly diatomic mixture)	20.8	29.1

✓**TIP**

For solids and liquids, specific heat is heat capacity per unit mass; for gases, it's per mole.

You'll notice a pattern in the tabulated values. For each of the monatomic gases listed, $c_V = 12.5 \, \text{J}/(\text{mol} \cdot {}^\circ\text{C})$. This isn't a coincidence, as we'll show shortly. For the diatomic gases, c_V varies over a narrow range, from $20.4 \, \text{J}/(\text{mol} \cdot {}^\circ\text{C})$ to $20.9 \, \text{J}/(\text{mol} \cdot {}^\circ\text{C})$. Figure 13.6 shows why the specific heat of diatomic gases is larger than that of monatomic gases. When a monatomic gas absorbs heat, the energy shows up in the translational kinetic energy of individual molecules (Figure 13.6a). In a diatomic gas (Figure 13.6b), energy goes into both translational and rotational kinetic energy. Because temperature measures the average translational kinetic energy, this means more heat is needed for the same temperature rise in a diatomic gas compared with a monatomic gas.

Why is $12.5 \, \text{J}/(\text{mol} \cdot {}^\circ\text{C})$ the specific heat of a monatomic gas at constant volume? For a monatomic gas, all the heat Q goes into molecular kinetic energy, and Equation 12.7 gives the total energy as $E_{\text{th}} = \frac{3}{2}Nk_BT$. Therefore, adding heat Q changes the thermal energy by $\frac{3}{2}Nk_B\Delta T$, so Q and ΔT are related by $Q = \frac{3}{2}Nk_B\Delta T$. In Chapter 12 we found that the number of molecules N is nN_A, where N_A is Avogadro's number, and that $N_Ak_B = R$, the molar gas constant. So our Q vs. ΔT relation becomes $Q = \frac{3}{2}Rn\Delta T$. Comparison with Equation 13.3, $Q = nc_V\Delta T$, shows that $c_V = \frac{3}{2}R$. With $R = 8.315 \, \text{J}/(\text{mol} \cdot {}^\circ\text{C})$, that gives $c_V = 12.5 \, \text{J}/(\text{mol} \cdot {}^\circ\text{C})$.

Note also that c_P is larger than c_V for every gas. You can understand this by considering the difference between constant-volume and constant-pressure processes for an ideal gas obeying $PV = nRT$. In the constant-volume process, the heat absorbed increases the kinetic energy of individual molecules. This increased molecular energy corresponds to a temperature increase. In a constant-pressure process, the gas volume increases in proportion to the temperature increase, and the gas does work as it expands against the pressure of its surroundings. Additional energy is needed for that expansion (Figure 13.7). This means that more heat must be supplied at constant pressure to produce a given temperature increase as compared with the constant-volume process.

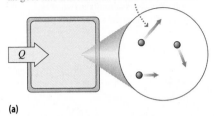

When a monatomic gas absorbs heat, the energy all goes into translational motion of the atoms.

(a)

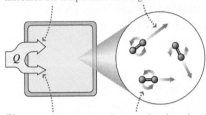

When a diatomic gas absorbs heat, some of the energy goes into translational motion, which increases the temperature of the gas . . .

(b) . . . and some goes into rotational motion of the molecules, which does not change the temperature.

FIGURE 13.6 Kinetic energy in monatomic and diatomic gases.

EXAMPLE 13.7   **A Cold Breath**

You inhale 4.0 L of 0°C air and hold it in your lungs. How much energy does your body supply to heat the air to your 37°C body temperature if you hold constant (a) your lung volume or (b) your lung pressure?

ORGANIZE AND PLAN Equations 13.3 and 13.4 describe constant-volume and constant-pressure processes. Table 13.2 gives the two specific heats for air, which are nearly the same as that for air's main component, nitrogen.

Known: $V = 4.0 \, \text{L}$, $\Delta T = 37{}^\circ\text{C}$.

SOLVE At standard conditions of 0°C and 1 atm, we found in Chapter 12 that a mole of ideal gas occupies 22.4 L. So your 4.0-L breath contains 4.0/22.4 moles, giving $n = 0.179 \, \text{mol}$. Then for the constant-volume process,

$$Q = nc_V\Delta T = (0.179 \, \text{mol})(20.8 \, \text{J}/(\text{mol} \cdot {}^\circ\text{C}))(37{}^\circ\text{C}) = 138 \, \text{J}$$

At constant pressure:

$$Q = nc_P\Delta T = (0.179 \, \text{mol})(29.1 \, \text{J}/(\text{mol} \cdot {}^\circ\text{C}))(37{}^\circ\text{C}) = 193 \, \text{J}$$

REFLECT More heat is required in the constant-pressure process. That's because only some of the energy goes into raising the temperature; the rest helps the gas expand.

MAKING THE CONNECTION   How much food energy would you expend in warming that breath?

ANSWER With one food Calorie equal to 4186 J, and assuming full conversion of food energy to heat, you'd only expend 0.03 to 0.04 Calories. This is not an effective way to lose weight!

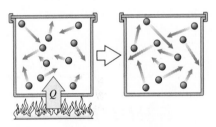

(a) When heat is added at constant volume, all the energy goes into thermal motion.

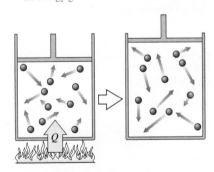

(b) When heat is added at constant pressure, some of the energy goes into thermal motion and some into expanding the container.

FIGURE 13.7 The difference between constant-volume and constant-pressure processes.

Rotation occurs around the two axes perpendicular to the molecular bond.

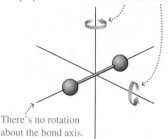

There's no rotation about the bond axis.

FIGURE 13.8 Rotational motion of a diatomic molecule.

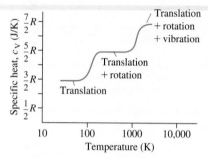

FIGURE 13.9 Specific heat of hydrogen (H_2) versus temperature. Below 20 K hydrogen is liquid, and above 3200 K it dissociates into individual atoms.

The atoms in a diatomic molecule can vibrate back and forth like masses joined by a spring.

FIGURE 13.10 Modeling vibration of a diatomic molecule.

Each atom can move in three directions (x, y, z) and has both kinetic and potential energy due to each direction's motion.

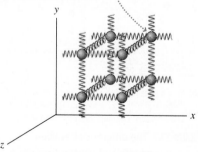

FIGURE 13.11 Atoms vibrating in a solid.

Equipartition

We've shown why the volume specific heat of monatomic gases has the value $\frac{3}{2}R$. A principle called the **equipartition theorem** lets us generalize this result for other gases.

> **Equipartition theorem:** The molar specific heat of a substance is $\frac{1}{2}R$ for each molecular degree of freedom.

What's a "degree of freedom"? It's an independent way for a molecule to have energy; mathematically, each quadratic (squared) term in the kinetic or potential energy of the molecule represents a degree of freedom. For example, the monatomic gas is free to move in the x-, y-, and z-directions, so its kinetic energy is $K = \frac{1}{2}mv^2 = \frac{1}{2}mv_x^2 + \frac{1}{2}mv_y^2 + \frac{1}{2}mv_z^2$. Thus there are 3 degrees of freedom, associated with the three velocity components. The equipartition theorem predicts that a monatomic gas has a molar heat capacity $c_V = 3 \times \frac{1}{2}R = \frac{3}{2}R$, as we found in the preceding section.

Now consider the diatomic gas, for which a crude model is two spheres (the atoms) connected by a solid rod (the molecular bonding forces), as shown in Figure 13.8. In addition to the three translational degrees of freedom, there are two rotational ones. Therefore, there are a total of $3 + 2 = 5$ degrees of freedom, and the equipartition theorem predicts a molar specific heat $c_V = 5 \times \frac{1}{2}R = \frac{5}{2}R = 20.8\ \text{J}/(\text{mol} \cdot °\text{C})$. This value closely matches Table 13.2's specific heats for diatomic gases.

Experiment shows that molar specific heat of a diatomic gas depends on temperature (Figure 13.9). Just above the boiling point, $c_V \approx \frac{3}{2}R$. This is a quantum-mechanical effect, and shows that rotations don't "turn on" until higher temperatures. For most diatomic gases, $c_V \approx \frac{5}{2}R$ at room temperature, showing that both translation and rotation occur. At still higher temperatures, the specific heat increases again. Figure 13.10 shows why: now vibrational motion has "turned on," adding 1 degree of freedom from the associated kinetic energy and 1 from the potential energy. That makes 7 degrees of freedom total, giving $c_V = 7 \times \frac{1}{2}R = \frac{7}{2}R$. In Figure 13.9, you see hydrogen approaching this value. But before it can be reached, thermal energy breaks the molecule into individual atoms.

Vibration also occurs in solids. Although individual atoms aren't free to translate or rotate, the "springs" that bind them permit oscillations in three independent directions (Figure 13.11). There are 2 degrees of freedom (kinetic plus potential) for each, giving $3 \times 2 = 6$ total degrees of freedom. The predicted molar specific heat of a cubic solid is then $c_V = 6 \times \frac{1}{2}R = 3R$. The actual molar specific heat of many cubic solids (for example, copper) is quite close to this value.

Why the name *equipartition*? At the microscopic level, the equipartition theorem states something very simple: random collisions share energy among molecules—and on average they share it equally among all the possible ways a molecule can have energy. Those ways are the degrees of freedom, so each degree of freedom gets, on average, the same energy. Each therefore absorbs heat equally, and so contributes equally to the specific heat.

Reviewing New Concepts: Specific Heats of Gases

- The specific heat of a gas is measured at either constant volume (c_V) or constant pressure (c_P).
- Specific heats are higher for diatomic gases than for monatomic gases.
- For any gas, c_P is larger than c_V by R, because of energy needed to expand the gas.
- The specific heats of monatomic and diatomic gases follow from the equipartition theorem.

GOT IT? Section 13.2   The molar specific heat of a diatomic gas is larger than the molar specific heat of a monatomic gas by about (a) $R/2$; (b) R; (c) $3R/2$; (d) $2R$.

13.3 Phase Changes

Here's another way to cool that hot drink: Drop in an ice cube. Even though the ice is only a few degrees colder than the milk you tried before, it results in much greater cooling. Why?

Heats of Transformation

Melting ice requires energy, to break the bonds joining neighboring H_2O molecules. The energy per unit mass required to melt a solid is the **heat of fusion** L_f. To melt a sample of mass m then requires heat Q, where

$$Q = mL_f \quad \text{(Heat of fusion; SI unit: J/kg)} \tag{13.5}$$

Similarly, it takes energy to turn liquid into gas, further separating the molecules. The **heat of vaporization**, L_v, gives the energy required per unit mass:

$$Q = mL_v \quad \text{(Heat of vaporization; SI unit: J/kg)} \tag{13.6}$$

Table 13.3 shows some values of L_f and L_v, which are collectively called **heats of transformation**. Notice that L_v is substantially larger than L_f. In the examples that follow, you'll see the physical significance of this fact.

Melting and vaporization are reversible, and require *removing* the corresponding amounts of heat. For that reason, heats of transformation are also called **latent heats** because the energy that goes into melting or vaporizing is "latent" in the new state and can be recovered by refreezing the liquid or condensing the solid. It's the release of that latent heat from moist tropical air that powers hurricanes.

Spring Skiing

Winter snow piles up in the mountains, but remains long after temperatures rise above freezing. It can take months to melt a large snowpack, because of water's large heat of fusion. Many communities in the western United States depend on runoff from melting snow for water and hydroelectric power into the dry summer season.

✓**TIP**

To change phase from a solid to a liquid or from a liquid to a gas, heat must be added; when going in the other direction, heat must be removed.

TABLE 13.3 Heats of Transformation at $P = 1$ atm

Substance	Melting point (°C)	Heat of fusion L_f (J/kg)	Boiling point (°C)	Heat of vaporization L_v (J/kg)
Copper	1084	2.05×10^5	2560	3.92×10^5
Ethanol	−114	1.04×10^5	78	8.52×10^5
Gold	1064	6.45×10^4	2650	1.57×10^6
Helium	N/A	No solid phase at $P = 1$ atm	−269	2.09×10^4
Lead	328	2.50×10^4	1740	8.66×10^5
Mercury	−39	1.22×10^4	358	2.67×10^5
Nitrogen	−210	2.57×10^4	−196	1.96×10^5
Oxygen	−218	1.38×10^4	−183	2.12×10^5
Tungsten	3400	1.82×10^5	5880	4.81×10^6
Uranium	1133	8.28×10^4	3818	1.88×10^6
Water	0	3.33×10^5	100	2.26×10^6

- The heat of transformation is the energy per unit mass needed to change a substance's phase, either by melting (L_f) or by vaporization (L_v).
- The heat Q required to melt or vaporize a mass m is $Q = mL_f$ for melting or $Q = mL_v$ for vaporization.
- In freezing or condensation, heats of transformation give the energy per unit mass that must be removed from the substance.

EXAMPLE 13.8   **Ice to Steam**

You have a 0.250-kg lump of ice at 0°C. (a) How much energy is required (a) to melt it, (b) to bring the liquid water from 0°C to 100°C, and (c) to boil it all to vapor once the water is at 100°C?

ORGANIZE AND PLAN   Table 13.3 gives the heats of fusion and vaporization; with those, we can use Equations 13.5 and 13.6 to find the energies involved in the phase changes. For the energy needed to raise the liquid's temperature, we'll use Equation 13.2, $Q = mc\Delta T$, and the specific heat from Table 13.1.

Known: $m = 0.250$ kg.

SOLVE   (a) To melt the ice, $Q = mL_f$. With $L_f = 333$ kJ/kg (Table 13.3),

$$Q = mL_f = (0.250 \text{ kg})(333 \text{ kJ/kg}) = 83.3 \text{ kJ}$$

(b) Table 13.1 gives specific heat of water: 4186 J/(kg·°C). Then the heat needed to raise the temperature from 0°C to 100°C is

$$Q = mc\Delta T = (0.250 \text{ kg})(4186 \text{ J/(kg·°C)})(100°C) = 105 \text{ kJ}$$

(c) Using $L_v = 2.26 \times 10^6$ J/kg from Table 13.3,

$$Q = mL_v = (0.250 \text{ kg})(2260 \text{ kJ/kg}) = 565 \text{ kJ}$$

REFLECT   It takes almost as much heat to melt the ice as to raise the temperature of the liquid water from 0°C to 100°C. The heat required to boil the water is much greater than the other two values. This should be consistent with your experience: Put a pot of water on the stove, and it takes a few minutes to reach boiling, but the water boils for a long time before the pan goes dry.

MAKING THE CONNECTION   If your heater puts energy into the water at the rate of 500 W, how much time does each of the above processes take?

ANSWER   (a) 2.8 min, (b) 3.5 min, (c) 19 min.

CONCEPTUAL EXAMPLE 13.9   **History of Water**

You start with the same 250-g ice block at 0°C and add heat at a constant rate of 500 W until it has all boiled away. Graph the temperature of the ice/water/steam versus time. Assume the temperature remains uniform throughout the sample at all times.

SOLVE   The preceding example tells what you need to draw the graph. While melting, the ice/water mixture holds at 0°C for 2.8 min. Then the liquid water warms steadily for 3.5 min until it reaches 100°C, at which point it takes 19 min to boil away. Our graph (Figure 13.12) shows that most of the time is spent boiling the water.

REFLECT   If you continue supplying heat after all the water has vaporized, the steam's temperature will rise above 100°C. The rate of increase will be faster than for the 0°C–100°C rise, because steam's specific heat is about half that of liquid water.

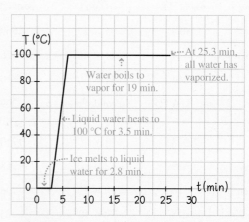

FIGURE 13.12 Temperature versus time for this experiment.

EXAMPLE 13.10 **Cooling Your Coffee**

Your 300-g cup of coffee is at 85°C—too hot to drink. You drop in a 42-g ice cube (at 0°C). What's the final temperature of the mixture, after thermal equilibrium is reached? Coffee's specific heat is essentially that of water.

ORGANIZE AND PLAN Consider two steps (Figure 13.13): (1) Heat flows from coffee to ice, melting it. (2) Heat continues to flow from the coffee into the melted ice, until equilibrium is reached. In step 1, $Q = mL_f$, for the heat transferred to the ice. In step 2, $Q = mc\Delta T$ for the heat that flows from the coffee to the water.

Known: $m_c = 0.300$ kg, $m_i = 0.042$ kg, where the subscripts denote coffee and ice.

SOLVE In step 1, the heat that flows from the coffee to the ice is

$$Q = m_i L_f = (0.042 \text{ kg})(333 \text{ kJ/kg}) = 14.0 \text{ kJ}$$

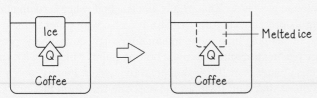

❶ Heat from the coffee melts the ice. ❷ More heat from the coffee warms the melted ice.

FIGURE 13.13 The two-step process: ice melts, and the melted ice warms.

This heat flows *from* the coffee, so $Q_c = -14.0$ kJ, and with $Q_c = mc\Delta T_c$,

$$\Delta T_c = \frac{Q}{m_c c} = \frac{-14.0 \text{ kJ}}{(0.300 \text{ kg})(4.186 \text{ kJ/(kg} \cdot {}^\circ\text{C}))} = -11.1{}^\circ\text{C}$$

Therefore, melting the ice reduces the coffee's temperature from 85°C to 73.9°C.

In step 2, we follow the procedure of Section 13.2, considering that all the energy leaving the coffee ends up in the water. The water's temperature increases from 0°C to a final temperature T_f, while the coffee drops from 73.9°C to T_f. Mathematically, $\Delta T_i = T_f - 0{}^\circ\text{C}$ and $\Delta T_c = T_f - 73.9{}^\circ\text{C}$. Using these in the heat flow equation gives $m_i c (T_f - 0{}^\circ\text{C}) + m_c c (T_f - 73.9{}^\circ\text{C}) = 0$.

The specific heat of water, c, cancels, and we solve for T_f:

$$T_f = \frac{m_c (73.9{}^\circ\text{C})}{m_i + m_c} = \frac{(0.300 \text{ kg})(73.9{}^\circ\text{C})}{0.042 \text{ kg} + 0.300 \text{ kg}} = 64.8{}^\circ\text{C}$$

REFLECT This seems reasonable, given the masses of coffee and ice. You might wonder how we knew there was enough energy available to melt the ice. We didn't, but the fact that we get a final temperature greater than 0°C confirms that there was enough. Put a huge chunk of ice in the coffee, though, and you might end up at 0°C with some ice unmelted. That's what you do with a cold drink.

MAKING THE CONNECTION How much water at 0°C would you have to add to 300 g of coffee to have the same cooling effect?

ANSWER Using water's specific heat, you'd have to add 94 g of water. That would dilute the coffee much more than the ice cube did!

PROBLEM-SOLVING STRATEGY 13.1 **Heating, Melting, and Vaporizing**

ORGANIZE AND PLAN
- Visualize the situation using a diagram.
- Identify the substance(s) involved.
- Identify the processes: are there temperature changes, phase changes, or both?
- If temperature(s) change, use the specific heat(s) to relate temperature change(s) and heat supplied.
- If two substances are coming into thermal equilibrium, use the same final temperature for both and write an equation expressing that the heat lost by one substance is gained by the other.
- For a single substance changing phase, use the appropriate heat of transformation to relate mass and heat involved.
- If two substances are coming to equilibrium and one undergoes a phase change, consider first the phase change, then any additional temperature change. Make sure the final answer is consistent with the final phase; if it isn't, then your final state is a mixture including both phases at the phase change temperature.
- Review the information you have, such as masses, temperature changes, and phase changes. Find the specific heats or heats of transformation for any known substances. Plan how to use that information to solve for the unknown(s).

SOLVE
- Gather given information and tabulated values.
- Combine and solve equations for the unknown quantity (or quantities), using appropriate units.

REFLECT
- Check the dimensions and units of the answer. Are they reasonable?
- If the problem relates to something familiar, consider whether the answer makes sense.

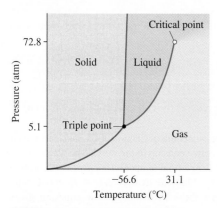

FIGURE 13.14 Phase diagram for carbon dioxide (not to scale).

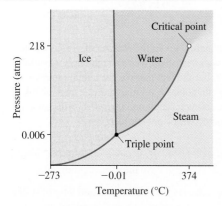

(a)

(b)

FIGURE 13.15 (a) Phase diagram for water (not to scale). (b) At sufficiently low pressure, water boils at room temperature.

Phase Diagrams

At a given temperature and pressure, a substance will generally be a solid, liquid, or gas. A **phase diagram**, a graph of pressure versus temperature, shows boundaries between phases. Figure 13.14 shows the phase diagram for carbon dioxide. The lines separating phases represent combinations of temperature and pressure where two phases can coexist. Note that there's one **triple point** where all three phases coexist. That's useful for temperature calibrations, because it defines unambiguously a unique temperature. Note also that the liquid-gas line ends at a **critical point**. Here liquid and gas have the same density, and at higher pressures they become indistinguishable. Instead of an abrupt change from liquid and gas with increasing temperature, there's a gradual transition. Figure 13.14 also shows what you already know: Under common conditions ($P = 1\,\text{atm}$, $T = 20°C$), carbon dioxide is a gas. You exhale it with every breath!

Figure 13.15a shows water's phase diagram. Note that water's triple point occurs at a low 0.006 atm. Above that pressure, water has its three familiar phases. If you start with liquid water at $P = 1\,\text{atm}$ and cool it without changing the pressure, you're moving leftward on the phase diagram. Eventually, you cross into the solid phase (ice). Similarly, heating moves you to the right on the diagram, eventually crossing into the gas phase (steam). But the phase diagram shows a less familiar approach to phase changes: reduce the pressure without changing water's temperature, and you move downward in the diagram, eventually crossing into the gas phase. That means you can boil water at room temperature simply by reducing its pressure (Figure 13.15b). It's also why water boils at a lower temperature when you're camping high in the mountains, where atmospheric pressure is lower. Conversely, increasing pressure raises the boiling point, which is why you can boil foods at temperatures greater than 100°C in a pressure cooker. Finally, note the solid-liquid line on water's phase diagram: It slopes the opposite way from that of CO_2 and most other substances. That's related to water's unusual thermal expansion, which we discussed in Chapter 12.

CONCEPTUAL EXAMPLE 13.11 **Dry Ice**

Solid CO_2 is called *dry ice*. What happens if you take dry ice from a freezer at −80°C and put it on a table at room temperature? Assume normal atmospheric pressure throughout.

SOLVE The CO_2 begins as a solid at $P = 1\,\text{atm}$ and $T = -80°C$. With constant pressure, heating to room temperature takes the CO_2 to the *right* on its phase diagram. Another look at the phase diagram (Figure 13.16) shows that at 1 atm pressure, we're well below the triple point (5.2 atm). That means there's no liquid phase at 1 atm, so as it warms the CO_2 goes directly from solid to gas—a process called **sublimation**.

REFLECT Dry ice is a popular refrigerant because it's much colder than water ice and doesn't form puddles when it warms (hence "*dry* ice"). The resulting CO_2 gas is harmless in small quantities.

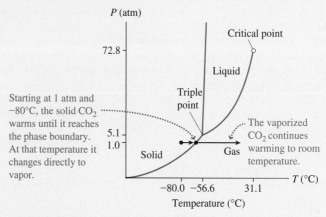

FIGURE 13.16 Phase diagram illustrating the sublimation of dry ice.

Evaporative Cooling

Turning liquid to gas takes energy. For that reason evaporation provides a means of cooling both biological and mechanical systems. Why do you sweat on a hot day or during vigorous activity? That's how your body transfers heat to the surrounding air. Vaporizing the water in sweat cools your skin (and leaves the salt behind). Refrigerators work similarly, evaporating a *working fluid* chosen for its appropriate temperature-pressure behavior.

Table 13.3 shows that water's heat of vaporization is $L_v = 2.26$ MJ/kg at 100°C. It's slightly larger, about 2.4 MJ/kg, at body temperature, 37°C. Suppose while exercising you lose 100 g of water by evaporating sweat. That requires energy:

$$Q = mL_v = (0.10 \text{ kg})(2.4 \times 10^6 \text{ J/kg}) = 2.4 \times 10^5 \text{ J}$$

Using an average body's specific heat (Table 13.1) of $c = 3.5$ kJ/(kg·°C), and with $Q = mc\Delta T = -2.4 \times 10^5$ J, a 70-kg person's temperature could change by

$$\Delta T = \frac{Q}{mc} = \frac{-2.4 \times 10^5 \text{ J}}{(70 \text{ kg})(3500 \text{ J}/(°C \cdot \text{kg}))} = -1.0°C$$

This estimate ignores sweat dripping off your body or absorbed in your clothes. It also neglects other heat-transfer mechanisms, which we'll discuss in the next section.

Dogs don't sweat, so to cool by evaporation they stick out their wet tongues. They also pant, rapidly exchanging hot air in their lungs for cool fresh air. That helps cool the blood in major vessels going to the head, and it enhances evaporation as air rushes over the wet tongue.

Why does water evaporate at temperatures below boiling? Because faster-moving molecules in the liquid escape into the air, and as long as the air isn't *saturated* with water vapor—that is, as long as the humidity is below 100%—there are more molecules leaving the liquid than returning. Hence, there's a net loss of liquid to the gas phase.

GOT IT? Section 13.3   For the same mass of water in its different phases, rank the energy needed to (a) raise the temperature of ice from −100°C to 0°C; (b) melt the ice at 0°C; (c) raise the temperature of water from 0°C to 100°C; (d) boil the water at 100°C.

13.4 Conduction, Convection, and Radiation

Heat, you've seen, is energy in transit because of a temperature difference. Here we introduce the three principal heat-transfer mechanisms: conduction, convection, and radiation.

 TIP

Conduction, convection, and radiation can occur singly or in combination.

Conduction

Conduction occurs when substances are in direct contact. Microscopically, conduction results from collisions that transfer energy from faster-moving particles in the hotter substance to slower-moving particles in the cooler substance. Put a frying pan on a hot stove burner, and conduction transfers heat to the bottom of the pan, then through the pan and into the food. Warm your home's interior on a winter day, and conduction transfers heat through the wall to the outside—resulting in expensive heating bills. Figure 13.17 emphasizes that conduction occurs whenever there's a temperature difference in matter.

Different materials conduct heat at different rates. Try paddling an aluminum canoe and a Kevlar one; the aluminum boat feels a lot cooler. Different conduction rates originate at the molecular and atomic level, where molecules and atoms conduct heat with greater or lesser ease depending on how they're bound to their neighbors. Free electrons, which move rapidly, greatly enhance thermal conductivity. That's why metals that are good

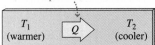

FIGURE 13.17 Heat conduction caused by a difference in temperature.

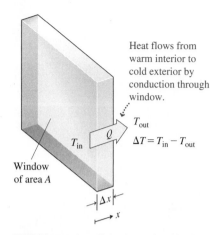

Heat flows from warm interior to cold exterior by conduction through window.

T_{out}

$\Delta T = T_{in} - T_{out}$

T_{in}

Q

Window of area A

Δx

x

FIGURE 13.18 Heat flow through a window, used to illustrate thermal conductivity.

TABLE 13.4 Thermal Conductivities

Substance	Thermal conductivity k $(W/(°C \cdot m))$
Metals	
Aluminum	240
Copper	390
Iron	52
Silver	420
Liquid	
Water	0.57
Gases	
Air	0.026
Hydrogen	0.17
Nitrogen	0.026
Oxygen	0.026
Other substances	
Brick	0.70
Concrete	1.28
Fiberglass	0.042
Glass (common)	0.80
Goose down	0.043
Human body (average)	0.20
Ice	2.2
Styrofoam	0.024
Wood (pine)	0.12

electrical conductors also tend to be good thermal conductors. (You'll study electrical conductivity in Chapter 17.) That aluminum canoe feels cooler because aluminum conducts heat away from your body more rapidly.

A common example—conduction through window glass—provides a model for quantitative analysis (Figure 13.18). You might guess that the heat-flow rate is proportional to the window area A, because a larger window means greater heat loss. The rate is also proportional to the temperature difference ΔT, because your home loses more heat on a colder day or when you crank up the thermostat. Finally, the heat-flow rate is inversely proportional to the window thickness Δx, because a thinner pane facilitates conduction. Putting together these results gives the heat-flow rate H:

$$H = kA\frac{\Delta T}{\Delta x} \quad \text{(Heat conduction: SI unit: W)} \quad (13.7)$$

Like other quantities involving energy rates, H is measured in joules per second, or watts. Here k is the **thermal conductivity**, a property of the conducting material (glass, in this case). Better thermal conductors have higher k values (Table 13.4).

EXAMPLE 13.12 **Heating Bills!**

A house has wooden siding 1.0 cm thick, with total surface area of 275 m^2. Suppose it's 19°C inside and 1°C outside. (a) What's the rate of energy loss through the walls? (b) What's the daily heating cost with energy at $0.10 per kWh?

ORGANIZE AND PLAN Equation 13.7 gives the heat-flow rate. The energy lost in a day is the rate times the time, 24 h = 86,400 s. With 3600 s in 1 h, 1 kWh is 3.6×10^6 J.

Known: $A = 275$ m^2, $\Delta x = 1.0$ cm, $\Delta T = 18$°C. Table 13.4 gives the thermal conductivity for pine as 0.12 W/(°C·m).

SOLVE (a) Using the parameters given, the rate of energy loss is

$$H = kA\frac{\Delta T}{\Delta x} = (0.12 \text{ W/(°C·m)})(275 \text{ m}^2)\frac{18°C}{0.01 \text{ m}} = 59.4 \text{ kW}$$

(b) The total energy lost in 1 day is then

$$Q = Ht = (5.94 \times 10^4 \text{ J/s})(86,400 \text{ s}) = 5.13 \times 10^9 \text{ J} = 5.13 \text{ GJ}$$

At $0.10 per kWh, this energy costs

$$5.13 \times 10^9 \text{ J} \times \frac{1 \text{ kWh}}{3.6 \times 10^6 \text{ J}} \times \frac{\$0.10}{\text{kWh}} = \$142.50$$

REFLECT That can't be right! But it is: That's what it *would* cost if only 1.0 cm of wood stood between you and the outside air. Fortunately, homes are much better insulated than that. There's plaster or drywall, then fiberglass or foam insulation, then the wood siding. That insulation is crucial if you're to avoid astronomical energy bills. By the way, just because we priced energy in kWh doesn't mean it's necessarily electricity. That 5 GJ could equally well come from about a barrel of oil (check the current price!), 400 pounds of coal, or 5000 cubic feet of natural gas. Or it could come from direct sunlight shining on 150 m^2 for 10 hours.

MAKING THE CONNECTION What thickness of fiberglass would provide the same insulating effect as the wood in the example?

ANSWER Table 13.4 lists fiberglass's thermal conductivity as lower than wood's by a factor of about 3, so you'd need only about 3 mm of fiberglass. Of course, you'd use a lot more; in the northern United States, 14 cm of fiberglass insulation is standard.

Convection

Convection is heat transfer through the bulk motion of a fluid. In mechanical systems, convection is often *forced* using a fan or pump to expedite heat transfer. Many home-heating furnaces use forced-air convection, with a fan blowing heated air through ducts into living areas. Your car's cooling system circulates coolant through the engine, where it absorbs heat, and then through the radiator, where it's transferred to the air. Animals, including humans, use forced convection in respiration. Air you inhale is usually cooler than air you exhale; the result is a net heat transfer from your body. Over the course of a day you can lose a lot of energy this way, particularly outdoors in cold weather.

Convection also occurs naturally, as a warmed fluid becomes less dense and rises. That's why the upper floors of your house tend to be warmer. Outdoors, sunlight warms the ground, and the warm air rises in *convection currents* (also, *thermals*) that can carry birds and hang gliders to high altitudes.

Convection also occurs in liquids. Put a pot of water on the stove, and the water at the bottom is warmed through conduction from the pan. This warmer liquid is less dense and rises, and it's replaced by cooler, denser liquid from above (Figure 13.19a). Striking geometric patterns, called convection cells, often appear in a heated liquid (Figure 13.19b). Convection cells in the oceans help to generate surface currents, and they contribute to the seasonal turnover of water in lakes. Convection in Earth's liquid core helps create our planet's magnetism (more on magnetism in Chapter 18).

Radiation

Radiation is energy transfer by electromagnetic waves. Radiation carries energy from the hot Sun to the cooler Earth to power essentially all life. It cooks your marshmallows over a campfire, and it keeps Earth from overheating by returning to space the energy gained from the Sun.

Turn an electric stove burner to "high" and you'll soon see it glowing red-orange. That shows it's emitting electromagnetic radiation in the form of visible light. Even with the burner on "low," you can still feel the radiated heat. That's invisible infrared radiation, like visible light, but with longer wavelengths. You'll study electromagnetic waves in Chapter 20, where you'll see that there's a wide spectrum of such waves, distinguished by wavelength; it runs from radio waves through infrared and visible light to ultraviolet, x rays, and gamma rays. Hot objects emit a range of wavelengths, with the predominant wavelengths depending on temperature: The higher the temperature, the shorter the predominant wavelengths. The Sun, at 5800 K, emits about half visible light and half infrared, along with a small amount of ultraviolet. Your stove burner is cooler than the Sun, and it emits a greater portion of its radiation as infrared. Your own body emits infrared of longer wavelengths. The ear thermometer in Chapter 12 measures infrared radiation to deduce temperature. Some stars are so hot they emit predominantly ultraviolet or even x rays, and the entire universe—at an average temperature around 2.7 K—emits predominantly radio waves.

Radiation can occur simultaneously with conduction or convection. Stand before a roaring fire and you see visible radiation from the glowing coals and feel the infrared radiation. The fire also causes convection in the air, which can help warm the room, although a lot of convective heat is lost up the chimney. There's also conduction through the air, although air's low thermal conductivity limits this effect.

You'll stay cooler in summer wearing light-colored clothing than you will in dark colors. That's because lighter colors reflect much of the incident radiation, while dark objects absorb most of it. A material that's a good absorber of radiation at some wavelength is necessarily also a good emitter at that wavelength. A perfect absorber/emitter—one that absorbs all radiation at all wavelengths—is called a **blackbody**, because it would appear completely black.

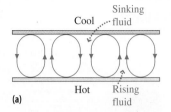

FIGURE 13.19 (a) Convection between two plates at different temperatures. (b) Top view of convection cells in a laboratory experiment. Fluid rises at the center and sinks at the edges of the cells.

APPLICATION

Earth's Climate

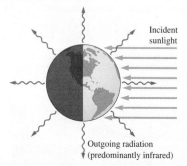

Incident sunlight

Outgoing radiation (predominantly infrared)

Solar energy reaching Earth's surface averages about 240 W for every square meter. The warmed Earth in turn emits infrared radiation according to Equation 13.8. A constant average temperature requires the rate at which infrared radiation escapes to equal the rate of solar energy input. Setting $e\sigma T^4$ equal to 240 W/m² then provides a rough estimate of Earth's surface temperature (see Problem 99). This simple picture is complicated by convective energy flows and by the *greenhouse effect*—the absorption of outgoing infrared by carbon dioxide and other gases. By burning fossil fuels and thus adding more carbon dioxide, we're increasing that infrared absorption, requiring a higher surface temperature T in the Stefan-Boltzmann law. Equation 13.8 is thus at the heart of our century's greatest environmental challenge: global climate change.

The **Stefan-Boltzmann law** gives the rate at which an object at temperature T radiates energy:

$$P = e\sigma AT^4 \qquad \text{(Stefan-Boltzmann law; SI unit: W)} \qquad (13.8)$$

Here A is the object's surface area, and e its **emissivity**—a measure of how good an absorber/emitter it is. For a perfect blackbody $e = 1$, and $e = 0$ for a perfectly reflective surface. The quantity σ is the **Stefan-Boltzmann constant**, with numerical value $\sigma = 5.67 \times 10^{-8}$ W/(m²·K⁴). The temperature in Equation 13.8 must be in kelvins, and thus the equation shows that any object above absolute zero emits radiation. The object also receives radiation from its environment, giving a net radiated power $P_{net} = e\sigma A(T^4 - T_e^4)$. This assumes that the object's entire surroundings are at the environmental temperature T_e, so the object has radiation incident on its entire surface.

One application of radiation you may have seen is a **thermogram**—an image mapping an object's surface temperature, as measured by the infrared radiation it emits. Physicians use thermograms to detect tumors noninvasively; this works because a concentration of cancer cells tends to be warmer than surrounding tissue. Energy conservation specialists use thermograms to pinpoint heat losses from buildings, and infrared imaging from satellites helps assess Earth's ecological health.

EXAMPLE 13.13   **Human Radiation**

Your body has a surface area around 1.0 m² and emissivity 0.75. If the ambient temperature is 20°C, find (a) your body's net radiated power and (b) the total energy lost in 1 day.

ORGANIZE AND PLAN The net power radiated is $P = e\sigma A(T^4 - T_e^4)$. Then the total energy lost in a day is the power times the time, 24 h = 86,400 s.

Known: $A = 1.0$ m², $e = 0.75$. Body temperature $T = 37°C = 310$ K, and environmental temperature $T_e = 20°C = 293$ K. The value of the Stefan-Boltzmann constant is $\sigma = 5.67 \times 10^{-8}$ W/(m²·K⁴).

SOLVE (a) Using the parameters given, the net radiated power is

$$P_{net} = e\sigma A(T^4 - T_e^4)$$
$$= (0.75)(5.67 \times 10^{-8}\text{ W/(m}^2\cdot\text{K}^4))(1.0\text{ m}^2)((310\text{ K})^4 - (293\text{ K})^4)$$
$$= 79\text{ W}$$

(b) The total energy loss in 1 day is then

$$Q = Pt = (79\text{ J/s})(86,400\text{ s}) = 6.8 \times 10^6\text{ J}$$

REFLECT Is that reasonable? You consume about 2000 food calories per day; that's about 8.4×10^6 J. So our answer is the right order of magnitude, but it doesn't give you much energy to spare. And there are also conductive and convective losses. Clothing reduces all these losses, leaving you plenty of energy for muscle and brain activity.

MAKING THE CONNECTION   How does the answer for net radiated power change on a hot day at 30°C?

ANSWER The radiated power drops to just 34 W. The exponent in T^4 makes a huge difference here.

CONCEPTUAL EXAMPLE 13.14 **Thermos Bottle**

How does a thermos bottle limit heat transfer by conduction, convection, and radiation?

SOLVE Convection is easiest to understand. The screw top (Figure 13.20) makes a tight seal, preventing convective loss of hot vapors. To limit conduction, there's a good vacuum between the inner bottle and outer surface; that makes thermal conductivity nearly zero. Finally, the bottle's walls are highly reflective, giving them very low emissivity. This limits heat transfer by radiation.

REFLECT Scientists use *Dewar flasks*—essentially large thermos bottles—to store liquid nitrogen, helium, and other substances that are gases at normal temperatures. Nitrogen boils at 77 K, helium at 4.2 K. High-quality Dewars store these liquids for long periods with only minor loss due to boiling. Liquid nitrogen and liquid helium are used by scientists to cool materials, so their low-temperature properties can be studied.

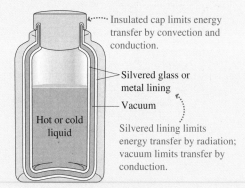

Insulated cap limits energy transfer by convection and conduction.

Silvered glass or metal lining

Vacuum

Hot or cold liquid

Silvered lining limits energy transfer by radiation; vacuum limits transfer by conduction.

FIGURE 13.20 A thermos bottle.

GOT IT? Section 13.4 For the same surface area and thickness, rank in increasing order the rate of heat flow through the following: (a) air; (b) wood; (c) glass; (d) water.

Chapter 13 in Context

This chapter began by defining *heat* as a temperature-driven energy flow, distinct from temperature and thermal energy introduced in Chapter 12. We explored *heat capacity* and *specific heat*, which determine temperature changes resulting from heat flow. We then showed how the *equipartition theorem* predicts specific heats of some gases and solids. Heat can also cause *phase changes*, and we explored energies associated with such changes and saw regions of different *phase diagrams* that relate temperature, pressure, and phase. Finally, we discussed heat transfer by *conduction, convection*, and *radiation*.

Looking Ahead In Chapter 14 you'll learn the first and second laws of thermodynamics, building on concepts of temperature and heat from Chapters 12 and 13. The first law extends conservation of energy to include heat as well as work. The second law limits the ability to convert thermal energy into mechanical work and establishes a direction for thermal processes—an "arrow of time." Later, in Chapter 24, we'll study further the blackbody radiation introduced here in Section 13.4.

CHAPTER 13 SUMMARY

Heat and Thermal Energy

(Section 13.1) **Heat** is energy in transit as a result of a temperature difference. It's measured in joules, although commonly used alternatives are calories and food calories.

Calories and joules: $1 \text{ cal} = 4.186 \text{ J}$

Food calories: $1 \text{ food calorie} = 1 \text{ Cal} = 1 \text{ kcal} = 1000 \text{ cal} = 4186 \text{ J}$

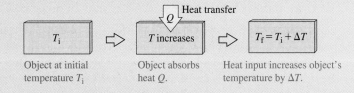

Heat transfer

Object at initial temperature T_i ⟹ Object absorbs heat Q. T increases ⟹ Heat input increases object's temperature by ΔT. $T_f = T_i + \Delta T$

Heat Capacity and Specific Heat

(Section 13.2) Heat flows between two objects in thermal contact until their temperatures are equal; then they're in **thermal equilibrium**.

Heat capacity relates heat and temperature change. **Specific heat** is heat capacity per unit mass. Specific heats of gases are measured at constant volume or constant pressure.

The **equipartition theorem** predicts specific heats of some gases and solids.

Heat capacity: $Q = C\Delta T$

Specific heat: $Q = mc\Delta T$

Molar specific heat (gases, constant volume): $Q = nc_V\Delta T$

Molar specific heat (gases, constant pressure): $Q = nc_P\Delta T$

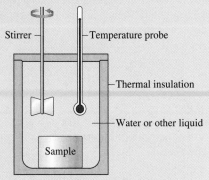

A colorimeter, used to measure specific heats

Phase Changes

(Section 13.3) **Heats of transformation** describe the energy per unit mass needed for phase changes: the **heat of fusion** for melting and the **heat of vaporization** for vaporizing.

Heat of fusion: $Q = mL_f$

Heat of vaporization: $Q = mL_v$

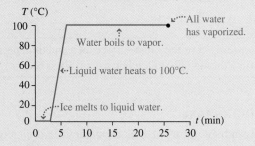

Conduction, Convection, and Radiation

(Section 13.4) **Conduction** is transfer of heat by direct contact and involves collisions among molecules, atoms, and electrons. **Thermal conductivity** quantifies a material's heat conduction capability. **Convection** is the bulk motion of fluid, carrying thermal energy. **Radiation** is energy transfer by electromagnetic waves. Any object above absolute zero radiates power given by the **Stefan-Boltzmann law**.

Thermal conductivity: $\dfrac{Q}{t} = kA\dfrac{\Delta T}{\Delta x}$

Stefan-Boltzmann law: $P = e\sigma AT^4$

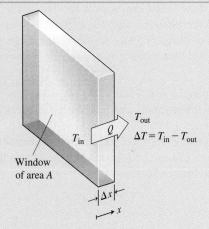

NOTE: Problem difficulty is labeled as ■ straightforward to ■■■ challenging. Problems labeled BIO are of biological or medical interest.

Conceptual Questions

1. Distinguish heat from temperature.
2. How is heat related to temperature?
3. Why is it incorrect to say that a substance contains some amount of heat?
4. Distinguish heat capacity and specific heat.
5. If equal amounts of heat are supplied to equal masses of aluminum and iron, which experiences the greatest temperature increase?
6. Explain what is meant by the "mechanical equivalent of heat."
7. On a hot day in early summer a lake may be cooler than on a cool day in late summer. Why?
8. Is specific heat ever negative?
9. You have two objects of different mass made from the same material. If they're initially at different temperatures, which one's temperature changes more when they're in thermal contact?
10. Why does an igloo, made from thick blocks of snow, offer good protection in an Arctic climate?
11. On rare occasions when subfreezing temperatures are predicted, Florida citrus farmers sometimes spray water on the fruit. How does this help protect the fruit?
12. A 20°C day is mild enough for you to wear light clothing, but a plunge into 20°C water feels very cold. Why the difference?
13. If you want to increase a gas's temperature quickly, should you apply heat at constant volume or constant pressure?
14. Why do you feel colder on a windy day than on a calm day at the same temperature?
15. Discuss the ways a long-distance runner maintains a constant body temperature.
16. Suppose you start with water at 1 atm and steadily increase the pressure. What happens to the melting and boiling points? Would these changes continue indefinitely?
17. Why is a double-paned window better than one with a single pane? And why is it better if the air gap between the two panes is small?
18. Snow is made of ice crystals. However, the thermal conductivity of snow is smaller than that of ice. Why?
19. In winter, why does your house lose less energy if you keep the curtains closed?
20. Critique the statement "This pot of hot water contains a lot of heat." What's a better way to put it?
21. To measure your body temperature, you place the bulb of a thermometer under your tongue. How can the thermometer read accurately when most of it is sticking out of your mouth?

Multiple-Choice Problems

22. A candy bar containing 200 food calories has, equivalently, (a) 837 J; (b) 200 kJ; (c) 418 kJ; (d) 837 kJ.
23. How many food calories would be required just to change the gravitational potential energy of a 60-kg hiker climbing a 1200-m mountain? (a) 169 Cal; (b) 215 Cal; (c) 276 Cal; (d) 313 Cal.
24. What energy is required to raise the temperature of 450 g of water by 6°C? (a) 4200 J; (b) 5200 kJ; (c) 6600 kJ; (d) 11,300 J.
25. 100 cal of heat will raise the temperature of 25 g of water by (a) 2500°C; (b) 4°C; (c) 2°C; (d) 1°C.
26. 1 kJ of heat will raise the temperature of a 1.0-kg piece of aluminum by (a) 0.9°C; (b) 1.1°C; (c) 1.4°C; (d) 1.7°C.

27. If you mix 2.0 kg of water at 45°C with 1.2 kg of water at 10°C, the final temperature is (a) 32°C; (b) 34°C; (c) 36°C; (d) 38°C.
28. How much heat is required to melt a 200-g ice cube at 0°C? (a) 33.3 kJ; (b) 66.6 kJ; (c) 99.9 kJ; (d) 66,600 kJ.
29. If 6.5 kJ completely melts a block of ice, the energy required to completely vaporize the equivalent water at 100°C is (a) 6.5 kJ; (b) 13 kJ; (c) 18 kJ; (d) 44 kJ.
30. If you drop a 30-g ice cube at 0°C into 320 g of tea (essentially water) at 75°C, what's the temperature after thermal equilibrium is reached? (a) 50°C; (b) 54°C; (c) 58°C; (d) 62°C.
31. A star with temperature 7200 K has radius 1.62×10^9 m. Treating the star as a blackbody, at what rate does it radiate energy? (a) 5×10^{27} W; (b) 8×10^{27} W; (c) 2×10^{28} W; (d) 6×10^{28} W.

Problems

Section 13.1 Heat and Thermal Energy

32. ■ Find the daily energy intake in joules for (a) an individual on a 1500-Cal/day weight-loss diet; (b) an athletic individual who consumes 2600 Cal/day; (c) a Tour de France competitor consuming 6000 Cal/day.
33. ■ How many joules are in a 280-Cal candy bar?
34. ■ A 1120-kg car is going 60 mph. Find the thermal energy created in the car's brakes when stopping the car.
35. ■ Repeat the previous problem for a 34,000-kg semi-trailer truck with the same speed.
36. ■■ How many food calories are converted to gravitational potential energy when a 70-kg hiker climbs a 2200-m mountain?
37. ■■ Suppose a 65-kg person can run 1000 m in 5 min. Assume that each stride is 1.5 m long, and with each stride the runner must supply an amount of energy equal to her kinetic energy. How much energy (in food calories) is required to run 1 km? Is your answer realistic?
38. ■■ Repeat the previous problem for the same person, walking 1000 m in 12 min with an 85-cm stride.
39. ■■ You're burning calories by weight lifting. If you lift a 75-kg barbell 1.9 m, and repeat 20 times, how much energy (in J and Cal) do you expend?

Section 13.2 Heat Capacity and Specific Heat

40. ■ The British thermal unit (Btu) is the energy required to raise the temperature of 1 pound of water by 1°F. What's 1 Btu in (a) joules; (b) calories?
41. ■ Natural gas is often sold in *therms*, where 1 therm = 10^5 Btu (see preceding problem). If a home uses 92 therms in a given month, what's the corresponding average power use in watts?
42. ■■ Use the information in the preceding two problems to rank the following from smallest to largest: (a) 1 therm; (b) 1 cal; (c) 1 J; (d) 1 Btu.
43. ■ A piece of metal absorbs 2.48 kJ of heat, increasing its temperature by 25°C. (a) What's its heat capacity? (b) How much heat is required to increase its temperature by 200°C?
44. ■ A 0.450-kg rock is dropped 10.0 m into 2.5 kg of water. If all the rock's kinetic energy is converted to thermal energy in the water, what's the water's temperature increase?
45. ■■ If you mix 18 kg of water at 25°C with 6 kg of water at 2.0°C, what's the final temperature?

46. ■■ An electric water heater holds 189 L. If electricity costs $0.12 per kWh, how much does it cost to raise the water temperature from 10°C to 60°C?

47. ■■ You have 300 g of coffee at 55°C. Coffee has the same specific heat as water. How much 10°C water do you need to add in order to reduce the coffee's temperature to a more bearable 49°C?

48. ■■ Consider Joule's paddle wheel apparatus described in Section 13.1. Joule claimed that a weight of 817 pounds falling 1 foot was the energy equivalent needed to raise the temperature of 1 pound of water by 1°F. (a) What's the percentage difference between those two supposedly equivalent values? (b) What should have been the weight, in pounds?

49. ■■ Consider again Joule's apparatus from the preceding problem. If you wanted to state similar results in SI units, what mass of water falling 1 m would produce the energy needed to raise the temperature of 1 kg of water by 1 K?

50. ■■ A 0.35-kg piece of copper at a temperature of 150°C is dumped into 500 mL of water at a temperature of 25°C. (a) What's the final equilibrium temperature? (b) How much heat flowed in this process?

51. ■■ The electric heater in a tea kettle delivers 1250 W to the water. If the kettle contains 1.0 L of water initially at room temperature (20°C), what's the time until the water begins boiling?

52. ■■ Capacities of industrial refrigerators and air conditioning units are often measured in *tons*, where a 1-ton unit can, in 1 day, provide the same refrigeration as melting 1 ton of ice at its freezing point. Express in watts the rate at which a 25-ton air conditioning system can remove energy from a building.

53. ■■ You drop 25.0 g of unknown material at 34.5°C into a calorimeter containing 125 g of 18°C water and measure the equilibrium temperature to be 22.9°C. What's the specific heat of the unknown material?

54. ■■ A 210-g sample of unknown, silvery metal at 115.4°C is dropped into a calorimeter containing 250 g of 15.5°C water. If the equilibrium temperature is 20.0°C, which material in Table 13.1 is the unknown likely to be?

55. ■■ A mercury thermometer contains 2.30 mL of liquid mercury at 0°C. How much heat must it absorb in order to reach 100°C?

56. ■■■ The specific heat of copper is close to the $3R$ per mole predicted by the equipartition theorem. Given that one mole of copper has mass 63.6 g, convert this molar specific heat into the more familiar J/(kg · °C), and compare with the value in Table 13.1.

57. ■■ If 56 g of nitrogen gas at 25°C is combined with 12 g of helium gas at 45°C, what's the mixture's equilibrium temperature?

58. ■■ You measure a heat capacity of 37.5 J/°C at constant volume for a sample of monatomic gas. (a) How many moles are in the sample? (b) What's the heat capacity of this sample at constant pressure?

59. ■■■ A house has floor area of 190 m² and 2.3-m ceilings. Assume air contains 79% nitrogen and 21% oxygen. (a) What energy is required to raise the air temperature throughout the house by 1°C? (b) At 16 ¢/kWh, what's the cost to heat this air by 1°C? (Your heating system isn't 100% efficient, so your answer underestimates the cost.)

Section 13.3 Phase Changes

60. ■ How much energy is required to melt a 120-g ice cube at 0°C?

61. ■■ How much energy is required to melt a 120-g ice cube initially at −25°C?

62. ■ You have a 1.0-kg lead block at room temperature. Find the energy required (a) to raise the lead's temperature to its melting point and (b) to melt the lead once it's reached the melting point.

63. ■■ How much energy must be removed from a 15.0-g sample of liquid copper initially at its melting point in order to turn it into a solid at 600°C?

64. ■■ On a −15°C morning, you find ice 1.5 mm thick on your car's rear window. If the window is 1.4 m wide and 0.65 m high, at what rate must the window heater supply energy to melt all the ice in 5 min?

65. ■■ How long will it take a 120-kW industrial furnace to melt a 52-kg copper ingot initially at 20°C?

66. ■■ You have 325 g of water at 12°C. What is the most ice (at 0°C) that you could put into this water and have it all melt?

67. ■■ You place a tray containing 410 g of water at 11°C in your freezer. How much energy must be removed to turn the water to ice at −14°C?

68. ■■ Liquid helium, which boils at 4.2 K, is used to cool materials to very low temperatures. (a) How much energy does 250 g of liquid helium absorb in boiling completely? (b) If that amount of energy is then added to the gas, what's its final temperature?

69. ■■ You start with 0.50 kg of ice at −10°C and begin heating it at the constant rate of 1000 W. Plot temperature versus time until the ice has all become water vapor.

70. ■■ It takes 9.53 kJ to melt some ice at 0°C. (a) How much heat is required to boil the same mass of water at 100°C? (b) What's that mass?

71. ■■ 150 g of water in a calorimeter is initially at 30°C. You add some steam at 100°C, which condenses to water. After reaching thermal equilibrium, the water temperature is 42°C. How much steam was added?

72. ■■■ A 4.5-cm-thick ice sheet covers a lake in early spring. The ice density is 910 kg/m³. If the spring sun radiates an average of 150 W/m² onto the ice, how much time will it take to melt all the ice?

73. BIO ■■ **Cooling by evaporation.** How much sweat must evaporate from a 90-kg athlete's body to cool the body by an average of 1°C?

74. BIO ■■ **Cooling by liquid intake.** How much 2°C water must a 62-kg person drink to lower her body temperature by 1°C?

Section 13.4 Conduction, Convection, and Radiation

75. ■■ An iron cylinder with radius 1.0 cm and length 25 cm has one end held at 250°C and the other at 20°C. Find the heat-flow rate through the cylinder.

76. ■■■ A window with area 1.7 m² is made from a single pane of 3.2-mm-thick glass. (a) If it's 15°C colder outdoors than inside, what's the heat-flow rate through the window? (b) Repeat for a double-pane window made from the same glass, with a 1.0-mm air gap between the panes.

77. ■■ Rework Example 13.12 to find the daily energy cost if the walls also include a 6.0-cm-thick layer of Styrofoam insulation.

The following four problems involve "R-value" (R for resistance), a measure of insulating quality used in the construction and home heating businesses. It's defined as $R = \Delta x/k$, where k is the thermal conductivity and Δx the thickness of a material.

78. ■■ If you want to save energy and money, should you opt for a larger or a smaller R? Explain.

79. ■■ Calculate and compare the R-values of 3.2-mm thick (a) glass, (b) wood, and (c) Styrofoam.

80. ■■■ Show that for two different materials sandwiched together, the net R-value is the sum of the individual R-values.

81. ■■■ Using the results of the preceding two problems, compute the R-value of a double-pane window with 3.2-mm-thick glass panes and a 2.0-mm air gap. Compare with standard R-19 wall construction.

82. ■■ At what rate does the Sun radiate energy, given that it's essentially a spherical blackbody with radius 6.96×10^8 m and surface temperature 5800 K?

83. ■■■ (a) Use the result of the preceding problem, along with appropriate astronomical data, to find the solar power per unit area at Earth's orbit. (b) If an orbiting solar power station uses 20% efficient solar panels to capture this energy, how large an area would be required to replace a 1.0-GW power plant?

84. ■■ An aluminum cylinder with radius 2.0 cm and length 12 cm is heated to 450°C. Find the net power this cylinder radiates into a room at 25°C.

General Problems

85. ■■ Rework Example 13.10, this time with 400 g of ice dumped into the coffee.

86. ■■ Building heat loss in the United States is usually expressed in Btu/h. What is 1 Btu/h in SI?

87. ■■ You drop 250 g of ice at 0°C into 250 g of water at 25°C. (a) Once equilibrium is reached, do you have all water or an ice-water mixture? (b) If the former, find the temperature; if the latter, find the masses of water and ice.

88. **BIO** ■■ **Food intake and temperature.** Suppose your daily food intake amounts to 1800 Cal. If all this energy were released through metabolism, and if your body had no way to lose energy to your surroundings, by how much would your temperature increase in a day? Assume your mass is 65 kg.

89. ■■ (a) How much heat does it take to bring a 3.4-kg iron skillet from 20°C to 130°C? (b) If a stove burner supplies the heat at a rate of 2.0 kW, how much time does it take to heat the pan?

90. ■■ You bring a 1.25-kg wrench into the house from your car. The house is at 22°C, your car at 7.0°C. The wrench absorbs 9.0 kJ in warming to room temperature. Find (a) its heat capacity, (b) its specific heat, and (c) the metal it's likely made of.

91. ■■ In a nuclear accident, a reactor shuts down but its **2.5 $\times 10^5$-kg** uranium core continues to produce energy at the rate of 120 MW due to radioactive decay. Once it reaches the melting point, how long will it take the core to melt?

92. ■■ Find the heat-loss rate through 1.0-m² slabs of (a) wood and (b) Styrofoam, each 2.0 cm thick, with a 30°C temperature difference between their faces.

93. ■■■ The top of a steel wood stove measures 90 cm by 40 cm and is 0.45 cm thick. The fire maintains the inside of the stovetop at 310°C, while the outside surface is 295°C. (a) Find the rate of heat conduction through the stovetop. (b) Suppose that the entire stove heats the air in the room at three times the rate you found in part (a). The room measures 8.6 m by 6.5 m by 2.8 m. Assuming the room is at $T = 20$°C and $P = 1$ atm, at what rate does the room's temperature increase?

94. ■■ How thick a concrete wall would be needed to give the same insulating value as 1.8 cm of wood?

95. ■■ An 8.0 m by 12 m house is built on a concrete slab 23 cm thick. What is the heat-loss rate through the floor if the interior is at 18°C while the ground is at 10°C?

96. ■■ A horseshoe has a surface area of 50 cm², and a blacksmith heats it to a red-hot 810°C. At what rate does it radiate energy?

97. ■■■ A 1500-kg car traveling at 32 m/s comes to a sudden stop. If all of the car's energy is dissipated in its four 5.0-kg steel-disk brakes, what's the temperature rise of the brakes?

98. ■■ What's the power of a microwave oven that can heat 430 g of water from 15°C to boiling in 5.0 min?

99. ■■ Follow the suggestion in the application "Earth's Climate" to estimate Earth's average surface temperature. Take $e = 1$ for

Earth's infrared emissivity. Your result is low because it neglects the greenhouse effect.

100. **BIO** ■■■ **Warm breaths.** Suppose you take 10 breaths per minute, each containing 4.0 L of 20°C air with density 1.29 kg/m³. How much energy do you expend each day warming all this air to your 37°C body temperature? Answer in joules and food calories (Cal).

101. ■■ A circular lake 1.0 km in diameter is 10 m deep. Solar energy is incident on the lake at an average rate of 200 W/m². If the lake absorbs all this energy and doesn't exchange energy with its surroundings, how much time is required to raise the temperature from 5°C to 20°C?

102. ■■ Two neighbors return from a tropical vacation to find their houses at a frigid 2°C. Each house has a furnace that outputs 10^5 Btu/h. One house is made of steel and has mass 75,000 kg, the other of wood with mass 15,000 kg. Neglecting heat loss, find the time required to bring each house to 18°C.

103. **BIO** ■■ **Keeping warm!** The average human body produces heat at the rate of 100 W and has surface area of 1.5 m². What is the coldest outdoor temperature in which a sleeping bag insulated with 4.0-cm-thick goose down can be used without the body's temperature dropping below 37°C?

104. **BIO** ■■ **Keeping cool!** A 65-kg runner needs to dissipate excess energy at the rate of 400 W. (a) At what rate would the runner's temperature increase if this energy weren't dissipated? (b) Suppose the energy is dissipated by evaporating sweat, with $L_f = 2.4 \times 10^6$ J/kg at body temperature. What mass of sweat is lost each minute? (c) How often do you need to drink 500 mL of water to replace that liquid?

105. ■■ You're the public affairs director for an electric utility. A nuclear plant has been refueled and needs to power up again. The news media want to know how much time will pass before the reactor is on line again. The reactor needs to heat 5.4×10^6 kg of water from 10°C to 350°C. (It's a pressurized-water reactor, so the water remains liquid.) How much time will this take, if the reactor's thermal power output is 1.42 GW? Ignore the heat capacity of the reactor vessel and plumbing.

106. ■■■ When a nuclear power plant's reactor is shut down, radioactive decay continues to produce heat at 10% the rate of the reactor's normal 3.0-GW output. In a major accident, a pipe breaks and all the cooling water is lost. The reactor is immediately shut down, the break sealed, and 420 m³ of 20°C water injected into the reactor. If the water were not actively cooled, how much time would it take (a) to reach boiling and (b) to boil away completely?

107. ■■■ When the reactor in the preceding problem is operating normally, it produces thermal energy at the rate of 3.0 GW, of which 1.0 GW is converted to electrical energy. The remaining 2.0 GW is waste heat that must be dissipated using water from a nearby river. Suppose that the river averages 250 m wide by 3.0 m deep and flows at 1.5 m/s. What's the rise in the river's temperature due to the reactor?

Answers to Chapter Questions

Answer to Chapter-Opening Question

Water vapor, evaporated from the warm tropical ocean, condenses and gives up its latent heat, which in turn drives the hurricane's intense winds.

Answers to GOT IT? Questions

Section 13.2 (b) R

Section 13.3 (a) < (b) < (c) < (d)

Section 13.4 (a) Air < (b) wood < (d) water < (c) glass

14 The Laws of Thermodynamics

■ Power plants provide the world's electrical energy. But most of the energy released from their fuels is dumped into the environment as waste heat, here through huge cooling towers. Is this waste just the result of poor engineering?

To Learn

By the end of this chapter you should be able to

- Define internal energy.
- Describe how the first law of thermodynamics relates internal energy, heat, and work.
- Find the work done in different thermodynamic processes (constant-temperature, constant-pressure, constant-volume, and adiabatic).
- Apply the first law to human metabolism.
- Express the second law of thermodynamics in several different ways.
- Describe entropy and its relation to heat flows and the second law.
- Describe energy flows in a heat engine.
- Discuss the Carnot cycle and the limit it places on engine efficiencies.
- Describe energy flows in a refrigerator and define COP.
- Explain the relation between probability, disorder, and entropy.

The laws of thermodynamics govern all systems that rely on heat flows for energy. That includes planet Earth, whose energy source is largely solar radiation, and virtually all Earth's natural and technological subsystems: living things; weather, winds and moving water; and the engines that power our cars, airplanes, industries, and electric power plants.

The first law of thermodynamics is about energy conservation. It relates energy contained in a system to exchanges of work and heat with the system's surroundings. The second law describes the tendency to evolve from more ordered to less ordered states. Among its important consequences are that heat does not flow spontaneously from cooler to hotter objects; that it's impossible to build an engine that converts thermal energy to mechanical work with 100% efficiency; and that not all joules are created equal—meaning energy has quality as well as quantity. We'll introduce the concept of entropy as a way of measuring energy quality. After examining the implications of the second law and energy quality for our modern energy-intensive society, we'll end with a statistical look at the meaning of the second law and entropy.

14.1 The First Law of Thermodynamics

As you saw in Chapter 13, you can transfer energy to a gas by heating it (see Figure 13.8). Another way is to do work on the gas, as shown in Figure 14.1. The work you do ends up as thermal energy of the gas. Energy transfer by these two processes—heat and work—is what the first law of thermodynamics is all about. Before stating that law, we need to be precise in defining the energy contained in a system.

Internal Energy

We defined *thermal energy* in Chapter 12 as the kinetic and potential energy associated with individual molecules. Here we expand that concept, defining **internal energy** to include also any potential energy associated with interactions among molecules, such as the energy of bonds that get broken during phase changes. Physicists commonly use U for internal energy. That's the same symbol we've used for potential energy, but you needn't be confused because the two meanings won't occur in the same context.

 TIP

In this chapter, U is internal energy, not potential energy.

The First Law: Heat and Work

Heat a gas or compress it as in Figure 14.1, and the result is the same: The gas inside gets warmer, showing that its internal energy has increased. That's essentially the **first law of thermodynamics**, which states that the change ΔU in a system's internal energy is the sum of the heat Q transferred to it and the work W done on it:

$$\Delta U = Q + W \quad \text{(First law of thermodynamics; SI unit: J)} \tag{14.1}$$

The first law extends the familiar conservation-of-energy principle to include internal energy and heat. We'll see how this makes many physical processes and applications understandable in terms of energy conservation.

In most systems we'll consider, a change in internal energy is equivalent to a change in thermal energy. For example, in Chapter 12 you learned (Equation 12.7) that the thermal energy of a monatomic ideal gas containing N molecules is $E_{th} = \frac{3}{2}Nk_BT$. Ideal-gas molecules don't interact, and that makes a change in the gas's thermal energy equal to the change in its internal energy. Thus

$$\Delta U = \Delta E_{th} = \frac{3}{2}Nk_B\Delta T \quad \text{(Monotomic gas)}$$

This equation makes clear that a change ΔU in internal energy corresponds directly to a temperature change ΔT.

For a diatomic gas, you saw in Chapter 13 that the thermal energy is $E_{th} = \frac{5}{2}Nk_BT$— larger for a given temperature because diatomic molecules have rotational as well as translational energy. Therefore, the relationship between internal energy and temperature for a diatomic gas is

$$\Delta U = \Delta E_{th} = \frac{5}{2}Nk_B\Delta T \quad \text{(Diatomic gas)}$$

Compressing a Gas

The first law involves the work W done on a system. Details of that work depend on whether the system is solid, liquid, or gas. A gas readily changes volume; solids and liquids don't. Since work is essentially force times displacement (Section 5.1), the larger displacements possible for gases mean work is usually more significant for gases than for liquids and solids.

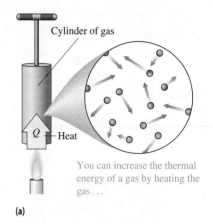

You can increase the thermal energy of a gas by heating the gas . . .

(a)

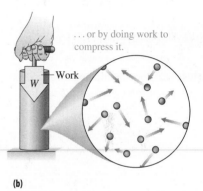

. . . or by doing work to compress it.

(b)

FIGURE 14.1 Adding energy to a gas as (a) heat and (b) work.

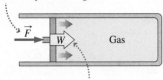

Piston compresses the gas.

Force and displacement are in the same direction, so the piston does positive work on the gas.

(a) Gas is compressed: work is positive

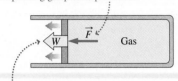

Expanding gas pushes piston out.

Because the gas does positive work on the piston, the work done *on the gas* is *negative*.

(b) Gas expands: work is negative

FIGURE 14.2 Positive and negative work on a gas in a piston-cylinder system.

The work done on a gas can be positive or negative, depending on whether volume decreases or increases. Figure 14.2 makes this point for a gas in a cylinder with a movable piston. It's important to associate the proper sign with a volume change:

Sign rule for work: When a gas is compressed, the work W done on it is positive. When a gas expands, the work W done on it is negative.

The question of how much work is done in compressing (or expanding) a gas isn't a simple one, and it depends on the process involved. Consider n moles of gas obeying the ideal-gas law $PV = nRT$. Change V, and gas pressure P and temperature T are both subject to change. You could hold one of these constant, or both might vary in the compression or expansion process. We'll consider these possibilities in the next section. In any case, the work done in compressing a gas depends on the process and follows from the ideal-gas law.

The first law alone provides some insight into what can happen when a gas is compressed or expanded. Solving the first law (Equation 14.1) for W gives $W = \Delta U - Q$. Compressing the gas means $W > 0$. By the first law, then either there must be an increase in internal energy ($\Delta U > 0$), or heat must flow out ($Q < 0$), or both. You can analyze other situations similarly.

✓TIP

Compression is only one kind of work. Any energy transfer that isn't heat—that isn't driven by a temperature difference—is included in the work W.

CONCEPTUAL EXAMPLE 14.1 **Heat without ΔT?**

You apply a flame to a gas. Use the first law of thermodynamics to describe how this *could* occur without changing the gas's temperature.

SOLVE The first law of thermodynamics says that $\Delta U = Q + W$. For an ideal gas, internal energy U is proportional to temperature. Therefore, $\Delta T = 0$ implies that $\Delta U = 0$, so for this process the first law reads $\Delta U = Q + W = 0$, or $W = -Q$.

That flame delivers heat, so $Q > 0$. Therefore, $W = -Q < 0$. The work done on a gas is negative when the gas expands. Thus, the only way for the gas's temperature to remain constant is for it to expand while being heated.

REFLECT The heating processes in Chapter 13 assumed a container of fixed size. In that case $W = 0$, and the first law says that $\Delta U = Q$. Heating a gas with fixed volume must result in a temperature increase.

EXAMPLE 14.2 **Big Balloon!**

The largest balloons carry multi-ton scientific payloads to the edge of space. As they rise, decreasing atmospheric pressure causes them to expand more than 200 times in volume. A balloon containing 350 kmol of helium does 270 MJ of work as it expands, and its temperature drops from 10°C at launch to −25°C at altitude. Find the heat that flows into or out of the balloon.

ORGANIZE AND PLAN The first law of thermodynamics, $\Delta U = Q + W$, relates heat, work, and internal energy. You're given the work done *by* the balloon: 270 MJ. Since W in the first law is the work done *on* a system, we take $W = -270$ MJ. For an ideal monatomic gas, we've just seen that $\Delta U = \frac{3}{2}Nk_B\Delta T$; in molar terms, that's $\Delta U = \frac{3}{2}nR\Delta T$.

Known: $n = 350$ kmol; $W = -270$ MJ; $\Delta T = -25°C - 10°C = -35°C$.

SOLVE We solve the first law for Q:

$$Q = \Delta U - W = \frac{3}{2}nR\Delta T - W$$

$$= \frac{3}{2}(350 \times 10^3 \text{ mol})(8.315 \text{ J/K} \cdot \text{mol})(-35 \text{ K})$$

$$- (-270 \times 10^6 \text{ J}) = 120 \text{ MJ}$$

cont'd.

REFLECT That's heat flowing *into* the gas ($Q > 0$). The gas did so much work that it needed this additional 120 MJ to keep the temperature from falling even lower. Note that we use °C and K interchangeably when we're dealing with a temperature *difference*.

MAKING THE CONNECTION How low would the balloon's temperature get if there were no heat flow?

ANSWER With $Q = 0$, the first law gives $\Delta U = W$. Solving for ΔT in that case gives $\Delta T = -62°C$, for a final temperature of $-52°C$.

Reviewing New Concepts: The First Law of Thermodynamics

- Internal energy U includes thermal energy—the random motions and potential energies within molecules—as well as energy associated with molecular interactions.
- The first law of thermodynamics states $\Delta U = Q + W$.
- Heat flows into the system when $Q > 0$ and out when $Q < 0$.
- $W > 0$ when volume decreases, and vice versa.

GOT IT? Section 14.1 For each of the following situations involving an ideal gas, tell whether the gas's temperature increases or decreases, or whether this cannot be determined from the information given. (a) No heat is transferred while the gas expands. (b) Heat is removed from the gas while it's compressed. (c) Heat is removed from the gas while it expands. (d) Heat is added to the gas while it's compressed. (e) Heat is added to the gas while it expands.

14.2 Thermodynamic Processes

You've just seen that the work involved in compressing or expanding a gas depends on temperature change and heat flow. Here we'll focus on specific processes and use the first law of thermodynamics to explore the relation between heat and work.

Constant-Pressure Processes

The simplest process to understand involves a gas at constant pressure, as shown in Figure 14.3 (such a process is also called **isobaric**). With constant pressure, the work W done on the gas follows from the definition of work done by a constant force in one dimension (Equation 5.1): $W = F_x \Delta x$, where the force F moves the piston through a displacement Δx (Figure 14.3). Pressure is force per unit area, so for this cylinder with cross-sectional area A, $F_x = PA$. Therefore, $W = F_x \Delta x = PA\Delta x$. The cylinder's volume is $V = Ax$, and Figure 14.3 shows that a positive Δx decreases the volume, so $\Delta V = -A\Delta x$. Therefore, the work done on the gas is

$$W = -P\Delta V \quad \text{(Work done on a gas at constant pressure; SI unit: J)} \quad (14.2)$$

Does the negative sign here make sense? Yes: When the gas is compressed, $V_f < V_i$, which makes $\Delta V = V_f - V_i$ negative. Then Equation 14.2 shows that positive work is done on the gas, as is expected for compression. When the gas expands, ΔV is positive, and the work done on it is negative.

A constant force is applied and heat is allowed to escape, so the pressure remains constant as the gas is compressed.

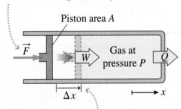

The piston moves through displacement Δx, resulting in work $W = F_x \Delta x = PA\Delta x$.

FIGURE 14.3 Compression at constant pressure.

EXAMPLE 14.3 **Tire Pump 1**

Your tire pump contains 0.0020 mol of air at 22°C, but its hose is clogged, so air can't escape. You push on the handle, maintaining a constant 1-atm pressure while decreasing the volume by half. (a) What's the initial gas volume? (b) How much work do you do?

ORGANIZE AND PLAN You can get the volume from the ideal-gas law, $PV = nRT$. The work done in compressing the air then follows from Equation 14.2, $W = -P\Delta V$.

cont'd.

Known: $n = 0.0020$ mol, $P = 1$ atm $= 1.013 \times 10^5$ Pa, initial $T = 22°C = 295$ K.

SOLVE (a) Solving the ideal-gas law for V,

$$V = \frac{nRT}{P} = \frac{(0.0020 \text{ mol})(8.315 \text{ J/(mol} \cdot \text{K}))(295 \text{ K})}{1.013 \times 10^5 \text{ Pa}}$$

$$= 4.84 \times 10^{-5} \text{ m}^3$$

(b) We *reduce* the volume by half this amount, so $\Delta V = -V/2$, giving

$$W = -P\Delta V = -(1.013 \times 10^5 \text{ Pa})(-4.84 \times 10^{-5} \text{ m}^3/2) = 2.5 \text{ J}$$

REFLECT That's positive work done *on* the gas, as expected for a compression.

MAKING THE CONNECTION What happens to the air's temperature in this process?

ANSWER It drops. With $PV = nRT$ and constant P, temperature, like volume, drops to half its initial value (in kelvins). That makes this example unrealistic: You'd need to put the pump in a very cold environment so that heat could continue to flow out even as the temperature dropped. Otherwise, the pressure would rise and you'd have to push harder—and that wouldn't be a constant-pressure process. Our next tire pump examples will be more realistic.

In an isothermal process, the temperature of the gas is held constant while the gas is compressed (or expands).

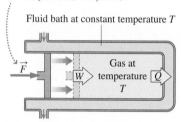

FIGURE 14.4 Isothermal compression.

Constant-Temperature Processes

A process that takes place at constant temperature is **isothermal**. One way to achieve an isothermal process is to immerse our piston-cylinder system in a large reservoir of water or other fluid at constant temperature T (Figure 14.4). Moving the piston slowly enough that the gas remains in equilibrium with the fluid ensures that the gas temperature remains constant. Then the ideal-gas law dictates that pressure varies inversely with volume: $P = nRT/V$.

What's the work done in an isothermal process? We've answered this question for a constant-pressure process: the magnitude of the work is $P\Delta V$. Figure 14.5a shows that this work is the rectangular area under the pressure-versus-volume curve, in this case a horizontal line. The same is true for any process, including the isothermal process whose pressure volume curve is shown in Figure 14.5b: **The magnitude of the work done is the area under the curve**. In the isothermal case, calculus shows that

$$W = nRT \ln\left(\frac{V_i}{V_f}\right) \quad \text{(Work done on gas at constant temperature; SI unit: J)} \tag{14.3}$$

where V_i and V_f are the initial and final volumes.

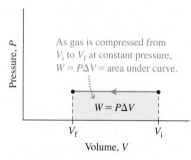

As gas is compressed from V_i to V_f at constant pressure, $W = P\Delta V =$ area under curve.

$W = P\Delta V$

(a) Compression at constant pressure

✓TIP

Use pressure-versus-volume graphs to describe ideal-gas processes and visualize the work involved.

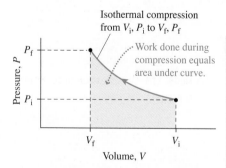

Isothermal compression from V_i, P_i to V_f, P_f

Work done during compression equals area under curve.

(b) Isothermal compression

FIGURE 14.5 Compression of a gas at (a) constant pressure and (b) constant temperature.

CONCEPTUAL EXAMPLE 14.4 **Work: Positive or Negative?**

Show that the sign of the work done in isothermal processes makes sense.

SOLVE For compression, $V_i > V_f$. Therefore, the ratio V_i/V_f in Equation 14.3 is greater than 1, and the natural logarithm of a number larger than 1 is positive. The quantities n, R, and T are also positive, so the work done in compressing a gas isothermally is positive.

For $V_i < V_f$, now $V_i/V_f < 1$. Since the logarithm of a number smaller than 1 is negative, the work done during isothermal expansion is negative.

REFLECT These results, coming from analysis of Equation 14.3, are consistent with what you already know about compression and expansion. It takes positive work to compress a gas, but the gas does work—that is, negative work is done on it—during expansion.

Temperature and internal energy are proportional in an ideal gas, so $\Delta U = 0$ for an isothermal process. Therefore, the first law becomes $Q + W = 0$, or by Equation 14.3,

$$Q = -W = -nRT \ln\left(\frac{V_i}{V_f}\right)$$

This means that $Q < 0$ when the gas is compressed, and $Q > 0$ when the gas expands. This makes sense: When the gas is compressed, its temperature tends to rise because of the work being done on it. In order for temperature to remain constant, heat must flow out. Conversely, an expanding gas tends to cool, so heat must flow in to maintain constant temperature.

EXAMPLE 14.5 **Tire Pump 2**

Consider again the tire pump of Example 14.3. This time you halve the air's volume slowly enough that it stays in equilibrium with its 22°C surroundings. Now how much work do you do?

ORGANIZE AND PLAN This is an isothermal process, so we'll use Equation 14.3 for the work. The volume is halved, so $V_i/V_f = 2$.

Known: $n = 0.0020\,\text{mol}, T = 22°C = 295\,\text{K}, V_i/V_f = 2$.

SOLVE Equation 14.3 gives

$$W = nRT \ln\left(\frac{V_i}{V_f}\right)$$

$$= (0.0020\,\text{mol})(8.315\,\text{J/(mol}\cdot\text{K}))(295\,\text{K})(\ln 2) = 3.4\,\text{J}$$

REFLECT This is more than the 2.5 J we found for the constant-pressure process in Example 14.3. Comparing pressure-volume curves for the two processes (Figure 14.6) shows why the isothermal process takes more work: This process increases the pressure, requiring a greater force and therefore more work.

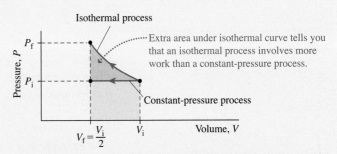

FIGURE 14.6 Isothermal and constant-pressure compression.

MAKING THE CONNECTION Discuss the heat flow in this process.

ANSWER You know that $Q = -W$ for an isothermal process. Therefore, 3.4 J of heat leaves the gas during compression.

Constant-Volume Processes

Many processes, like the calorimetry experiments in Section 13.2, take place in sealed containers with fixed volume. Since work requires displacement ($W = F\Delta x$), the work done in a **constant-volume process** is zero. Figure 14.7 confirms this, showing zero area under the pressure-versus-volume graph for this process. With $W = 0$, the first law of thermodynamics gives $\Delta U = Q$ for a constant-volume process. Therefore, any change in internal energy requires a heat flow. As you saw in the calorimetry examples of Chapter 13, heat flow into the system ($Q > 0$) increases the internal energy, while outflow ($Q < 0$) decreases it. Changing internal energy causes either a temperature change or a phase change.

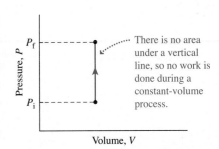

FIGURE 14.7 A constant-volume process.

Adiabatic Processes

In an **adiabatic process** there's no heat transfer: $Q = 0$. One way to make a process adiabatic is to insulate the system (Figure 14.8). Even without insulation, some processes are very nearly adiabatic because they happen so quickly that there's not time for significant heat flow. Compression and expansion of gases in the cylinders of your car's engine—processes that repeat hundreds of times each second—are essentially adiabatic for this reason.

For an adiabatic process $Q = 0$, so the first law of thermodynamics says $\Delta U = W$. In other words, internal energy changes solely because of the work done. For an ideal gas, an internal-energy change implies a temperature change. An adiabatic compression increases temperature, while an expansion decreases it. Figure 14.9 (next page) is the pressure-versus-volume graph for an adiabatic process and shows that the work done is greater than that for an isothermal process with the same volume change.

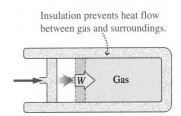

FIGURE 14.8 An adiabatic process can take place as shown, in an insulated system, or suddenly, so there's no time for heat to flow.

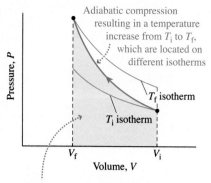

Adiabatic compression resulting in a temperature increase from T_i to T_f, which are located on different isotherms

T_f isotherm

T_i isotherm

Pressure, P

V_f V_i

Volume, V

Area under adiabatic curve is greater than area under T_i isotherm, so adiabatic compression requires more work than equivalent isothermal compression.

FIGURE 14.9 Pressure-volume diagram for an adiabatic compression.

The exact relationship between pressure and volume in an adiabatic process is more complex than for the other processes we've described, and depends on the type of gas. A general way of stating this relationship is

$$PV^\gamma = \text{constant} \qquad (14.4)$$

where the **adiabatic exponent** is $\gamma = c_P/c_V$, the ratio of the specific heats at constant pressure and constant volume. Fortunately, that ratio is known for many gases. For example, recall from Chapter 13 that a monatomic gas has molar specific heats $c_P = 5R/2$ and $c_V = 3R/2$. Therefore, $\gamma = 5/3$ for a monatomic gas. Similarly, a diatomic gas has $\gamma = 7/5$. Calculus shows that the work done in an adiabatic process, taking a gas from pressure P_i and volume V_i to pressure P_f and volume V_f, is

$$W = \frac{P_f V_f - P_i V_i}{\gamma - 1} \qquad \text{(Work done in an adiabatic process; SI unit: J)} \qquad (14.5)$$

✓ **TIP**

For gas molecules more complicated than diatomic, the ratio c_P/c_V is determined experimentally.

An example of a nearly adiabatic process in nature is cloud formation. As warm, moist air rises, the lower density and pressure of the air aloft allows the rising air to expand. The expansion is fast enough to be considered adiabatic. Because a gas cools during adiabatic expansion, it can reach a temperature at which water vapor condenses into droplets, forming a cloud.

EXAMPLE 14.6   **Tire Pump 3**

You now compress your tire pump so rapidly that the process is essentially adiabatic. For the tire pump described in Example 14.3 (again assuming compression in volume by a factor of 2 from the original state), find (a) the final pressure, (b) the final temperature, and (c) the work done.

ORGANIZE AND PLAN Equation 14.4 is the pressure-volume relation for an adiabatic process, which we can express as a relation between initial and final values of the quantity PV^γ. Air is essentially diatomic, so $\gamma = 7/5$. Once we know the final pressure, we can use the ideal-gas law to find the temperature, and Equation 14.5 for the work. We'll also need the initial and final volumes; we found V_i in Example 14.3, and $V_f = V_i/2$. We start at 1 atm pressure.

Known: $P_i = 1$ atm $= 1.013 \times 10^5$ Pa, $V_i = 4.84 \times 10^{-5}$ m^3 (from Example 14.3), $V_f = V_i/2$.

SOLVE (a) In terms of initial and final values, Equation 14.4 reads $P_i V_i^\gamma = P_f V_f^\gamma$. Therefore, the final pressure is $P_f = P_i V_i^\gamma/V_f^\gamma = P_i(V_i/V_f)^\gamma$. Since $V_i = 2V_f$, $P_f = P_i(V_i/V_f)^\gamma = (1.0 \text{ atm})(2)^{7/5} = 2.64$ atm, or 2.67×10^5 Pa.

(b) Solving the ideal-gas law for temperature gives

$$T_f = \frac{P_f V_f}{nR} = \frac{(2.67 \times 10^5 \text{ Pa})(2.42 \times 10^{-5} \text{ m}^3)}{(0.0020 \text{ mol})(8.315 \text{ J/K} \cdot \text{mol})} = 389 \text{ K}$$

or 116°C.

(c) Equation 14.5 gives the work done in the adiabatic process:

$$W = \frac{P_f V_f - P_i V_i}{\gamma - 1}$$

$$= \frac{(2.67 \times 10^5 \text{ Pa})(2.42 \times 10^{-5} \text{ m}^3) - (1.013 \times 10^5 \text{ Pa})(4.84 \times 10^{-5} \text{ m}^3)}{7/5 - 1} = 3.9 \text{ J}$$

cont'd.

REFLECT That's more than the 3.4 J we found in Example 14.5 for the isothermal process—consistent with our expectation from Figure 14.9.

MAKING THE CONNECTION How much work is required to compress this gas adiabatically by another factor of 2, from $2.42 \times 10^{-5}\,\mathrm{m^3}$ to $1.21 \times 10^{-5}\,\mathrm{m^3}$?

ANSWER Using the same procedure as in Example 14.6, the work required is 5.2 J. The more you compress, the harder it gets—consistent with Figure 14.9, which shows the pressure-volume graph getting steeper as the volume decreases.

Reviewing New Concepts: The First Law of Thermodynamics in Thermal Processes

Process	Work W	First law accounting, with $\Delta U = Q + W$
Constant pressure	$W = -P\Delta V$	$\Delta U = Q - P\Delta V$
Constant temperature (isothermal)	$W = nRT\ln\left(\dfrac{V_i}{V_f}\right)$	$\Delta U = 0$ $Q = -W = -nRT\ln\left(\dfrac{V_i}{V_f}\right)$
Constant volume	$W = 0$	$\Delta U = Q$
Adiabatic ($Q = 0$)	$W = \dfrac{P_f V_f - P_i V_i}{\gamma - 1}$	$\Delta U = W = \dfrac{P_f V_f - P_i V_i}{\gamma - 1}$

Metabolic Processes

The first law is a general statement governing thermodynamic processes throughout the universe. We've focused on gases, because it's easier to understand a process when one variable (such as pressure, volume, or temperature) is constant.

However, first-law applications are everywhere, including our own bodies. You learned about energy and metabolic rates in Chapter 5; the first law extends that discussion. Your body has internal energy U, the energy of all your molecules. You convert some of that energy into useful work, and some leaves your body as heat. You consume food to replenish your internal energy.

Some of the work your body does is obvious, moving muscles to walk, run, swim, lift, and climb. Less obvious is the work done to pump blood and other fluids throughout your body. Even your brain does work, sending electrical energy coursing through your neural system. Chemical reactions within cells convert internal energy to work as motor proteins drag cell constituents about in the fundamental processes of life. As far as the first law is concerned, all these processes involve work. In general, anything that isn't heat—the spontaneous flow of energy from hotter to cooler—counts as work. In accounting for human metabolic work in the first law, it's important to note that W is negative for all these metabolic processes. Remember that we defined W as positive for work done *on* a system. Metabolic processes constitute work done *by* the system, and hence W in the first law is negative. This makes sense: With $\Delta U = Q + W$, your internal energy decreases when you do work.

What about heat? Generally heat flows *from* your body, which is warmer than its surroundings. When you exercise, the heat-flow rate increases to keep your body temperature from rising. Because it's an outflow, heat Q, like work W, is generally negative in your body's first-law accounting. Your food intake establishes the rate at which you can expend energy, and any energy you lose as heat is unavailable to do work.

TACTIC 14.1 **The First Law of Thermodynamics**

The first law of thermodynamics states:

$$\Delta U = Q + W$$

Solving first-law problems requires keeping track of which terms in this equation change and which don't. Specifically:

In an isothermal (constant-temperature) process involving an ideal gas, the internal energy U is constant, so $\Delta U = 0$.

In a constant-volume process, the work done is zero ($W = 0$).

In an adiabatic process, the heat flow is zero ($Q = 0$).

In other processes, the quantities in $\Delta U = Q + W$ can all change simultaneously.

The work done depends on the type of process but is always equal to the area under the pressure-volume graph. Work is positive for compression and negative for expansion.

GOT IT? Section 14.2   Heat flows out of a system in (a) an adiabatic compression; (b) an isothermal compression; (c) an isothermal expansion; (d) an adiabatic expansion.

14.3 The Second Law of Thermodynamics

Leave your hot coffee on the kitchen counter, and an hour later it's cooled to room temperature. Of course: You know that heat flows from hotter to cooler objects, and you've learned how to use tools such as specific heat, thermal conductivity, and the first law of thermodynamics to explore and quantify that heat flow.

But the first law doesn't say anything about heat *having* to flow from hotter to cooler; it only says that energy is conserved. Nor do the laws of classical mechanics—including Newton's second law of motion, momentum conservation, and energy conservation. Yet mechanical systems seem to "know" that many processes have a preferred direction. Drop a ball: It bounces a few times and comes to rest, having transferred its energy to the floor's internal energy. Energy conservation alone doesn't prohibit the floor from giving energy back to the ball, causing it to jump spontaneously upward, but you never see that happen.

These examples illustrate what physicists call **time's arrow**—the notion that many processes have a preferred direction, even though it wouldn't violate other laws of physics if those processes were reversed. The "other law of physics" we'll focus on here is energy conservation, because so much of thermal physics is about energy flows and using energy to do work.

The **second law of thermodynamics** sets the preferred direction for many physical processes, including heat flow. Here we'll explore the second law and show why it limits our ability to turn thermal energy into useful work—a limit that's at the heart of society's growing energy crisis. The second law of thermodynamics is unusual among physical laws in that there are multiple ways to state it. Different second-law statements seem quite distinct, but actually they're equivalent. Each is useful in a different context. We'll start with the familiar example of heat flow:

Second law of thermodynamics: Heat flows spontaneously only from hotter to cooler objects.

This is familiar, so we won't pursue it with examples here. Later you'll see how this statement of the second law relates to other versions.

✓**TIP**

The second law of thermodynamics prohibits certain processes that are nevertheless allowed by the first law.

Entropy

The second law of thermodynamics may seem rather qualitative and imprecise, but the concept of **entropy** (symbol S) makes it quantitative. Entropy is a state variable—a term introduced in Chapter 12—so a thermodynamic system with state variables T, P, and V also has a definite entropy S. It's often difficult to compute entropy, and we won't pursue such calculations here. What's more important is the *change* in entropy that takes place when heat flows. That change is

$$\Delta S = \frac{Q}{T} \qquad \text{(Entropy change; SI unit: J/K)} \qquad (14.6)$$

where Q is the heat that flows into a system at absolute temperature T, with zero work done. Because the absolute temperature is always positive, the entropy change ΔS can be positive or negative, depending on whether heat flows in $(Q > 0)$ or out $(Q < 0)$, respectively.

You can use Equation 14.6 to relate entropy and the second law of thermodynamics. Suppose you have two large objects in thermal contact, one at $T_1 = 300$ K and the other at $T_2 = 400$ K (Figure 14.10). The second law of thermodynamics says that heat flows from the hotter to the cooler object. Suppose that after a short time, 24 kJ of heat has flowed. We'll assume both objects have heat capacities large enough that this heat flow doesn't significantly change their temperatures. In that case Equation 14.6 gives the entropy changes of the two objects:

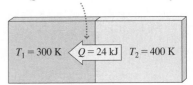

Heat Q flows from warmer to cooler object.

FIGURE 14.10 Heat flow between objects at different temperatures, used to compute entropy changes.

$$\Delta S_{cool} = \frac{Q}{T_{cool}} = \frac{24{,}000 \text{ J}}{300 \text{ K}} = 80 \text{ J/K}$$

Heat flows *out* of the hot object, so $Q = -24{,}000$ J, and

$$\Delta S_{hot} = \frac{Q}{T_{hot}} = \frac{-24{,}000 \text{ J}}{400 \text{ K}} = -60 \text{ J/K}$$

The net entropy change of the whole system is

$$\Delta S_{total} = \Delta S_{cool} + \Delta S_{hot} = 80 \text{ J/K} - 60 \text{ J/K} = +20 \text{ J/K}$$

This illustrates a general rule: Heat flow from a hotter to a cooler body is accompanied by an increase in the system's total entropy. If you think about Equation 14.6 as applied to two objects at different temperatures, you'll see that there's no way to avoid that increase, no matter what temperatures you use. This leads to another statement of the second law of thermodynamics, equivalent to the one we gave earlier:

Second law of thermodynamics: Heat flow is accompanied by an increase in the entropy of the universe.

You may sense that entropy is an unusual quantity. Unlike energy, entropy is not conserved, because the entropy of the universe grows every time there's a flow of heat. Notice how entropy fits with the notion of time's arrow: As time passes, entropy can only increase, never decrease. This leads to yet another statement of the second law:

Second law of thermodynamics: Natural processes evolve toward a state of maximum entropy.

Our entropy example assumed a heat flow so small that temperatures didn't change significantly, so we could apply Equation 14.6 to each object. Leave the objects in contact, of course, and they'll eventually reach a common equilibrium temperature. We'd then need calculus to find the entropy change with continually changing temperatures—but the result would be the same: a net increase in entropy.

✓ **TIP**

Entropy is *not* a conserved quantity.

EXAMPLE 14.7 **Melting Ice**

You put a 40-g ice cube at 0°C in a 20°C room and it melts. Find the entropy change of the ice, the air in the room, and the net entropy change of the system. (Assume the melted ice remains at 0°C and the air at 20°C.)

ORGANIZE AND PLAN From Chapter 13, the heat needed to melt ice is $Q = mL_f$, where m is the mass of ice and $L_f = 333$ kJ/kg is the heat of fusion. That heat flows *from* the air *to* the ice. We'll use Equation 14.6 to find the entropy changes, with Q positive for the ice and negative for the air.

Known: $m = 40$ g, $L_f = 333$ kJ/kg, $T_{ice} = 0°C = 273$ K, $T_{air} = 20°C = 293$ K.

SOLVE The heat that flows from the air to the ice is

$$Q = mL_f = (0.040 \text{ kg})(333 \text{ kJ/kg}) = 13.3 \text{ kJ}$$

Therefore, for the ice $Q = +13.3$ kJ, and

$$\Delta S_{ice} = \frac{Q}{T_{ice}} = \frac{13.3 \text{ kJ}}{273 \text{ K}} = 48.7 \text{ J/K}$$

For the air $Q = -13.3$ kJ, and

$$\Delta S_{air} = \frac{Q}{T_{air}} = \frac{-13.3 \text{ kJ}}{293 \text{ K}} = -45.4 \text{ J/K}$$

Then the net entropy change is

$$\Delta S_{net} = \Delta S_{ice} + \Delta S_{air} = 48.7 \text{ J/K} - 45.4 \text{ J/K} = +3.3 \text{ J/K}$$

REFLECT Again, heat flow results in a net entropy increase, with the ice's entropy gain slightly exceeding the entropy loss from the warmer air. This is consistent with the second law.

MAKING THE CONNECTION How would your answers differ on a hotter day, with air temperature 30°C?

ANSWER The ice's entropy gain is the same, because it takes the same amount of heat to melt it. However, the air's temperature means a smaller entropy decrease. For the air at $T = 303$ K, $\Delta S = Q/T = -43.9$ J/K, giving a net entropy change of 4.8 J/K. A larger temperature difference results in a larger entropy increase, even though the heat flow is the same!

CONCEPTUAL EXAMPLE 14.8 **Entropy of Freezing Water**

Using the results of the preceding example, what's the entropy change when 40 g of water at 0°C freezes to ice? Does this violate the second law?

SOLVE In the preceding example, melting the ice required a 13.3-kJ heat flow *to* the ice, resulting in a 48.7-J/K entropy increase. Freezing the same amount of water requires that 13.3 kJ flow *out* of the water. With $Q < 0$, there's an entropy *decrease*: $\Delta S = -48.7$ J/K.

How can you reconcile this entropy decrease with the second law of thermodynamics? The second law requires an entropy increase for the entire system, which certainly includes the refrigerator that makes the ice. But the refrigerator won't operate without electrical energy, and its operation generates heat, which is expelled into the room. This heat flow increases entropy, more than offsetting the entropy decrease of the freezing water.

REFLECT It's the entropy of a *closed* system that increases—one with no inflows or outflows of matter or energy. No system is truly closed, so ultimately it's the entropy of the universe that increases. Any time you see an entropy decrease, you can be sure that somewhere else there's a more-than-compensating entropy increase. In Section 14.4 we'll discuss the working of refrigerators, and you'll see more precisely how the second law governs their operation.

GOT IT? Section 14.3   What is the entropy increase when 1 kg of water at 100°C turns to steam? (a) 6.06 kJ; (b) 8.37 kJ; (c) 22.6 kJ; (d) 2.26 MJ.

14.4 Heat Engines and Refrigerators

What makes your car go? It starts with potential energy locked in gasoline molecules that is then released when the gasoline burns in your engine's cylinders. The hot combustion gases expand, pushing on the pistons. Gears and cranks transfer the resulting mechanical energy to the wheels, and away you go.

But not all the thermal energy released in combustion ends up as mechanical energy. Hot gases carry energy out of the exhaust. The cooling system circulates water through the engine, then to the radiator, where it's transferred to the surrounding air. Energy leaving your car as heat isn't available to do work. The bottom line is that only some of the gasoline's energy ends up moving your car, and the rest is lost as heat. This isn't a matter of poor engineering. Rather, it's a fundamental limitation set by the second law of thermodynamics. We'll explore this limitation through **heat engines**, which include automobile engines, jet aircraft engines, and power plants.

Figure 14.11 diagrams the energy flows in a heat engine. This model roughly describes your car's engine, as some of the energy from a hot substance (the combusted gasoline) is used to do mechanical work W, while the rest is expelled to the environment as heat Q_C. Another example is the steam engine, in which thermal energy boils water, and the resulting steam does mechanical work. Electric power plants that run on coal, oil, natural gas, or nuclear fission are steam engines, in which steam turns a turbine connected to a generator that transforms mechanical into electrical energy (more in Chapter 19).

Our simplified heat engine includes a *hot reservoir*, which represents thermal energy produced, for example, by burning a fuel, the fission process in uranium (discussed in Chapter 25), or even concentrating sunlight. There's also a *cold reservoir*, usually the ambient air or a river used to cool a power plant. Heat Q_H flows from the hot reservoir into the engine, which converts some of it to useful work W, while the rest flows as heat to the cold reservoir. Energy is conserved, so $W = Q_H - Q_C$.

The engine's efficiency e is the ratio of the work done to the energy input Q_H; that is $e = W/Q_H$. With $W = Q_H - Q_C$, the efficiency becomes $e = (Q_H - Q_C)/Q_H$, or

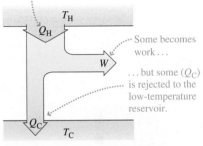

Extract heat Q_H from the high-temperature reservoir of a real heat engine.

Some becomes work ...

... but some (Q_C) is rejected to the low-temperature reservoir.

FIGURE 14.11 Energy flows in a heat engine.

$$e = 1 - \frac{Q_C}{Q_H} \quad \text{(Efficiency of a heat engine)} \quad (14.7)$$

The efficiency is between 0 and 1, with 1 representing 100% conversion of thermal energy to work. An efficiency of 0.34, typical of a nuclear power plant, means that only 34% of the thermal energy released from the fuel ends up as useful work. Equation 14.7 shows that you want Q_C as small as possible for maximum efficiency.

✓ **TIP**

In Equation 14.7, both Q_H and Q_C are taken as positive, with the direction of heat flow defined in Figure 14.11.

EXAMPLE 14.9 | **Gas Guzzler!**

A car delivers 105 horsepower to the wheels. Its engine's efficiency is 24%. In each second, (a) how much energy is released from its fuel and (b) how much heat is dumped to the environment?

ORGANIZE AND PLAN From Chapter 5, one horsepower (hp) is 754.7 W or 754.7 J/s. The energy delivered to the car's wheels is mechanical energy W, and the energy released from fuel is Q_H. With the efficiency e known, you can solve for Q_H using $e = W/Q_H$. The waste heat is then the difference $Q_C = Q_H - W$.

Known: $W = 105$ hp, $e = 0.24$.

SOLVE (a) Converting 105 hp to watts gives the rate at which the engine supplies energy to the wheels: $(105 \text{ hp})(754.7 \text{ W/hp}) = 79.2 \text{ kW}$; equivalently, that's 79.2 kJ of work each second. With $e = W/Q_H$, the energy released each second from burning gasoline is

$$Q_H = \frac{W}{e} = \frac{79.2 \text{ kJ}}{0.24} = 330 \text{ kJ}$$

cont'd.

(b) Then the heat flowing to the environment each second is

$$Q_C = Q_H - W = 330 \text{ kJ} - 79.2 \text{ kJ} = 251 \text{ kJ}$$

REFLECT The 24% efficiency is typical for automobile engines. That means more than three-fourths of the fuel energy is wasted! Despite decades of engineering, the internal combustion engine remains remarkably wasteful.

MAKING THE CONNECTION How would a 1% efficiency gain change the car's horsepower rating, assuming the same energy from fuel?

ANSWER With 25% efficiency, the same fuel energy Q_H (330 kJ) yields 82.5 kJ of work, an increase of 3.3 kJ or 4.2%. That raises the horsepower rating from 105 to 109.

The Carnot Engine

Example 14.9 suggests that a car engine is around 25% efficient. Including frictional losses, the energy delivered to the wheels is actually more like 15% of the fuel energy. Similarly, coal-fired and nuclear power plants usually operate at 30–40% efficiency, while modern gas plants approach 60%.

Why can't we do better? Since the first steam engine, over 200 years ago, scientists and engineers have worked to develop more efficient engines. In the early 19th century, the French physicist Sadi Carnot showed that the most efficient engine is one that operates between two fixed temperatures, T_H (hot) and T_C (cold), following a four-step process called the **Carnot cycle** (Figure 14.12).

The efficiency of the Carnot cycle depends on the two temperatures:

$$e_{\text{Carnot}} = 1 - \frac{T_C}{T_H} \quad \text{(Carnot efficiency)} \tag{14.8}$$

Both temperatures in Equation 14.8 must be on the absolute (kelvin) scale. Note that the Carnot efficiency depends on the ratio of T_C and T_H. If they're close, then T_C/T_H is nearly 1, and the efficiency is close to 0. Raising efficiency means increasing the hot-to-cold temperature ratio. The surrounding environment usually establishes T_C, so the way to raise efficiency is to increase T_H. But T_H is limited by properties of the materials used in an engine, and real engines are therefore a compromise between thermal efficiency, material properties, and economics. Although the cold temperature T_C is out of our control, its seasonal variation means that engines are generally more efficient in cold weather. Power plants, in particular, have higher power ratings in winter than in summer.

✓ **TIP**

No real engine is a Carnot engine, and a real engine's efficiency is always less than the Carnot efficiency.

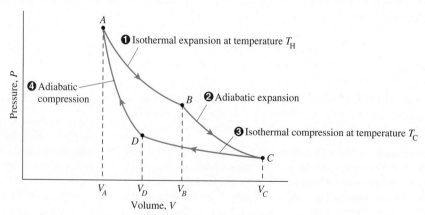

FIGURE 14.12 Pressure-volume diagram for a Carnot cycle.

EXAMPLE 14.10   **Solar-Thermal Power**

The electric generating station shown in Figure 14.13 uses parabolic reflectors to concentrate sunlight, heating a fluid to 288°C. The fluid transfers its energy to water, which boils in a steam-turbine power system. If the system rejects heat to the environment at 35°C, what's the limit on its maximum possible efficiency?

ORGANIZE AND PLAN No engine can be more efficient than a Carnot engine, so the efficiency limit is

$$e_{Carnot} = 1 - T_C/T_H$$

Known: $T_C = 35°C = 308$ K, $T_H = 288°C = 561$ K.

SOLVE With the given values, we have

$$e_{Carnot} = 1 - \frac{T_C}{T_H} = 1 - \frac{308 \text{ K}}{561 \text{ K}} = 0.45$$

REFLECT This is the Carnot efficiency, a theoretical upper limit. The actual efficiency of the generating station shown here is around 29%.

FIGURE 14.13 Solar electric generating station in California.

MAKING THE CONNECTION   Why not use a refrigerator to lower T_C as a means to higher efficiency?

ANSWER As you'll see, a refrigerator is just a heat engine in reverse, and it's subject to similar thermodynamic limitations. You're generally stuck with the ambient environment as the lowest possible T_C.

The Carnot engine is conceptually useful, but it's not a practical device. One reason is that the four steps in Figure 14.12 must occur slowly enough so that the gas is never far from equilibrium. By contrast, real engines repeat their cycles many times each second. The idealized Carnot cycle also ignores losses due to friction as well as thermal losses that effectively lower T_H. Again, the Carnot engine is an idealization that sets an upper limit on the efficiency of *any* heat engine. You'll now see how the second law predicts that even the best engineered, well-oiled, well-insulated engine cannot work with perfect efficiency, consistent with the Carnot limit in Equation 14.8.

Real Engines and the Second Law

Any heat engine has an efficiency given by Equation 14.7: $e = 1 - Q_C/Q_H$, where Q_H is the heat flow into the engine from the hot reservoir at temperature T_H, and Q_C is the flow from the engine to the cold reservoir at T_C. These heat flows remove entropy Q_H/T_H from the hot reservoir, and add entropy Q_C/T_C to the cold reservoir. The second law of thermodynamics requires a net entropy gain, so $Q_C/T_C > Q_H/T_H$. Rearranging this inequality gives $Q_C/Q_H > T_C/T_H$. Therefore, the engine's efficiency satisfies

$$e < 1 - \frac{T_C}{T_H}$$

Since $1 - T_C/T_H$ is the Carnot efficiency (Equation 14.8), this says that no engine can be more efficient than a Carnot engine. Friction, heat loss, and thermodynamic disequilibrium all prevent real engines from achieving the Carnot limit. Our analysis of heat-engine efficiency suggests another statement of the second law:

Second law of thermodynamics: It is impossible to convert thermal energy entirely to work.

This means that mechanical and thermal energy aren't equivalent in their ability to do work. For example, a billiard ball with 4 J of kinetic energy can strike another ball elastically and transfer all 4 J of its energy (Chapter 6). However, a heat engine cannot transform 4 J of thermal energy into work, because the second law requires that some energy leave the engine as heat. In that sense, mechanical energy is a higher-quality form of energy.

APPLICATION

Cogeneration

The second law requires that engines convert some of their fuel energy into heat, often called "waste heat." But it needn't be wasted; although the heat rejected to the environment can't do work, it can still warm buildings. In today's energy-strapped world, it's smart to use the waste heat from electric power plants for just that purpose. Industries, educational institutions, and even whole cities increasingly use such *cogeneration*, thus benefiting from all the energy in the fuels they burn. The photo shows a 600-kW cogenerating unit in the heating plant at Middlebury College.

This inequity between mechanical and thermal energy is related to the arrow of time. Whenever mechanical energy is converted to thermal energy, some of its ability to do useful work is lost. As time goes by, less of the universe's energy is of the higher-quality type, and more is thermal energy. Extrapolating far into the future, the universe may eventually suffer a "heat death," in which all energy has degraded to thermal energy, leaving no way to do work.

We can also relate energy quality to entropy. As that billiard ball rolls along, its molecules share a common translational motion. Friction slows and eventually stops the ball, converting its kinetic energy to thermal energy in itself and the table. That thermal energy is associated with random motions, and is therefore less ordered than the kinetic energy of the rolling ball. As you'll learn in Section 14.5, entropy is associated with disorder. According to the second law of thermodynamics, the entropy of the universe increases as time passes. Thus, increasing entropy goes along with the degradation of energy from more to less ordered forms.

Refrigerators

Run a heat engine backwards, and you have a **refrigerator**. We'll refer to all cooling machines as refrigerators, including air conditioners and heat pumps. As Figure 14.14 shows, a refrigerator pulls heat Q_C from a cold reservoir (the contents of your kitchen fridge) and dumps heat Q_H into a hot reservoir (the kitchen air). But that's not the way heat flows naturally. So you have to supply work W to drive the heat flow. That's usually from electrical energy—you have to plug in your refrigerator.

The measure of a refrigerator's efficiency is, as for an engine, the ratio of the benefit to the energy cost. This ratio is the **coefficient of performance** (COP). The benefit is the amount of cooling, Q_C, and the cost is the energy W. Therefore,

$$\text{COP} = \frac{Q_C}{W} \qquad \text{(Coefficient of performance)} \qquad (14.9)$$

Typical refrigerator COPs range from about 2 to 4. A COP of 4 means that for every joule of electrical energy it uses, the refrigerator removes 4 J of thermal energy from its contents. By conservation of energy, $W = Q_H - Q_C$, so an alternate expression for COP is

$$\text{COP} = \frac{Q_C}{Q_H - Q_C} = \frac{1}{Q_H/Q_C - 1}$$

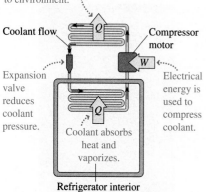

Heat flows from cold to hot...

... but this requires work W. By conservation of energy, $W = Q_H - Q_C$.

FIGURE 14.14 Energy flows in a refrigerator.

Real Refrigerators

Coolant condenses in coils on outside of refrigerator, releasing heat of vaporization to environment.

Coolant flow

Compressor motor

Expansion valve reduces coolant pressure.

Electrical energy is used to compress coolant.

Coolant absorbs heat and vaporizes.

Refrigerator interior

In a real refrigerator, fluid with a low boiling point cycles through a closed system. A motor-driven compressor brings the gas phase to high pressure. It then condenses to liquid, giving up its heat of vaporization to the surrounding environment. The liquid then expands, dropping its pressure. It passes through coils in contact with the refrigerator's interior, where it vaporizes. The energy required—the heat of vaporization—comes from the refrigerator's contents. The gas then flows to the compressor, and the cycle repeats. The net effect is to transfer heat from the refrigerator's interior to its surroundings. Mechanical work is required in the process, work that comes from the electrical energy supplied to the compressor motor.

EXAMPLE 14.11 **The Best Refrigerator**

Use the second law of thermodynamics to express the maximum COP for a refrigerator in terms of the temperatures T_C and T_H. Evaluate for a refrigerator maintaining 4°C in a 20°C room.

ORGANIZE AND PLAN Heat is being removed from the refrigerator's cold interior, so its entropy decreases by Q_C/T. Heat is dumped into the warmer room, so its entropy increases by Q_H/T. The second law requires a net entropy increase, so $Q_H/T_H > Q_C/T_C$. We've just seen the COP written as

$$\text{COP} = \frac{1}{Q_H/Q_C - 1}$$

Combining these two expressions will give the COP in terms of the temperatures.

Known: $T_C = 4°C = 277$ K, $T_H = 20°C = 293$ K.

SOLVE Rearranging the inequality above gives $Q_H/Q_C > T_H/T_C$. Substituting into the COP expression yields

$$\text{COP} < \frac{1}{T_H/T_C - 1} = \frac{1}{293 \text{ K}/277 \text{ K} - 1} = 17$$

cont'd.

REFLECT A COP of 17 would be fantastic, but it's unrealistic. The maximum COP allowed by the second law can only be achieved in a Carnot cycle, and a real refrigerator doesn't run on a Carnot cycle. The compressor that drives the refrigeration cycle isn't perfectly efficient, and there's plenty of heat leakage into the refrigerator. Opening the door further compromises the average COP. But refrigerators are better than they used to be; today, the average household refrigerator consumes only one-third the energy of a 1970s model.

MAKING THE CONNECTION What's the maximum COP of this refrigerator's freezer unit, which maintains a temperature of $-18°C$?

ANSWER Reworking with $T_C = -18°C = 255$ K, the maximum COP is 6.7.

CONCEPTUAL EXAMPLE 14.12 **Running Hot and Cold**

Is a refrigerator's COP higher when it's first plugged in or after its interior has cooled?

SOLVE The preceding example gave the maximum possible COP:

$$COP < \frac{1}{T_H/T_C - 1}$$

When the refrigerator is first plugged in, the two temperatures are equal, and the COP is unrestricted. In essence, you get the first little bit of cooling almost for free. (We say "almost," because it takes some

energy to start the compressor.) But as T_C drops, the denominator $T_H/T_C - 1$ grows, and the COP decreases. The colder the refrigerator becomes, the more work is required for further cooling.

REFLECT Helium refrigerators liquefy helium at 4.2 K. In a room-temperature environment, this gives a maximum COP of only 0.0145, which means a lot of work to move a little heat! Quantitatively, that means at least 69 J of work to remove 1 J of thermal energy. To make things easier, the helium is surrounded by liquid nitrogen at 77 K, raising the maximum COP to a more reasonable 0.058.

Air Conditioners and Heat Pumps

An **air conditioner** is similar to a refrigerator, because it removes heat from a building's interior and dumps it to the higher-temperature environment outside. As for a refrigerator, this requires work W. In effect, the cold reservoir is the space being cooled, and the hot reservoir is the outdoors. Like the refrigerator, an air conditioner has to work harder to maintain a larger temperature difference, so it consumes the most energy on hot days.

A **heat pump** is like an air conditioner run backwards: It removes energy from the environment and deposits it in a space you're trying to heat. Figure 14.15 shows the advantage of the heat pump: It requires only energy W to deliver a larger quantity of heat Q_H. In northern climates, the reservoir Q_C is usually groundwater at around 10°C, and the pump might deliver 3–4 J of heat for every J of electrical energy W used to run the device. Whether the heat pump is more efficient than burning a fuel directly depends on the efficiency of the power plant used to generate the electricity. For hydropower and wind, which are mechanical energy sources that don't require fossil or nuclear fuel, a heat pump "leverages" a small amount of high-quality electrical energy to move a larger quantity of heat. For thermal power plants with typical efficiencies between 30 and 50%, the heat pump may or may not make thermodynamic sense, because every unit of electrical energy generated requires 2–3 units of fuel energy. But a heat pump is always far more efficient than electric heat, which turns each joule of high-quality electrical energy into 1 joule of low-grade heat.

As usual, the COP of a heat pump is the ratio of the benefit to the energy cost. But the heat pump's purpose is heating, not cooling, so the benefit is Q_H rather than Q_C as for a refrigerator. Thus COP $= Q_H/W$ for a heat pump. With $Q_H = W + Q_C$, this ensures that COP > 1 for a heat pump.

Heat pumps can be made with reversible connections, so they bring heat into the house in winter and remove it in summer. Thus one device serves as both heater and air conditioner!

Heat pump uses energy W to extract heat Q_C from outdoors and deposit heat Q_H indoors. By conservation of energy, $Q_H = W + Q_C$.

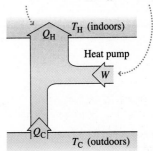

FIGURE 14.15 Energy flows in a heat pump.

Reviewing New Concepts: Heat Engines and Refrigerators

- A heat engine does work by drawing heat from a hot reservoir, and ejecting heat to a cold reservoir.
- The efficiency of a heat engine is $e = 1 - \dfrac{Q_C}{Q_H}$.
- The Carnot cycle gives the maximum theoretical efficiency for a heat engine: $e_{Carnot} = 1 - \dfrac{T_C}{T_H}$.
- A refrigerator operates like a heat engine in reverse, with coefficient of performance given by $COP = \dfrac{Q_C}{W}$. For heat pumps, $COP = \dfrac{Q_H}{W}$.

EXAMPLE 14.13 **Cutting Heat Bills**

A home heat pump with COP = 4.0 uses electrical energy at a rate of 2.5 kW. (a) At what rate does it supply heat to the house? (b) With electricity at 14¢/kWh, what's the cost per MJ of heat?

ORGANIZE AND PLAN The high-quality energy input (W) to the heat pump is 2.5 kW or 2.5 kJ/s. With $COP = Q_H/W$, we can find the heat delivered. Then we'll be able to determine the price per MJ.

Known: COP = 4.0, W = 2.5 kJ every second.

SOLVE (a) Solving for Q_H gives

$$Q_H = (COP)W = (4.0)(2.5\ kJ) = 10\ kJ$$

of heat delivered each second, or a rate of 10 kW.

   (b) At 10 kJ/s it will take 100 s to deliver 1 MJ. With COP = 4, the electrical energy consumed is one-fourth of the heat delivered, so the pump uses 250 kJ while delivering 1 MJ of heat. With 1 kWh = 3.6 MJ, the pump uses 0.25 MJ/(3.6 MJ/kWh) = 0.0694 kWh. At 14¢/kWh, that costs 0.97¢—just under a penny.

REFLECT The savings obtained with heat pumps have to be weighed against their high initial cost, as well as the overall thermodynamic situation including the power plant. Still, cost savings can be significant; with oil at $4 per gallon, and burned in an 85% efficient furnace, that 1 MJ of heat would cost about 3¢.

MAKING THE CONNECTION If the electricity in this example comes from a 33% efficient coal-fired power plant, find the ratio of heat delivered to fuel energy released.

ANSWER At 33% efficiency, each unit of electrical energy produced requires three units of fuel energy. So it takes 0.75 MJ of coal energy to produce the 0.25 MJ of electricity that runs the heat pump to deliver 1 MJ of heat to the house. That's a ratio of 1/0.75 = 1.3. You're still ahead, because the best you could do by burning the coal directly is 0.75 MJ, and that assumes 100% efficient combustion.

GOT IT? Section 14.4 What's the maximum efficiency of a heat engine operating between 25°C and 300°C? (a) 36%; (b) 48%; (c) 71%; (d) 92%.

14.5 Statistical Interpretation of Entropy

In Section 14.3 we noted that higher entropy is associated with greater disorder. How does disorder—a seemingly subjective quality—connect to a physical quantity such as entropy? Here we'll establish that connection by relating entropy increase and the second law of thermodynamics.

An Ideal Gas

Consider a box containing an ideal gas with just two identical molecules. One measure of order in this system is to count the molecules in each half of the box. Table 14.1 shows that there are four different possible configurations for those two molecules. Each molecule is as likely to be in the left half as the right, so all four combinations are equally likely.

   The term **microstate** describes a specific configuration, giving the position of each molecule. A **macrostate** is a less detailed description that doesn't include positions of individual molecules. Here the macrostate description means listing the number of molecules on each side of the box: 0, 1, or 2.

   What's the **probability** of each macrostate? In general, that probability is the number of microstates with that number of molecules on the left, divided by the total number of

TABLE 14.1 Macrostates for a Two-Molecule Gas

Microstates (ways of distributing the two atoms in the two halves of the box)	Macrostates (number of atoms in each half)
	2 | 0
	1 | 1
	0 | 2

TABLE 14.2 Macrostates and Probabilities for a Four-Molecule Gas

Microstates (16 total)	Macrostates	Probability of macrostate
	4 : 0	$\frac{1}{16} = 0.06$
	3 : 1	$\frac{4}{16} = 0.25$
	2 : 2	$\frac{6}{16} = 0.38$
	1 : 3	$\frac{4}{16} = 0.25$
	0 : 4	$\frac{1}{16} = 0.06$

microstates. Thus, the probability of the macrostate with two molecules on the left is $\frac{1}{4}$, because only one of the four microstates has this configuration. Similarly, one configuration has zero molecules on the left, so the probability of this macrostate is also $\frac{1}{4}$. On the other hand, two microstates have one molecule on the left. Therefore, the probability of the macrostate with one molecule on the left is $2/4 = 1/2$. Note that the sum of all the probabilities equals 1 ($= \frac{1}{4} + \frac{1}{4} + \frac{1}{2}$), because the two molecules have to exist in one of the four allowed microstates.

Next, consider what happens when the number of molecules increases to four (Table 14.2). There are now 16 microstates, which group into five macrostates. The last column in Table 14.2 shows the probability of each macrostate, still defined as the number of microstates with that number of molecules on the left, divided by the total number of microstates. Notice that the probability is highest for the "2" macrostate and lowest for the states with all the molecules on one side of the box.

Now increase to 100 molecules. There are 101 macrostates (0, 1, ...99, 100, as shown in Table 14.3). We can't list all the microstates, because they now number 2^{100} or about 1.27×10^{30}!

Figure 14.16a (next page) graphs these probabilities. Notice how sharply peaked the graph is. The most probable macrostates are those that have nearly equal numbers of molecules on left and right. The probability of finding only a few molecules (say 10 or fewer) on one side is ridiculously small—it will almost never happen! The probabilities become even more skewed in a real gas, which might contain some 10^{23} molecules (Figure 14.16b, next page). Your experience confirms this: In a room full of gas molecules, the chances of a sudden drop in the gas density in one part of the room are impossibly small.

TABLE 14.3 Macrostates and Probabilities for the 100-Molecule Gas

Macrostate (number of molecules on left side)	Number of microstates in this macrostate	Probability of this macrostate*
0	1	7.89×10^{-31}
1	100	1.73×10^{-29}
10	1.73×10^{13}	1.36×10^{-17}
40	1.37×10^{28}	0.011
50	1.01×10^{29}	0.080
60	1.37×10^{28}	0.011
90	1.73×10^{13}	1.36×10^{-17}
99	100	1.73×10^{-29}
100	1	7.89×10^{-31}

*Probability of macrostate = number of microstates here divided by total number of microstates (1.27×10^{30}).

Probability, Entropy, and Order

Based on these ideal-gas examples, we can now relate disorder and entropy. In the late 19th century, Ludwig Boltzmann showed that the entropy of a macrostate comprising Ω microstates is

$$S = k_B \ln \Omega \qquad \text{(Boltzmann entropy formula; SI unit: J/K)} \qquad (14.10)$$

where Boltzmann's constant is $k_B = 1.38 \times 10^{-23}$ J/K, as introduced in Chapter 12.

If you apply Boltzmann's formula to the 100-molecule gas (Table 14.3), the results are striking. The entropy of the state with 50 molecules on each side is

$$S = k_B \ln \Omega = (1.38 \times 10^{-23} \text{ J/K}) \ln (1.01 \times 10^{29}) = 9.22 \times 10^{-22} \text{ J/K}$$

The entropy of the state with one molecule on the left is

$$S = k_B \ln \Omega = (1.38 \times 10^{-23} \text{ J/K}) \ln (100) = 6.36 \times 10^{-23} \text{ J/K}$$

while the state with all the molecules on one side has

$$S = k_B \ln \Omega = (1.38 \times 10^{-23} \text{ J/K}) \ln (1) = 0$$

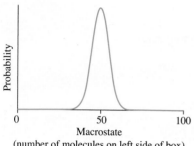

(a)

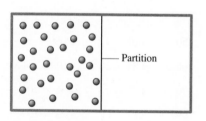

For a 10^{23}-molecule gas, the peak is *much* sharper.

(b)

FIGURE 14.16 Probability distributions for a gas of (a) 100 molecules and (b) 10^{23} molecules.

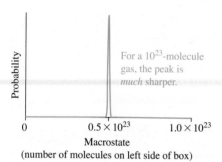

— Partition

When the partition is removed, the molecules spread evenly through the container.

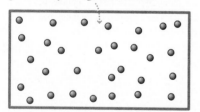

FIGURE 14.17 Opening the partition allows the gas to expand, increasing its entropy.

So here's the connection between entropy and order: The more disordered states, with the greatest mixing of molecules, have the highest entropy. The more ordered states, with most of the molecules on one side, have lower entropy.

Recall our earlier statement of the second law:

> **Second law of thermodynamics:** Natural processes tend to evolve toward a state of maximum entropy.

Figure 14.17 shows a box divided by a removable partition. With the partition in place, fill one side with gas and leave the other side empty. At the moment you open the partition, the entropy is zero, because all the molecules are on one side. Quickly molecules flood the other side, and in a short time you've got equilibrium, with roughly equal numbers of molecules in both sides. Our ideal-gas examples show that this redistribution of molecules is accompanied by an entropy increase. The system has evolved from lower to higher entropy, in accordance with the second law of thermodynamics.

Once equilibrium is established, it's unlikely to be reversed. It would take outside work, such as running a vacuum pump, to get all the gas back to one side. Running the pump would result in another entropy increase (for example, by heat flowing from the pump's motor to the environment) that would offset the decrease associated with rearranging the molecules. As you saw in Section 14.3, an entropy decrease must be accompanied by an equal or larger entropy increase elsewhere.

Entropy, Heat Flow, and the Second Law

We can extend these statistical ideas to heat flow. Consider again the two-chambered box, this time with a hot gas on one side and a cold gas on the other (Figure 14.18). Recall from Chapter 12 that the speeds of gas molecules follow the Maxwell distribution (Figure 12.11), with a hot gas having a higher rms speed than a cold one. Now open the partition, so the hot and cold gases mix. At the moment the partition is opened, the system's entropy is low. That's because with faster molecules on one side and slower molecules on the other, the system is well ordered. Once the gases mix, there's a heat flow from hot to cold, and soon the temperature throughout the box reaches an intermediate equilibrium. Now faster and slower molecules are distributed throughout the box. The system has become more disordered, so entropy has increased. Thus, heat flow results in entropy increase, as in the examples of Section 14.3. Once again, evolution toward higher entropy is consistent with the second law.

Life and the Second Law

The second law of thermodynamics is universal, applying even to living organisms. Even a single cell is a highly ordered system. The second law says that over time systems become more disordered. So how is it that living things are able to live, grow, and survive for years?

As in any thermodynamic system, an entropy decrease in one part of the system must be offset by an entropy increase elsewhere. That's just how we survive. We take energy from our environment, and use some of it to do work and build the organized structures of our bodies. But some energy is lost as heat. As you've seen, such heat flow results in an entropy gain, which offsets the entropy decrease associated with increasing organization.

The plant and animal products we eat consist of highly ordered molecules containing stored energy. After digestion, some of the energy refuels our cells, helping them maintain their order. But the digested food leaves our bodies in a more disordered state and with more entropy than it had when entering. We can trace all our food energy to sunlight, and ultimately the entropy-lowering organization of ourselves and even our societies is more than compensated by increasing entropy associated with nuclear fusion in the Sun's core.

Humans and other living things are improbably ordered systems. Yet we can't escape the second law of thermodynamics. We can work to sustain life and society, but in the universe at large entropy must continue to increase.

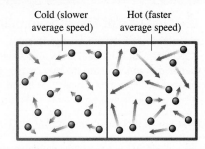

Cold (slower average speed)   Hot (faster average speed)

Chapter 14 in Context

This chapter builds on ideas of thermal physics introduced in Chapters 12 and 13. The *first law of thermodynamics* extends the concept of energy conservation, first discussed in Chapter 5 and used throughout mechanics. The *second law of thermodynamics* is special among the laws of physics in that it has many interpretations and because it establishes an arrow of time that doesn't appear in Newtonian dynamics. Together, the first and second laws of thermodynamics make it possible to understand the operation and limitations of practical devices such as *engines, refrigerators,* and *heat pumps.*

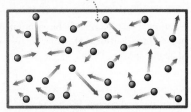

After mixing, average molecular speed is intermediate throughout the container.

Looking Ahead This chapter concludes our discussion of thermal physics. Although tools and concepts from thermal physics will appear later, we'll turn our attention now to other important areas of physics. The second half of this book introduces electricity and magnetism, optics, and topics from modern physics—relativity, quanta, atomic and nuclear physics, elementary particles, and cosmology. You'll see the ideas of work and energy—central to both classical dynamics and thermal physics—in other contexts, for example, in our discussion of electrical energy in Chapters 16 and 19.

FIGURE 14.18 Mixing hot and cold gas increases entropy.

CHAPTER 14 SUMMARY

The First Law of Thermodynamics

(Section 14.1) **Internal energy** is the random kinetic and potential energy associated with individual molecules and their interactions. The **first law of thermodynamics** relates internal energy (U), heat (Q), and work (W). Heat flows into the system when $Q > 0$ and out of the system when $Q < 0$. When volume decreases, $W > 0$, and when volume increases, $W < 0$.

First law of thermodynamics: $\Delta U = Q + W$

The change ΔU in a system's internal energy is the sum of the heat Q transferred to the system and the work W done on it.

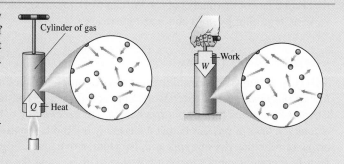

Thermodynamic Processes

(Section 14.2) Work can be computed for constant-pressure, constant-temperature (**isothermal**), constant-volume, and **adiabatic** processes. In an adiabatic process, there is no heat flow.

Metabolic processes can be understood using the first law. Humans replenish energy by consuming food, whose energy content is converted to the body's internal energy. We use that energy to do work, and release energy in the form of heat.

Work done in compressing a gas at constant pressure: $W = -P\Delta V$

Work done in compressing a gas at constant temperature:

$$W = nRT \ln\left(\frac{V_i}{V_f}\right)$$

Work done in an adiabatic process: $W = \dfrac{P_f V_f - P_i V_i}{\gamma - 1}$

No work is done in a constant-volume process.

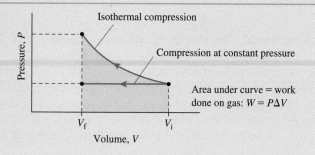

The Second Law of Thermodynamics

(Section 14.3) The **second law of thermodynamics** describes an **arrow of time** for physical processes: Heat flows spontaneously from hotter to cooler objects. **Entropy** is a measure of disorder, and heat flow is accompanied by an overall entropy increase. Natural processes evolve toward states of higher entropy.

Change in entropy at constant temperature: $\Delta S = \dfrac{Q}{T}$

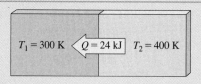

Heat Engines and Refrigerators

(Section 14.4) **Heat engines** convert thermal energy to work but cannot do so with 100% efficiency. The efficiency of an engine is the ratio of the work output to the thermal energy input. The **Carnot engine** has the maximum possible efficiency. **Refrigerators** (and air conditioners and **heat pumps**) remove heat from a cold body and transfer it to a warmer one, requiring work to do so.

Efficiency of a heat engine: $e = 1 - \dfrac{Q_C}{Q_H}$

Efficiency of a Carnot engine: $e_{Carnot} = 1 - \dfrac{T_C}{T_H}$

Coefficient of performance for a refrigerator: $COP = \dfrac{Q_C}{W}$

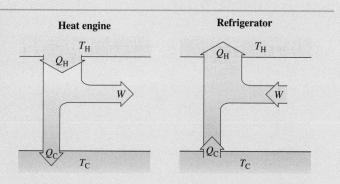

Statistical Interpretation of Entropy

(Section 14.5) Entropy is related to the statistical probability of a system's macroscopic state: the higher the probability, the higher the entropy. A **microstate** is a description of the specific configuration of molecules, noting the position of each. A **macrostate** is a description that does not detail the individual molecules.

Boltzmann (statistical) entropy: $S = k_B \ln \Omega$

More disordered states have higher entropy; more ordered states have lower entropy.

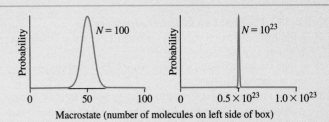

Macrostate (number of molecules on left side of box)

NOTE: Problem difficulty is labeled as ■straightforward to ■■■challenging. Problems labeled BIO are of biological or medical interest.

Conceptual Questions

1. Compare the work done in compressing a gas (a) at constant pressure, (b) adiabatically, and (c) isothermally.
2. After you use a hand pump to inflate a bicycle tire, why is the pump hot? Why does the tire also feel warm?
3. Does the temperature of a gas necessarily increase if simultaneously (a) heat flows in while the gas is compressed and (b) heat flows in while the gas expands?
4. What happens to a gas's temperature when it (a) expands adiabatically and (b) is compressed adiabatically?
5. Exercising vigorously, you do work on your surroundings, so $W < 0$. You also transfer heat to your surroundings, so $Q < 0$. With $W < 0$ and $Q < 0$, why doesn't your body temperature drop?
6. A heat pump uses electrical energy at a rate of 500 W. How can it deliver more than 500 W of heat without violating some law of thermodynamics?
7. Why are air conditioners put in windows instead of in the middle of rooms?
8. In a Carnot engine, the temperature of the hot reservoir is increased. Does this increase or decrease the engine's efficiency?
9. Why can't you cool the room by leaving your refrigerator door open?
10. Assume your house is warmer in summer than in winter. Does it take more or less energy to run your refrigerator in summer?
11. Which has more entropy, a piece of silver when it's solid or when it's liquid?
12. If you mix hot water and cold water, the mixture comes to equilibrium at some intermediate temperature. Has the entropy of the system increased?
13. Suppose the two-chambered box in Section 14.5 contains 50 molecules in each side. As the molecules move randomly, a short time later there are 48 on one side and 52 on the other. Does this constitute a violation of the second law?
14. In an example of *free expansion*, a gas doubles its volume by expanding into a vacuum without pushing on anything. Discuss whether each of the following quantities increase, decrease, or remain the same: pressure, temperature, and entropy.
15. Give an example not mentioned in this chapter in which the first law of thermodynamics is satisfied but the second law would be violated.

Multiple-Choice Problems

16. Heat flows into a system during (a) adiabatic expansion; (b) adiabatic compression; (c) isothermal compression; (d) isothermal expansion.
17. A monatomic ideal gas occupies 23.0 L at 1 atm pressure. Its internal energy is (a) 2300 J; (b) 2800 J; (c) 3100 J; (d) 3500 J.
18. A cylinder with a movable piston encloses 0.25 mole of ideal gas at 300 K. How much heat is required to increase its volume isothermally by a factor of 2? (a) 430 J; (b) 620 J; (c) 900 J; (d) 1730 J.
19. In an adiabatic compression, a gas's temperature (a) increases; (b) decreases; (c) remains the same; (d) cannot be determined from the information given.
20. 2.5 moles of neon gas expand adiabatically and in the process do 930 J of work. The gas's temperature drops by (a) 10°C; (b) 30°C; (c) 45°C; (d) 60°C.
21. Consider a closed cycle on a pressure-versus-volume graph—that is, a cycle in which a gas returns to its initial state, as shown in Figure MC14.21. The work done on the gas is positive (a) if the path is traversed clockwise; (b) if the path is traversed counterclockwise; (c) always; (d) never.

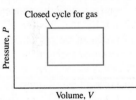

FIGURE MC14.21

22. What's the entropy change of a 50-g ice cube that melts at 0°C? (a) −14 J/K; (b) 42 J/K; (c) 61 J/K; (d) 85 J/K.
23. What's the entropy change of 40 g of water that freezes at 0°C? (a) −39 J/K; (b) −49 J/K; (c) −59 J/K; (d) −69 J/K.
24. A heat engine does 800 J of work and dumps 1400 J of heat to the cold reservoir. What's its efficiency? (a) 0.22; (b) 0.36; (c) 0.57; (d) 0.64.
25. A refrigerator with COP = 3.5 draws 800 W of electric power. At what rate can this refrigerator remove heat from its interior? (a) 229 W; (b) 2000 W; (c) 2800 W; (d) 3600 W.
26. What's the maximum efficiency for a heat engine operating between 273 K and 400 K? (a) 0.32; (b) 0.44; (c) 0.56; (d) 0.68.

Problems

Section 14.1 The First Law of Thermodynamics

27. ■ Find the internal-energy change when 1.25 mol of helium gas is warmed by 100°C.

28. ■ Find the internal-energy change when 1.25 mol of oxygen (O_2) gas is warmed by 100°C.

29. ■ Find the rate of heat flow into a system whose internal energy is increasing at the rate of 45 W, given that the system is doing work at a rate of 165 W.

30. ■■ In a perfectly insulated container, 1.0 kg of water is stirred vigorously until its temperature rises by 7.0°C. How much work was done on the water?

31. ■■ Krypton is a monatomic ideal gas. Suppose you have 4.0 moles of krypton at 293 K and 1 atm. After absorbing 1830 J of heat, the gas temperature has increased by 45°C. (a) How much work is involved in this process? (b) Does the gas volume increase or decrease? (c) Repeat parts (a) and (b) if the same heat flow results in a temperature increase of only 15°C.

32. ■■ Show that $U = \frac{3}{2}PV$ for a monatomic ideal gas.

33. ■■ Find the internal energy of 2.0 mol of argon at 273 K and 1.0 atm. If the gas pressure and volume both double, what's the change in internal energy?

34. ■■ The combustion of 1 L of gasoline releases about 3.4×10^7 J of energy. In one car's engine, about 80% of that energy is lost as heat. (a) How much work is done by the engine for each liter of gasoline burned? (b) If the car requires 4.5×10^5 J of work to travel 1 km, find the fuel efficiency in km/L.

35. ■■ An 8.5-L nitrogen-filled balloon is heated in a warm oven and absorbs 860 J of heat. If its temperature increases by 82°C, how much work is done on the gas?

Section 14.2 Thermodynamic Processes

36. ■ A balloon contains 2.5 L of oxygen gas at 1.0 atm and 20 °C. How much work is done in compressing the balloon at constant pressure until its volume drops by 10%?

37. ■■ For the situation of the preceding problem, find (a) the increase in the gas temperature and (b) the heat that must then be removed to return the gas to its original temperature.

38. ■■ A balloon contains 0.30 mol of helium. It rises, maintaining a constant 300-K temperature, to an altitude where its volume has expanded five times. How much work is done by the gas during this isothermal expansion? Neglect tension forces in the balloon.

39. ■■ The balloon in the preceding problem starts at a pressure of 100 kPa and rises to an altitude where the pressure is 75 kPa, maintaining a constant 300 K temperature. (a) By what factor does its volume increase? (b) How much work does the gas do?

40. ■■ A balloon contains 2.0 mol of helium at 1.0 atm, initially at 200°C. (a) What's the initial volume? (b) What's the volume after the gas cools at constant pressure to 20°C? (c) How much work does the gas do in this process?

41. ■■ Diesel engines rely on compression to raise air in the cylinders to the fuel ignition temperature. In a particular diesel engine, 0.60 L of air initially at 293 K and 1 atm is compressed adiabatically to one-twentieth of its initial volume. Find (a) the final pressure, (b) the work done, and (c) the final temperature.

42. ■■ A piston-cylinder system contains 0.10 mole of ideal gas at 300 K. (a) If the gas expands isothermally, does heat flow into the gas or out? (b) How much heat is required to expand the gas volume by a factor of 3?

43. ■ A container of fixed size contains 8.0 g of helium at 273 K. How much heat is required to double the gas temperature?

44. ■■ Consider air to be a mixture of 80% nitrogen and 20% oxygen. How much work is required to raise the temperature of 1 mole of air by 1°C in an adiabatic process?

45. ■■ You have an expandable container with 0.50 mole of oxygen gas at 273 K and 1 atm. (a) What's the gas volume? (b) If the gas expands adiabatically to twice this volume, find the final pressure and the work done.

46. ■■ Undergoing adiabatic expansion, 2.0 moles of hydrogen gas (H_2) do 750 J of work. (a) Does the gas temperature increase or decrease? Why? (b) Find the temperature change.

47. ■■ Consider the four-step process shown in Figure P14.47. Find (a) the work done in each step and (b) the net work done over the cycle.

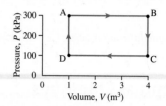

FIGURE P14.47

48. ■■ Repeat the preceding problem if all the arrows are reversed.

49. ■■ A gas is compressed along path AB in Figure P14.49. (a) Does its internal energy increase or decrease in this process? (b) Find the work done in this process.

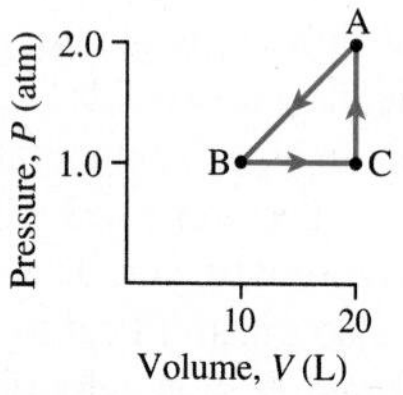

FIGURE P14.49

50. ■■■ Consider the three-step cycle described by the complete triangle in Figure P14.49. (a) Is the net work done over the cycle positive or negative? Why? (b) Find the net work.

51. ■■ A high-performance gasoline engine has a compression ratio of 10:1, meaning that air entering the cylinders is compressed to one-tenth its volume as the piston rises. If air enters this engine's cylinders at 290 K and 1 atm, find (a) the pressure and (b) the temperature of the air at full compression. Assume adiabatic compression.

52. ■■■ The pressure on a sample of water is increased from 1 atm to 200 atm, reducing its volume by 1%. (a) Draw the pressure-versus-volume graph for this process. (b) Estimate the work required. (c) Assume that the process takes place quickly enough so that no heat flows out of the water. By how much does the water's temperature increase?

53. BIO ■■ **Food, heat, and work.** A runner uses about 80 kcal of internal energy to run each kilometer. (a) If the runner loses 120 g of perspiration while running 1 km, find the heat lost in the process. (b) How much work does the runner do in running that kilometer?

54. BIO ■■ **Heat and physical activity.** In a bench press, a weight lifter lifts 125 kg 42 cm. In so doing, the athlete expends 1.3 kJ. (a) What's the heat flow involved in this process? (b) How many repetitions does the weight lifter need to make in order to "burn" 100 kcal?

Section 14.3 The Second Law of Thermodynamics

55. ■ Find the entropy change when water at 25°C (a) absorbs 1000 J of heat and (b) loses 1000 J of heat. Assume there's enough water that its temperature change is negligible.

56. ■ ■ Find the entropy change when 75 g of steam condenses to water at 100°C.

57. ■ ■ Find the entropy created when 100 g of ice melts at 0°C. Compare with the entropy created when the same mass of water boils at 100°C.

58. ■ ■ Find the net entropy change when a 50-g ice cube initially at 0°C melts in (a) a 10°C room and (b) a 35°C room.

59. ■ ■ 150 g of water at 0°C are placed in a freezer at −4°C. Find the net entropy change after all the water has turned to ice but is still at 0°C.

60. ■ ■ ■ Using data from Chapter 13, find the entropy increase in 1.5 kg of mercury when it melts at −39°C and 1 atm pressure.

61. ■ ■ ■ Using data from Chapter 13, find the entropy increase when 100 g of nitrogen boil at atmospheric pressure.

62. ■ ■ ■ Earth receives energy from the Sun at an average rate of about 240 W/m². Given Earth's average temperature of 15°C, find the rate of entropy increase associated with this energy flow.

Section 14.4 Heat Engines and Refrigerators

63. ■ A heat engine does 650 J of work and dumps 1270 J of heat to its cold reservoir. What's its efficiency?

64. ■ ■ A 40% efficient heat engine does mechanical work at the rate of 400 W. If it runs for an entire day, find the heat removed from the hot reservoir and the heat transferred to the cold reservoir.

65. ■ ■ Each second a nuclear power plant extracts 1700 MJ of thermal energy from its uranium fuel and dumps 1100 MJ of waste heat into a river. (a) What's its efficiency? (b) At what rate does it produce electrical energy, assuming all the work it does ends up as electricity?

66. ■ ■ An automobile engine with an efficiency 0.21 dumps heat to the environment at the rate of 330 kW. What's its mechanical power output, in W and in hp?

67. ■ Find the maximum possible efficiency of a heat engine operating between the freezing and boiling points of water.

68. ■ Find the maximum temperature in a Carnot cycle operating in a room-temperature (20°C) environment if its Carnot efficiency is 0.5.

69. ■ ■ ■ One alternative energy source is the temperature difference between surface and deep water in the tropical oceans. Suppose a heat engine uses 25°C surface water as its hot reservoir and 4°C deep water as its cold reservoir. (a) Find the maximum efficiency of such an engine. (b) How much heat would have to be removed from the surface water each day in order to produce energy at the rate of a typical large power plant, about 1000 MW?

70. ■ ■ ■ A power plant produces electrical energy at the rate of 1300 MW with an efficiency of 0.31. The excess heat is dumped into a river that carries 1500 m³/s of water. How much does the river's temperature increase?

71. ■ A refrigerator with COP = 3.8 consumes electrical energy at the rate of 600 W. At what rate can this refrigerator remove heat from its interior?

72. ■ ■ A window air conditioner consumes 1200 W of electric power and has COP = 3.2. If it runs continuously for 24 h, how much heat gets removed from the house?

73. ■ Find the maximum possible COP for a refrigerator that liquefies nitrogen at 77 K in a lab at 22°C.

74. ■ Find the maximum possible COP for a freezer that maintains −5°C when the surrounding room temperature is 25°C.

75. ■ ■ ■ Winter heating of an office building requires an average of 250 kBtu/h. The heating system uses a heat pump with COP = 3.0. (a) If the energy supplied to the heat pump costs $0.09/kWh,

what's the cost per day to heat the building? (b) What's the daily monetary savings, compared with heat from oil at $4.20 per gallon? The oil contains 40 kWh per gallon and burns at 87% efficiency.

Section 14.5 Statistical Interpretation of Entropy

76. ■ Construct tables similar to Table 14.2 for (a) five and (b) six molecules.

77. ■ ■ For both cases in the preceding problem, show that the probabilities of all the macrostates sum to 1.

78. ■ ■ Compute the entropy of each macrostate in the gases of Problem 76.

79. ■ ■ ■ For an N-molecule gas, show that the number of microstates with n molecules on the left side of the box is $N!/[n!(N - n)!]$. *Hint:* Consider the pattern established by the cases of two, three, four, and five molecules.

80. ■ ■ ■ You flip a coin 50 times. (a) How many different microstates are there, counting as microstates each distinct ordering of heads and tails in the sequence? (b) Find the probability of getting heads 0 times, 10 times, 25 times, 40 times, and 50 times.

81. ■ ■ ■ A standard deck of cards contains 52 different cards. A poker hand consists of five cards, chosen randomly. How many different poker hands are there?

82. ■ ■ ■ In the game of poker (see the preceding problem), three of a kind (for example, three queens) beats two pair (for example, two aces and two fives). The better hand is the one with the lower probability. Show that three of a kind is less probable than two pair.

General Problems

83. ■ ■ An ideal gas expands 10-fold in volume, maintaining a constant 440-K temperature. If the gas does 3.3 kJ of work, (a) how much heat does it absorb and (b) how many moles of gas are there?

84. ■ It takes 600 J to compress a gas isothermally to half its initial volume. How much work would it take to compress it by a factor of 10 from its original volume?

85. ■ ■ A gas undergoes an adiabatic compression that halves its volume. If its pressure increases by a factor of 2.55, what's the specific-heat ratio γ?

86. ■ ■ A gas undergoes the cyclic process ABCA shown in Figure GP14.86, where AB lies on an isotherm. The pressure at point A is 60 kPa. Find (a) the pressure at B and (b) the net work done on the gas.

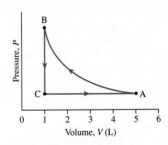

FIGURE GP14.86

87. ■ ■ Repeat the preceding problem if AB is an adiabat and $\gamma = 1.4$.

88. ■ ■ The McNeil Generating Station in Burlington, Vermont is one of the world's largest wood-fired electric power plants. The plant produces steam at 950°F to drive its turbines, and condensed steam returns to the boiler at 90°F. (Note the temperatures in Fahrenheit, used in U.S. engineering situations.) Find McNeil's maximum thermodynamic efficiency and compare with its actual efficiency of 25%. *Note:* Some of the difference comes from having to evaporate moisture out of the wood-chip fuel.

89. ■ ■ A nuclear power plant has a maximum steam temperature (T_H) of 310°C. It produces 650 MW of electric power in the winter, when its T_C is effectively 0°C. (a) Find its maximum winter efficiency. (b) If its summertime T_C is 38°C, what's its summertime electric power output, assuming nothing else changes?

90. BIO ■ ■ **Blood flow.** As the heart beats, blood pressure in an artery varies from 80 mm Hg to 125 mm Hg (gauge pressures). An air bubble trapped in the artery has a diameter of 1.52 mm at the minimum pressure. (a) What will its diameter be at the maximum pressure? (b) How much work does the blood (and ultimately the heart) do in compressing the bubble? Assume a constant 37°C temperature.

91. ■ ■ A gas with $\gamma = 1.4$ occupies 5.0 L at 100 kPa pressure. (a) What's the gas's final pressure? (b) How much work is needed to compress the gas adiabatically to 2.5 L?

92. ■ ■ ■ Two moles of ideal gas with molar specific heat $c_V = \frac{5}{2}R$ are at $T = 300$ K and $P = 100$ kPa. Determine the final temperature and the work done on the gas when 1.5 kJ of heat is added (a) isothermally, (b) at constant volume, and (c) at constant pressure.

93. ■ ■ ■ Warm winds called Chinooks sweep across the plains just east of the Rocky Mountains. They carry air from high in the mountains down to the plains so fast that the air has no time to exchange significant heat with its surroundings (Figure GP14.93). On the day of a Chinook, pressure and temperature high in the Alberta Rockies are 62.0 kPa and −11.0°C, respectively. (a) What will the air temperature be once it's reached the city of Calgary, where the pressure is 86.5 kPa? (b) How much work is done on what was initially a cubic meter of air as it descends to the plain?

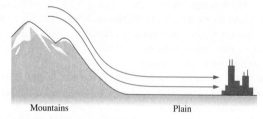

Mountains   Plain

FIGURE GP14.93

94. ■ ■ It costs $180 to heat a home with electricity in a typical winter month. (An electric furnace converts all the electrical energy to heat.) What's the monthly heating bill following conversion to an electrically powered heat pump with COP = 3.1?

95. ■ ■ A 4.0-L sample of water at 9.0°C is put into a refrigerator. The refrigerator's 130-W motor runs for 4.0 min to cool the water to the refrigerator's low temperature of 1.0°C. (a) What's the COP of the refrigerator? How does this compare with the maximum possible COP if the refrigerator exhausts heat to the kitchen at 25°C?

96. ■ A jet aircraft engine is a *gas turbine,* using hot combustion gases directly to spin a turbine. Gas turbines operate at high temperatures (T_H), but they aren't very efficient because their exhaust temperature (T_C) is also high. Find the maximum possible efficiency of a gas turbine operating between 1050°C and 590°C.

97. ■ ■ Despite its inefficiency, the gas turbine enables our most efficient fossil-fueled power plant, the *combined-cycle gas turbine.* This system burns natural gas in a gas turbine, then uses the turbine's exhaust to drive a conventional steam cycle. Consider a system using the gas turbine of the preceding problem, with the steam cycle then operating between the turbine's 590°C exhaust and discharging waste heat at 42°C. Find (a) the maximum efficiency of the steam cycle alone and (b) the maximum efficiency of the combined cycle.

98. BIO ■ ■ ■ **Human entropy production.** An athlete with a daily food intake of 2800 kcal uses only about 25% of that energy for vital body functions, with the rest lost as heat. The heat is conducted from the body's core at 37°C to the outer skin, which has an average temperature of 30°C. (a) How much entropy is produced each day in this process? (b) How much additional entropy is produced when the heat is radiated away to the surrounding air, at 20°C?

99. ■ ■ ■ An ideal gas with $\gamma = 5/3$ starts at point A in Figure GP14.99, where its volume and pressure are 1.00 m^3 and 250 kPa, respectively. It undergoes adiabatic expansion that triples its volume, ending at point B. It's then heated at constant volume to point C, and finally is compressed isothermally back to point A. Find (a) the pressure of the gas at B, (b) the pressure at C, and (c) the net work done on the gas.

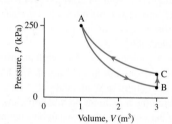

FIGURE GP14.99

Answers to Chapter Questions

Answer to Chapter-Opening Question
It's not poor engineering. It's a fundamental limitation on our ability to convert thermal energy to mechanical work, embodied in the second law of thermodynamics.

Answers to GOT IT? Questions
Section 14.1 (a) Decreases; (b) can't be determined; (c) decreases; (d) increases; (e) can't be determined
Section 14.2 (b) an isothermal compression
Section 14.3 (a) 6.06 kJ
Section 14.4 (b) 48%

Quadratic Formula

If $ax^2 + bx + c = 0$, then $x = \dfrac{-b \pm \sqrt{b^2 - 4ac}}{2a}$.

Circumference, Area, Volume

Where $\pi \simeq 3.14159\ldots$

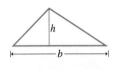

circumference of circle	$2\pi r$
area of circle	πr^2
surface area of sphere	$4\pi r^2$
volume of sphere	$\frac{4}{3}\pi r^3$
area of triangle	$\frac{1}{2}bh$
volume of cylinder	$\pi r^2 l$

Trigonometry

definition of angle (in radians): $\theta = \dfrac{s}{r}$

2π radians in complete circle

1 radian $\simeq 57.3°$

Trigonometric Functions

$\sin\theta = \dfrac{y}{r}$

$\cos\theta = \dfrac{x}{r}$

$\tan\theta = \dfrac{\sin\theta}{\cos\theta} = \dfrac{y}{x}$

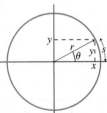

Values at Selected Angles

$\theta \rightarrow$	0	$\dfrac{\pi}{6}$ (30°)	$\dfrac{\pi}{4}$ (45°)	$\dfrac{\pi}{3}$ (60°)	$\dfrac{\pi}{2}$ (90°)
$\sin\theta$	0	$\dfrac{1}{2}$	$\dfrac{\sqrt{2}}{2}$	$\dfrac{\sqrt{3}}{2}$	1
$\cos\theta$	1	$\dfrac{\sqrt{3}}{2}$	$\dfrac{\sqrt{2}}{2}$	$\dfrac{1}{2}$	0
$\tan\theta$	0	$\dfrac{\sqrt{3}}{3}$	1	$\sqrt{3}$	∞

Graphs of Trigonometric Functions

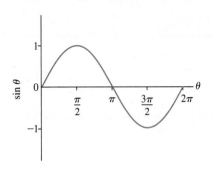

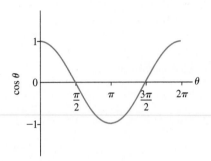

Trigonometric Identities

$\sin(-\theta) = -\sin\theta$

$\cos(-\theta) = \cos\theta$

$\sin\left(\theta \pm \dfrac{\pi}{2}\right) = \pm\cos\theta$

$\cos\left(\theta \pm \dfrac{\pi}{2}\right) = \mp\sin\theta$

$\sin^2\theta + \cos^2\theta = 1$

$\sin 2\theta = 2\sin\theta\cos\theta$

Laws of Cosines and Sines

Where A, B, C are the sides of an arbitrary triangle and α, β, γ are the angles opposite those sides:

Law of cosines

$C^2 = A^2 + B^2 - 2AB\cos\gamma$

Law of sines

$\dfrac{\sin\alpha}{A} = \dfrac{\sin\beta}{B} = \dfrac{\sin\gamma}{C}$

Exponentials and Logarithms

$e^{\ln x} = x \quad \ln e^x = x \quad e = 2.71828\ldots$

$a^x = e^{x \ln a}$ $\qquad \ln(xy) = \ln x + \ln y$

$a^x a^y = a^{x+y}$ $\qquad \ln\left(\dfrac{x}{y}\right) = \ln x - \ln y$

$(a^x)^y = a^{xy}$ $\qquad \ln\left(\dfrac{1}{x}\right) = -\ln x$

$\qquad\qquad\qquad 10^{\log x} = x$

Approximations

For $|x| \ll 1$, the following expressions provide good approximations to common functions:

$e^x \simeq 1 + x$

$\sin x \simeq x$

$\cos x \simeq 1 - \frac{1}{2}x^2$

$\ln(1 + x) \simeq x$

$(1 + x)^p \simeq 1 + px \qquad$ (binomial approximation)

Expressions that don't have the forms shown may often be put in the appropriate form. For example:

$$\frac{1}{\sqrt{a^2 + y^2}} = \frac{1}{a\sqrt{1 + \dfrac{y^2}{a^2}}} = \frac{1}{a}\left(1 + \frac{y^2}{a^2}\right)^{-1/2} \approx \frac{1}{a}\left(1 - \frac{y^2}{2a}\right)$$

for $y^2/a^2 \ll 1$, or $y^2 \ll a^2$.

Unit Vector Notation

An arbitrary vector $\vec{A}$ may be written in terms of its components A_x, A_y, A_z and the unit vectors $\hat{\imath}, \hat{\jmath}, \hat{k}$ that have length 1 and lie along the x-, y-, z-axes:

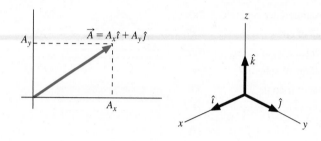

The International System of Units (SI)

This material is from the U.S. edition of the English translation of the seventh edition of "Le Système International d'Unités (SI)," the definitive publication in the French language issued in 1991 by the International Bureau of Weights and Measures (BIPM). The year the definition was adopted is given in parentheses.

length (meter): The meter is the length of the path traveled by light in vacuum during a time interval of 1/299 792 458 of a second. (1983)

mass (kilogram): The kilogram is equal to the mass of the international prototype of the kilogram. (1889)

time (second): The second is the duration of 9 192 631 770 periods of the radiation corresponding to the transition between the two hyperfine levels of the ground state of the cesium-133 atom. (1967)

electric current (ampere): The ampere is that constant current which, if maintained in two straight parallel conductors of infinite length, of negligible circular cross section, and placed 1 meter apart in vacuum, would produce between these conductors a force equal to 2×10^{-7} newton per meter of length. (1948)

temperature (kelvin): The kelvin, unit of thermodynamic temperature, is the fraction 1/273.16 of the thermodynamic temperature of the triple point of water. (1967)

amount of substance (mole): The mole is the amount of substance of a system that contains as many elementary entities as there are atoms in 0.012 kilogram of carbon-12. (1971)

luminous intensity (candela): The candela is the luminous intensity, in a given direction, of a source that emits monochromatic radiation of frequency 540×10^{12} hertz and that has a radiant intensity in that direction of (1/683) watt per steradian. (1979)

SI Base and Supplementary Units

Quantity	SI Unit Name	SI Unit Symbol
Base Unit		
Length	meter	m
Mass	kilogram	kg
Time	second	s
Electric current	ampere	A
Thermodynamic temperature	kelvin	K
Amount of substance	mole	mol
Luminous intensity	candela	cd
Supplementary Units		
Plane angle	radian	rad
Solid angle	steardian	sr

SI Prefixes

Factor	Prefix	Symbol
10^{24}	yotta	Y
10^{21}	zetta	Z
10^{18}	exa	E
10^{15}	peta	P
10^{12}	tera	T
10^{9}	giga	G
10^{6}	mega	M
10^{3}	kilo	k
10^{2}	hecto	h
10^{1}	deka	da
10^{0}	—	—
10^{-1}	deei	d
10^{-2}	centi	c
10^{-3}	milli	m
10^{-6}	micro	μ
10^{-9}	nano	n
10^{-12}	pico	p
10^{-15}	femto	f
10^{-18}	atto	a
10^{-21}	zepto	z
10^{-24}	yocto	y

Some SI Derived Units with Special Names

Quantity	Name	Symbol	SI Unit Expression in Terms of Other Units	SI Unit Expression in Term of SI Base Units
Frequency	hertz	Hz		s^{-1}
Force	newton	N		$m \cdot kg \cdot s^{-2}$
Pressure, stress	pascal	Pa	N/m^2	$m^{-1} \cdot kg \cdot s^{-2}$
Energy, work, heat	joule	J	$N \cdot m$	$m^2 \cdot kg \cdot s^{-2}$
Power	watt	W	J/s	$m^2 \cdot kg \cdot s^{-3}$
Electric charge	coulomb	C		$s \cdot A$
Electric potential, potential difference, electromotive force	volt	V	J/C	$m^2 \cdot kg \cdot s^{-3} \cdot A^{-1}$
Capacitance	farad	F	C/V	$m^{-2} \cdot kg^{-1} \cdot s^4 \cdot A^2$
Electric resistance	ohm	Ω	V/A	$m^2 \cdot kg \cdot s^{-3} \cdot A^{-2}$
Magnetic flux	weber	Wb	$V \cdot s$	$m^2 \cdot kg \cdot s^{-2} \cdot A^{-1}$
Magnetic field	tesla	T	Wb/m^2	$kg \cdot s^{-2} \cdot A^{-1}$
Inductance	henry	H	Wb/A	$m^2 \cdot kg \cdot s^{-2} \cdot A^{-2}$
Radioactivity	becquerel	Bq	1 decay/s	s^{-1}
Absorbed radiation dose	gray	Gy	J/Kg, 100 rad	$m^2 \cdot s^{-2}$
Radiation dose equivalent	sievert	Sv	J/Kg, 100 rem	$m^2 \cdot s^{-2}$

Conversion Factors

The listings below give the SI equivalents of non-SI units. To convert from the units shown to SI, multiply by the factor given; to convert the other way, divide. For conversions within the SI system, see the table of SI prefixes in Appendix B, Chapter 1, or the inside front cover. Conversions that are not exact by definition are given to, at most, four significant figures.

Length

1 inch (in) $= 0.0254$ m

1 foot (ft) $= 0.3048$ m

1 yard (yd) $= 0.9144$ m

1 mile (mi) $= 1609$ m

1 nautical mile $= 1852$ m

1 angstrom (Å) $= 10^{-10}$ m

1 light year (ly) $= 9.46 \times 10^{15}$ m

1 astronomical unit (AU) $= 1.5 \times 10^{11}$ m

1 parsec $= 3.09 \times 10^{16}$ m

1 fermi $= 10^{-15}$ m $= 1$ fm

Mass

1 slug $= 14.59$ kg

1 metric ton (tonne; t) $= 1000$ kg

1 unified mass unit (u) $= 1.661 \times 10^{-27}$ kg

Force units in the English system are sometimes used (incorrectly) for mass. The units given below are actually equal to the number of kilograms multiplied by g, the acceleration of gravity.

1 pound (lb) $=$ weight of 0.454 kg

1 ton $= 2000$ lb $=$ weight of 908 kg

1 ounce (oz) $=$ weight of 0.02835 kg

Time

1 minute (min) $= 60$ s

1 hour (h) $= 60$ min $= 3600$ s

1 day (d) $= 24$h $= 86{,}400$ s

1 year (y) $= 365.2422$ d $= 3.156 \times 10^{7}$ s

Area

1 hectare (ha) $= 10^{4}$ m^2

1 square inch (in^2) $= 6.452 \times 10^{-4}$ m^2

1 square foot (ft^2) $= 9.290 \times 10^{-2}$ m^2

1 acre $= 4047$ m^2

1 barn $= 10^{-28}$ m^2

1 shed $= 10^{-30}$ m^2

Volume

1 liter (L) $= 1000$ cm^3 $= 10^{-3}$ m^3

1 cubic foot (ft^3) $= 2.832 \times 10^{-2}$ m^3

1 cubic inch (in^3) $= 1.639 \times 10^{-5}$ m^3

1 fluid ounce $= 1/128$ gal $= 2.957 \times 10^{-5}$ m^3

1 barrel $= 42$ gal $= 0.1590$ m^3

1 gallon (U.S.; gal) $= 3.785 \times 10^{-3}$ m^3

1 gallon (British) $= 4.546 \times 10^{-3}$ m^3

Angle, Phase

1 degree (°) $= \pi/180$ rad $= 1.745 \times 10^{-2}$ rad

1 revolution (rev) $= 360° = 2\pi$ rad

1 cycle $= 360° = 2\pi$ rad

*The length of the year changes very slowly with changes in Earth's orbital period.

Speed, Velocity

$1 \text{ km/h} = (1/3.6) \text{ m/s} = 0.2778 \text{ m/s}$ $1 \text{ ft/s} = 0.3048 \text{ m/s}$

$1 \text{ mi/h (mph)} = 0.4470 \text{ m/s}$ $1 \text{ ly/y} = 3.00 \times 10^8 \text{ m/s}$

Angular Speed, Angular Velocity, Frequency, and Angular Frequency

$1 \text{ rev/s} = 2\pi \text{ rad/s} = 6.283 \text{ rad/s } (\text{s}^{-1})$ $1 \text{ rev/min (rpm)} = 0.1047 \text{ rad/s } (\text{s}^{-1})$

$1 \text{ Hz} = 1 \text{ cycle/s} = 2\pi \text{ s}^{-1}$

Force

$1 \text{ dyne} = 10^{-5} \text{ N}$ $1 \text{ pound (lb)} = 4.448 \text{ N}$

Pressure

$1 \text{ dyne/cm}^2 = 0.10 \text{ Pa}$ $1 \text{ lb/in}^2 \text{ (psi)} = 6.895 \times 10^3 \text{ Pa}$

$1 \text{ atmosphere (atm)} = 1.013 \times 10^5 \text{ Pa}$ $1 \text{ in } H_2O \text{ (60°F)} = 248.8 \text{ Pa}$

$1 \text{ torr} = 1 \text{ mm Hg at } 0°C = 133.3 \text{ Pa}$ $1 \text{ in Hg (60°F)} = 3.377 \times 10^3 \text{ Pa}$

$1 \text{ bar} = 10^5 \text{ Pa} = 0.987 \text{ atm}$

Energy, Work, Heat

$1 \text{ erg} = 10^{-7} \text{ J}$ $1 \text{ Btu}^* = 1.054 \times 10^3 \text{ J}$

$1 \text{ calorie}^* \text{(cal)} = 4.184 \text{ J}$ $1 \text{ kWh} = 3.6 \times 10^6 \text{ J}$

$1 \text{ electronvolt (eV)} = 1.602 \times 10^{-19} \text{ J}$ $1 \text{ megaton (explosive yield: Mt)}$

$1 \text{ foot-pound (ft·lb)} = 1.356 \text{ J}$ $= 4.18 \times 10^{15} \text{ J}$

Power

$1 \text{ erg/s} = 10^{-7} \text{ W}$ $1 \text{ Btu/h (Btuh)} = 0.293 \text{ W}$

$1 \text{ horsepower (hp)} = 746 \text{ W}$ $1 \text{ ft·lb/s} = 1.356 \text{ W}$

Magnetic Field

$1 \text{ gauss (G)} = 10^{-4} \text{ T}$

Radiation

$1 \text{ curie (ci)} = 3.7 \times 10^{10} \text{ Bq}$ $1 \text{ rad} = 10^{-2} \text{ Gy}$

$1 \text{ rem} = 10^{-2} \text{ Sv}$

Energy Content of Fuels

Energy Source	Energy Content
Coal	$29 \text{ MJ/kg} = 7300 \text{ kWh/ton} = 25 \times 10^6 \text{ Btu/ton}$
Oil	$43 \text{ MJ/kg} = 39 \text{ kWh/gal} = 1.3 \times 10^5 \text{ Btu/gal}$
Gasoline	$44 \text{ MJ/kg} = 36 \text{ kWh/gal} = 1.2 \times 10^5 \text{ Btu/gal}$
Natural gas	$55 \text{ MJ/kg} = 30 \text{ kWh/100 ft}^3 = 1000 \text{ Btu/ft}^3$
Uranium (fission)	
Normal abundance	$5.8 \times 10^{11} \text{ J/kg} = 1.6 \times 10^5 \text{ kWh/kg}$
Pure U-235	$8.2 \times 10^{13} \text{ J/kg} = 2.3 \times 10^7 \text{ kWh/kg}$
Hydrogen (fusion)	
Normal abundance	$7 \times 10^{11} \text{ J/kg} = 3.0 \times 10^4 \text{ kWh/kg}$
Pure deuterium	$3.3 \times 10^{14} \text{ J/kg} = 9.2 \times 10^7 \text{ kWh/kg}$
Water	$1.2 \times 10^{10} \text{ J/kg} = 1.3 \times 10^4 \text{ kWh/gal} = 340 \text{ gal gasoline/gal}$
H_2O	
100% conversion, matter to energy	$9.0 \times 10^{16} \text{ J/kg} = 931 \text{ MeV/u} = 2.5 \times 10^{10} \text{ kWh/kg}$

*Values based on the thermochemical calorie; other definitions vary slightly.

Properties of Selected Isotopes

Subatomic Particle Masses

Particle	Mass (u)	Mass (kg)
Electron	$5.48\,580 \times 10^{-4}$	9.1094×10^{-31}
Proton	1.007 276	1.6726×10^{-27}
Neutron	1.008 665	1.6749×10^{-27}
Alpha particle	4.001 506	6.6447×10^{-27}

Atomic Number (Z)	Element	Symbol	Mass Number (A)	Atomic Mass* (u)	Abundance (%) or Decay Mode [†] (if Radioactive)	Half-Life (if Radioactive)
0	(Neutron)	n	1	1.008 665	β^-	10.6 min
1	Hydrogen	H	1	1.007 825	99.985	
	Deuterium	D	2	2.014 102	0.015	
	Tritium	T	3	3.016 049	β^-	12.33 y
2	Helium	He	3	3.016 029	0.00014	
			4	4.002 603	≈ 100	
3	Lithium	Li	6	6.015 123	7.5	
			7	7.016 005	92.5	
4	Beryllium	Be	7	7.016 930	EC, γ	53.3 d
			8	8.005 305	2α	6.7×10^{-17} s
			9	9.012 183	100	
5	Boron	B	10	10.012 938	19.8	
			11	11.009 305	80.2	
			12	12.014 353	β^-	20.4 ms
6	Carbon	C	11	11.011 433	β^+, EC	20.4 ms
			12	12.000 000	98.89	
			13	13.003 355	1.11	
			14	14.003 242	β^-	5730 y
7	Nitrogen	N	13	13.005 739	β^-	9.96 min
			14	14.003 074	99.63	
			15	15.000 109	0.37	
8	Oxygen	O	15	15.003 065	β^+, EC	122 s
			16	15.994 915	99.76	
			18	17.999 159	0.204	
9	Fluorine	F	19	18.998 403	100	
10	Neon	Ne	20	19.992 439	90.51	
			22	21.991 384	9.22	
11	Sodium	Na	22	21.994 435	β^+, EC, γ	2.602 y
			23	22.989 770	100	
			24	23.990 964	β^-, γ	15.0 h
12	Magnesium	Mg	24	23.985 045	78.99	
13	Aluminum	Al	27	26.981 541	100	
14	Silicon	Si	28	27.976 928	92.23	
			31	30.975 364	β^-, γ	2.62 h
15	Phosphorus	P	31	30.973 763	100	
			32	31.973 908	β^-	14.28 d
16	Sulfur	S	32	31.972 072	95.0	
			35	34.969 033	β^-	87.4 d

cont'd.

Atomic Number (Z)	Element	Symbol	Mass Number (A)	Atomic Mass* (u)	Abundance (%) or Decay Mode † (if Radioactive)	Half-Life (if Radioactive)
17	Chlorine	Cl	35	34.968 853	75.77	
			37	36.965 903	24.23	
18	Argon	Ar	40	39.962 383	99.60	
19	Potassium	K	39	38.963 708	93.26	
			40	39.964 000	β^-, EC, γ, β^+	1.28×10^9 y
20	Calcium	Ca	40	39.962 591	96.94	
24	Chromium	Cr	52	51.940 510	83.79	
25	Manganese	Mn	55	54.938 046	100	
26	Iron	Fe	56	55.934 939	91.8	
27	Cobalt	Co	59	58.933 198	100	
			60	59.933 820	β^-, γ	5.271 y
28	Nickel	Ni	58	57.935 347	68.3	
			59	58.934 352	β^+, EC	7.6×10^4 y
			60	59.930 789	26.1	
			64	63.927 968	0.91	
29	Copper	Cu	63	62.929 599	69.2	
			64	63.929 766	β^-, β^+	12.7 h
			65	64.927 792	30.8	
30	Zinc	Zn	64	63.929 145	48.6	
			66	65.926 035	27.9	
33	Arsenic	As	75	74.921 596	100	
35	Bromine	Br	79	78.918 336	50.69	
36	Krypton	Kr	84	83.911 506	57.0	
			89	88.917 563	β^-	3.2 min
38	Strontium	Sr	86	85.909 273	9.8	
			88	87.905 625	82.6	
			90	89.907 746	β^-	28.8 y
39	Yttrium	Y	89	88.905 848	100	
43	Technetium	Tc	98	97.907 210	β^-, γ	4.2×10^6 y
47	Silver	Ag	107	106.905 095	51.83	
			109	108.904 754	48.17	
48	Cadmium	Cd	114	113.903 361	28.7	
49	Indium	In	115	114.903 88	95.7; β^-	5.1×10^{14} y
50	Tin	Sn	120	119.902 199	32.4	
53	Iodine	I	127	126.904 477	100	
			131	130.906 118	β^-, γ	8.04 d
54	Xenon	Xe	132	131.904 15	26.9	
			136	135.907 22	8.9	
55	Cesium	Cs	133	132.905 43	100	
56	Barium	Ba	137	136.905 82	11.2	
			138	137.905 24	71.7	
			144	143.922 73	β^-	11.9 s
61	Promethium	Pm	145	144.912 75	EC, α, γ	17.7 y
74	Tungsten	W	184	183.950 95	30.7	
76	Osmium	Os	191	190.960 94	β^-, γ	15.4 d
			192	191.961 49	41.0	
78	Platinum	Pt	195	194.964 79	33.8	
79	Gold	Au	197	196.966 56	100	
80	Mercury	Hg	202	201.970 63	29.8	
81	Thallium	Tl	205	204.974 41	70.5	
			210	209.990 069	β^-	1.3 min
82	Lead	Pb	204	203.974 044	β^-, 1.48	1.4×10^{17} y
			206	205.974 46	24.1	
			207	206.975 89	22.1	
			208	207.976 64	52.3	
			210	209.984 18	α, β^-, γ	22.3 y
			211	210.988 74	β^-, γ	36.1 min
			212	211.991 88	β^-, γ	10.64 h
			214	213.999 80	β^-, γ	26.8 min

cont'd.

Atomic Number (Z)	Element	Symbol	Mass Number (A)	Atomic Mass* (u)	Abundance (%) or Decay Mode † (if Radioactive)	Half-Life (if Radioactive)
83	Bismuth	Bi	209	208.980 39	100	
			211	210.987 26	α, β^-, γ	2.15 min
84	Polonium	Po	210	209.982 86	α, γ	138.38 d
			214	213.995 19	α, γ	164 μs
86	Radon	Rn	222	222.017 574	α, β	3.8235 d
87	Francium	Fr	223	223.019 734	α, β^-, γ	21.8 min
88	Radium	Ra	226	226.025 406	α, γ	1.60×10^3 y
			228	228.031 069	β^-	5.76 y
89	Actinium	Ac	227	227.027 751	α, β^-, γ	21.773 y
90	Thorium	Th	228	228.028 73	α, γ	1.9131 y
			232	232.038 054	100; α, γ	1.41×10^{10} y
92	Uranium	U	232	232.037 14	α, γ	72 y
			233	233.039 629	α, γ	1.592×10^5 y
			235	235.043 923	0.72; α, γ	7.038×10^8 y
			236	236.045 563	α, γ	2.342×10^7 y
			238	238.050 786	99.275; α, γ	4.468×10^9 y
			239	239.054 291	β^-, γ	23.5 min
93	Neptunium	Np	239	239.052 932	β^-, γ	2.35 d
94	Plutonium	Pu	239	239.052 158	α, γ	2.41×10^4 y
95	Americium	Am	243	243.061 374	α, γ	7.37×10^3 y
96	Curium	Cm	245	245.065 487	α, γ	8.5×10^3 y
97	Berkelium	Bk	247	247.070 03	α, γ	1.4×10^3 y
98	Californium	Cf	249	249.074 849	α, γ	351 y
99	Einsteinium	Es	254	254.088 02	α, γ, β^-	276 d
100	Fermium	Fm	253	253.085 18	EC, α, γ	3.0 d

*The masses given throughout this table are those for the neutral atom, including the Z electrons.
†"EC" stands for electron capture.

Astrophysical Data

Sun, Planets, Principal Satellites

Body	Mass (10²⁴ kg)	Mean Radius (10⁶ m Except as Noted)	Surface Gravity (m/s²)	Escape Speed (km/s)	Sidereal Rotation Period* (days)	Mean Distance from Central Body† (10⁶ km)	Orbital Period	Orbital Speed (km/s)	Eccentricity	Semimajor Axis (10⁹ m)
Sun	1.99×10^6	696	274	618	36 at poles 27 at equator	2.6×10^{11}	200 My	250		
Mercury	0.330	2.44	3.70	4.25	58.6	57.6	88.0 d	48	0.2056	57.6
Venus	4.87	6.05	8.87	10.4	−243	108	225 d	35	0.0068	108
Earth	5.97	6.37	9.81	11.2	0.997	150	365.3 d	30	0.0167	149.6
Moon	0.0735	1.74	1.62	2.38	27.3	0.385	27.3 d	1.0		
Mars	0.642	3.38	3.74	5.03	1.03	228	1.88 y	24.1	0.0934	228
Phobos	9.6×10^{-9}	9–13 km	0.001	0.008	0.32	9.4×10^{-3}	0.32 d	2.1		
Dcimos	2×10^{-9}	5–8 km	0.001	0.005	1.3	23×10^{-3}	1.3 d	1.3		
Jupiter	1.90×10^3	69.1	26.5	60.6	0.414	778	11.9 y	13.0	0.0483	778
Io	0.0888	1.82	1.8	2.6	1.77	0.422	1.77 d	17		
Europa	0.479	1.57	1.3	2.0	3.55	0.671	3.55 d	14		
Ganymede	0.148	2.63	1.4	2.7	7.15	1.07	7.15 d	11		
Callisto	0.107	2.40	1.2	2.4	16.7	1.88	16.7 d	8.2		
and 13 smaller satellites										
Saturn	569	56.8	11.8	36.6	0.438	1.43×10^3	29.5 y	9.65	0.0560	1430
Tethys	0.0007	0.53	0.2	0.4	1.89	0.294	1.89 d	11.3		
Dione	0.00015	0.56	0.3	0.6	2.74	0.377	2.74 d	10.0		
Rhea	0.0025	0.77	0.3	0.5	4.52	0.527	4.52 d	8.5		
Titan	0.135	2.58	1.4	2.6	15.9	1.22	15.9 d	5.6		
and 12 smaller satellites										
Uranus	86.6	25.0	9.23	21.5	−0.65	2.87×10^3	84.1 y	6.79	0.0461	2870
Ariel	0.0013	0.58	0.3	0.4	2.52	0.19	2.52 d	5.5		
Umbriel	0.0013	0.59	0.3	0.4	4.14	0.27	4.14 d	4.7		
Titania	0.0018	0.81	0.2	0.5	8.70	0.44	8.70 d	3.7		
Oberon	0.0017	0.78	0.2	0.5	13.5	0.58	13.5 d	3.1		
and 11 smaller satellites										
Neptune	103	24.0	11.9	23.9	0.768	4.50×10^3	165 y	5.43	0.0100	4500
Triton	0.134	1.9	2.5	3.1	5.88	0.354	5.88 d	4.4		
and 7 smaller satellites										

*Negative rotation period indicates retrograde motion, in opposite sense from orbital motion. Periods are sidereal, meaning the time for the body to return to the same orientation relative to the distant stars rather than the Sun.

†Central body is galactic center for Sun, Sun for planets, and planet for satellites.

Answers to Odd-Numbered Problems

Chapter 1
Answers to odd-numbered multiple-choice problems
9. (b)
11. (c)
13. (a)
15. (c)
17. (a)

Answers to odd-numbered problems
19. (a) 1.395×10^4 m;
(b) 2.46×10^{-5} kg;
(c) 3.49×10^{-8} s;
(d) 1.28×10^9 s
21. 10^9 kg
23. (a) 1.083×10^{21} m³;
(b) 5.51×10^3 kg/m³, about 5.5 times the density of water
25. 1.389×10^{-3}
27. 31.29 m/s
29. 2.29 m
31. 5.08 m
33. (a) 30.14; (b) 11740
35. 1.993×10^{-26} kg
37. (a) 1 mi = 1.609 km;
(b) 1 kg = $10^9 \mu$g;
(c) 1 km/h = 0.278 m/s;
(d) 1 ft³ = 0.0283 m³
39. 7.842×10^3 m/s
41. (a) 0.385 AU; (b) 1.52 AU;
(c) 5.20 AU; (d) 30.1 AU
43. (a) 4; (b) 8
45. (a) 1.283 s; (b) 499 s; (c) 1.50×10^4 s
47. T: $\sqrt{m/k}$
49. v: $\sqrt{gh}$
51. (a) 1; (b) 3; (c) 3; (d) 5
53. 1.50×10^2 cm²
55. 2.700 g/cm³
57. 3×10^9
59. 0.1 mm
61. (a) Earth; (b) $\rho_E/\rho_V = 1.063$
63. (a) 619.54 kg/m³;
(b) The average density is only about 60% of the density of water
65. (a) 16.76 m/s; (b) 1.64
67. about 5×10^{25} atoms
69. (a) 2.563 kg; (b) 4.826×10^{25}

Chapter 2
Answers to odd-numbered multiple-choice problems
13. (c)
15. (a)
17. (b)
19. (c)
21. (c)
23. (c)
25. (b)

Answers to odd-numbered problems
27. $\Delta x = 0$, and s (my notation for total distance) = 400 m
29. (a) $\Delta x = 0$, $s = 960$ km;
(b) $\Delta x = 160$ km, $s = 1120$ km;
(c) $\Delta x = 80$ km, $s = 1200$ km.
31. 500 s or 8 min 20 s
33. 4.44 m/s
35. (a) 204 min = 3.40 h;
(b) 485.3 km/h
37. $\vec{v} = 58$ km/h, $+x$; average speed is 194 km/h
39. 3.96 m/s
41. (a) -3.16 mi/h;
(b) 55.38 s for each mile
43. 2.2 s
45. $0.755c$
47. $v(t) = -(55/12) + (5/3)t$
49. (i) $0 - 7.5$ s: 2.33 m/s²;
(ii) $7.5 - 12.5$ s: 0 m/s²;
(iii) $12.5 - 20$ s: -2.67 m/s²

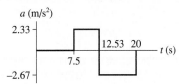

51. (a) a is greatest in the interval $0 - 7.5$ s;
(b) a is smallest in the interval $12.5 - 20$ s;
(c) a is zero in the interval $7.5 - 12.5$ s;
(d) maximum: 2.33 m/s², min: -2.67 m/s²
53. (a) 26.8 m/s; (b) 6.7 m/s²
55. 7.8 m/s
57.

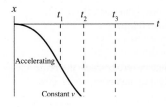

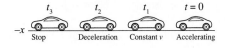

59. (a) -2.41 m/s²;
(b) 5.76 s; yes
61. (a) yes;
(b) 0.332 m/s
63. (a) 25.28 m/s;
(b) 27.27 s;
(c) 386.5 m
65. (a) -9.61×10^5 m/s²;
(b) -9.36×10^5 m/s²
67. 2.014 m/s²
69. -3.71 m/s²
71. (a) 9.58 m/s²;
(b) 37.56 m
73. 13.65 m/s
75. 2.286 s, -22.4 m/s
77.

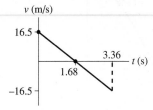

79. 3.60 m/s²
81. (a) 0.948 s;
(b)

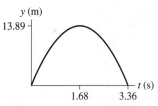

(c) The player spends about 70% of the time at or above half of the maximum height. This gives the appearance that the player is "hanging" in the air.
83. (a) 138.6 m/s; (b) 1742.4 m;
(c) 184.8 m/s; (d) 44 s
85. -4.95 m/s²

87. $y = 20 + 4t - 4.9t^2$; $v_y = 4 - 9.8t$

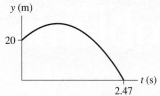

89. (a) 11.2 m/s; (b) -448 m/s^2
91. (a) -5.99 m/s^2; (b) 2.24 s; (c) -12.9 m/s^2
93. $x = 25t - 1.25t^2$; $v_x = 25 - 2.5t$
    $x(0) = 0, x(2) = 45, x(4) = 80,$
    $x(6) = 105, x(8) = 118.75,$
    $x(10) = 125$ (in meters)

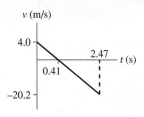

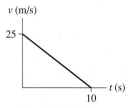

95. (a) 0.554 s, 1.915 s;
    (b) 6.67 m/s, -6.67 m/s
97. 31.25 m from the left, assuming the moving sidewalk moves to the right.
99. 90 s
101. 14.67 s, 161.3 m
103. 3.57 s; yes, flight time is 3.57 s so the opponent still has time.

Chapter 3
Answers to odd-numbered multiple-choice problems
15. (b)
17. (c)
19. (a)
21. (c)
23. (a)
25. (b)

Answers to odd-numbered problems
27. 53.1°, 36.9°, 90°
29. 1.60 m

31. (a) 15.3 cm;
    (b) 52.3°, 37.6°, 90°
33. 0.575 km
35. (a) 117.1°; (b) 199°; (c) $-73.3°$
37. (i) for $\vec{r}_1$, 5.61 m, $-64.76°$;
    (ii) for $\vec{r}_2$, 3.69 m, 164.6°
39. $(5.95 \text{ m})\hat{\imath} - (6.05 \text{ m})\hat{\jmath}$
41. 130.6°
43. (a) (1.11 m, 6.3 m); (b) (11.26 m, 6.5 m);
    (c) (0, -10 m)
45. $\vec{T} = \vec{R} + \vec{S} = (10.5 \text{ m})\hat{\imath} + (7.8 \text{ m})\hat{\jmath}$;
47. $(58.52 \text{ m})\hat{\imath} - (2.93 \text{ m})\hat{\jmath}$
49. (a) 28.56 m/s; (b) $-(18.18 \text{ m/s})\hat{\imath}$;
    (c) $(28.56 \text{ m/s})\hat{\jmath}$
51. $(9.33 \times 10^{-4} \text{ m/s})\hat{\imath} +$
    $(7.17 \times 10^{-4} \text{ m/s})\hat{\jmath}, 1.18 \times 10^{-3} \text{ m/s}$
53. $(3.54 \text{ m/s}^2)\hat{\imath} - (8.54 \text{ m/s}^2)\hat{\jmath}$
55. (a) (245.9 m/s, 917.6 m/s);
    (b) $(4.47 \text{ m/s}^2)\hat{\imath} + (16.68 \text{ m/s}^2)\hat{\jmath}$
57. (a) $-(68 \text{ m/s})\hat{\imath}$;
    (b) $-(9.07 \times 10^4 \text{ m/s}^2)\hat{\imath}$,
       magnitude: $9.07 \times 10^4 \text{ m/s}^2$, direction: $-\hat{\imath}$, opposite of the initial direction of the ball's velocity
59. (a) $(1.14 \text{ m/s}^2, -0.15 \text{ m/s}^2)$;
    (b) $(11.4 \text{ m/s})\hat{\imath} - (1.5 \text{ m/s})\hat{\jmath}$, speed: 11.5 m/s
61. (a) $\Delta\vec{v} = +(2.546 \text{ m/s})\hat{\jmath}$
    (b) $\Delta\vec{v}' = +(0.142 \text{ m/s})\hat{\imath} + (2.403 \text{ m/s})\hat{\jmath}$
63. (a) 74.4 m; (b) 3.896 s; (c) 18.6 m
65. (a) 1.2 m; (b) 0.30 m
67. (a) 18.66 m/s; (b) 1.392 s;
    (c) $+(13.4 \text{ m/s})\hat{\imath} - (13.64 \text{ m/s})\hat{\jmath}$
69. 24.5 m
71. (a) 383.4 m/s; (b) 55.3 s
73. 720 m
75. 9.073 m/s
77. Between 7.98° and 20.76°
79. Between 20.4° and 26.57°
81. 0.0265 m/s^2; the ratio is given by $\cos 38° / \cos 0° = 0.788$
83. (a) At top of the loop, gravity provides the source of centripetal force,
       $mg = mv^2/r$, so $a_r = g$;
    (b) 8.46 m/s
85. (a) 2.51×10^{-5} s; (b) 7.5×10^{13} m/s^2
87. 630.6 m/s^2
89. 4.74×10^5 m/s^2
91. (d)
93. (a) 400 m;
    (b) With A = (0, 0), we have
       $\Delta\vec{v} = (80 \text{ m})\hat{\imath} - (76.4 \text{ m})\hat{\jmath}$
95. (a) $(21.2 \text{ m/s})\hat{\imath}$;
    (b) $(21.2 \text{ m/s})\hat{\imath} - (17.8 \text{ m/s})\hat{\jmath}$;
    (c) 67.2 m
97. (a) 38.7°, north due east; (b) 0.32 h;
    (c) (10 km, 8.0 km)
99. (a) 11.2 m/s; (b) 1.62 s
101. (a) 45°; (b) 68°;
    (c) the range is zero because the projectile goes straight up and comes back down

105. 2.72×10^{-3} m/s^2, about 2.78×10^{-4} g
107. 0.125 m

Chapter 4
Answers to odd-numbered multiple-choice problems
17. (a)
19. (c)
21. (c)
23. (a)
25. (c)
27. (c) under the assumption of 25 m/s for the speed of the car.

Answers to odd-numbered problems
29. 2.2 N to the right
31. 17.43 m/s^2, upward
33. 0.184 N, to the left
35. 0.0698 N, 111.4°
37.

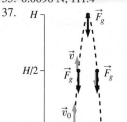

39. 1.64×10^{-21} m/s^2
41. 0.361 m/s
43. 7.9×10^5 N
45. 75 m/s
47. 15.77 kg
49. $(-156.1 \text{ N})\hat{\imath} + (108 \text{ N})\hat{\jmath}$
51. 31.98 m/s
53. (a) 477.8 N on hand;
    (b) 19.11 N on head;
    (c) 54.9 N on each leg
55. 2.97 m/s^2
57. (a) 291.2 N; (b) 2184 N
59. 7.82°
61. 113 N
63. (a) 12 m/s^2;
    (b) 18 N, 12 N, 6.0 N;
    (c) $F_{12} = 18$ N, $F_{23} = 6$ N
65. 1.1×10^4 N
67. (a) 286.7 N; (b) 296.8 N
69. 2.2 m/s^2
71. (a) -0.441 m/s^2; (b) 6.8 m
73. (a) -0.0396 m/s^2; (b) 37.87 s;
    (c) 0.0040
75. (a)

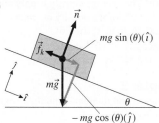

(b) 0.0868

77. (a)

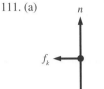

(b) $mg = 17.15$ N, $f = 4.44$ N,
   $n = 16.57$ N
79. 427.9 m
81. (a) 39.37 m for both car and truck;
   (b) 39.37 m for car going at 50 km/h;
      157.47 m for car going at 100 km/h
83. (a) 116.7 N; (b) 0.594
85. (a) 0.0711;
   (b) the system remains at rest
87. (a)

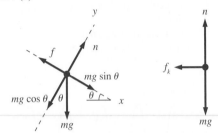

(b) 7.463 m/s; (c) 47.36 m;
   (d) 23.4 s
89. (a) $f_s = 966.4$ N $> F_{max} = 900$ N;
   (b) 8.57°
91. (a) 3.72 m/s²; (b) 2.94 m/s²
95. (a) 2.01×10^{20} N; (b) 3.53×10^{22} N;
   $\dfrac{F_{moon\text{-}Earth}}{F_{Earth\text{-}Sun}} = 5.7 \times 10^{-3}$
97. (a)

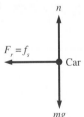

(b) 37.86 m/s
99. 224 days, very close
101. (a) 826.5 m; (b) 2 mg
103. (a) 617.4 N; (b) 2498 N; (c) 1263 N
105. (a) $\dfrac{mv^2_{max}}{L} = mg \Rightarrow v_{max} = \sqrt{Lg}$;
   (b) assuming $L = 1.2$ m, $v_{max} = 3.4$ m/s
107. (a)

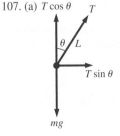

(b) $2\pi\sqrt{\dfrac{L\cos\theta}{g}}$;
(c) as $\theta \to 0$, $T \to 2\pi\sqrt{\dfrac{L}{g}}$
109. 0.069
111. (a)

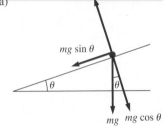

(b) 2.79×10^4 N $= 37$ mg; (c) 0.788 s
113. (a)

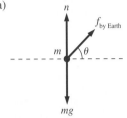

(b) $(1.484 \times 10^3$ m/s², 2.37×10^3 m/s²$)$;
(c) $F_x = 1.83 \times 10^{-2}$ N $= 151.4$ mg,
   $F_y = 2.92 \times 10^{-2}$ N $= 242.2$ mg
115. (a)

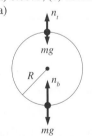

(b) 0.915 m; (c) 1.463 s; (d) 1.25 m/s;
(e)
117. (a) 0.0073; (b) 186.15 N
119. (a)

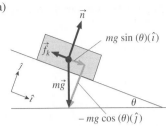

(b) $n_t = 570$ N, $n_b = 606$ N

Chapter 5
Answers to odd-numbered multiple-choice problems
15. (c)
17. (a)
19. (a)
21. (c)
23. (d)
25. (c)
27. (c)

Answers to odd-numbered problems
29. 1890 J
31. (a) -2104 N; (b) -3.05×10^5 J

33. -2.59 J
35. (a) 5.85 J; (b) -5.85 J; (c) 8.5 J
37. (a) 4620 J; (b) -1813 J; (c) 2807 J
39. (a) 3.17 N; (b) 1.9 J; (c) -1.9 J;
   (d) 0
41. (a) 3.92 m/s²;
   (b) 0.294 J on mass 1, and 0.196 J on
      mass 2;
43. (a) 11.76 N/m; (b) 0.833 m
45. 4.77×10^4 N/m
47. 6.0 J
49. (a) 1.75 J; (b) 1.3125 J;
   (c) 3.5 J; (d) -1.75 J
51. (a) -20 J; (b) 20 J; (c) 80 J;
   (d) 80 J; (e) 20 J
53. 5.39 cm
55. (a) 0.0392 m; (b) 0.0784 m;
   (c) 0.136 m
57. (a) 3.97 kg; (b) 1220 J; (c) 76.25 J
59. (a) 1.64×10^8 J; (b) 1.366×10^5 N;
   (c) yes
61. (a) -15.3 J; (b) -0.83 N
63. (a) 34.45 J; (b) -34.45 J; (c) 25.45 m;
   (d) 1.705 J; (e) 22.33 m/s
65. (a) 14 m/s; (b) 7.5 m
67. 8008 N
69. (a) 6.5×10^4 J; (b) 7.30×10^4 J
71. (a) 1.219 m/s; (b) 1.716 m/s;
   (c) 2.211 m/s
73. 2.58×10^6 J
75. 0
77. (a) 5.02×10^5 J;
   (b) 0.14 glass
79. about 9300 times; no
81. (a) 32.87 J; (b) 37.84 m/s
83. (a) 5.25 m/s; (b) 69 J
85. 11.38 m/s
87. 0.486 m
89. (a) 6.86 m/s; (b) 1.8 m
91. (a) -1.0×10^4 J; (b) 2.1×10^4 J;
   (c) 4.747 m/s
93. 1.014 m
95. 4.915 m/s
97. 2990 N/m
99. 2.55×10^5 J
101. 8.983×10^{12} W
103. (a) 2058 J on person, 603.68 on chair;
   (b) 171.5 W
105. 22.5 m/s
107. 0.617 MJ
109. (a) 3.822 J;
   (b) a straight line that starts at (0,0) and
      ends at (0.728, 10.5);
   (c) 3.82 J
111. (a) 95.2 J; (b) 1.586 W;
   (c) Blood is viscous and the passage
      through the blood vessels is much longer
      than the height of the person; some energy
      turns into thermal energy and is not
      recoverable.
113. 1.34 miles
115. (a) 24 J; (b) 12 J
117. (a) 0.288 m; (b) 11.66 N/m
119. (a) $x = 1$ m; (b) $x = 2.917$ m
121. (a) 17.96 m; (b) 20.02 m/s

123. -18.29 J
125. (a) 2.217 m/s; (b) 1.097 m;
(c) 5.33 m/s

Chapter 6

Answers to odd-numbered multiple-choice problems
19. (a)
21. (b)
23. (d)
25. (a)
27. (d)
29. (c)
31. (b)

Answers to odd-numbered problems
33. 196.875 N, 1575 kg·m/s
35. 467.2 kg·m/s
37. 10.687 kg·m/s
39. (a) $1\,N\cdot s = 1\,(kg\cdot m/s^2)\cdot s = 1\,kg\cdot m/s$;
(b) 53.57 m/s
41. (a) 0.2156 N;
(b) 2.59×10^{-4} kg·m/s;
(c) 2.59×10^{4} kg·m/s
43. (a) $(1.53\,kg\cdot m/s)\hat{\imath} + (1.08\,kg\cdot m/s)\hat{\jmath}$;
(b) $(2.02\,kg\cdot m/s)\hat{\imath} + (1.31\,kg\cdot m/s)\hat{\jmath}$
45. (a) 28.95 kg·m/s, 86.84 kg·m/s;
(b) 279.3 J, 837.9 J
47. (a) 678.4 kg·m/s; (b) 479.7 kg·m/s;
(c) 0
49. (a) Let $\vec{v}_i = v_i(\cos\theta\hat{\imath} - \sin\theta\hat{\jmath})$,
$\vec{v}_f = v_f(\cos\theta\,\hat{\imath} + \sin\theta\,\hat{\jmath})$;
$\Delta\vec{p} = (0.267\,kg\cdot m/s)\hat{\jmath}$;
(b) $(-0.0346\,kg\cdot m/s)\hat{\imath} + (0.247\,kg\cdot m/s)\hat{\jmath}$;
(c) $(10.68\,N)\hat{\jmath}$ for (a) and $(-1.384\,N)\hat{\imath} + (9.88\,N)\,\hat{\jmath}$ for (b)
51. (a) 1.5 kg·m/s for 0 to 0.5 s, and 3.0 kg·m/s for 0.5 to 1.0 s;
(b) 37.5 m/s
53. (a) 1.46 kg·m/s; (b) 919.3 kg
55. 0.477 m/s, in the $-x$-direction
57. 17.33 m/s
59. 1.61 m/s
61. $-0.186\,v_0$ for 60-kg person, and $0.814\,v_0$ for the 87.5-kg person (need to know v_0 to have numerical answers)
65. $v_f = v_i/2$; $K_i = \dfrac{1}{2} mv_i^2$, and
$K_f = \dfrac{1}{2}(2m)v_f^2 = \dfrac{1}{4} mv_i^2 = \dfrac{K_i}{2} \neq K_i$,
energy is not conserved
67. When the bullet hits the block; as the bullet + block system swings to height h.
69. (a) $mv/(m + M)$; (b) $\dfrac{M + m}{m}\sqrt{2gh}$
71. (a) 48.79 n/s;
(b) 45.45%;
(c) 47.88 m/s
73. Before: 9.70 m/s; after: -1.70 m/s
75. 0.551
77. 0.05 m/s, and 0.90 m/s

79. $(0.725\,m/s)(\hat{\imath} + \hat{\jmath})$
81. 15.4 m/s or 34.5 mi/h, which exceeds 30-mi/h speed limit
83. One ball 1.167 m/s, 45° above x-axis; other ball 1.167 m/s, 45° below x-axis
85. $(8.2 \times 10^{-22}\,kg\cdot m/s)\hat{\imath} + (3.1 \times 10^{-28}\,kg\cdot m/s)\hat{\jmath}$
87. $(11.5\,m/s)\hat{\imath} - (23\,m/s)\hat{\jmath}$
89. 0.40 m
91. 7.42×10^{8} m, outside Sun's radius
93. 2.063 m from the pivot
95. (a) 0.30 m/s;
(b) velocities are exchanged;
(c) the CM velocity remains the same
97. $(x_{cm}, y_{cm}) = (54.16\,m, 28.0\,m)$
99. (a) 36.39 cm; (b) 12.55 cm;
(c) your muscles contract, pulling the tendons
103. $(-3.01 \times 10^{-24}\,N)\hat{\imath} - (4.62 \times 10^{-24}\,N)\hat{\jmath}$
105. (a) $(101.8\,N)\hat{\imath}$; (b) 0.042 m/s
107. 62.2 cm/s, in the opposite direction
109. (a) 3.27×10^{-4} s; (b) 2.30×10^{4} N;
(c) 1.875 m/s; (d) 0.221
111. 9.6 km/h
113. (a) 4.276 m/s; (b) 5.55 m/s
115. (a) 323 s $\approx$ 5.4 min;
(b) 838 s $\approx$ 14 min
117. (a) $(x_{cm}, y_{cm}) = (0, 100.8\,cm)$;
(b) $(0, 108.9\,cm)$
119. (a) before: $(0, 98.4\,cm)$, after: $(0, 100.8\,cm)$;
(b) 2.4 cm
123. (a) $-v_i/3$; (b) 5

Chapter 7

Answers to odd-numbered multiple-choice problems
17. (c)
19. (b)
21. (b)
23. (c)
25. (d)
27. (a)
29. (d)

Answers to odd-numbered problems
31. (a) periodic; (b) 3.17×10^{-8} Hz
33. 0.375 s
35. (a) 6.25×10^{-5} s; (b) 1.01×10^{5} rad/s
37. $T \sim \sqrt{m/k}$
39. 0.067 N/m
41. (a) 4.9 m; (b) 4.62 m; (c) 3.54 m
43. 154 N/m
45. 0.088 kg
47. $x(t) = (0.5\,m)\cos(8.71t)$, with $T = 0.721$ s

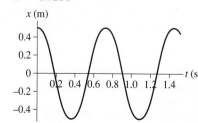

49. (a) $x(t) = A\cos(\omega t + \pi/2)$ (plot with $A = 1$, $\omega = 1$)

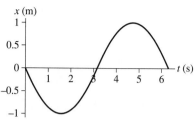

51. 0.268 m
53. (a) 0.733 s; (b) 0.0187 m
55. $T/8$
57. 0.58 m
59. (a) 14.74 J; (b) 4.89 m/s; (c) 0.203
61. (a) 10 s; (b) 10.13 kg;
(c) $v_{max} = 0.47$ m/s, $a_{max} = 0.296\,m/s^2$
63. (a) 0.295 m; (b) 0.609 J; (c) 1.32 m/s
65. 5.83 Hz
67. (a) 10 Hz; (b) 62.83 rad/s
69. (a) 4.17 m; (b) no change
71. (a) 1.28 s; (b) 0.407 m
73. $\theta(t) = (5°)\cos(2.556t)$

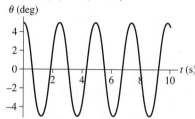

75. 2.16 s
77. 0.007 s
79. lightly damped since ω_{damped} is real and positive
81. (a) $\omega_{damped} = \sqrt{\dfrac{k}{m} - \left(\dfrac{b}{2m}\right)^2} = 6.778$ rad/s > 0;
(b) $T_{damped} = 0.927$ s, $T_0 = 0.91$ s;
(c) Since $e^{-bT_{damped}/2m} = 0.3$, the amplitude will damp below $A/2$ in one oscillation
83. 0.045 kg/s
85. (a) 196.5 N/m; (b) 21.6 s
87. (a) 6.67×10^{-5} N/m;
(b) 3.446×10^{-10} kg
89. (a) $T = 0.555$ s;
(b) $T = T_1/\sqrt{2}$, where $T_1 = 0.785$ is the period with one spring
91. (a) $f_1 = 2f_2 \Rightarrow \omega_1 = 2\omega_2$;
(b) $a_{1,max} = 4a_{2,max}$
93. (a) 0.144 m;
(b) when the normal force on the block is zero (when the acceleration exceeds g)
95. (a) $v_{max} = 754$ m/s, $a_{max} = 5.68 \times 10^{16}\,m/s^2$;
(b) 5.70×10^{-21} J
97. (a) 0.306 m; (b) A = 0.222 m;
$T = 0.583$ s

Chapter 8

Answers to odd-numbered multiple-choice problems

15. (d)
17. (a)
19. (b)
21. (a)
23. (d)
25. (c)
27. (b)
29. (a)

Answers to odd-numbered problems

31. 1.267×10^4 m
33. 314 rad/s, about half that of the *E. Coli's* flagellum.
35. 245 rad/s = 2340 rpm
37. -4.12×10^{-22} rad/s^2
39. 1757.4 turns
41. (a) -119.7 rad/s^2; (b) 408 rev; (c) 241 m
43. -6.13×10^{-22} rad/s^2
45. 5.02 rad/s
47. (a) 97.1 rad/s^2; (b) 84.9 rad/s
49. (a) 16.82 m/s; (b) 0.128 rad/s^2
51. (a) 1.09×10^4 m; (b) 4.74 GB
53. (a) 1.047 rad/s; (b) 1.1 m/s
55. 2.57×10^{29} J
57. 11 J
59. (a) 7.94×10^{-5} kg m^2; (b) trans: 35.1 J; rotation: 0.627 J; $K_{trans}/K_{rot} \approx 56$
61. 40.26 rad/s
65. 102.1 rad/s
67. $\dfrac{K_{trans}}{K_{tot}} = \dfrac{2}{3}$, $\dfrac{K_{rot}}{K_{tot}} = \dfrac{1}{3}$
69. 5.42 rad/s
71. $\dfrac{5}{7} g \sin\theta < g \sin\theta$
73. (a) 0.933 m; (b) the θ dependence cancels out (i.e, the ratio of the times taken to reach the bottom by the two objects does not depend on θ)
75. 22.75 N·m
77. 33.33 pN
79. (a) 83.3 N·m; (b) 0.346 rad/s^2
81. 61.67 cm
83. (a) 130.95 N; (b) 194.2 N
85. (a) 0.702 kg; (b) 10.9 N
87. 2.66×10^{40} kg·m^2/s, 10^6 times greater than that due to rotation
89. 2.22 rad/s
91. (a) 45.5 kg·m^2/s, into the page as viewed from above; (b) 4.55 N·m, out of the page, as viewed from above
93. $-z$-direction.
95. -5.96×10^{16} N·m, down along the axis of rotation
97. (a) 2.94 Nm; (b) 752.6 rad/s^2
99. (a) 52.5°; (b) -260.66 J; (c) -284.5 N·m
101. (a) solid sphere wins; (b) solid cylinder: 1.4 m; hollow ball: 1.26 m; hollow cylindrical shell: 1.05 m
103. 1663 N·m

Chapter 9

Answers to odd-numbered multiple-choice problems

17. (a)
19. (d)
21. (b)
23. (b)
25. (b)
27. (b)

Answers to odd-numbered problems

29. 1.41×10^{-9} N
31. (a) 1.347×10^{26} N; (b) $F_{Earth-Sun} = 3.548 \times 10^{22}$ N, $F/F_{Earth-Sun} = 3.8 \times 10^4$
33. 3.63×10^{-47} N
35. Saturn: 11.1 m/s^2; Jupiter: 25.94 m/s^2
37. 6.48×10^{23} kg; very close to the given value of 6.42×10^{23} kg
39. (a) $F_g = 1.27 \times 10^{-10}$ N; (b) 6.17×10^{-10} N·m
41. (a) 3187 m; (b) 32090 m; (c) 3.446×10^5 m
43. 0.866
45. 2.12×10^{10} m
47. approximately 3.36×10^{18} m^3/s^2 for Mercury, Venus and Jupiter
49. $T_A/T_B = 2.828$
51. (a) 1.43×10^4 s; (b) $h = 2.175 R_E$.
53. 2.732 years
55. (a) -3.24×10^{35} J; (b) -5.27×10^{34} J; (a)/(b) = 6.15
57. (a) 443 m/s; (b) 4120 m/s; (c) 8747 m/s
59. 1.775×10^{32} J; 996 m/s
61. -1.267×10^{10} J
63. 2375 m/s
65. 6.71×10^7 m
67. 6160 s = 1.7 h
69. (a) 2439 m/s; (b) 2.1×10^8 m/s
71. 8.84×10^7 m = 51 R_{moon}
75. (a) 3.8×10^{28} J; (b) -7.6×10^{28} J; (c) -3.8×10^{28} J
77. (a) 4.0×10^{10} J; (b) 1478 kg; (c) 7360 m/s
79. 7060 s
81. 1.65×10^{12} m/s^2
83. (a) 3.44×10^{-5} m/s^2; (b) 3.22×10^{-5} m/s^2; (c) 2.2×10^{-6} m/s^2
85. 0.1796 m/s^2
87. b, d, e, a, c
89. 6.2 m; 22.8 m
91. (a) 4.597×10^{10} m/s; (b) 0.2065
93. (a) 2068 km; (b) 804 km
95. 3.78 m/century
97. 2200 m
99. (a) 2950 m; (b) 2.95×10^{14} m

Chapter 10

Answers to odd-numbered multiple-choice problems

15. (d)
17. (b)
19. (c)
21. (d)
23. (c)

Answers to odd-numbered problems

25. 1.24 m^3
27. 1.69
29. 0.657 L
31. 1.1×10^{-7} m
33. 7.69 kg
35. 0.31 m
37. 10.34 m, not practical
39. (a) 5.81×10^7 Pa; (b) -3.63×10^{-4}
41. 20.1 m
43. 23.8 km
45. 0.898 m
47. 2.29×10^3 kg
49. 1.87×10^6 N
51. $F_b = 0.836$ N; $W = mg = 686$ N, $F_b << W$
53. (a) 63700 N; (b) 6.31 m^3; (c) 0.904
55. 500 kg/m^3
57. 1033.4 kg/m^3
59. (a) 3.31 N; (b) 2.08 N
61. 17.2%
63. 0.11 m/s^2
65. 2.67×10^{-4} m^3/s
67. (a) 1.27×10^{-4} m^3/s; (b) 18 m/s
69. 6.0×10^{-5} m^3/s
71. 113.6 kPa
73. 3.83 m/s
75. 1800 N, outward
77. decreases by 2.6%
79. 111.4 Pa
81. 0.27 m^2
83. (a) 143 kg; (b) 0.50 m; (c) 7170 Pa
85. 2.28×10^4 N; not likely
87. (a) 1.96 N; (b) 1.33×10^{-2} m/s; (c) 5.89 m/s
89. (a) 309 m^3; (b) 8.4 m
91. 1.5×10^4 Pa
93. (a) 23.7 N; (b) 23.6 N
95. 7.196×10^{-6} m^3/s

Chapter 11

Answers to odd-numbered multiple-choice problems

15. (d)
17. (b)
19. (b)
21. (b)
23. (b)
25. (d)

Answers to odd-numbered problems

27. (a) 0.566 m/s; (b) 1.13 m/s
29. 288 km
31. (d) $v = \omega/k$
37. 21.5 cm
39. (a) 235.2 m/s; (b) 8.86×10^{-4} kg/m
41. G: 392 Hz, 588 Hz; D: 588 Hz, 882 Hz; A: 880 Hz, 1320 Hz; E: 1318 Hz, 1977 Hz
43. (a) 192.96 m/s
45. 493.8 Hz 4 times every 10 seconds
47. 701.4 m
49. (a) 12005 m; (b) 36.27 s.
51. 2.48×10^{-8} W
53. $\frac{1}{2}$
55. (a) 89.18 dB; (b) 3.16%

57. (a) -6.02 dB; (b) -20 dB; (c) -40 dB
59. (a) 89.2 km; (b) 892 m.
61. 5.01
63. 238.15 Hz
65. 19.94 Hz, 59.83 Hz, 99.71 Hz, 139.6 Hz
67. (a) 1.53 m; (b) 0.327 m; (c) 0.164 m; (d) 0.071 m
69. 2048 Hz
71. 6.82 m/s
73. (a) 247.3 Hz; (b) 238 Hz
75. 381.4 Hz
77. (a) 1.2×10^5 Hz, 2.86×10^{-3} m; (b) 603 Hz, 0.569 m
79. 343 m/s
83. (a) a factor of 4; (b) -6.0 dB
85. 31.7 dB, louder than a quiet whisper at 20 dB
87. 54.3 kHz
89. 0.138 m
91. 0.80 W
93. 0.0148 m/s

Chapter 12

Answers to odd-numbered multiple-choice problems
15. (d)
17. (b)
19. (b)
21. (c)
23. (b)

Answers to odd-numbered problems
25. $-15°C$
27. (a) $-320.8°F$, 77.15 K; (b) $620.6°F$, 600.15 K
29. $5.4°F$
31. (a) $59.4°F$; (b) $-18°C$, $-0.4°F$
33. $100.76°F$
35. (a) $-40°C = -40°F$; (b) 233.15 K
37. (a) 50.0072 m; (b) 49.982 m
39. (a) 1.26×10^{-4} m; 4.87×10^{-6} m
41. -2.3×10^{-9} m
43. 5.2 s
47. 0.2532 m^2
49. (a) 39.95 g; (b) 11 g; (c) 52.47 g; (d) 528 g
51. 3.90×10^{27}
53. (a) Decrease; (b) 8.43 cm
55. (a) in order He, Ne, Ar, Kr, Xe, Rn: 0.164 kg/m^3, 0.818 kg/m^3, 1.635 kg/m^3, 3.426 kg/m^3, 5.37 kg/m^3, 9.08 kg/m^3; (b) only He and Ne
57. (a) 264.36 kPa; (b) 253.7 kPa
59. (a) 1.31 g; (b) 1.253 g
61. 56.14 cm^3
63. (a) less; (b) 1.9×10^4 kg
65. (a) 1845 m/s; (b) 764.3 m/s
67. 1172 K
69. 1.098
71. (a) 1.2×10^{-19} J; (b) $E/E_{ion} = 0.055$
75. -0.138
77. (a) 5.52×10^{-3} m; (b) 1.392 cm
79. (a) 19.3633 m; (b) 19.3721 m
81. 532 kPa
83. 162.3 atm
85. v238/v235 = 0.996
87. 0.266 m

Chapter 13

Answers to odd-numbered multiple-choice problems
23. (a)
25. (b)
27. (a)
29. (d)
31. (a)

Answers to odd-numbered problems
33. 1.17×10^6 J
35. 1.22×10^7 J
37. 57.5 Cal
39. 27930 J, or 6.67 Cal
41. 3.74 kW
43. (a) 99.2 J/°C; (b) 19.84 kJ
45. 19.25 °C
47. 46 g (46 mL)
49. 427 kg
51. 268 s (=4 min 28 s)
53. 8841 J/kg
55. 438 J
57. 34.5 °C, at constant volume
59. (a) 0.147 kWh; (b) 2.35 cents
61. 4.62×10^4 J
63. -5900 J
65. 266 s
67. 1.67×10^5 J
71. 3.0 g
73. 0.131 kg
75. 15 W
77. $4.6
79. (a) 0.004 m^2°C/W; (b) 0.0267 m^2°C/W; (c) 0.133 m^2°C/W; $R_{Styrofoam} > R_{wood} > R_{glass}$
81. 0.085 m^2°C/W; smaller than R-19 which has a value of 3.346 m^2°C/W
83. (a) 1387 W/m^2; (b) 3.6×10^6 m^2
85. 0°C, 80 g of ice
87. (a) ice-water mixture; (b) 328.6 g of water and 171.4 g of ice
89. (a) 1.68×10^5 J; (b) 84 s
91. 172 s
93. (a) 4.8×10^4 W (using $k = 40$ W/°C m for steel, helpful if given in Table 13.4); (b) 1.06 °C/s at constant volume
95. 4.27×10^3 W
97. 76.8 K
99. 255 K
101. 3.14×10^6 s
103. $-25°C$
105. 90 minutes
107. 0.425 °C

Chapter 14

Answers to odd-numbered multiple-choice problems
17. (d)
19. (a)
21. (b)
23. (b)
25. (c)

Answers to odd-numbered problems
27. 64.86 J
29. 210 W

31. (a) 415 J; (b) volume decreases; (c) -1081.65 J, volume increases
33. 6810 J; 20430 J
35. -258 J
37. (a) 11.7 °C, assuming at room temperature initially; (b) 25.3 J
39. (a) 4/3; (b) 215.3 J
41. (a) 32.42 atm; (b) 258.6 J; (c) 791.7 K
43. 1565.5 J
45. (a) 11.21 L; (b) 0.379 atm; 684.5 J
47. (a) $W_{A \to B} = -9.0 \times 10^5$ J; $W_{B \to C} = 0$; $W_{C \to D} = +3.0 \times 10^5$ J; $W_{D \to A} = 0$; (b) $W_{net} = -6.0 \times 10^5$ J;
49. (a) decreases; (b) 1519.5 J
51. (a) 25.12 atm; (b) 728.4 K
53. (a) 288 kJ; (b) 46.9 kJ
55. (a) 3.36 J/K; (b) -3.36 J/K
57. Melting: 121.98 J/K; boiling: 605.9 J/K
59. 2.72 J/K
61. 254.5 J/K
63. 0.34
65. (a) 0.353; (b) 600 MJ
67. 0.268
69. 1.23×1015 J
71. 2280 W
73. 0.353
75. (a) $52.7; (b) $17.94 saved
81. $C(52, 5) = 2.6 \times 10^6$
83. (a) 3.3 kJ; (b) 0.392 mol
85. 1.35
87. (a) 571.1 kPa; (b) 437.74 J
89. (a) 0.532; (b) 570 MW
91. (a) 399.4 J; (b) 2.65×10^5 Pa
93. (a) 288.2 K; (b) 15.5 kJ
95. (a) 4.29; (b) $(COP)_{max} = 11.42$
97. (a) 0.635; (b) 0.762
99. (a) 40 kPa; (b) 83.33 kPa; (c) about 80 kJ

Chapter 15

Answers to odd-numbered multiple-choice problems
15. (b)
17. (b)
19. (c)
21. (c)
23. (d)
25. (d)

Answers to odd-numbered problems
27. (a) 1.76×10^{11} C; (b) 9.58×10^7 C
29. (a) 3×10^{26} electrons
31. 4.8 mC
33. 0.38 m
35. (a) 10^{10} N directed straight up; (b) The weight is 1.64×10^{-26} N, insignificant compared to the electrical force.
37. 36.4 cm
39. 5 m. This means that gravity can be ignored in atomic or quantum calculations.
41. $\tan \theta \sin^2 \theta = \dfrac{kQ^2}{4L^2 mg}$

43. (a) Between the charges, 36.6 cm from the smaller charge. (b) Both magnitude and sign of the third charge cancel out of the equations for the Coloumb force when those forces are set equal to each other.
45. 2.77 N directed away from the center of triangle.
47. 8.6×10^{-6} N directed away from the center of the square.
49. (1.78 m, 2.29 m)
51. 3.44×10^{-8} N
53. 0.0014 N/C
55. a, d, b, c
57. (a) 8600 N/C in the $+x$-direction; (b) 3200 N/C in the $-x$-direction
59. 11°
61. 940,000
63. 8.85×10^{-8} C/m^2
65. (a) 5.15×10^{11} N/C; (b) 8.24×10^{-8} N
67. $x = 0.53$ m
69. (a) 23.6 kN/C in the $-x$-direction; (b) 38.0 kN/C at an angle 71.0° below the $+x$ axis
71. 5500 N/C
73. 185 N/C, 30° below the $-x$-axis
75. (a) Down; (b) $r = 523$ nm $(d = 1.05\,\mu m)$
77. 4.13×10^6 m/s, 19.1° below horizontal
79. (a) 9.22×10^6 m/s; (b) 2.54×10^{-15} m. This is on the order of the diameter of a small nucleus.
81. (a) 136 N/C; (b) The same direction as the electron's original motion.
83. 4.6×10^{13}
85. (a) Place the third charge at $y = 2.72$ m; (b) 0
87. (a) 3.33×10^{-12} kg; (b) (i) Up; (ii) Down
89. (a) 3.4×10^8; (b) 6.6×10^{-9} N, directed vertically upward; (c) The bee's weight is 1.2×10^{-3} N, roughly 180,000 times the electrical force.
91. $\{+40.3\,\mu C, -6.91\,\mu C\}$ or $\{-40.3\,\mu C, +6.91\,\mu C\}$
93. (a) 2.64 mm; (b) 1.32 mm
95. $4.3\,\mu m$ compression

Chapter 16
Answers to odd-numbered multiple-choice problems
19. (b)
21. (d)
23. (c)
25. (b)
27. (d)
29. (c)
31. (c)

Answers to odd-numbered problems
33. 3.15×10^{14} J
35. (a) earth: 4.4×10^4 C, moon: 546 C; (b) earth: 8.7×10^{11} C/m^2, moon: 1.4×10^{11} C/m^2
37. -1.99×10^7 J

39. 15.82 J
41. (b) 5.25 m/s
43. (a) 140 V; (b) 3.11×10^{-8} C, positive
45. (a) 3.56×10^{-10} C; (b) 2.2×10^9
47. 15.49 m/s
49. 1.15×10^5 V
51. (a) 3.2×10^{-19} C; (b) 3.2×10^{-19} C
53. 2.03×10^3 V/m
55. 1.496×10^6 m/s
57. (a) 3.85×10^4 V/m; (b) -6.15×10^{-15} N
59. -12.65 nC
61. (a) negative; (b) 1.8×10^4 V/m; (c) 4.8×10^{-19} C; (d) 3
63. 2.80 cm
65. (a) 0.70 F; (b) let $C_1 = 0.25$ F, $C_2 = 0.45$ F total Q: 16.8 C, $Q_1 = 6.0$ C, $Q_2 = 10.8$ C
67. (a) $17.14\,\mu c$; (b) 8.57 V; (c) $V_1 = V_2 = 3.43$ V; (d) $Q_1 = 3.43\,\mu C, Q_2 = 13.72\,\mu C$;
69. $9.0\,\mu F$
71. $1.0\,\mu F$
73. $2.0\,\mu F$
77. (a) 648 J; (b) 0.259 MW
79. (a) 0.443 nF; (b) 22.125 nC; 553 nJ
81. 1250 nF, in parallel
83. (a) 8.87 pF; (b) 106.42 pC; (c) 1920 V
85. 0.0127 mm
87. 2.475 mC
93. (a) 10^5 J; (b) 2.127×10^4 V
97. (a)

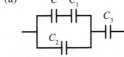

(b) $4.27\,\mu F$

Chapter 17
Answers to odd-numbered multiple-choice problems
19. (b)
21. (c)
23. (c)
25. (c)
27. (d)
29. (b)
31. (a)

Answers to odd-numbered problems
33. 1.29×10^{-5} m/s
35. 8.75×10^{-5} Ω
37. 7.78 kA
39. (a) 8.1 Ω; (b) 0.081 Ω; (c) 8.1×10^{-4} Ω
41. Within an uncertainty of $\Delta R = 0.0228$ Ω
43. 4
45. (a) 125 s; (b) 12 J
47. 4.48 Ω
49. (a) 3.6 C; (b) 5040 C; (c) 6048 J
51. 3.6 V, 1.94 Ω
53. (a) 149 Ω; (b) 0.048 A (in 250-Ω resistor) and 0.032 A (in 370-Ω resistor)

59. 40 Ω
61. 75 V
63. (a) 2.5 V; (b) let $(R_1, R_2, R_3) = (20\,k\Omega, 30\,k\Omega, 75\,k\Omega)$, $(I_1, I_2, I_3) = (0.125$ mA, 0.0833 mA, 0.0333 mA)
65. 230.4 J
67. 0.042 A
69. (a) 14.58 A; (b) 1.05 MJ = 0.292 kWh; (c) \$1.31
71. 12
73. (a) 870 W; (b) 1.81°C
75. (a) 0.034 nC; (b) 0.316 nC; (c) 1.69 nC
77. (a) 0.693; (b) 1.228
79. (a) 2.4 mC; (b) 0.635 mA
81. (a) 0.48 C; (b) 0.0229 C more
83. (a) 5.735 mC, 5.479 mA; (b) 39.4 mC, 2.417 mA
87. 20.8 Ω, 1.84 V
89. $3.367\,\mu F$
91. (a) 6800 Ω; (b) 1.632 s
95. (a) 4800; (b) 1.0 A

Chapter 18
Answers to odd-numbered multiple-choice problems
23. (b)
25. (d)
27. (d)
29. (c)
31. (a)
33. (d)

Answers to odd-numbered problems
35. (a) $0.1995\,\mu N$, down; (b) $0.1995\,\mu N$, up
37. (a) 5.30 N, vertically downward; (b) 5.30 N, vertically upward; (c) 5.30 N, vertically upward; (d) 5.30 N, vertically downward; (e) 0
39. (a) 1.50×10^{-18} N, west; (b) 9.0×10^8 m/s^2, west
41. (a) 0.1088 pN, $-z$-direction; (b) 6.8×10^5 N/C, $-z$-direction
43. (2025 N/C), $-x$-direction
45. 1.87×10^{-28} kg; $m_\mu/m_p = 0.112$, $m_\mu/m_e = 205$
47. (a) 1.885×10^5 m/s; (b) 0.052 T
49. 0.69 mT
51. (a) $85.37\,\mu m$; (b) 3.655 mm; (c) 3.644 mm
53. 0.0876 T
55. 38.59 mA
57. (a) 54208 T; (b) 54.208 T
59. assuming current flows eastward, $B = 73.3\,\mu T$, north
61. (a) 0; (b) $-(0.09$ N·m$)\hat{j}$; (c) $(0.0636$ N·m$)(\hat{i} - \hat{j})$
63. $(0.01875$ N$)\hat{j}$ on segment along x, $(0.01875$ N$)\hat{i}$ on segment along y, and $-(0.01875$ N$)(\hat{i} + \hat{j})$ along diagonal; the net force is zero

65. (a) 70.5 μN, west;
    (b) 75 μN, 70° north of up;
67. 11.04 A · m^2
69. R = 1.99 cm, 12 turns
71. 0.0768 T
73. 2.92 μT
75. (a) 133.3 μT; (b) 16.67 μT;
    (c) 83.33 N/m, repulsive
77. 3.35 × 10^{-3} N/m, toward each other,
    force attractive
79. (a) 0;
    (b) 1.326 × 10^{-4} N/m, toward the center
    of square
81. (a) 2.356 × 10^{-4} T; (b) 0
83. (a) (8.6 × 10^{-7} N), toward the wire;
    (b) (8.6 × 10^{-7} N), toward the loop;
    (c) (8.6 × 10^{-7} N), away from the wire;
    (d) (8.6 × 10^{-7} N), away from the loop
85. (a) 0.102 T; (b) 0.02356 T;
    (c) 0.0628 T in both cases
87. $-(147.6 \text{ m/s})\hat{\imath}$
89. $(0.06 \text{ N})\hat{\imath}$
91. 0.726 T
93. 22.62°, west of magnetic north
95. 1000 m

Chapter 19

Answers to odd-numbered multiple-choice problems
17. (a)
19. (c)
21. (a)
23. (a)
25. (b)
26. (c)

Answers to odd-numbered problems
27. 0.0736 Wb
29. 0.0117 Wb
31. 0.2873 A
33. (a) Clockwise;
    (b) 0.6136 mA; 0.3375 mV
35. 35
37. (a) 0.0751, clockwise; (b) 12.98 mW
39. (a) 1.764 V; (b) 3.528 V
41. (b) 1.225 V
45. (a) 0.01178 V; (b) 1508 V
47. 50
49. (a) 17.5 Ω;
    (b) All power would be lost.
51. 0.06 s
53. (a) 11.5 mH; (b) 9.325 A;
55. (a) 4.8 A; (b) 4.15 A; (c) 0.03475 s;
    (d) 4.15 A
57. Between 2516 and 3559 Hz
59. Increase C fourfold
61. (a) 0.0393 s; (b) 0.0731 J;
    (c) 8.55 mC
63. (a) 0.0687 A; (b) 16.51 W
65. (a) 280 V; (b) 560 W
67. (a) 169.7 V; (b) 49 Hz;
    (c) $\mathcal{E}(t) = (169.7 \text{ V}) \sin(308t)$
69. (a) 31.83 Hz; (b) 79.58 Hz, 150 Ω
71. 129.5 Ω
73. (a) 0.33929 A; (b) 0.33938 A

75. (a) Capacitive reactance: 1061 Ω, induc-
    tive reactance: 84.82 Ω;
    (b) 1666 Ω; (c) 31.27 μF
77. (a) −27.7°; (b) 0.885; (c) 75.3 W
79. (a) 516.4 rad/s; (b) 10.87 W
81. 42
83. 1102.9 V
85. (a) 3.573 A/s; (b) Opposite
87. (a) 5.07 × 10^{16} J;
    (b) 3380 s, or 56.3 minutes

Chapter 20

Answers to odd-numbered multiple-choice problems
13. (d)
15. (d)
17. (b)
19. (d)
21. (d)
23. (a)

Answers to odd-numbered problems
25. 4.74 × 10^{14} Hz
27. 273 nm
29. 468 nm
31. (a) 0.50 μT; (b) 7.5 N/C
33. (a) 1.18 × 10^{-4} Pa; (b) 5.9 × 10^{-5} Pa
35. (a) 0.15 MHz, radio waves;
    (b) 150 MHz, radio waves;
    (c) 150 GHz, microwaves;
    (d) 1.5 × 10^{14} Hz, IR;
    (e) 1.5 × 10^{17} Hz, UV
37. 10^9 − 10^{12} Hz
39. 1.50 m
41. 3 × 10^8 m/s
43. $t_{perp} = 35.36$ s; $t_{perp} < t_{parallel}$
45. (a) Earth: 133.3 s, ship: 133.3267 s;
    (b) Earth: 4.44 s, ship: 4.24 s;
    (c) Earth: 1.778 s, ship: 1.176 s;
    (d) Earth: 1.3468 s, ship: 0.190 s;
47. (a) 10 ns; (b) 11.547 ns;
    (c) 3.464 m; (d) 3 × 10^8 m/s
49. 0.141 c
51. 59.945 s
53. 0.9999995 c
55. 709 m
57. (a) Yes, 771.442 m < 75 m; (b) 0.033 s
59. 0.9803 c
61. 0.9396 c
63. (a) 534.6 nm; (b) 488.45 nm;
    (c) 311.77 nm; (d) 123.88 nm;
    (e) 38.28 nm
65. 0.47 c, to the right
67. $f_0 \sqrt{\dfrac{1 + v/c}{1 - v/c}}$; 2.29 × 10^{15} Hz
69. 34.4/min
71. 0.999949 c
75. 1.022 Mev
77. 0.866 c
79. (a) 0.14 c; (b) 0.910527 c
81. (a) 2.733 × 10^{-20} kg · m/s,
    $E = 51.1$ Me V; (c) 100
83. 0.995 c
85. 7.94 MW
87. 1.067 × 10^{-16} kg

89. 0.96 c
91. 0.93237 c

Chapter 21

Answers to odd-numbered multiple-choice problems
17. (b)
19. (c)
21. (a)
23. (c)
25. (b)

Answers to odd-numbered problems
27. 6 m
29. (a) 4.0 m; (b) 0.50 m/s
31. (a)

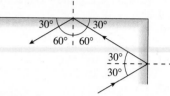

    (b) 24 cm
33. (a) 116.7 cm; (b) 53.85 cm
39. (a) 187.5 cm; (b) real; (c) −1.5
41. di = 52.2 cm; M = −1.5, image inverted
45. (a) Upright;
    (b) 2.25 cm from center of the ball;
    (c) 1.5 cm
47. (a) 40 cm;
    (b) image located at di = −15 cm,
    upright, reduced with M = 0.25
49. Diamond with $n = 2.42$, $v = 0.413 c =$
    1.28 × 10^8 m/s
51. (a) 2.29 × 10^8 m/s; (b) 1.948 × 10^8 m/s;
    (c) 1.27 × 10^8 m/s; (d) 1.24 × 10^8 m/s
53. (a) 14.9° (b) 12.92°
55. 0.55 m
57. 0.93 m
59. $\theta_c = 24.4°$
61. (a) 22.48°; (b) 0.469 mm
67. (a) $d_o = 60$ cm, di = 30 cm;
    (b) $d_o = 30$ cm, di = 60 cm
59. (a) Convex; (b) Inverted; (c) do = 15 cm
71. (a) 4.91 mm; (b) 9.0 mm; (c) 54 mm
73. 4.89 cm
75. (a) 6.56; (b) 17.67
77. (a) 0.172°; (b) 1.72°
79. 1.87 cm
85. (a) 23.0 cm; (b) Switching the two values
    with $R_1 = 20$ cm, and $R_2 = -40$ cm
    gives the same result.
87. −2.67 diopters
89. 2.03 diopters, assuming glasses 2 cm
    from eyes
91. (a) $f = \dfrac{R}{n - 1}$; (b) 13 cm
93. $n_2 = 1 + \dfrac{1}{3}(n_1 - 1)$
95. (a) 19.95 cm;
    (b) M = −0.5536, image size reduced
97. (a) 72 cm;
    (b) di = −4.5 cm, or 4.5 cm inside the
    surface

99. (i) for $M = -1/2$: (a) do $= 56$ cm,
    di $= 28$ cm; (b) image real;
    (c) f $= 18.67$ cm;
    (ii) for $M = -2$: (a) do $= 28$ cm,
    di $= 56$ cm; (b) image real;
    (c) f $= 18.67$ cm;
101. converging lens with P $= +3.33$ diopters
103. 7.56 mm

Chapter 22
Answers to odd-numbered multiple-choice problems
15. (d)
17. (d)
19. (b)
21. (b)
23. (c)

Answers to odd-numbered problems
25. (a) 76.16 nm; (b) 96.03 nm;
    (c) 105.96 nm
27. (a) No violet light;
    (b) Violet light
29. 542.64 nm, green light
31. 176.5 nm
33. 2945 nm
35. (a) 2.10 mm; (b) 3.155 mm;
37. (a) $0.364°, 0.728°$ and $1.092°$; (b) Yes
39. 526.45 nm
41. 25
43. 11.81 mm
45. 5.33 μm, with $\lambda = 1.0$ nm
47. (a) 550 nm; (b) yes, $61.63°$
49. (a) $12.37°$; (b) $23.13°$; (c) $40°$
51. (a) Entire visible spectrum;
    (b) $500 - 700$ nm (green to violet);
    (c) No part of visible spectrum
53. (a) $0.02967°$; (b) $0.1253°$
55. (a) 68.76μm; (b) 99.3μm
57. 0.812 mm, using $\sin \theta_2 = 2.23 \dfrac{\lambda}{D}$
59. 1.82 cm
61. (a) 2.796×10^{-7} rad;
    (b) 2.645×10^{10} m, or about $0.176\, d_{SE}$,
    the Earth-Sun distance
63. 549.7 nm
65. 0.67
67. (a) 5.0 mW; (b) 3.75 mW
69. (a) 0.1875; (b) 0.1875; (c) 3.045×10^{-4}
71. (a) 0.25; (b) 0
73. 1.34 m, with $\lambda = 550$ nm
75. 0.33 m
77. 1.98 s
79. 0.079
81. 210 nm

Chapter 23
Answers to odd-numbered multiple-choice problems
17. (b)
19. (c)
21. (d)
23. (d)
25. (d)
27. (d)

Answers to odd-numbered problems
29. 6.25×10^{18}
31. (a) 139.6 g;
    (b) 2.4×10^{-7} C, too small to be noticed
33. 2282 K
35. 7431 K
37. (a) 1369 nm, infrared;
    (b) 2.18×10^{-5} m^2
39. 2.73 K
41. 951 K
43. 1.028×10^{15} Hz, 292 nm
45. (a) 1.12×10^{15} Hz, 267.5 nm; (b) 4.56 V
47. (a) 0.448 V; (b) 3.97×10^5 m/s
49. (a) 2.32 V; (b) 3.926×10^{-19} J
51. (a) 6.272×10^{-34} J·s, -5.34%;
    (b) 7.2×10^{14} Hz, 4.5×10^{-19} J
53. (a) 6.626×10^{-25} J;
    (b) 1.988×10^{-22} J
55. (a) 8.28×10^{-20} J; (b) 1.21×10^{20}
57. (a) UVB: 1.56×10^{18}, visible:
    2.77×10^{18}
59. (a) 4.53×10^{-10}; (b) 1387 W/m^2
61. $90°$
63. $108.5°$
65. 97.2 pm
67. (a) 1876.5 MeV;
    (b) Each photon has an energy of
    938.27 MeV
69. (a) 7.27×10^5 m/s; (b) 546 fm
71. 2.91×10^6 m/s
73. (a) 16.73 eV; (b) 1.46×10^{-21} J;
    (c) 3.65×10^{-22} J
75. 27.82 pm, very unlikely
77. (a) 1.506×10^{-24} J;
    (b) 4.96×10^{-19} J
79. (a) 5.273×10^{-25} kg·m/s;
    (b) 1.526×10^{-19} J; (c) 1.3μm
81. (a) 24.1 pm; (b) 4.2 pm; (c) 0.345 pm
83. 0.02478 fm, much smaller than the diameter of a proton
85. (a) 1.32 fm; (b) 9.3827×10^8 eV
87. 1.988×10^{-26} m
89. 294 eV
91. (a) 2.77×10^{20}; (b) 6.1×10^9

Chapter 24
Answers to odd-numbered multiple-choice problems
17. (c)
19. (a)
21. (d)
23. (b)
25. (d)

Answers to odd-numbered problems
27. 1.92×10^7 m/s
29. 4.86 fm with Al target, much shorter than 29.6 fm found in Ex. 24.1 for Au target
31. (a) Atom: 6.236×10^{-31} m^3, nucleus:
    7.238×10^{-45} m^3;
    (b) 2.68×10^3 kg/m^3, about 2.7 times
    the density of water;
    (c) 2.31×10^{17} kg/m^3, much greater than
    that found in atom for part (b)
33. (a) 13; (b) 28; (c) 41;

35. $2279 - 7458$ nm
37. (a) 3.4 eV; (b) 365 nm
39. (a) 2.19×10^6 m/s; (b) 1.09×10^6 m/s
41. (a) 486 nm; (b) 0.82 m/s
43. (a) 1.89 eV; (b) absorption; (c) 656 nm
45. 69
49. (a) -54.4 eV; (b) -122.4 eV
51. (a) $2s$; (b) $4p$; (c) $4f$; (d) $3d$
53. (a) Allowed, 656 nm;
    (b) Forbidden;
    (c) Allowed, 1875 nm;
    (d) Allowed, 102.6 nm
55. (a) $l = 1$;
    (b) $m_l = -1, 0, +1$;
    (c) $m_s = \pm\frac{1}{2}$
57. (a) $\dfrac{L_f}{L_i} = \dfrac{1}{\sqrt{3}}$; (b) $\dfrac{L_f}{L_i} = \sqrt{3}$;
    (c) $\dfrac{L_f}{L_i} = \dfrac{1}{\sqrt{2}}$
63. (a) Holes; (b) Electrons
65. 40 kV
67. 0.0248 fm
69. 6895 V
71. 2.33 eV
73. (a) 532 nm; (b) 1.68 W
75. $n = 5 \rightarrow 3$
79. (a) $35.3°, 65.9°, 90°, 114.1°, 144.7°$;
    (b) $L_z = m_l \dfrac{h}{2\pi}$, $m_l = 2, 1, 0, -1, -2$
81. 3.02×10^{17} per second

Chapter 25
Answers to odd-numbered multiple-choice problems
17. (a)
19. (b)
12. (d)
23. (b)
25. (b)
27. (c)

Answers to odd-numbered problems
29. (a) ^{49}Cr; (b) ^{94}Tc; (c) ^{190}Pb;
31. For ^{80}Br: $Z = 35, N = 45$;
    ^{80}Kr: $Z = 36, N = 44$;
33. (a) 2.3 fm; (b) 3.26 fm;
    (c) 6.13 fm; (d) 7.45 fm
35. Any nuclide with A $= 216$ (e.g., ^{216}At)
45. (a) 2.66×10^{33} J; (b) 2.66×10^{33} J;
    (c) 2.93×10^{16} kg
47. (a) ^{67}Cu $\rightarrow \beta^- + ^{67}$Zn;
    (b) ^{85}Kr $\rightarrow \beta^- + ^{85}$Rb;
    (c) ^{112}Pd $\rightarrow \beta^- + ^{112}$Ag
49. (a) ^{158}Er; (b) β^+; (c) ^{4}He; (d) ^{4}He
51. ^{211}Pb $\rightarrow ^{211}$Bi $+ \beta^-$, 1.38 MeV
53. (a) 0.148 Bq/L;
    (b) 1.28×10^4/L, assuming constant rate
55. Electron: 0.45 mm; alpha: 19.3 mm
57. (a) 0.325 mg; (b) 0.081 mg
59. 9.32 mg
61. (a) 1.42×10^{11} Bq;
    (b) 6.16×10^8 Bq, with 0.108μg
    remained

63. 8.92 kg
65. About 6.3×10^9 years
67. (a) 3.11 MeV released;
    (b) $Q = -1.64$ MeV, $K_{min} = 11.9$ MeV;
    (c) 2.22 MeV released
69. (a) $Q = -0.59$ MeV;
    (b) $Q = -12.7$ MeV;
    (c) $Q = -6.09$ MeV
71. (a) ^{89}Kr; (b) ^{143}La; (c) ^{96}Zr
73. (a) ^{239}Np; (b) 2n; (c) ^{120}Ag
75. 3.5 kg/day
77. 0.776 kg, much smaller than the 50 kg used in early bomb
79. (a) $Q = -22.37$ MeV;
    (b) $Q = -7.46$ MeV;
    (c) 17.6 MeV
81. (a) 12.86 MeV; (b) 17.35 MeV;
    (c) 18.35 MeV
83. 0.0307 mol each (0.09 g of ^{3}H and 0.06g of ^{2}H)
85. 11.45 MeV, 0.11 pm
87. (a) 2.5×10^{-31} kg; (b) 9.74×10^{10} Bq;
    (c) After 1 week: 1.92×10^{-15} g; after 30 days: 4.32×10^{-43} g
89. Flight: 2.5×10^{-6}, or 1 in 400,000; PET scan: 5×10^{-4}, or 1 in 2,000
91. ^{198}Hg + ^{1}n → ^{197}Ag + ^{2}H
93. Yes; takes only 4.31 days to drop to 2000 Bq/L

Chapter 26
Answers to odd-numbered multiple-choice problems
13. (c)
15. (c)
17. (b)
19. (c)
21. (c)

Answers to odd-numbered problems
23. 0.511 MeV, 2.43 pm
25. 0.866c
27. 1250 kg total (625 kg of matter and 625 kg of anti-matter)
29. 1.32×10^{-18} m
31. 0.252 fm
33. (a) 1.24 TeV;
    (b) 1.24 TeV
35. (a) $\mu^- \rightarrow e^- + \bar{\nu}_e + \nu_\mu$;
    (b) $\mu^+ \rightarrow e^+ + \nu_e + \bar{\nu}_\mu$
37. 105.66 MeV, 11.7 fm
39. (a) $\bar{\nu}_e + p \rightarrow n + e^+$;
    (b) $K^+ \rightarrow \mu^+ + \nu_\mu$
41. (a) Allowed;
    (b) Not allowed, violation of baryon number conservation and muon lepton number conservation;
    (c) Not allowed, electric charge not conserved

45. (a) 62 MeV;
    (b) K^{-1} gets more
47. $\overline{sss}$
49. (a) $\bar{c}u$;
    (b) Same mass and lifetime, different charm number and decay products
55. (a) uuu, ccc, ttt; (b) $\overline{uuu}, \overline{ccc}, \overline{ttt}$
57. (a) ve = 0.999 999 87 c;
    (b) vp = 0.875 c, ve/vp = 1.143
59. (a) 2.67×10^{-6} s
61. (a) Colliding beam: 2.0 GeV, with stationary target: 2.32 GeV;
    (b) Colliding beam: 20 GeV, with stationary target: 4.72 GeV;
    (c) Colliding beam: 200 GeV, with stationary target: 13.82 GeV
63. 5.42×10^{16} K
65. 2.73 K
67. 2.87×10^8 m/s
69. (a) 1.29 MeV;
    (b) 9.98×10^9 K; $< 10^{-3}$ s after Big Bang
71. No, baryon number not conserved
73. 7.0×10^{19} m
75. (a) 7462;
    (b) $v = c(1 - 8.98 \times 10^{-9})$
77. (a) 2.556 fm; (b) -2.815 keV

Credits

A